Optimization for Learning and Control

Optimization for Learning and Control

Anders Hansson
Linköping University
Linköping
Sweden

Martin Andersen
Technical University of Denmark
Kongens Lyngby
Denmark

Library of Congress Cataloging-in-Publication Data

Names: Hansson, Anders, author. | Andersen, Martin, author.
Title: Optimization for learning and control / Anders Hansson, Linköping University, Linköping Sweden, Martin Andersen, Technical University of Denmark.
Description: First edition. | Hoboken, NJ, USA : Wiley, [2023] | Includes index.
Identifiers: LCCN 2023002568 (print) | LCCN 2023002569 (ebook) | ISBN 9781119809135 (cloth) | ISBN 9781119809142 (adobe pdf) | ISBN 9781119809173 (epub)
Subjects: LCSH: System analysis–Mathematics. | Mathematical optimization. | Machine learning–Mathematics. | Signal processing–Mathematics. | MATLAB.
Classification: LCC T57.62 .H43 2023 (print) | LCC T57.62 (ebook) | DDC 004.2/10151–dc23/eng/20230202
LC record available at https://lccn.loc.gov/2023002568
LC ebook record available at https://lccn.loc.gov/2023002569

Cover Design: Wiley
Cover Image: © pluie_r/Shutterstock

Set in 9.5/12.5pt STIXTwoText by Straive, Chennai, India

To Erik.
To Cassie, Maxwell, and Patrick.

Contents

Preface

This is a book about optimization for learning and control. The literature and the techniques for learning are vast, but we will here not focus on all possible learning methods. Instead we will discuss some of them, and especially the ones that result in optimization problems. We will also discuss what optimization methods are relevant to use for these optimization problems. The book is primarily intended for graduate students with a background in science or engineering and who want to learn more about what optimization methods are suitable for learning problems. It is also useful for those who want to study optimal control. Very limited knowledge of optimization, control, or learning is needed as a background. The book is accompanied with a large number of exercises, many of which involve computer tools in order for the students to obtain hands-on experience.

The topics in learning span a wide range from classical statistical learning problems like regression and maximum likelihood estimation to more recent problems like deep learning using, e.g. recurrent neural networks. Regarding optimization methods, we cover methods from simple gradient methods to more advanced interior-point methods for conic optimization. A special emphasis is on stochastic methods applicable to the training of neural networks. We also put a special emphasis on nondifferentiable problems for which we discuss subgradient methods and proximal methods. We cover second-order methods, variable-metric methods, and augmented Lagrangian methods. Regarding applications to system identification, we discuss identification both for input–output models as well as for state-space models. Recurrent neural networks and temporal convolutional networks are naturally introduced as ways of modeling nonlinear dynamical systems. We also cover calculus of variations and dynamic programming in detail, and its generalization to reinforcement learning.

The book can be used to teach several different courses. One could be an introductory course in optimization based on Chapters 4–6. Another course could be on optimal control covering Chapters 7–8, and possibly also Chapter 11. Another course could be on learning covering Chapters 9–10 and perhaps Chapter 12. There is of course also the possibility to combine more chapters, and a course that has been taught at Linköping University for PhD students covers all but the material for optimal control.

Linköping and Kongens Lyngby
November 2022

Anders Hansson and Martin Andersen

Acknowledgments

We would like to thank Andrea Garulli at University of Siena who invited Anders Hansson to give a course on optimization for learning in the spring of 2019. The experience from teaching that course provided most valuable inspiration for writing this book. Daniel Cederberg, Markus Fritzsche, and Magnus Malmström are gratefully acknowledged for having proofread some of the chapters.

Anders Hansson and Martin Andersen

Glossary

Sets

$\mathbb{N}$	set of natural numbers
$\mathbb{N}_k$	set $\{1, 2, \ldots, k\}$
$\mathbb{Z}$	set of integers
$\mathbb{Z}_k$	set $\{0, 1, \ldots, k\}$
$\mathbb{Z}_+$	set of nonnegative integer numbers
$\mathbb{R}$	set of real numbers
$\mathbb{R}_+$	set of nonnegative real numbers
$\mathbb{R}_{++}$	set of positive real numbers
$\bar{\mathbb{R}}$	set of extended real numbers
$\bar{\mathbb{R}}_+$	set of nonnegative extended real numbers
$\bar{\mathbb{R}}_{++}$	set of positive extended real numbers
$\mathbb{C}$	set of complex numbers
$\mathbb{S}^n$	set of symmetric real-valued matrices of order n
$\mathbb{S}^n_+$	set of positive semidefinite real-valued matrices of order n
$\mathbb{S}^n_{++}$	set of positive definite real-valued matrices of order n
$\mathbb{Q}^n$	quadratic cone of dimension n
Δ^n	probability simplex of dimension $n - 1$
$\emptyset$	empty set

Numbers, Vectors, and Matrices

e	Euler's number
π	Archimedes' constant
$\mathbb{1}$	vector of ones
I	identity matrix

Elementary Functions

exp	natural exponential function
log	logarithm function
ln	natural logarithm function
$\log_2$	logarithm function, base 2

sin	sine function
cos	cosine function
tan	tangent function
sgn	sign function
I_A	indicator function for set A

Operations on Sets or Functions

aff	affine hull of set
argmin	minimum argument
argmax	maximum argument
cl	closure of set
conv	convex hull of set
cone	conic hull of set
dom	effective domain
epi	epigraph of function
inf	infimum of set or function
int	interior of set
max	maximum of set or function
min	minimum of set or function
prox	proximal operator
relint	relative interior of set
sup	supremum of set or function

Operations on Vectors, Vector Spaces or Matrices

adj	adjugate of matrix
blkdiag	block diagonal matrix from matrices
det	determinant of matrix
diag	diagonal matrix from vector
dim	dimension of vector space or convex set
$\mathcal{N}$	nullspace of matrix
nnz	number of nonzero entries
nullity	nullity of matrix
$\mathcal{R}$	range of matrix
rank	rank of matrix
span	span of vectors
svec	symmetric vectorization of matrix
tr	trace of matrix
vec	vectorization of matrix

Probability Spaces

$\mathbb{E}$	expectation functional
$\mathcal{N}$	normal probability density function
$\mathbb{P}$	probability measure
Var	variance functional

Symbols

$\sum$	summation
$\prod$	product
$\int$	integral
$\oint$	contour integral
∞	infinity
$\in$	belongs to
$\notin$	does not belong to
$\subset$	proper subset of
$\subseteq$	subset of
$\supset$	proper superset of
$\supseteq$	superset of
$\not\subset$	not proper subset of
$\nsubseteq$	not subset of
$\not\supset$	not proper superset of
$\nsupseteq$	not superset of
$\cup$	set union
$\cap$	set intersection
$\setminus$	set difference
$+$	plus
$-$	minus
$\pm$	plus or minus
$\times$	multiplied by
$\otimes$	Kronecker product
$\circ$	Hadamard product or composition of functions
$=$	is equal to
$<$	is less than
$\leq$	is less than or equal to
$>$	is greater than
$\geq$	is greater than or equal to
$\neq$	is not equal to
$\not<$	is not less than
$\nleq$	is neither less than nor equal to
$\not>$	is not greater than
$\ngeq$	is neither greater than nor equal to
$\ll$	much smaller than
$\gg$	much greater than
$\approx$	is approximately equal to
$\sim$	asymptotically equivalent to
$\propto$	proportional to
$\prec$	precedes
$\preceq$	precedes or equals
$\succ$	succeeds
$\succeq$	succeeds or equals
$\nprec$	does not precede
$\npreceq$	neither precedes nor equals
$\nsucc$	does not succeed
$\nsucceq$	neither succeeds nor equals

$\exists$	there exists
$\nexists$	there is no
$\forall$	for all
$\neg$	logical not
$\wedge$	logical and
$\vee$	logical or
$\implies$	implies
$\impliedby$	is implied by
$\iff$	is equivalent to
$\to$	to or tends toward
$\leftrightarrow$	corresponds to
$\searrow$	tends toward from above
$\mapsto$	maps to
$\perp$	is perpendicular to
$\mid$	such that or given
$:$	such that

Acronyms

AdaGrad	adaptive gradient method
Adam	adaptive moment estimation
ADMM	alternating direction method of multipliers
a.e.	almost everywhere
ANN	artificial neural network
ARE	algebraic Riccati equation
ARMAX	auto-regressive-moving-average with exogenous terms
ARX	auto-regressive with exogenous terms
a.s.	almost surely
BB	Barzilai–Borwein
BFGS	Broyden, Fletcher, Goldfarb, and Shanno
BM	Boltzmann machine
DAE	differential algebraic equation
df	distribution function
DFP	Davidon, Fletcher, and Powell
DC	diagonal/correlated
EM	expectation maximization
EVD	eigenvalue decomposition
FIR	finite impulse response
FLOP	floating point operation
GN	Gauss–Newton
GMM	Gaussian mixture model
i.i.d.	independent, identically distributed
HMM	hidden Markov model
IFT	iterative feedback tuning
ILC	iterative learning control
IP	interior point
IPG	incremental proximal gradient
KKT	Karush–Kuhn–Tucker
LICQ	linear independence constraint qualification
LM	Levenberg–Marquardt
LMI	linear matrix inequality
LP	linear program
LQ	linear quadratic
LS	least-squares
LSTM	long short-term memory

MA	moving average
MAP	maximum a posteriori
MDP	Markov decision process
ML	maximum likelihood
MPC	model predictive control
m.s.	mean square
MSE	mean square error
NP	nondeterministic polynomial time
ODE	ordinary differential equation
OE	output error
PCA	principal component analysis
pdf	probability density function
pf	probability function
PI	policy iteration
PMP	Pontryagin maximum principle
QP	quadratic program
RBM	restricted Boltzmann machine
RMSprop	root mean square propagation
RNN	recurrent neural network
ReLU	rectified linear unit
SA	stochastic approximation
SAA	stochastic average approximation
SARSA	state-action-reward-state-action
SG	stochastic gradient
SGD	stochastic gradient descent
SMW	Sherman–Morrison–Woodbury
SNR	signal-to-noise ratio
SR1	symmetric rank-1
SVD	singular value decomposition
SVM	support vector machine
SVRG	stochastic variance-reduced gradient
TCM	temporal convolutional network
TPBVP	two-point boundary value problem
VI	value iteration
w.p.1	with probability one

About the Companion Website

This book is accompanied by a companion website.

www.wiley.com/go/opt4lc

This website includes:

- Data files and templates for solving the exercises
- Solutions to the exercise in the book in terms of a pdf-document (Instructors only)
- MATLAB solution files (Instructors only)

Part I

Introductory Part

1

Introduction

This book will take you on a journey through the fascinating field of optimization, where we explore techniques for designing algorithms that can learn and adapt to complex systems. Whether you are an engineer, a scientist, or simply curious about the world of optimization, this book is for you. We will start with the basics of optimization and gradually build up to advanced techniques for learning and control. By the end of this book, you will have a solid foundation in optimization theory and practical tools to apply to real-world problems. In this opening, we informally introduce problems and concepts, and we will explain their close interplay with simple formulations and examples. Chapters 2–13 will explore the topic with more rigor, and we end this chapter with an outline of the remaining content of the book.

1.1 Optimization

Optimization is about choosing a best option from a set of available alternatives based on a specific criterion. This concept applies to a range of disciplines, including computer science, engineering, operations research, and economics, and has a long history of conceptual and methodological development. One of the most common optimization problems is of the form

$$\text{minimize} \sum_{k=1}^{m} f_k(x)^2, \tag{1.1}$$

with variable $x \in \mathbb{R}^n$. This is called a nonlinear *least-squares* problem, since we are minimizing the squares of the possibly nonlinear functions f_k. We will see that the least-squares problem and its generalizations have many applications to learning and control. It is also the backbone of several optimization methods for solving more complicated optimization problems. In Chapter 4, we set the foundations for optimization theory, in Chapter 5, we cover different classes of optimization problems that are relevant for learning and control, and in Chapter 6, we discuss different methods for solving optimization problems numerically.

1.2 Unsupervised Learning

Learning is often about finding low-dimensional descriptions of data that provide insight and are easy to understand or interpret. A very simple example is the repeated measurement of the same real-valued quantity $x \in \mathbb{R}$. This could be the length of a piece of wood. Each measurement may be assumed to be somewhat different because of random measurement errors, i.e. we have

Optimization for Learning and Control, First Edition. Anders Hansson and Martin Andersen.
© 2023 John Wiley & Sons, Inc. Published 2023 by John Wiley & Sons, Inc.
Companion Website: www.wiley.com/go/opt4lc

a sequence of, say, m measurements $y_1, \ldots, y_m$. Clearly, a reasonable estimate of x would be the average of the m measurements, i.e.

$$\hat{x} = \frac{1}{m} \sum_{k=1}^{m} y_k.$$

This is a scalar representation of the m measurements, and hence of lower dimension, which is a reasonable representation of the length of the piece of wood. Here we have estimated the length of the piece of wood by averaging the measurements. The learning we have discussed here is called *unsupervised learning*, and it is discussed in more detail in Chapter 9.

1.3 Supervised Learning

Let us consider another simple problem where we are driving a car with constant speed v. We start at an unknown distance y_0 from a city toward which we are driving, and we receive distances information y_k to the city from road signs corresponding to having traveled for a time of $t_k, k = 1, \ldots, m$. The following relation should hold

$$y_k = y_0 - t_k v = a_k^T x, \quad k = 1, \ldots, m,$$

where $a_k = (1, -t_k)$ and $x = (y_0, v)$. We assume that we do not know x and we would like to learn it. It is enough to have $m = 2$, i.e. two linear equations, to solve for x, and this solution will be unique since the a_ks will be linearly independent unless $t_1 = t_2$. However, in an application like this, it is unlikely that the above equations hold exactly due to measurement errors e_k, i.e.

$$y_k = a_k^T x + e_k, \quad k = 1, \ldots, m,$$

is a more appropriate description, and hence, it might be more suitable to have m larger than 2 to average out the effect of the measurement errors. This can be accomplished by solving

$$\text{minimize} \sum_{k=1}^{m} \left(a_k^T x - y_k \right)^2,$$

with variable x. This is an example of a least-squares problem for which $f_k(x) = a_k^T x - y_k$ is an affine function. Hence, this is often called a linear least-squares problem. Typically, m is much larger than n, and therefore, the optimal solution $x^\star$ is of lower dimension than the measurements. If we later get a new value of a, we may predict the value of the corresponding measurement as $a^T x^\star$ without performing the measurement. For our application, this means that we can estimate the distance to the city by just checking how long we have been traveling. We do not have to wait for a new sign to appear. This is a so-called *supervised learning* problem, since for each a_k, we know the corresponding y_k. For learning the length of the piece of wood the data did not come in pairs, but instead, we just had one stream of data y_k. We learned the mean of the data. That is the reason for the name unsupervised learning for such a learning problem. We will discuss supervised learning in more detail in Chapter 10.

1.4 System Identification

Many physical phenomena can be described with *dynamical systems*, which in mathematical terms are given by *differential equations* or *difference equations*. A simple example is the difference equation

$$y_{k+1} = a y_k + b u_k, \quad k = 1, \ldots, m, \tag{1.2}$$

where k denotes time, $y_k \in \mathbb{R}$ is a physical phenomenon that we can observe. It is influenced by both its previous value and by a quantity denoted by $u_k \in \mathbb{R}$. Depending on which field is studying dynamical systems, the pairs (u_k, y_k) are called *stimuli and response* or *input and output*, respectively. Sometimes the word *signal* is added at the end for each of the four words. Often, the above equation does not hold exactly due to measurement errors e_k, and hence, it is more appropriate to consider

$$y_{k+1} = ay_k + bu_k + e_k, \quad k = 1, \ldots, m.$$

When we do not know the *parameters* $(a, b) \in \mathbb{R} \times \mathbb{R}$, we can use supervised learning to learn the values assuming that we are given pairs of data (u_k, y_k) for $1 \leq k \leq m + 1$. The following linear least-squares problem

$$\text{minimize} \sum_{k=1}^{m} \left(y_{k+1} - ay_k - bu_k\right)^2,$$

with variables (a, b) will provide an estimate of the parameters. Learning for dynamical systems is called *system identification*, and it will be discussed in more detail in Chapter 12.

1.5 Control

Control is about making dynamical systems behave in a way we find desirable. Let us again consider the dynamical system in (1.2), where we are going to influence the behavior by manipulating the input signal u_k. In the context of control, we also often call it the *control signal*. We assume that the *initial value* y_0 and the parameters (a, b) are known, and our objective is to make y_k for $1 \leq k \leq m$ small. We can make y_1 equal to zero by taking $u_0 = -ay_0/b$, and then we can make all future values of y_k equal to zero by taking all future values of u_k equal to zero. However, it could be that the value of u_0 is large, and in applications, this could be costly. Hence, we are interested in finding a trade-off between how large the values of u_k are in comparison to the values of y_k. This can be accomplished by solving

$$\text{minimize} \sum_{k=1}^{m} \left(y_k^2 + \rho u_k^2\right), \tag{1.3}$$
$$\text{subject to } y_{k+1} = ay_k + bu_k, \quad k = 1, \ldots, m - 1,$$

with variables $(u_1, y_2, \ldots, u_m, y_m)$. This is an equality *constrained* linear least-squares problem. The parameter $\rho > 0$ can be used to trade-off how small $|y_k|$ should be as compared to $|u_k|$. We will cover control of dynamical systems in Chapter 7 for continuous time, and in Chapter 8 for discrete time.

1.6 Reinforcement Learning

When we discussed control above we had to know the values of the parameters (a, b). Had they not been known, we could have used system identification to estimate them, and then we could have used the estimated parameters for solving the control problem. However, sometimes it is desirable to skip the system identification step and do control without knowing the parameter values. One way to accomplishing this for the formulation in (1.3) is called *reinforcement learning*. This will be discussed in more detail in Chapter 11.

1.7 Outline

The book is organized as follows: first, we give a background on linear algebra and probabilities in Chapters 2 and 3. Background on optimization is given in Chapter 4. We will cover both convex and nonconvex optimization. Chapter 5 introduces different classes of optimization problems that we will encounter in the learning chapters later on. In Chapter 6, we discuss different optimization methods that are suitable for solving learning problems. After this, we discuss calculus of variations in Chapter 7 and dynamic programming in Chapter 8. We then cover unsupervised learning in Chapter 9, supervised learning in Chapter 10, and reinforcement learning in Chapter 11. Finally, we discuss system identification in Chapter 12. For information about notation, basic definitions, and software useful for optimization and for the applications we consider, see the Appendix.

2

Linear Algebra

Linear algebra is the study of vector spaces and linear maps on such spaces. It constitutes a fundamental building block in optimization and is used extensively for theoretical analysis and derivations as well as numerical computations.

A procedure for solving systems of simultaneous linear equations appeared already in an
ancient Chinese mathematical text. Systems of linear equations were introduced in Europe
in the seventeenth century by René Descartes in order to represent lines and planes by linear
equations and to compute their intersections. Gauss developed the method of elimination. Further
important developments were done by Gottfried Wilhelm von Leibniz, Gabriel Cramer, Hermann
Grassmann, and James Joseph Sylvester, the latter introducing the term "matrix."

The purpose of this chapter is to review key concepts from linear algebra and calculus in
finite-dimensional vector spaces as well as a number of useful identities that will be used
throughout the book. We also discuss some computational aspects, including a number of matrix
factorizations and their application to solving systems of linear equations.

2.1 Vectors and Matrices

We start by introducing vectors and matrices. A vector x of length n is an ordered collection of n
numbers,

$$x = (x_1, x_2, \ldots, x_n),$$

where x_i is the ith element or entry of the vector x. The n-dimensional real space, denoted $\mathbb{R}^n$, is
the set of real-valued n-vectors, i.e. vectors of length n whose elements are all real. The product
of a scalar $t \in \mathbb{R}$ and a vector $x \in \mathbb{R}^n$ is defined as $tx = (tx_1, \ldots, tx_n)$, the sum of two real-valued
n-vectors a and b is the vector $a + b = (a_1 + b_1, \ldots, a_n + b_n)$, and the Euclidean *inner product* or *dot
product* of a and b is the real number

$$\langle a, b \rangle = \sum_{i=1}^{n} a_i b_i = a_1 b_1 + a_2 b_2 + \cdots + a_n b_n. \tag{2.1}$$

The vectors a and b are said to be *orthogonal* if $\langle a, b \rangle = 0$.

A matrix A of size m-by-n, also written as $m \times n$, is an ordered rectangular array that consists of mn elements arranged in m rows and n columns, i.e.

$$A = \begin{bmatrix} a_{11} & a_{12} & \cdots & a_{1n} \\ a_{21} & a_{22} & \cdots & a_{2n} \\ \vdots & \vdots & \ddots & \vdots \\ a_{m1} & a_{m2} & \cdots & a_{mn} \end{bmatrix},$$

where a_{ij} denotes the element of A in its ith row and jth column. The set of m-by-n matrices with real-valued elements is denoted by $\mathbb{R}^{m \times n}$. The transpose of A is the $n \times m$ matrix defined as

$$A^T = \begin{bmatrix} a_{11} & a_{21} & \cdots & a_{m1} \\ a_{12} & a_{22} & \cdots & a_{m2} \\ \vdots & \vdots & \ddots & \vdots \\ a_{1n} & a_{2n} & \cdots & a_{mn} \end{bmatrix},$$

i.e. the (i,j)th element of A^T is the (j,i)th element of A.

It is often convenient to think of a vector as a matrix with a single row or column. For example, when interpreted as a matrix of size $1 \times n$, the vector $a \in \mathbb{R}^n$ is referred to as a *row vector*, and similarly, when interpreted as a matrix of size $n \times 1$, a is referred to as a *column vector*. In this book, we use the convention that all vectors are column vectors. Thus, a vector $x \in \mathbb{R}^n$ is always interpreted as the column vector

$$x = \begin{bmatrix} x_1 \\ x_2 \\ \vdots \\ x_n \end{bmatrix},$$

and hence, x^T is interpreted as the row vector $\begin{bmatrix} x_1 & x_2 & \cdots & x_n \end{bmatrix}$. Similarly, to refer to the columns of a matrix $A \in \mathbb{R}^{m \times n}$, we will sometimes use the notation

$$A = \begin{bmatrix} a_1 & a_2 & \cdots & a_n \end{bmatrix},$$

where $a_1, a_2, \ldots, a_n \in \mathbb{R}^m$. When referring to the rows of A, we will define

$$A^T = \begin{bmatrix} a_1 & a_2 & \cdots & a_m \end{bmatrix},$$

where $a_1, a_2, \ldots, a_m \in \mathbb{R}^n$ such that A is the matrix with rows $a_1^T, a_2^T, \ldots, a_m^T$. The notation is somewhat ambiguous because a_i may refer to the ith element of a vector a or the ith column of either A or A^T, but the meaning will be clear from the context and follows from our convention that vectors are column vectors.

Given two vectors $x, y \in \mathbb{R}^n$, the inner product $\langle x, y \rangle$ can also be expressed as the product

$$x^T y = \begin{bmatrix} x_1 & x_2 & \cdots & x_n \end{bmatrix} \begin{bmatrix} y_1 \\ y_2 \\ \vdots \\ y_n \end{bmatrix} = \sum_{i=1}^{n} x_i y_i.$$

In contrast, the outer product of two vectors $u \in \mathbb{R}^m$ and $v \in \mathbb{R}^n$, not necessarily of the same length, is defined as the $m \times n$ matrix

$$uv^T = \begin{bmatrix} u_1 v_1 & u_1 v_2 & \cdots & u_1 v_n \\ u_2 v_1 & u_2 v_2 & \cdots & u_2 v_n \\ \vdots & \vdots & \ddots & \vdots \\ u_m v_1 & u_m v_2 & \cdots & u_m v_n \end{bmatrix}.$$

The product of a matrix $A \in \mathbb{R}^{m \times n}$, with columns $a_1, \ldots, a_n \in \mathbb{R}^m$, and a vector $x \in \mathbb{R}^n$ is the linear combination

$$y = a_1 x_1 + a_2 x_2 + \cdots + a_n x_n.$$

Equivalently, the ith element of the vector $y = Ax$ is the inner product of x and the ith row of A.

The vector inner and outer products and matrix–vector multiplication are special cases of matrix multiplication. Two matrices A and B are said to be *conformable for multiplication* if the number of columns in A is equal to the number of rows in B. Given two such matrices $A \in \mathbb{R}^{m \times p}$ and $B \in \mathbb{R}^{p \times n}$, the product $C = AB$ is the $m \times n$ matrix with elements

$$c_{ij} = \sum_{k=1}^{p} a_{ik} b_{kj}, \quad i \in \mathbb{N}_m, \; j \in \mathbb{N}_n.$$

Note that c_{ij} is the inner product of the ith row of A and the jth column of B. As a result, $C = AB$ may be expressed as

$$C = \begin{bmatrix} a_1^T \\ a_2^T \\ \vdots \\ a_m^T \end{bmatrix} \begin{bmatrix} b_1 & b_2 & \cdots & b_n \end{bmatrix} = \begin{bmatrix} a_1^T b_1 & a_1^T b_2 & \cdots & a_1^T b_n \\ a_2^T b_1 & a_2^T b_2 & \cdots & a_2^T b_n \\ \vdots & \vdots & \ddots & \vdots \\ a_m^T b_1 & a_m^T b_2 & \cdots & a_m^T b_n \end{bmatrix},$$

where $a_1^T, \ldots, a_m^T$ are the rows of A and $b_1, \ldots, b_n$ are the columns of B. Equivalently, by expressing A in terms of its columns and B in terms of its rows, C may also be written as the sum of p outer products

$$C = \begin{bmatrix} a_1 & a_2 & \cdots & a_p \end{bmatrix} \begin{bmatrix} b_1^T \\ b_2^T \\ \vdots \\ b_p^T \end{bmatrix} = \sum_{i=1}^{p} a_i b_i^T,$$

where $a_1, \ldots, a_p$ are the columns of A and $b_1^T, \ldots, b_p^T$ are the rows of B.

It is important to note that matrix multiplication is associative, but unlike scalar multiplication, it is not commutative. In other words, the associative property $(AB)C = A(BC)$ holds, provided that A, B, and C are conformable for multiplication, but the identity $AB = BA$ does not hold in general.

The Frobenius inner product of two matrices $A, B \in \mathbb{R}^{m \times n}$ is defined as

$$\langle A, B \rangle = \sum_{i=1}^{m} \sum_{j=1}^{n} a_{ij} b_{ij}. \tag{2.2}$$

This is also called the trace inner product of A and B since

$$\langle A, B \rangle = \text{tr}(A^T B),$$

where the trace of a square matrix $C \in \mathbb{R}^{n \times n}$ is defined as

$$\text{tr}(C) = \sum_{i=1}^{n} c_{ii}.$$

The inner product $\langle A, B \rangle$ may also be written as $\text{vec}(A)^T \text{vec}(B)$, where $\text{vec}(A)$ maps a matrix $A \in \mathbb{R}^{m \times n}$ to a vector of length mn by stacking the columns of A, i.e.

$$\text{vec}(A) = \text{vec}\left(\begin{bmatrix} a_1 & \cdots & a_n \end{bmatrix}\right) = \begin{bmatrix} a_1 \\ \vdots \\ a_n \end{bmatrix} \in \mathbb{R}^{mn}.$$

2.2 Linear Maps and Subspaces

The span of n vectors $v_1, \dots, v_n \in \mathbb{R}^m$ is the set of all linear combinations of these vectors, i.e.

$$\text{span}(v_1, \dots, v_n) = \{\alpha_1 v_1 + \dots + \alpha_n v_n \mid \alpha_1, \dots, \alpha_n \in \mathbb{R}\}.$$

The set of vectors $v_1, \dots, v_n$ is said to be *linearly independent* if

$$\alpha_1 v_1 + \dots + \alpha_n v_n = 0 \iff \alpha_1 = \dots = \alpha_n = 0,$$

and otherwise, the set is said to be *linearly dependent*. Equivalently, the set $v_1, \dots, v_n$ is linearly independent if and only if all vectors in $\text{span}(v_1, \dots, v_n)$ have a unique representation as a linear combination of $v_1, \dots, v_n$. The span of a set of k linearly independent vectors $v_1, \dots, v_k \in \mathbb{R}^m$ is a k-dimensional *subspace* of $\mathbb{R}^m$, and $v_1, \dots, v_k$ is a *basis* for the subspace. In other words, the *dimension* of a subspace is equal to the number of linearly independent vectors that span the subspace. The vectors $v_1, \dots, v_n$ are said to be *orthonormal* if they are mutually orthogonal and of unit length, i.e. $v_i^T v_j = 0$ if $i \neq j$ and $v_i^T v_i = 1$ for $i \in \mathbb{N}_n$. The *standard basis* or *natural basis* for $\mathbb{R}^m$ is the orthonormal basis $e_1, \dots, e_m$, where $e_i \in \mathbb{R}^m$ is the unit vector whose ith element is equal to 1, and the rest are zero.

The *range* of a matrix $A \in \mathbb{R}^{m \times n}$, denoted $\mathcal{R}(A)$, is the span of its columns. This is also referred to as the *column space* of A, whereas $\mathcal{R}(A^T)$ is referred to as the *row space* of A. The dimension of $\mathcal{R}(A)$ is called the *rank* of A, denoted rank(A). The *null space* of $A \in \mathbb{R}^{m \times n}$, denoted $\mathcal{N}(A)$, consists of all vectors v such that $Av = 0$, i.e.

$$\mathcal{N}(A) = \{v \in \mathbb{R}^n \mid Av = 0\}.$$

The dimension of $\mathcal{N}(A)$ is called the *nullity* of A and is denoted nullity(A). The null space of A is said to be *trivial* if $\mathcal{N}(A) = \{0\}$ in which case nullity$(A) = 0$.

2.2.1 Four Fundamental Subspaces

It follows from the definition of the range and nullspace of a matrix A that

$$\mathcal{R}(A) = \mathcal{N}(A^T)^\perp, \quad \mathcal{R}(A^T) = \mathcal{N}(A)^\perp. \tag{2.3}$$

To see that $\mathcal{R}(A) = \mathcal{N}(A^T)^\perp$, note that for every $y \in \mathcal{N}(A^T)$, we have that

$$(A^T y)^T x = y^T A x = 0, \quad \forall x \in \mathbb{R}^n,$$

or equivalently, if we let $u = Ax$, we see that $y^T u = 0$ for all $u \in \mathcal{R}(A)$. This shows that $\mathcal{R}(A)$ and $\mathcal{N}(A^T)$, which are both subspaces of $\mathbb{R}^m$, are orthogonal complements. Similarly, for every $x \in \mathcal{N}(A)$, it immediately follows that $y^T A x = 0$ for all $y \in \mathbb{R}^m$, and hence, $\mathcal{R}(A^T)$ is the orthogonal complement of $\mathcal{N}(A)$.

The result (2.3) can be used to derive the so-called *rank-nullity* theorem which states that

$$\text{rank}(A) + \text{nullity}(A) = n. \tag{2.4}$$

To see this, note that the identity $\mathcal{R}(A^T) = \mathcal{N}(A)^\perp$ combined with the fact that

$$\dim(\mathcal{N}(A)^\perp) = n - \text{nullity}(A),$$

implies that rank$(A^T) = n - \text{nullity}(A)$. The rank-nullity theorem follows from the identity

$$\text{rank}(A) = \text{rank}(A^T), \tag{2.5}$$

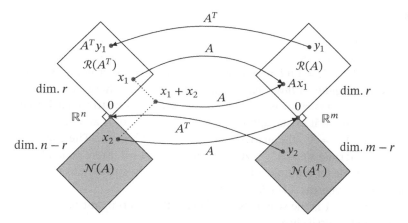

Figure 2.1 The four subspaces associated with an $m \times n$ matrix A with rank r.

which we will now derive. First, suppose that $\text{rank}(A^T) = r$, and let $v_1, \ldots, v_r \in \mathbb{R}^n$ be a linearly independent set of vectors that span $\mathcal{R}(A^T)$. It follows that the set of vectors $Av_1, \ldots, Av_r$ is linearly independent since

$$\alpha_1 A v_1 + \cdots + \alpha_r A v_r = A(\alpha_1 v_1 + \cdots + \alpha_r v_r) = 0 \iff \alpha_1 = \cdots = \alpha_r = 0.$$

As a result, we have that $\text{rank}(A) \geq \text{rank}(A^T)$. Applying the same inequality to $B = A^T$ implies that $\text{rank}(A) = \text{rank}(A^T)$. A direct consequence of this identity is that $\text{rank}(A) \leq \min{(m, n)}$. We say that A has *full rank* if $\text{rank}(A) = \min{(m, n)}$, and we will use the term *full row rank* when $r = m$ and the term *full column rank* when $r = n$. The four subspaces $\mathcal{R}(A)$, $\mathcal{N}(A^T)$, $\mathcal{R}(A^T)$, and $\mathcal{N}(A)$ and their dimensions are illustrated in Figure 2.1.

If two matrices A and B are conformable for multiplication, then

$$\text{rank}(AB) \leq \min{(\text{rank}(A), \text{rank}(B))}. \tag{2.6}$$

This follows from the fact that $\mathcal{R}(AB) \subseteq \mathcal{R}(A)$ and $\mathcal{R}(B^T A^T) \subseteq \mathcal{R}(B^T)$ which implies that $\text{rank}(AB) \leq \text{rank}(A)$ and $\text{rank}(B^T A^T) \leq \text{rank}(B^T) = \text{rank}(B)$. Furthermore, we have that $\mathcal{R}(AB) = \mathcal{R}(A)$ when B has full row rank, in which case $\text{rank}(AB) = \text{rank}(A)$.

If A and B are conformable for addition, then

$$\text{rank}(A + B) \leq \text{rank}(A) + \text{rank}(B). \tag{2.7}$$

This means that rank is *subadditive*, and it follows from the fact that $\mathcal{R}(A + B) \subseteq \mathcal{R}(A) + \mathcal{R}(B)$ where

$$\mathcal{R}(A) + \mathcal{R}(B) = \{u + v \mid u \in \mathcal{R}(A), v \in \mathcal{R}(B)\}$$

is the *Minkowski sum* of $\mathcal{R}(A)$ and $\mathcal{R}(B)$. Subadditivity implies that a rank k matrix can be decomposed as the sum of k or more rank 1 matrices. Thus, a rank k matrix $A \in \mathbb{R}^{m \times n}$ can be factorized as

$$A = CR^T = \sum_{i=1}^{k} c_i r_i^T,$$

where $c_1, \ldots, c_k$ and $r_1, \ldots, r_k$ are the columns of $C \in \mathbb{R}^{m \times k}$ and $R \in \mathbb{R}^{n \times k}$, respectively, and $\mathcal{R}(C) = \mathcal{R}(A)$ and $\mathcal{R}(R) = \mathcal{R}(A^T)$.

2.2.2 Square Matrices

A matrix with an equal number of rows and columns is called a *square matrix*, and the *order* of such a matrix refers to its size. The square matrix with ones on the main diagonal and zeros elsewhere is an *identity* matrix and denoted by I, i.e. $I_{ij} = 1$ if $i = j$ and otherwise, $I_{ij} = 0$. Furthermore, a square matrix of order n is said to be *nonsingular* or *invertible* if it has full rank, and otherwise, it is said to be *singular*. Equivalently, a square matrix A is invertible if there exists a matrix B such that $AB = BA = I$. Such a matrix is unique if it exists, and it is called the *inverse* of A and is denoted by A^{-1}. In the special case where $A^{-1} = A^T$, the matrix A is said to be *orthogonal*. The columns of an orthogonal matrix are orthonormal since $A^{-1} = A^T$ implies that $A^T A = I$. More generally, if two matrices A and B are conformable for multiplication, i.e. not necessarily square matrices, and $AB = I$, then B is a *right inverse* of A whereas A is a *left inverse* of B.

The *determinant* of a scalar matrix A is the scalar itself, whereas the determinant of a square matrix A of order n may be defined recursively as

$$\det(A) = \sum_{i=1}^{n} (-1)^{i+j} a_{ij} M_{ij}, \tag{2.8}$$

where M_{ij} denotes the *minor* of a_{ij} which is the determinant of the $(n-1) \times (n-1)$ matrix that is obtained by removing the ith row and jth column of A. This expression is a so-called *Laplace expansion* of the determinant along the jth column of A, and it holds for every $j \in \mathbb{N}_n$. As a special case, the determinant of a 2×2 matrix A may be expressed as

$$\det(A) = a_{11}a_{22} - a_{12}a_{21},$$

and its absolute value may be interpreted as the area of a parallelogram defined by the columns of A, as illustrated in Figure 2.2. More generally, the absolute value of the determinant of an $n \times n$ matrix A is the volume of the parallelotope defined by the columns of A, i.e. the set

$$\{Ax \in \mathbb{R}^n \mid 0 \leq x_i \leq 1 \quad \forall i \in \mathbb{N}_n\}.$$

As a result, $\det(A) \neq 0$ if and only if A has full rank.

The term $(-1)^{i+j} M_{ij}$ is called the *cofactor* of a_{ij}. The $n \times n$ matrix composed of all the cofactors, i.e. the matrix C with elements

$$c_{ij} = (-1)^{i+j} M_{ij}, \quad i, j \in \mathbb{N}_n,$$

is called the *cofactor matrix*. Expressed in terms of the cofactors, the Laplace expansion (2.8) may be written as the inner product of the jth column of A and C, i.e.

$$\det(A) = \sum_{k=1}^{n} a_{kj} c_{kj}.$$

Furthermore, since the Laplace expansion is valid for any $j \in \mathbb{N}_n$, the diagonal elements of the matrix $C^T A$ are all equal to $\det(A)$. In fact, it can be shown that

$$C^T A = \det(A) I,$$

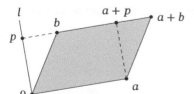

Figure 2.2 Parallelogram defined by two vectors a and b in $\mathbb{R}^2$. The area is given by $|\det([a\ b])| = \|p\|_2 \|a\|_2$, where p is the projection of b onto the line l with normal a.

and if A is nonsingular, this identity implies that

$$A^{-1} = \frac{1}{\det(A)} C^T,$$ (2.9)

where C^T, the transpose of the cofactor matrix, is known as the *adjugate matrix* which is denoted by adj(A).

2.2.3 Affine Sets

Given two points $x, y \in \mathbb{R}^n$ and $\theta \in \mathbb{R}$, the point

$$\theta x + (1 - \theta) y$$

is a so-called *affine combination* of x and y. The set of all such affine combinations of x and y, which may be expressed as

$$\{\theta x + (1 - \theta) y \mid \theta \in \mathbb{R}\}$$

is a line in $\mathbb{R}^n$ if $x \neq y$. This is illustrated in Figure 2.3. More generally, an affine combination of k points $x_1, \ldots, x_k \in \mathbb{R}^n$ is a linear combination

$$\theta_1 x_1 + \cdots + \theta_k x_k = \sum_{i=1}^{k} \theta_i x_i,$$

where $\theta_1 + \cdots + \theta_k = 1$. A set $\mathcal{A} \subseteq \mathbb{R}^n$ is an *affine set* if and only if it contains all affine combinations of its points. Equivalently, $\mathcal{A}$ is affine if for every pair of points $x, y \in \mathcal{A}$,

$$\{\theta x + (1 - \theta) y \mid \theta \in \mathbb{R}\} \subseteq \mathcal{A}.$$

An example of an affine set is the set of solutions to the system of equation $Ax = b$ with $A \in \mathbb{R}^{m \times n}$ and $b \in \mathbb{R}^m$, i.e.

$$\mathcal{A} = \{x \in \mathbb{R}^n \mid Ax = b\}.$$

In fact, every affine subset of $\mathbb{R}^n$ may be expressed in this form. The *affine hull* of a set $S \subseteq \mathbb{R}^n$, which we denote aff$S$, is the smallest possible affine set that contains S.

2.3 Norms

Equipped with the Euclidean inner product (2.1), the set $\mathbb{R}^n$ is a Euclidean vector space of dimension n with the induced norm

$$\|a\|_2 = \sqrt{\langle a, a \rangle} = \left(a_1^2 + a_2^2 + \cdots + a_n^2\right)^{1/2}.$$ (2.10)

More generally, a norm of a vector $x \in \mathbb{R}^n$, denoted $\|x\|$, is a function with the following defining properties:

Figure 2.3 Affine combinations $\theta x + (1 - \theta) y$ of two points x and y.

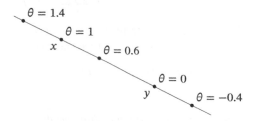

1. Subadditive[1]:

$$\|u + v\| \leq \|u\| + \|v\|, \quad \forall u, v \in \mathbb{R}^n.$$

2. Absolutely homogeneous:

$$\|\alpha x\| = |\alpha| \|x\|, \quad \forall x \in \mathbb{R}^n, \ \forall \alpha \in \mathbb{R}.$$

3. Positive definite:

$$\|x\| = 0 \text{ if and only if } x = 0.$$

4. Nonnegative:

$$\|x\| \geq 0, \quad \forall x \in \mathbb{R}^n.$$

Such a function is not unique, so to distinguish between different norms, a subscript is typically used to denote specific norms. For example, the 2-norm or Euclidean norm of a vector x, defined in (2.10), is denoted $\|x\|_2$. Other examples are the 1-norm and the infinity norm, defined as

$$\|x\|_1 = \sum_{i=1}^{p} |x_i| \quad \text{and} \quad \|x\|_\infty = \max \{|x_1|, \ldots, |x_n|\}.$$

These are special cases of the more general vector p-norm

$$\|x\|_p = \left(\sum_{i=1}^{n} |x_i|^p \right)^{1/p}, \tag{2.11}$$

which is defined for $p \geq 1$.

A norm on $\mathbb{R}^n$ is said to be *orthogonally invariant* if $\|x\| = \|Qx\|$ for all x and all orthogonal matrices Q. The 2-norm is orthogonally invariant, which follows by noting that $\|x\|_2^2 = x^T x = x^T Q^T Q x = \|Qx\|_2^2$ if Q is an orthogonal matrix.

The following inequality, known as the *Cauchy–Schwarz inequality*, is used extensively in linear algebra and optimization:

$$|x^T y| \leq \|x\|_2 \|y\|_2. \tag{2.12}$$

The inequality can be derived by noting that

$$\left\| \|y\|_2^2 x - x^T y y \right\|_2^2 = \|y\|_2^2 \left(\|y\|_2^2 \|x\|_2^2 - 2(x^T y)^2 + (x^T y)^2 \right) \geq 0,$$

where the inequality follows from the nonnegativity property of the norm. An immediate consequence is that $\|y\|_2^2 \|x\|_2^2 - (x^T y)^2 \geq 0$ for all x and y, from which the Cauchy–Schwarz inequality (2.12) follows by rearranging the terms and taking the square root.

The *Frobenius norm* of $A \in \mathbb{R}^{m \times n}$ is a vector norm on $\mathbb{R}^{m \times n}$ induced by the Frobenius inner product, i.e.

$$\|A\|_F = \sqrt{\text{tr}(A^T A)} = \left(\sum_{i=1}^{m} \sum_{j=1}^{n} |a_{ij}|^2 \right)^{1/2}. \tag{2.13}$$

This is sometimes referred to as an *entrywise* norm. In contrast, given both vector norms on $\mathbb{R}^m$ and $\mathbb{R}^n$, the *operator norm* or *induced norm* on $\mathbb{R}^{m \times n}$ may be defined as

$$\|A\| = \inf_{t \geq 0} \{t \mid \|Ax\| \leq t \|x\| \ \forall x \in \mathbb{R}^n\}.$$

---

1 This is also known as the *triangle inequality*.

It follows directly from this definition that $\|Ax\| \leq \|A\| \|x\|$ for all $x \in \mathbb{R}^n$ and that $\|I\| = 1$. The induced norm may also be expressed as

$$\|A\| = \sup_{x \neq 0} \left\{ \frac{\|Ax\|}{\|x\|} \right\}, \tag{2.14}$$

from which the submultiplicative property $\|AB\| \leq \|A\| \|B\|$ follows:

$$\begin{aligned}
\|AB\| &= \sup_{x \neq 0} \left\{ \frac{\|ABx\|}{\|x\|} \right\} \\
&= \sup_{Bx \neq 0} \left\{ \frac{\|ABx\|}{\|x\|} \right\} = \sup_{Bx \neq 0} \left\{ \frac{\|ABx\|}{\|Bx\|} \frac{\|Bx\|}{\|x\|} \right\} \\
&\leq \sup_{y \neq 0} \left\{ \frac{\|Ay\|}{\|y\|} \right\} \sup_{x \neq 0} \left\{ \frac{\|Bx\|}{\|x\|} \right\} = \|A\| \|B\|.
\end{aligned}$$

The matrix p-norm of A is the norm induced by the vector p-norm, and it is denoted $\|A\|_p$. The next example considers the matrix 1-norm, and the matrix infinity norm is treated in Exercise 2.3.

Example 2.1   The matrix 1-norm on $\mathbb{R}^{m \times n}$ is the induced norm defined as

$$\|A\|_1 = \sup_{x \neq 0} \left\{ \frac{\|Ax\|_1}{\|x\|_1} \right\} = \sup_{\|x\|_1 = 1} \left\{ \|Ax\|_1 \right\} = \max_{i = \mathbb{N}_n} \|a_i\|_1,$$

where $a_1, \ldots, a_n$ denote the n columns of A. To verify the third equality, first note that

$$\sup_{\|x\|_1 = 1} \left\{ \|Ax\|_1 \right\} \geq \|Ae_j\|_1 = \|a_j\|_1, \quad j \in \mathbb{N}_n,$$

where $e_1, \ldots, e_n$ are the columns of the identity matrix of order n. Moreover, subadditivity, i.e. the triangle inequality, implies that

$$\|A\|_1 = \sup_{\|x\|_1 = 1} \left\{ \left\| \sum_{j=1}^n a_j x_j \right\|_1 \right\} \leq \sup_{\|x\|_1 = 1} \left\{ \sum_{j=1}^n |x_j| \|a_j\|_1 \right\} = \max_{j \in \mathbb{N}_n} \|a_j\|_1,$$

which shows that $\|A\|_1 = \max_{i \in \mathbb{N}_n} \|a_i\|_1$.

2.4 Algorithm Complexity

The number of arithmetic operations on floating-point numbers required by an algorithm is often used as a rough measure of its complexity. One *floating-point operation* or "FLOP" refers to a single floating-point addition, subtraction, multiplication, or division, and an algorithm's FLOP count is the total number of FLOPs. When expressed in terms of the problem size, the FLOP count often provides a useful characterization of the asymptotic growth of the running time as a function of the problem size. However, FLOP counts can generally not be used to accurately predict computation time on modern computers. Table 2.1 shows the FLOP counts for some basic operations involving vectors and/or matrices. Since we are mostly interested in the asymptotic growth rate as one or several problem dimensions increase, it is customary to use the so-called "Big O" notation. A function $f(m, n)$ is said to be "big O" of $g(m, n)$, which is expressed mathematically as $f(m, n) = O(g(m, n))$, if there exists scalars $c > 0$, m_0, and n_0 such that

$$f(m, n) \leq c\, g(m, n), \quad \text{for all } m \geq m_0, n \geq n_0. \tag{2.15}$$

Table 2.1 FLOP counts for basic matrix and vector operations: α denotes a scalar, x and y are n-vectors, and $A \in \mathbb{R}^{m \times n}$ and $B \in \mathbb{R}^{n \times p}$ are matrices.

Operation		FLOPs
Scaling	αx	n
Vector addition/subtraction	$x \pm y$	n
Inner product	$x^T y$	$2n - 1$
Matrix–vector multiplication	Ax	$m(2n - 1)$
Matrix–matrix multiplication	AB	$mp(2n - 1)$

The function g provides an upper bound on the growth rate of f as the parameters m and/or n increase. For example, the FLOP count for the matrix–vector product Ax is $O(mn)$ since $m(2n - 1) \leq 2 \, mn$ for all $m \geq 1$, $n \geq 1$. Similarly, the function $f(m, n) = m^2 + 10m + n^3$ satisfies $f(m, n) = O(m^2 + n^3)$ since $m^2 + 10m + n^3 \leq 2 \, (m^2 + n^3)$ for $m \geq 10, n \geq 1$.

Example 2.2 Recall that matrix multiplication is associative, i.e. given three matrices A, B, and C that are conformable for multiplication, we have that $A(BC) = (AB)C$. This implies that the product of three or more matrices $A_1 A_2 \dots A_k$ can be evaluated in several ways, each of which may have a different FLOP count. For example, the product ABC with $A \in \mathbb{R}^{m \times n}$, $B \in \mathbb{R}^{n \times p}$, and $C \in \mathbb{R}^{p \times q}$ may be evaluated left-to-right by first computing $L = AB$ and then LC, or right-to-left by computing $R = BC$ and then AR. The first approach requires $O(mp(n + q))$ FLOPs, whereas the second approach requires $O(nq(m + p))$ FLOPs. A special case is the product $ab^T x$ with $a \in \mathbb{R}^m$, $b \in \mathbb{R}^p$, and $x \in \mathbb{R}^p$ which requires $O(mp)$ FLOPs when evaluated left-to-right, but only $O(m + p)$ FLOPs when evaluated right-to-left. More generally, the product $A_1 A_2 \dots A_k$ of k matrices requires $k - 1$ matrix–matrix multiplications which can be carried out in any order since matrix multiplication is associative. The problem of finding the order that yields the lowest FLOP count is a combinatorial optimization problem, which is known as the *matrix chain ordering* problem, see Exercise 5.7. This problem can be solved by means of dynamic programming, which we introduce in Section 5.5 and discuss in more detail in Chapter 8.

2.5 Matrices with Structure

Matrices are sometimes classified according to their mathematical properties and structure, which are often of interest from a computational point of view. This section provides a brief review of some of the most frequently encountered kinds of structure in this book.

2.5.1 Diagonal Matrices

A *diagonal* matrix is a square matrix in which all off-diagonal entries are equal to zero. Such a matrix may be stored as a vector $d = (d_1, \dots, d_n)$ of length n rather than as a general square matrix with n^2 entries, which amounts to significant saving in terms of storage when n is large. We use the notation

$$\text{diag}(d) = \text{diag}(d_1, \dots, d_n),$$

when referring to the diagonal matrix with the elements of d on its diagonal. Diagonal matrices are also attractive from a computational point of view. For example, a matrix–vector product of the form $\mathrm{diag}(d)x = (d_1 x_1, \ldots, d_n x_n)$ involves only n FLOPs, whereas a matrix–vector product Ax with a general square matrix A requires $O(n^2)$ FLOPs. Similarly, the inverse of a nonsingular diagonal matrix $\mathrm{diag}(d)$ is also diagonal, i.e. $\mathrm{diag}(d)^{-1} = \mathrm{diag}(1/d_1, \ldots, 1/d_n)$.

2.5.2 Orthogonal Matrices

Recall that a square matrix Q of order n is said to be *orthogonal* if $Q^T = Q^{-1}$, or equivalently, $Q^T Q = QQ^T = I$. The product of two orthogonal matrices Q_1 and Q_2 of order n is itself orthogonal, which follows by noting that $(Q_1 Q_2)^{-1} = Q_2^{-1} Q_1^{-1} = Q_2^T Q_1^T = (Q_1 Q_2)^T$.

A basic example of an orthogonal matrix of order 2 is a *rotation matrix* of the form

$$R = \begin{bmatrix} \cos(\theta) & -\sin(\theta) \\ \sin(\theta) & \cos(\theta) \end{bmatrix}. \tag{2.16}$$

The action of such a matrix corresponds to the counterclockwise rotation by an angle θ about the origin in $\mathbb{R}^2$. More generally, a *Givens rotation* in $\mathbb{R}^n$ is a rotation in a two-dimensional plane defined by two coordinate axes, say, i and j. Such a transformation can be represented by an orthogonal matrix of order n

$$G(i, j, \theta) = \begin{bmatrix} 1 & \ldots & 0 & \ldots & 0 & \ldots & 0 \\ \vdots & \ddots & \vdots & & \vdots & & \vdots \\ 0 & \ldots & c & \ldots & -s & \ldots & 0 \\ \vdots & & \vdots & \ddots & \vdots & & \vdots \\ 0 & \ldots & s & \ldots & c & \ldots & 0 \\ \vdots & & \vdots & & \vdots & \ddots & \vdots \\ 0 & \ldots & 0 & \ldots & 0 & \ldots & 1 \end{bmatrix}, \quad i < j, \tag{2.17}$$

where $c = \cos(\theta)$ and $s = \sin(\theta)$ such that the principal submatrix defined by i and j is of the form (2.16). Givens rotations are used frequently in linear algebra as a means to introduce zeros in a vector or a matrix. For example, a nonzero vector (a, b) can be transformed into the vector $(r, 0)$ with $r = \sqrt{a^2 + b^2}$ by means of a rotation. Indeed, it is easy to verify that

$$\begin{bmatrix} c & -s \\ s & c \end{bmatrix} \begin{bmatrix} a \\ b \end{bmatrix} = \begin{bmatrix} r \\ 0 \end{bmatrix}, \quad \begin{bmatrix} c & -s \\ s & c \end{bmatrix}^T \begin{bmatrix} c & -s \\ s & c \end{bmatrix} = I,$$

if we let $c = a/r$ and $s = -b/r$.

A *Householder transformation* or *Householder reflection* in $\mathbb{R}^n$ is a linear transformation that corresponds to the reflection about a hyperplane that contains the origin. Such a transformation can be represented in terms of an orthogonal matrix of the form

$$H = I - 2 \frac{vv^T}{\|v\|_2^2}, \quad v \in \mathbb{R}^n, \tag{2.18}$$

where $v \neq 0$ is normal to the reflection plane. This is illustrated in Figure 2.4. Matrices of the form (2.18) are referred to as *Householder matrices*, and they are typically stored implicitly. In other words, only the vector v, which uniquely determines H, is stored rather than storing H directly. This requires $O(n)$ storage rather than $O(n^2)$. Moreover, matrix–vector products with a Householder matrix can be computed in $O(n)$ FLOPs by noting that the operation Hx amounts to a vector operation, i.e. $Hx = x - \alpha v$ with $\alpha = 2\frac{v^T x}{\|v\|_2^2}$. A Householder transformation may be used to simultaneously introduce up to $n - 1$ zeros in a vector of length n, see Exercise 2.4.

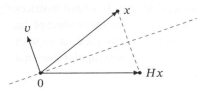

Figure 2.4 Householder reflection in $\mathbb{R}^2$.

Yet another example of a family of orthogonal matrices are *permutation matrices*. A permutation of the elements of a vector $x \in \mathbb{R}^n$ is a reordering of the elements. This can be expressed as a map of the form

$$(x_1, \ldots, x_n) \mapsto (x_{\pi(1)}, \ldots, x_{\pi(n)}), \tag{2.19}$$

where the function $\pi : \mathbb{N}_n \to \mathbb{N}_n$ defines the permutation and satisfies $\pi(i) \neq \pi(j) \iff i \neq j$. This map can be expressed as a matrix–vector product Px, where $P \in \mathbb{R}^{n \times n}$ is the permutation matrix defined as

$$P = \begin{bmatrix} e_{\pi(1)}^T \\ \vdots \\ e_{\pi(n)}^T \end{bmatrix}. \tag{2.20}$$

It is easy to verify that $PP^T = I$ since $\pi(i) \neq \pi(j)$ whenever $i \neq j$, which implies that P is orthogonal. It follows that the map (2.19) is invertible, and $P^{-1} = P^T$ is the permutation matrix that corresponds to the inverse permutation π^{-1}. The special case where $\pi(i) = n + 1 - i$ corresponds to reversing the order of n elements and leads to the permutation matrix

$$J = \begin{bmatrix} e_n^T \\ \vdots \\ e_1^T \end{bmatrix} = \begin{bmatrix} & & 1 \\ & \cdot^{\cdot^{\cdot}} & \\ 1 & & \end{bmatrix},$$

which is the so-called *exchange matrix* of order n.

2.5.3 Triangular Matrices

A square matrix in which the elements above or below the main diagonal are zero is called a *triangular matrix*. A matrix $L \in \mathbb{R}^{n \times n}$ is said to be *lower triangular* if $1 \leq i < j \leq n \implies l_{ij} = 0$, whereas $U \in \mathbb{R}^{n \times n}$ is said to be *upper triangular* if $1 \leq j < i \leq n \implies u_{ij} = 0$. Furthermore, a triangular matrix with ones on its diagonal is said to be *unit triangular*. The determinant of a triangular matrix T of order n may be expressed as

$$\det(T) = \prod_{i=1}^{n} T_{ii},$$

which follows from the Laplace expansion (2.8). A direct consequence is that T is nonsingular if and only if all of the diagonal entries of T are nonzero. The inverse of a nonsingular lower (upper) triangular matrix T is another lower (upper) triangular matrix. This follows from the identity (2.9) by noting that the cofactor matrix associated with T is itself triangular. To see this, first recall that the minor of t_{ij} is the determinant of the $(n-1) \times (n-1)$ matrix obtained by deleting the ith row and the jth column of T. This implies that if, say, T is lower triangular and $i > j$, then the effect of deleting the ith row and the jth column of T is a triangular matrix of order $n-1$. This matrix will have at least one diagonal entry that is equal to zero, and hence, it is singular, which implies that the minor of t_{ij} is zero. We note that the product of two lower (upper) triangular matrices is another lower (upper) triangular matrix.

Given a nonsingular triangular matrix T, it is possible to compute matrix–vector products of the form $y = T^{-1}x$ and $y = T^{-T}x$ without computing T^{-1} first. For example, if T is lower triangular, then $y = T^{-1}x$ can be computed by means of *forward substitution*, i.e.

$$y_1 \leftarrow T_{11}^{-1}x_1, \quad y_k \leftarrow T_{kk}^{-1}\left(x_k - \sum_{i=1}^{k-1} T_{ki}y_i\right), \quad k = 2, \ldots, n.$$

This requires approximately n^2 FLOPs. Similarly, if T is an upper triangular matrix, then $y = T^{-1}x$ can be computed using *backward substitution*, i.e.

$$y_n \leftarrow T_{nn}^{-1}x_n, \quad y_k \leftarrow T_{kk}^{-1}\left(x_k - \sum_{i=k+1}^{n} T_{ki}y_i\right), \quad k = n-1, \ldots, 1,$$

which also requires approximately n^2 FLOPs.

2.5.4 Symmetric and Skew-Symmetric Matrices

A square matrix X is said to be *symmetric* if $X = X^T$, and

$$\mathbb{S}^n = \{X \in \mathbb{R}^{n \times n} \mid X = X^T\} \tag{2.21}$$

is the set of symmetric matrices of order n. Similarly, a square matrix Y is *skew-symmetric* if $Y = -Y^T$, which implies that the diagonal elements of a skew-symmetric matrix are equal to zero. A square matrix $A \in \mathbb{R}^{n \times n}$ can always be decomposed into the sum of a symmetric and a skew-symmetric matrix, i.e. A can be expressed as $A = X + Y$, where $X = (1/2)(A + A^T)$ is symmetric and $Y = \frac{1}{2}(A - A^T)$ is skew-symmetric. We note that this is an orthogonal decomposition of A since the Frobenius inner product of a symmetric and a skew-symmetric matrix is always zero.

Equipped with the Frobenius inner product, which is defined in (2.2), the set $\mathbb{S}^n$ is a Euclidean vector space of dimension $n(n+1)/2$. It is sometimes convenient to work with a vector representation of a symmetric matrix, and to this end, we introduce the bijective linear map svec : $\mathbb{S}^n \to \mathbb{R}^{n(n+1)/2}$ defined as

$$\text{svec}(A) = \left(a_{11}, \sqrt{2}a_{21}, \ldots, \sqrt{2}a_{n1}, a_{22}, \sqrt{2}a_{32}, \ldots, \sqrt{2}a_{n2}, \ldots, a_{nn}\right), \tag{2.22}$$

i.e. svec stacks the lower triangular part of the columns of A and scales the off-diagonal entries by $\sqrt{2}$. It is straightforward to verify that this definition implies that for all $X, Y \in \mathbb{S}^n$,

$$x = \text{svec}(X), \ y = \text{svec}(Y) \implies x^Ty = \text{tr}(X^TY),$$

which means that the svec transformation preserves inner products.

2.5.5 Toeplitz and Hankel Matrices

A Toeplitz matrix of order n is a square matrix of the form

$$T = \begin{bmatrix} a_0 & a_{-1} & \cdots & a_{-(n-1)} \\ a_1 & a_0 & \ddots & \vdots \\ \vdots & \ddots & \ddots & a_{-1} \\ a_{n-1} & \cdots & a_1 & a_0 \end{bmatrix}, \tag{2.23}$$

where the first row and column of T define the subdiagonal, diagonal, and superdiagonal elements. Such a matrix is sometimes referred to as a *diagonal-constant* matrix, and it is uniquely determined by $2n - 1$ numbers in contrast to a general $n \times n$ matrix, which has n^2 degrees of freedom. A special case of (2.23) is the so-called *lower-shift matrix* of order n, which has ones on the first subdiagonal and zeros elsewhere, i.e.

$$S = \begin{bmatrix} 0 & & & \\ 1 & \ddots & & \\ & \ddots & \ddots & \\ & & 1 & 0 \end{bmatrix}. \tag{2.24}$$

Similarly, S^T is sometimes referred to as the *upper-shift matrix* of order n. It is easy to verify that S^k with $k \in \mathbb{N}_{n-1}$ is the matrix with ones on the kth subdiagonal and zeros elsewhere. Using the convention that $S^0 = I$, we may express any Toeplitz matrix of the form (2.23) as

$$T = \sum_{k=0}^{n-1} a_k S^k + \left(\sum_{k=1}^{n-1} a_{-k} S^k \right)^T. \tag{2.25}$$

A lower triangular Toeplitz matrix whose first column is given by $a_0, \ldots, a_{n-1}$ can be expressed as

$$L = \sum_{k=0}^{n-1} a_k S^k. \tag{2.26}$$

It then follows from the fact that $\det(L) = a_0^n$ that L is nonsingular if and only if $a_0 \neq 0$. Moreover, a matrix of the form (2.26) commutes with S, which follows from the fact that S commutes with S^k, i.e. $SS^k = S^k S$. It can also be shown that the inverse of a nonsingular lower triangular Toeplitz matrix is another lower triangular Toeplitz matrix, see Exercise 2.6.

Toeplitz matrices are closely related to another class of matrices called *Hankel* matrices. Like a Toeplitz matrix, a Hankel matrix is a square matrix that is uniquely determined by its first row and column. However, unlike a Toeplitz matrix, a Hankel matrix has constant entries along the anti-diagonal, the anti-subdiagonals, and anti-superdiagonals. As a consequence, the exchange matrix J maps a Toeplitz matrix to a Hankel matrix and vice versa, i.e. if T is a Toeplitz matrix, then JT and TJ are both Hankel matrices. We note that Hankel matrices are symmetric, whereas Toeplitz matrices are *persymmetric*, i.e. symmetric about the anti-diagonal. Finally, we note that the notion of Toeplitz (Hankel) matrices can be extended to include nonsquare matrices with diagonal-constant (anti-diagonal-constant) structure, and such matrices may be viewed as submatrices of square Toeplitz (Hankel) matrices.

2.5.6 Sparse Matrices

A matrix $A \in \mathbb{R}^{m \times n}$ is said to be *sparse* or *sparsely populated* if a large number of its entries are equal to zero. In contrast, a matrix with few or no zero entries is said to be *dense*. We will use nnz(A) to denote the number of nonzero entries in a matrix A. Sparse matrices can be stored efficiently by storing only the nonzero entries. For example, storing a sparse matrix A as a list of nnz(A) triplets of the form (i, j, A_{ij}) requires $O(\text{nnz}(A))$ rather than $O(mn)$ storage. Similarly, a matrix–vector product Ax requires only $O(\text{nnz}(A))$ FLOPs if sparsity is exploited as opposed to $O(mn)$ FLOPs if A is a general dense matrix.

2.5.7 Band Matrices

A sparse matrix $B \in \mathbb{R}^{m \times n}$ is said to be *banded* if all its nonzero entries are located within a band about the main diagonal. The *bandwidth* of B is the smallest integer k such that $b_{ij} = 0$ if $|i - j| > k$. In other words, all entries below the kth subdiagonal or above the k superdiagonal are equal to zero. The *lower bandwidth* of B is a nonnegative integer l such that $b_{ij} = 0$ if $j < i - l$, and similarly, the *upper bandwidth* of B is a nonnegative integer u such that $b_{ij} = 0$ if $j > i + u$. A bandwidth of 0 corresponds to a diagonal matrix, whereas a bandwidth of 1 is a *tridiagonal* matrix or, if $l = 0$ or $u = 0$, an upper or a lower *bidiagonal* matrix.

2.6 Quadratic Forms and Definiteness

A real n-ary *quadratic form* is a homogeneous quadratic polynomial in n variables, or equivalently, a function from $\mathbb{R}^n$ to $\mathbb{R}$ of the form

$$f(x) = x^T A x = \sum_{i=1}^{n} \sum_{j=1}^{n} A_{ij} x_i x_j,$$

for some $A \in \mathbb{R}^{n \times n}$. It is straightforward to verify that

$$x^T A x = x^T A^T x = (1/2) x^T (A + A^T) x,$$

which implies that only the symmetric part of A contributes to the value of $x^T A x$. We therefore limit our attention to the case where $A \in \mathbb{S}^n$.

A matrix $A \in \mathbb{S}^n$ is *positive semidefinite* if and only if $x^T A x \geq 0$ for all $x \in \mathbb{R}^n$, and it is *positive definite* if and only if $x^T A x > 0$ for all $x \neq 0$. Similarly, A is negative (semi)definite if $-A$ is positive (semi)definite, and it is *indefinite* if it is neither positive semidefinite nor negative semidefinite. We will use the notation $\mathbb{S}^n_+$ for the set of positive semidefinite matrices in $\mathbb{S}^n$, the interior of which is the set of positive definite matrices, denoted by $\mathbb{S}^n_{++}$.

Given two matrices $A, B \in \mathbb{S}^n$, the *generalized inequality* $A \succeq_{\mathbb{S}^n_+} B$, which is a partial ordering on $\mathbb{S}^n$, is defined as

$$A \succeq_{\mathbb{S}^n_+} B \iff A - B \in \mathbb{S}^n_+. \tag{2.27}$$

Similarly, the strict generalized inequality $A \succ_{\mathbb{S}^n_+} B$ is defined as

$$A \succ_{\mathbb{S}^n_+} B \iff A - B \in \mathbb{S}^n_{++}. \tag{2.28}$$

To simplify notation, we will omit the subscript $\mathbb{S}^n_+$ and simply write $A \geq B$ and $A > B$ when there is no danger of ambiguity. We return to generalized inequalities in Section 4.2.

We end this section by deriving some useful properties of positive semidefinite matrices. To this end, we consider a matrix $X \in \mathbb{S}^n_+$, which we partition as

$$X = \begin{bmatrix} A & B \\ B^T & C \end{bmatrix}, \tag{2.29}$$

where $A \in \mathbb{S}^{n_1}$, $B \in \mathbb{R}^{n_1 \times n_2}$, and $C \in \mathbb{S}^{n_2}$ with $n_1 + n_2 = n$. Positive semidefiniteness implies that $z^T X z \geq 0$ for all $z \in \mathbb{R}^n$ or, equivalently, for all $z = (u, v) \in \mathbb{R}^{n_1} \times \mathbb{R}^{n_2}$,

$$f(u, v) = u^T A u + v^T C v + 2 u^T B v \geq 0.$$

Thus, $f(u, 0) = u^T A u \geq 0$ for all u and $f(0, v) = v^T C v \geq 0$ for all v, so both A and C must be positive semidefinite. This holds for any partition of the form (2.29) and any symmetric permutation of X which, in turn, implies that every principal submatrix of X must be positive semidefinite.

Positive semidefiniteness of X also implies that

$$\mathcal{R}(B) \subseteq \mathcal{R}(A), \quad \mathcal{R}(B^T) \subseteq \mathcal{R}(C), \tag{2.30}$$

which is easily proven by contradiction: if $\mathcal{R}(B) \subsetneq \mathcal{R}(A)$, then there exists a vector v such that $Bv \neq 0$ and $Bv \notin \mathcal{R}(A)$, and hence, $f(-tBv, v) = v^T Cv - 2t\|Bv\|_2^2$ tends to $-\infty$ as $t \to \infty$. This is a contradiction since X is positive semidefinite. The condition $\mathcal{R}(B^T) \subseteq \mathcal{R}(C)$ can be proven in a similar manner. An immediate consequence of the range conditions (2.30) is that the ith row and column of X must be zero if $X_{ii} = 0$.

2.7 Spectral Decomposition

A symmetric matrix A of order n can be factorized as

$$A = Q\Lambda Q^T,$$

where Q is an orthogonal matrix and Λ is diagonal. This is a so-called *spectral decomposition* or *eigendecomposition* of the symmetric matrix A. The columns of Q are eigenvectors of A, and the diagonal entries of $\Lambda = \mathrm{diag}(\lambda_1, \dots, \lambda_n)$ are the associated eigenvalues, i.e. the ith column of Q satisfies the equation $Aq_i = \lambda_i q_i$. We will use the convention that the eigenvalues are ordered such that $\lambda_1 \geq \lambda_2 \geq \cdots \geq \lambda_n$. The definiteness of a symmetric matrix is related to the sign of its eigenvalues. Specifically, given an eigendecomposition $A = Q\Lambda Q^T$, we can express the quadratic form $x^T Ax$ as $y^T \Lambda y = \sum_{i=1}^{n} \lambda_i y_i^2$ by using the change of variables $y = Q^T x$. It follows that A is positive semidefinite if and only if its eigenvalues are nonnegative since

$$\sum_{i=1}^{n} \lambda_i y_i^2 \geq 0, \quad \forall y \in \mathbb{R}^n \iff \lambda_{\min}(A) \geq 0.$$

Analogously, A is positive definite if and only if $\lambda_{\min}(A) > 0$, which implies that A has full rank, and it is indefinite if it has both positive and negative eigenvalues. A positive definite matrix A of order n defines a *weighted* inner product $\langle y, x \rangle_A = \langle y, Ax \rangle = y^T Ax$, which induces the *quadratic norm*

$$\|x\|_A = \sqrt{\langle x, x \rangle_A} = \sqrt{x^T Ax}.$$

The *symmetric square root* of a matrix $A \in \mathbb{S}_+^n$ is the unique symmetric positive semidefinite matrix $A^{1/2}$ that satisfies $A = A^{1/2}A^{1/2}$. Given a spectral decomposition $A = Q\Lambda Q^T$, the symmetric square root of A may be expressed as

$$A^{1/2} = Q\Lambda^{1/2}Q^T = Q\mathrm{diag}\left(\lambda_1^{1/2}, \dots, \lambda_n^{1/2}\right)Q^T.$$

This implies that transformations of the form $A \mapsto F^T AF$ with $F \in \mathbb{R}^{n \times k}$ preserve positive semidefiniteness, i.e. we have that

$$x^T Ax \geq 0, \quad \forall x \in \mathbb{R}^n \implies y^T F^T AFy = \|A^{1/2}Fy\|_2^2 \geq 0, \quad \forall y \in \mathbb{R}^k.$$

Moreover, if A is positive definite and $\mathrm{rank}(F) = k$, then $F^T AF$ is also positive definite. We note that A and $B = F^T AF$ are said to be *congruent* if F is square and nonsingular.

The eigenvalues and eigenvectors of a symmetric matrix are related to the so-called *Rayleigh quotient* which, for a given matrix $A \in \mathbb{S}^n$ and a nonzero vector $x \in \mathbb{R}^n$, is defined as

$$R_A(x) = \frac{x^T Ax}{x^T x}, \quad x \neq 0. \tag{2.31}$$

A stationary point of R_A must satisfy

$$\nabla R_A(x) = 0 \iff Ax = \frac{x^T Ax}{x^T x}x,$$

which implies that x is a stationary point of R_A if and only if x is a real-valued eigenvector of A associated with the eigenvalue $R_A(x)$. This observation implies that we may express the largest and smallest eigenvalues of A as

$$\lambda_{\max}(A) = \max_{x \neq 0} \{R_A(x)\}, \quad \lambda_{\min}(A) = \min_{x \neq 0} \{R_A(x)\}.$$

2.8 Singular Value Decomposition

A *singular value decomposition* (SVD) of a matrix $A \in \mathbb{R}^{m \times n}$ is a factorization of the form $A = U\Sigma V^T$, where $U \in \mathbb{R}^{m \times m}$ and $V \in \mathbb{R}^{n \times n}$ are orthogonal matrices, and $\Sigma \in \mathbb{R}^{m \times n}$ is a matrix with the *singular values* of A on its main diagonal and zeros elsewhere, i.e. the diagonal entries of Σ are $\Sigma_{ii} = \sigma_i$, $i \in \mathbb{N}_{\min(m,n)}$ where we use the convention that $\sigma_1 \geq \sigma_2 \geq \cdots \geq 0$. If we let r denote the rank of A, then $\sigma_i = 0$ for $i > r$, and hence, we can partition an SVD of A as

$$A = \begin{bmatrix} U_1 & U_2 \end{bmatrix} \begin{bmatrix} S & 0 \\ 0 & 0 \end{bmatrix} \begin{bmatrix} V_1^T \\ V_2^T \end{bmatrix} = U_1 S V_1^T, \tag{2.32}$$

where $U_1 \in \mathbb{R}^{m \times r}$, $V_1 \in \mathbb{R}^{n \times r}$, and $S = \text{diag}(\sigma_1, \ldots, \sigma_r)$ is the square matrix with the nonzero singular values of A on its diagonal. This shows that an SVD is a so-called *rank-revealing* factorization, and $A = U_1 S V_1^T$ is commonly referred to as a *thin* or *reduced* SVD of A. We note that the largest integer k such that $\sigma_k > \epsilon$ for a given $\epsilon > 0$ is referred to as the *numerical rank* or the ϵ-rank of A. The ϵ-rank of A may also be defined as

$$\text{rank}(A, \epsilon) = \min_{\|E\|_2 \leq \epsilon} \text{rank}(A + E),$$

which allows us to interpret the numerical rank of A as the smallest attainable rank for matrices within a neighborhood of A.

The partition (2.32) can be linked to the four subspaces introduced in Section 2.2 and illustrated in Figure 2.1. Specifically, we have that

$$\mathcal{R}(A) = \mathcal{R}(U_1), \quad \mathcal{N}(A^T) = \mathcal{R}(U_2), \quad \mathcal{R}(A^T) = \mathcal{R}(V_1), \quad \mathcal{N}(A) = \mathcal{R}(V_2).$$

This allows us to express the projection of a vector $x \in \mathbb{R}^m$ onto $\mathcal{R}(A)$ as

$$\mathcal{P}_{\mathcal{R}(A)}(x) = \underset{u \in \mathcal{R}(a)}{\text{argmin}}\{\|u - x\|_2\} = U_1 U_1^T x.$$

The matrix $P = U_1 U_1^T$ is a *projection matrix*, and it is also an *idempotent matrix* since $P^2 = P$. Moreover, $I - P$ is also a projection matrix, which follows from the fact that $I - P = U_2 U_2^T$, and it defines a projection onto $\mathcal{N}(A^T) = \mathcal{R}(A)^\perp$.

An SVD can also be used to construct useful upper and lower bounds on the trace inner product of two matrices. A notable example is *von Neumann's trace inequality*, which states that

$$|\text{tr}(A^T B)| \leq \text{tr}(\Sigma^T \Gamma) = \sum_{i=1}^{\min(m,n)} \Sigma_{ii} \Gamma_{ii}, \tag{2.33}$$

where $A, B \in \mathbb{R}^{m \times n}$ are matrices with SVDs $A = U\Sigma V^T$ and $B = P\Gamma Q^T$, see, e.g. [76] for the a proof.

The singular values of an $m \times n$ matrix A can be used to define a family of matrix norms on $\mathbb{R}^{m \times n}$ that are known as *Schatten norms*. For $p \in [1, \infty)$, the Schatten p-norm is defined as

$$\|A\|_{(p)} = \left(\sum_{i=1}^{\min(m,n)} \sigma_i(A)^p \right)^{1/p}, \tag{2.34}$$

i.e. it may be viewed as the p-norm of a vector that contains the singular values of A. The parentheses in the subscript is not standard notation, but we include them here to avoid confusion with the induced matrix p-norm defined in (2.14). The Schatten 1-norm, which we will denote by $\|A\|_*$, is also known as the *nuclear norm* or the *trace norm*. It is straightforward to verify that the matrix norms $\|A\|_F$ and $\|A\|_2$ are both special cases of the Schatten p-norm, see Exercise 2.5.

2.9 Moore–Penrose Pseudoinverse

The *Moore–Penrose pseudoinverse* of a matrix $A \in \mathbb{R}^{m \times n}$ is a generalization of the inverse of a square matrix. It is denoted $A^\dagger$ and is the unique $n \times m$ matrix that satisfies the four so-called *Penrose conditions*

$$AA^\dagger A = A, \quad A^\dagger AA^\dagger = A^\dagger, \quad (AA^\dagger)^T = AA^\dagger, \quad (A^\dagger A)^T = A^\dagger A. \tag{2.35}$$

An SVD $A = U\Sigma V^T$ can be used to construct $A^\dagger$ as

$$A^\dagger = V\Sigma^\dagger U^T, \tag{2.36}$$

where $\Sigma^\dagger$ is obtained from Σ by taking the reciprocal of all the nonzero elements. Equivalently, using (2.32), we can write $A^\dagger$ as

$$A^\dagger = \begin{bmatrix} V_1 & V_2 \end{bmatrix} \begin{bmatrix} S^{-1} & 0 \\ 0 & 0 \end{bmatrix} \begin{bmatrix} U_1^T \\ U_2^T \end{bmatrix} = V_1 S^{-1} U_1^T, \tag{2.37}$$

from which it follows that $\mathcal{R}(A^\dagger) = \mathcal{R}(A^T)$ and $\mathcal{N}(A^\dagger) = \mathcal{N}(A^T)$. In the special case where A has full rank, the Moore–Penrose pseudoinverse can be expressed as

$$A^\dagger = \begin{cases} (A^T A)^{-1} A^T, & \operatorname{rank}(A) = n, \\ A^T (AA^T)^{-1}, & \operatorname{rank}(A) = m, \end{cases} \tag{2.38}$$

in which case $A^\dagger$ is a left inverse of A, if $\operatorname{rank}(A) = n$, and/or a right inverse of A, if $\operatorname{rank}(A) = m$. More generally, the Moore–Penrose pseudoinverse may be defined as the limit

$$A^\dagger = \lim_{\delta \searrow 0} (A^T A + \delta I)^{-1} A^T,$$

$$= \lim_{\delta \searrow 0} A^T (AAT + \delta I)^{-1}.$$

The Moore–Penrose pseudoinverse provides a convenient way to express projections onto the four subspaces $\mathcal{R}(A)$, $\mathcal{N}(A^T)$, $\mathcal{R}(A^T)$, and $\mathcal{N}(A)$. This follows from (2.32) and (2.36) by noting that $AA^\dagger$ and $A^\dagger A$ are projection matrices, i.e.

$$AA^\dagger = U_1 U_1^T, \quad A^\dagger A = V_1 V_1^T.$$

Thus, projections onto the four subspaces can be expressed in terms of the projection matrices included in Table 2.2.

Table 2.2 The four subspaces and the corresponding projection matrices.

Subspace	Projection matrix
$\mathcal{R}(A)$	$AA^\dagger = U_1 U_1^T$
$\mathcal{N}(A^T)$	$I - AA^\dagger = U_2 U_2^T$
$\mathcal{R}(A^T)$	$A^\dagger A = V_1 V_1^T$
$\mathcal{N}(A)$	$I - A^\dagger A = V_2 V_2^T$

2.10 Systems of Linear Equations

Consider a system of m linear equations in n unknowns, which may be expressed as

$$Ax = b, \tag{2.39}$$

where $A \in \mathbb{R}^{m \times n}$ is the *coefficient matrix*, $x \in \mathbb{R}^n$ is the vector of unknowns, and the vector $b \in \mathbb{R}^m$ is commonly referred to as the *right-hand side* of the system. Such systems are ubiquitous in numerical optimization and can be solved in many different ways, some of which we will review in this section.

The system (2.39) is said to be *consistent* if $b \in \mathcal{R}(A)$, in which case the solution set $\mathcal{X} = \{x \mid Ax = b\}$ is nonempty. Alternatively, the system is said to be *inconsistent* if $b \notin \mathcal{R}(A)$, which implies that there is no solution. A consistent system of the form (2.39) has a unique solution if and only if A has a trivial nullspace, or equivalently, if $\text{rank}(A) = n$. Indeed, if x and z are both solutions and $x \neq z$, then $A(x - z) = 0$ which implies that A has a nontrivial nullspace. Recall that $b = AA^\dagger b$ if $b \in \mathcal{R}(A)$, in which case $x = A^\dagger b$ satisfies $Ax = b$. This is the unique solution if $\text{rank}(A) = n$, and otherwise, there are infinitely many solutions. If $F \in \mathbb{R}^{n \times k}$ is a matrix such that $\mathcal{R}(F) = \mathcal{N}(A)$, then the solution set may be expressed as

$$\mathcal{X} = \{A^\dagger b + Fz \mid z \in \mathbb{R}^k\}. \tag{2.40}$$

The solution $x = A^\dagger b$ is the so-called *least-norm* solution, which is unique. To see this, note that $\|x\|_2^2$ can be expanded as

$$\begin{aligned}
\|x\|_2^2 &= \|A^\dagger b + Fz\|_2^2, \\
&= \|A^\dagger b\|_2^2 + \|Fz\|_2^2 + 2z^T F^T A^\dagger b, \\
&= \|A^\dagger b\|_2^2 + \|Fz\|_2^2,
\end{aligned}$$

where the last equality follows from the fact that $F^T A^\dagger = 0$ since $\mathcal{R}(A^\dagger) = \mathcal{R}(A^T) = \mathcal{N}(A)^\perp$.

2.10.1 Gaussian Elimination

We now turn to methods for solving a system of linear equations of the form $Ax = b$, where A is a nonsingular square matrix of order n. One such methods is *Gaussian elimination* with partial pivoting, which successively eliminates a variable from the system. To illustrate the basic principle behind variable elimination, we introduce a permutation matrix P and partition PA as

$$PA = \begin{bmatrix} \alpha & v^T \\ u & C \end{bmatrix},$$

such that α is a nonzero scalar. The element α is called the *pivot element*, and P is typically chosen such that α is an element of maximal absolute value from the first column of A. Introducing conformable partitions $x = (x_1, \tilde{x})$ and $Pb = (\beta, \tilde{b})$, the system $PAx = Pb$ can be expressed as

$$\alpha x_1 + v^T \tilde{x} = \beta, \quad u x_1 + C\tilde{x} = \tilde{b}.$$

Solving for x_1 in the first equation yields $x_1 = \alpha^{-1}(\beta - v^T \tilde{x})$, and substitution of this expression for x_1 in the second equations, we arrive at

$$(C - u\alpha^{-1}v^T)\tilde{x} = \tilde{b} - u\alpha^{-1}\beta,$$

which is a system of $n - 1$ equations in $n - 1$ unknowns. The coefficient matrix $C - u\alpha^{-1}v^T$ is the so-called *Schur complement* of α in PA, and it is nonsingular if and only if A is nonsingular. Thus, the variable elimination procedure may be repeated until we have one equation with a single unknown that is readily solved.

2.10.2 Block Elimination

It is sometimes convenient to eliminate several variables at the same time by means of *block elimination*. To illustrate the principle, we consider a system of equations $Mx = b$, where $M \in \mathbb{R}^{n \times n}$ is partitioned into four blocks

$$M = \begin{bmatrix} A & B \\ C & D \end{bmatrix}, \tag{2.41}$$

such that $A \in \mathbb{R}^{n_1 \times n_1}$ and $D \in \mathbb{R}^{n_2 \times n_2}$ are square matrices, and hence, $B \in \mathbb{R}^{n_1 \times n_2}$ and $C \in \mathbb{R}^{n_2 \times n_1}$. Introducing conformable partitions $x = (x_1, x_2)$ and $b = (b_1, b_2)$, the system $Mx = b$ may be expressed as

$$\begin{bmatrix} A & B \\ C & D \end{bmatrix} \begin{bmatrix} x_1 \\ x_2 \end{bmatrix} = \begin{bmatrix} b_1 \\ b_2 \end{bmatrix}. \tag{2.42}$$

Now, suppose A is nonsingular. The equation $Ax_1 + Bx_2 = b_1$ then allows us to express x_1 in terms of x_2 as

$$x_1 = A^{-1}(b_1 - Bx_2). \tag{2.43}$$

Substituting this expression for x_1 in the equation $Cx_1 + Dx_2 = b_2$ yields the equation

$$(D - CA^{-1}B)x_2 = b_2 - CA^{-1}b_1, \tag{2.44}$$

which only involves x_2. The matrix $D - CA^{-1}B$ is the Schur complement of A in M, which exists when A is nonsingular. Later in this section, we will see that this Schur complement is nonsingular if and only if M is nonsingular, provided that the Schur complement exists. As a consequence, we can solve the system $Mx = b$ in two steps when both M and A are nonsingular: first, we compute x_2 by solving the system (2.44), and then we compute x_1 using (2.43). This approach is often advantageous if A or its Schur complement $D - CA^{-1}B$ has some kind of structure that can be exploited.

2.11 Factorization Methods

Gaussian elimination and its block variant may be viewed as instances of a class of methods that solve $Ax = b$ in two steps: a *factorization* step and a *solve* step. The factorization step decomposes the coefficient matrix A into a product of factors, say, $A = A_1 \ldots A_k$, where k is typically equal to two or three. Moreover, each of the k factors generally has some kind of structure that can be exploited in the solve step. Typical examples include diagonal, orthogonal, and triangular factors. Using the factorization, the solve step amounts to solving

$$A_1 \ldots A_k x = b,$$

which is equivalent to k systems of equations

$$A_i x_i = x_{i-1}, \quad i \in \mathbb{N}_k,$$

where we define $x_0 = b$ and let $x = x_k$. The factorization step is typically the most costly, and the factorization can be reused if several systems that involve the same coefficient matrix need to be solved.

In what follows, we provide a review of some of the most common factorizations and their applications.

2.11.1 LU Factorization

A by-product of Gaussian elimination is a factorization of the form

$$A = PLU,$$

where A is nonsingular and square, P is a permutation matrix, L is unit lower triangular, and U is upper triangular. This is known as a *PLU factorization* or simply an *LU factorization* of A. The factorization requires roughly $(2/3)n^3$ FLOPs, and it only requires additional storage for a permutation vector if A is overwritten by the factors L and U. The factorization allows us to find the solution to $Ax = b$ by solving the three simpler systems of equations,

$$Pz = b, \quad Ly = z, \quad Ux = y.$$

The first system, $Pz = b$, has the solution $z = P^T b$, which is simply a permutation of b, and the two triangular systems can be solved by means of forward and backward substitution in roughly $2n^2$ FLOPs. Thus, the total cost of the factorization step and the solve step is roughly $(2/3)n^3 + 2n^2$ FLOPs. Note that the solve step is significantly cheaper than the factorization step, so it is generally advantageous to reuse the factorization of A if several systems with this coefficient matrix must be solved. We note that A^{-1} can be computed by solving the matrix equation $AX = I$, which is equivalent to solving the n systems $Ax_i = e_i$, $i \in \mathbb{N}_n$. This costs approximately $(2/3)n^3 + 2nn^2 = (8/3)n^3$ FLOPs if a single PLU factorization is computed and reused. Thus, the cost of solving $Ax = b$ by explicitly computing A^{-1} followed by the matrix–vector product $A^{-1}b$ is several times higher than that of the factor-solve approach.

2.11.2 Cholesky Factorization

The LU factorization may be simplified in the special case where the matrix A is symmetric and positive definite. Such matrices are strongly factorizable, which means that they can be factorized without pivoting. The resulting factorization is the *Cholesky factorization*, and it is the unique factorization of the form

$$A = LL^T, \tag{2.45}$$

where the Cholesky factor L is a lower triangular matrix with positive diagonal elements. The factorization is equivalent to the *LDL factorization*

$$A = \bar{L}D\bar{L}^T,$$

where $D = \mathrm{diag}(L_{11}^2, \dots, L_{nn}^2)$ and $\bar{L} = LD^{-1/2}$ is unit lower triangular. The cost of computing the Cholesky factorization is roughly $(1/3)n^3$ FLOPs.

2.11.3 Indefinite LDL Factorization

Unlike symmetric positive definite matrices, indefinite symmetric matrices are not strongly factorizable, and hence, it is generally not possible to factorize an indefinite symmetric matrix without pivoting. Moreover, symmetric pivoting of the form $A \leftarrow PAP^T$, where P is a permutation matrix is not sufficient to avoid zero pivots in general. To see this, note that the matrix

$$A = \begin{bmatrix} 0 & 1 \\ 1 & 0 \end{bmatrix}$$

is invariant under symmetric permutations. One approach to overcoming this issue, which due to [25], is to allow 1-by-1 and 2-by-2 pivot blocks. The resulting factorization is of the form

$$PAP^T = LDL^T, \tag{2.46}$$

where D is a block diagonal matrix that contains the pivot blocks, and the cost is roughly $(1/3)n^3$ FLOPs.

2.11.4 QR Factorization

A matrix $A \in \mathbb{R}^{m \times n}$ with linearly independent columns can be decomposed into the product of an orthogonal matrix $Q \in \mathbb{R}^{m \times m}$ and a matrix $R^{m \times n}$ with zeros below its diagonal, i.e.

$$A = QR = \begin{bmatrix} Q_1 & Q_2 \end{bmatrix} \begin{bmatrix} R_1 \\ 0 \end{bmatrix} = Q_1 R_1, \tag{2.47}$$

where $Q_1 \in \mathbb{R}^{m \times n}$ is the first n columns of Q, and $R_1 \in \mathbb{R}^{n \times n}$ is upper triangular with nonzero diagonal elements. Such a factorization is called a *QR factorization* and can be computed in several ways, e.g. by applying a sequence of $n - 1$ Householder transformations to A. The cost is approximately $2mn^2 - (2/3)n^3$ FLOPs or simply $O(mn^2)$, and it requires very little additional storage if A is overwritten by R and the vectors that define the $n - 1$ Householder transformations. Another benefit of storing Q implicitly in this way is that matrix–vector products with Q and Q^T can be computed in $O(mn)$ FLOPs rather than $O(m^2)$ FLOPs if Q is formed and stored explicitly. We note that $Q_1 R_1$ is referred to as a *thin* or *reduced* QR factorization of A, and the matrix R_1 is unique if we require the diagonal of R_1 to be positive.

A factorization of the form (2.47) yields an orthogonal basis for both the range of A and the nullspace of A^T. Specifically, the columns of Q_1 form an orthogonal basis for the range of A, whereas the columns of Q_2 form an orthogonal basis for the nullspace of A^T. This observation allows us to characterize the solution set to a system of underdetermined equations. Specifically, if $A \in \mathbb{R}^{m \times n}$ with rank$(A) = m$, then the set of solutions to $Ax = b$ may be expressed in terms of a QR factorization $A^T = QR$ as

$$\mathcal{X} = \{ Q_1 R_1^{-T} b + Q_2 z \mid z \in \mathbb{R}^{n-m} \}.$$

This follows directly from (2.40) by noting that

$$A^\dagger = A^T (AA^T)^{-1} = Q_1 R_1 (R_1^T R_1)^{-1} = Q_1 R_1^{-T}.$$

The QR factorization can also be used to compute a Cholesky factorization of a matrix of the form $A^T A$, where $A \in \mathbb{R}^{m \times n}$ and rank$(A) = n$. Indeed, if $A = Q_1 R_1$, then $A^T A = R_1^T R_1$ is the Cholesky factorization of $A^T A$, provided that R_1 is chosen such that its diagonal is positive.

Finally, by combining the QR factorization with *column pivoting*, it can be applied to general matrices $A \in \mathbb{R}^{m \times n}$ with linearly dependent columns. The result is a factorization of the form

$$A = QRP,$$

where $Q \in \mathbb{R}^{m \times m}$ is orthogonal, $R \in \mathbb{R}^{m \times n}$ is a matrix with zeros below the diagonal, and P is a permutation matrix, which is commonly chosen so that the diagonal elements of R are nonincreasing. The QR factorization with column pivoting is sometimes called a *QRP factorization*, and it is usually the first step of a so-called *rank-revealing QR factorization*, which can be used to compute the numerical rank of a matrix.

2.11.5 Sparse Factorizations

When the coefficient matrix $A \in \mathbb{R}^{n \times n}$ is large and sparse, it is often possible to reduce the cost of Gaussian elimination by using a *sparse LU factorization*. This is a factorization of the form

$$A = P_1 L U P_2,$$

Figure 2.5 The black squares in the sparsity pattern (left) denote nonzero entries. The sparsity graph (right) has a node for every row/column of the matrix, and each off-diagonal nonzero element in the sparsity pattern corresponds to an edge in the sparsity graph.

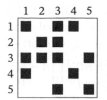

where the permutation matrices P_1 and P_2 affect both the stability of the factorization and the sparsity of the triangular matrices L and U. To illustrate the basic principle, we consider the somewhat simpler case where A is symmetric and positive definite. This implies that we can use a *sparse Cholesky factorization*

$$P^T A P = L L^T, \tag{2.48}$$

where P is a permutation matrix that determines the elimination order and affects the sparsity of L. Since $A \in \mathbb{S}^n_{++}$ is strongly factorizable, P can be chosen without taking numerical stability into account, and hence, it can be chosen based on the *sparsity pattern* of A. Figure 2.5 shows an example of a sparsity pattern and the corresponding *sparsity graph*, which is a graph with n nodes and an edge for every off-diagonal nonzero element.

The sparsity pattern of L can be determined by means of a *symbolic factorization* of the sparsity pattern of $P^T A P$. Only the location of the nonzero entries of $P^T A P$ are needed for this step. Nonzero entries in L that are not present in $P^T A P$ are referred to as *fill-in*. The example in Figure 2.6 illustrates this for two different symmetric permutations of an "arrow" sparsity pattern. It is clear from the figure that the elimination order has a significant effect on the amount of fill-in. A large amount of fill-in is undesirable, since additional nonzero entries in L lead to additional FLOPs. Unfortunately, the problem of computing the minimum fill-in is NP-complete , but several fill-in reducing heuristics exist that often work well in practice. We note that there exists a zero fill-in elimination order if and only if the sparsity graph is a so-called *chordal graph* [106]. An elimination order with zero fill-in is referred to as a *perfect elimination order*, and the corresponding symmetric permutation $P^T A P$ has the same sparsity pattern as that of $L + L^T$.

2.11.6 Block Factorization

The block variant of Gaussian elimination can also be expressed in terms of a block factorization of a matrix M of the form (2.41). Indeed, if the matrix A is nonsingular, then it is easy to verify that M may be expressed as

$$M = \begin{bmatrix} A & B \\ C & D \end{bmatrix} = \begin{bmatrix} I & 0 \\ CA^{-1} & I \end{bmatrix} \begin{bmatrix} A & 0 \\ 0 & D - CA^{-1}B \end{bmatrix} \begin{bmatrix} I & A^{-1}B \\ 0 & I \end{bmatrix}. \tag{2.49}$$

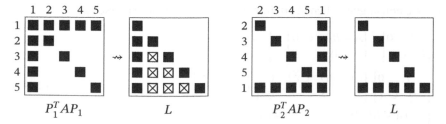

Figure 2.6 Symbolic Cholesky factorizations of two different symmetric permutations of A. The entries in L that are marked by ⊠ are fill-in.

This is a *block LDU factorization* of M, i.e. M is expressed as a a product of a block unit lower triangular matrix, a block diagonal matrix, and a block unit upper triangular matrix. Similarly, if D is nonsingular, then M may be expressed as

$$M = \begin{bmatrix} A & B \\ C & D \end{bmatrix} = \begin{bmatrix} I & BD^{-1} \\ 0 & I \end{bmatrix} \begin{bmatrix} A - BD^{-1}C & 0 \\ 0 & D \end{bmatrix} \begin{bmatrix} I & 0 \\ D^{-1}C & I \end{bmatrix}, \tag{2.50}$$

which is a *block UDL factorization*.

The determinant of M can be expressed in terms of the block factorizations (2.49) and (2.50) if A or D is nonsingular. Using the fact that the determinant of a triangular matrix with a unit diagonal is equal to 1 and the fact that $\det(\mathrm{bdiag}(A_1, A_2)) = \det(A_1)\det(A_2)$ if A_1 and A_2 are square matrices, we have that

$$A \text{ nonsingular} \implies \det(M) = \det(A)\det(D - CA^{-1}B), \tag{2.51a}$$

$$D \text{ nonsingular} \implies \det(M) = \det(D)\det(A - BD^{-1}C). \tag{2.51b}$$

It follows directly from (2.51a) that if A is nonsingular, then $D - CA^{-1}B$ is nonsingular if and only if M is nonsingular. Similarly, if D is nonsingular, then the Schur complement of D in M, $A - BD^{-1}C$, is nonsingular if and only if M is nonsingular. These observations can be used to derive the *Weinstein–Aronszajn identity*, also known as *Sylvester's determinant identity*, which states that

$$\det(I + BC) = \det(I + CB). \tag{2.52}$$

Indeed, this identity follows directly from (2.51) by letting $A = I$ and $D = -I$ such that

$$\det(M) = \det(A)\det(D - CA^{-1}B) = (-1)^{n_2}\det(I + CB)$$

and

$$\det(M) = \det(D)\det(A - BD^{-1}C) = (-1)^{n_2}\det(I + BC).$$

The identity is particularly useful if $n_1 \gg n_2$ or $n_2 \gg n_1$ so that either $I + CB$ or $I + BC$ is much smaller than the other. For example, in the special case, where $BC = uv^T$ is a rank-1 matrix, the identity reduces to $\det(I + uv^T) = 1 + v^T u$.

The block factorizations (2.49) and (2.50) can also be used to derive explicit expressions for the blocks of the inverse of M. It follows from (2.49) that if M and A are nonsingular, then M^{-1} is given by

$$
\begin{aligned}
\begin{bmatrix} A & B \\ C & D \end{bmatrix}^{-1} &= \begin{bmatrix} I & A^{-1}B \\ 0 & I \end{bmatrix}^{-1} \begin{bmatrix} A & 0 \\ 0 & D - CA^{-1}B \end{bmatrix}^{-1} \begin{bmatrix} I & 0 \\ CA^{-1} & I \end{bmatrix}^{-1}, \\
&= \begin{bmatrix} I & -A^{-1}B \\ 0 & I \end{bmatrix} \begin{bmatrix} A^{-1} & 0 \\ 0 & (D - CA^{-1}B)^{-1} \end{bmatrix} \begin{bmatrix} I & 0 \\ -CA^{-1} & I \end{bmatrix}, \\
&= \begin{bmatrix} A^{-1} + A^{-1}B(D - CA^{-1}B)^{-1}CA^{-1} & -A^{-1}B(D - CA^{-1}B)^{-1} \\ -(D - CA^{-1}B)^{-1}CA^{-1} & (D - CA^{-1}B)^{-1} \end{bmatrix}.
\end{aligned}
\tag{2.53}
$$

Similarly, if M and D are nonsingular, then (2.50) implies that M^{-1} can be expressed as

$$
\begin{aligned}
\begin{bmatrix} A & B \\ C & D \end{bmatrix}^{-1} &= \begin{bmatrix} I & 0 \\ D^{-1}C & I \end{bmatrix}^{-1} \begin{bmatrix} A - BD^{-1}C & 0 \\ 0 & D \end{bmatrix}^{-1} \begin{bmatrix} I & BD^{-1} \\ 0 & I \end{bmatrix}^{-1}, \\
&= \begin{bmatrix} I & 0 \\ -D^{-1}C & I \end{bmatrix} \begin{bmatrix} (A - BD^{-1}C)^{-1} & 0 \\ 0 & D^{-1} \end{bmatrix} \begin{bmatrix} I & -BD^{-1} \\ 0 & I \end{bmatrix}, \\
&= \begin{bmatrix} (A - BD^{-1}C)^{-1} & -(A - BD^{-1}C)^{-1}BD^{-1} \\ -D^{-1}C(A - BD^{-1}C)^{-1} & D^{-1} + D^{-1}C(A - BD^{-1}C)^{-1}BD^{-1} \end{bmatrix}.
\end{aligned}
\tag{2.54}
$$

Now, if both A and D are nonsingular, then the $(1,1)$ block of (2.53) and that of (2.54) must be equal, i.e.

$$(A - BD^{-1}C)^{-1} = A^{-1} + A^{-1}B(D - CA^{-1}B)^{-1}CA^{-1}. \qquad (2.55)$$

Substituting $(W, -U, V, Z^{-1})$ for (A, B, C, D), where W and Z are nonsingular, yields the so-called *Sherman–Morrison–Woodbury* (SMW) identity. This is also known as the *matrix inversion lemma*:

$$(W + UZV^T)^{-1} = W^{-1} - W^{-1}U(Z^{-1} + V^T W^{-1}U)^{-1}V^T W^{-1}. \qquad (2.56)$$

This identity is often useful when solving a system of equations of the form (2.44), where the coefficient matrix is a Schur complement. For example, using (2.56), the solution to the system $(A - BD^{-1}C)x = r$ can be expressed as

$$x = A^{-1}r + A^{-1}B(D - CA^{-1}B)^{-1}CA^{-1}r,$$

and can be computed as follows:

1. Compute u and Y by solving $Au = r$ and $AY = B$.
2. Form $S = D - CY$ and compute v by solving $Sv = Cu$.
3. Compute $x = u + Yv$.

This approach is particularly advantageous when the order of S is much smaller than that of A and A is simple or, cheap to factorize, e.g. a diagonal matrix.

2.11.7 Positive Semidefinite Block Factorization

Next, we consider the special case where M is symmetric and positive semidefinite, i.e.

$$M = \begin{bmatrix} A & B \\ B^T & D \end{bmatrix} \succeq 0,$$

with $A \in \mathbb{S}^{n_1}$ and $D \in \mathbb{S}^{n_2}$. Recall that $AA^\dagger$ defines a projection onto $\mathcal{R}(A)$ and $A^\dagger A$ defines a projection onto $\mathcal{R}(A^T) = \mathcal{R}(A)$. This means that the range conditions (2.30) can be expressed as

$$B = AA^\dagger B, \quad B^T = DD^\dagger B^T, \qquad (2.57)$$

and hence, we can express M as

$$\begin{bmatrix} A & B \\ B^T & D \end{bmatrix} = \begin{bmatrix} A & AA^\dagger B \\ B^T A^\dagger A & D \end{bmatrix},$$

$$= \begin{bmatrix} I & 0 \\ B^T A^\dagger & I \end{bmatrix} \begin{bmatrix} A & 0 \\ 0 & D - B^T A^\dagger B \end{bmatrix} \begin{bmatrix} I & A^\dagger B \\ 0 & I \end{bmatrix}.$$

It follows that M and the block-diagonal matrix in the center of the right-hand side are congruent. Thus, we can conclude that

$$M \succeq 0 \iff A \succeq 0, \quad D - B^T A^\dagger B \succeq 0, \quad B = AA^\dagger B. \qquad (2.58)$$

We note that $D - B^T A^\dagger B$ is a referred to as the *generalized* Schur complement of A in M. Finally, M can also be expressed as

$$\begin{bmatrix} A & B \\ B^T & D \end{bmatrix} = \begin{bmatrix} A & BD^\dagger D \\ DD^\dagger B^T & D \end{bmatrix},$$

$$= \begin{bmatrix} I & BD^\dagger \\ 0 & I \end{bmatrix} \begin{bmatrix} A - BD^\dagger B^T & 0 \\ 0 & D \end{bmatrix} \begin{bmatrix} I & 0 \\ D^\dagger B^T & I \end{bmatrix},$$

which implies that

$$M \succeq 0 \iff C \succeq 0, \quad A - BD^\dagger B^T \succeq 0, \quad B^T = DD^\dagger B^T. \qquad (2.59)$$

2.12 Saddle-Point Systems

A special case of the block system (2.42) is a symmetric system of the form

$$\begin{bmatrix} H & A^T \\ A & 0 \end{bmatrix} \begin{bmatrix} x \\ y \end{bmatrix} = \begin{bmatrix} -g \\ b \end{bmatrix}, \tag{2.60}$$

where $H \in \mathbb{S}^n$ and $A \in \mathbb{R}^{m \times n}$. This is an example of a so-called *saddle-point system* that arises frequently in numerical optimization, e.g. from the linearization of optimality conditions, as we will see in Chapter 6. The coefficient matrix

$$M = \begin{bmatrix} H & A^T \\ A & 0 \end{bmatrix}$$

is indefinite if $A \neq 0$, and if M is nonsingular, then it must hold that

$$\text{rank}(A) = m, \quad \text{rank}\left(\begin{bmatrix} H \\ A \end{bmatrix}\right) = n. \tag{2.61}$$

To see this, note that if $\text{rank}(A) < m$, then there exists a nonzero vector $y \in \mathcal{N}(A^T)$ such that $(0, y) \in \mathcal{N}(M)$, and if $\text{rank}\left(\begin{bmatrix} H & A^T \end{bmatrix}\right) < n$, then there exists a nonzero vector $x \in \mathcal{N}(H) \cap \mathcal{N}(A)$ such that $(x, 0) \in \mathcal{N}(M)$. The converse is generally not true: (2.61) does not imply that M is nonsingular without additional assumptions. In the special case, where H is positive semidefinite, the conditions (2.61) are both necessary and sufficient for M to be nonsingular. Indeed, if (x, y) belongs to the nullspace of M, i.e. $Hx + A^T y = 0$ and $Ax = 0$, then $x^T Hx + x^T A^T y = x^T Hx = 0$ from which it follows that $Hx = 0$ since H is positive semidefinite. Thus, we have that $y \in \mathcal{N}(A^T)$ and $x \in \mathcal{N}(H) \cap \mathcal{N}(A)$ which shows that (2.61) is sufficient for M to be nonsingular since it implies that $\mathcal{N}(A^T) = \{0\}$ and $\mathcal{N}(H) \cap \mathcal{N}(A) = \{0\}$. We note that the conditions (2.61) and positive semidefiniteness of H can be shown to imply that M has n positive eigenvalues and m negative ones; see, e.g. [13].

The system (2.60) can be solved in several ways. We now outline several approaches corresponding to different assumptions on H and assuming that the conditions in (2.61) hold.

2.12.1 *H* Positive Definite

The Cholesky factorization $H = LL^T$ allows us to express (2.60) as

$$LL^T x + A^T y = -g, \quad Ax = b.$$

Eliminating $x = -L^{-T} L^{-1} (g + A^T y)$ from the first equation and substituting for x in the second yields

$$AL^{-T} L^{-1} A^T y = -b - AL^{-T} L^{-1} g.$$

The matrix $AL^{-T} L^{-1} A^T$ is symmetric and positive definite because of our assumption that $\text{rank}(A) = m$, and hence, we can compute the Cholesky factorization $AL^{-T} L^{-1} A^T = \tilde{L} \tilde{L}^T$. This approach may be summarized as follows:

1. Compute Cholesky factorization $H = LL^T$.
2. Form $AL^{-T} L^{-1} A^T$ and compute its Cholesky factorization $\tilde{L} \tilde{L}^T$.
3. Solve $\tilde{L} \tilde{L}^T y = -b - AL^{-T} L^{-1} g$ and then $LL^T x = -g - A^T y$.

An alternative to the Cholesky factorization in step 2 is to compute a QR factorization $L^{-1} A^T = QR$, which leads to the following modification of steps 2 and 3:

2. Form $L^{-1} A^T$ and compute a QR factorization QR.
3. Solve $Ry = -R^{-T} b - Q^T L^{-1} g$ and then $L^T x = -L^{-1} g - QRy$.

The QR factorization approach is the preferred method from a numerical point of view.

The approach that we have just outlined relies on the assumption that H is positive definite, and hence, another approach is needed if H is only positive semidefinite.

2.12.2 *H* Positive Semidefinite

First recall that the conditions in (2.61) imply that $\mathcal{N}(H) \cap \mathcal{N}(A) = \{0\}$, and hence, $H + A^T A$ must be positive definite even if $\text{rank}(H) < n$. This observation allows us to rewrite the system (2.60) as

$$\begin{bmatrix} H + A^T A & A^T \\ A & 0 \end{bmatrix} \begin{bmatrix} x \\ y \end{bmatrix} = \begin{bmatrix} -g + A^T b \\ b \end{bmatrix}, \tag{2.62}$$

i.e. we have added $A^T b$ to the left- and right-hand side of $Hx + A^T y = -g$ and substituted Ax for b on left-hand side. The system (2.62) can be solved using the same approach as when H is positive definite.

2.13 Vector and Matrix Calculus

The analysis and solution of optimization problems often involve the partial derivatives of one or several differentiable multivariate functions. Here we introduce the notation used throughout the book and review some useful identities.

Given a real-valued differentiable function $f : \mathbb{R}^n \to \mathbb{R}^m$ and a vector $x \in \mathbb{R}^n$, we define $\frac{\partial f(x)}{\partial x^T}$ as the $m \times n$ matrix of partial derivatives

$$\frac{\partial f(x)}{\partial x^T} = \frac{\partial f}{\partial x^T} = \begin{bmatrix} \dfrac{\partial f_1}{\partial x_1} & \cdots & \dfrac{\partial f_1}{\partial x_n} \\ \vdots & \ddots & \vdots \\ \dfrac{\partial f_m}{\partial x_1} & \cdots & \dfrac{\partial f_m}{\partial x_n} \end{bmatrix}, \tag{2.63}$$

where $f(x) = (f_1(x), \ldots, f_m(x))$. The matrix $\frac{\partial f}{\partial x^T}$ is also known as the Jacobian of f. We will occasionally use the notation

$$\frac{\partial f^T}{\partial x} = \frac{\partial f(x)^T}{\partial x} = \left(\frac{\partial f}{\partial x^T} \right)^T, \tag{2.64}$$

which is an $n \times m$ matrix. In the special case, where f is real valued (i.e. $m = 1$), the vector of partial derivatives,

$$\nabla f(x) = \frac{\partial f}{\partial x} = \begin{bmatrix} \dfrac{\partial f}{\partial x_1} \\ \vdots \\ \dfrac{\partial f}{\partial x_n} \end{bmatrix} \tag{2.65}$$

is called the gradient of f. Moreover, if f is twice differentiable, then the $n \times n$ matrix of second-order partial derivatives,

$$\nabla^2 f(x) = \frac{\partial^2 f}{\partial x \partial x^T} = \begin{bmatrix} \dfrac{\partial^2 f}{\partial x_1 \partial x_1} & \cdots & \dfrac{\partial^2 f}{\partial x_1 \partial x_n} \\ \vdots & \ddots & \vdots \\ \dfrac{\partial^2 f}{\partial x_n \partial x_1} & \cdots & \dfrac{\partial^2 f}{\partial x_n \partial x_n} \end{bmatrix} \tag{2.66}$$

is the called the Hessian of f.

Now, given a matrix-valued function $F : \mathbb{R} \to \mathbb{R}^{m \times n}$ that is differentiable, we define the matrix of first-order derivatives

$$\frac{dF(x)}{dx} = \frac{dF}{dx} = \begin{bmatrix} \dfrac{dF_{11}}{dx} & \cdots & \dfrac{dF_{1n}}{dx} \\ \vdots & \ddots & \vdots \\ \dfrac{dF_{m1}}{dx} & \cdots & \dfrac{dF_{mn}}{dx} \end{bmatrix}, \tag{2.67}$$

where the function $F_{ij}(x)$ is the (i,j)th element of $F(x)$. Similarly, given a differentiable function $f : \mathbb{R}^{m \times n} \to \mathbb{R}$, we define the matrix of first-order partial derivatives

$$\frac{\partial f(X)}{\partial X} = \frac{\partial f}{\partial X} = \begin{bmatrix} \dfrac{\partial f}{\partial X_{11}} & \cdots & \dfrac{\partial f}{\partial X_{1n}} \\ \vdots & \ddots & \vdots \\ \dfrac{\partial f}{\partial X_{m1}} & \cdots & \dfrac{\partial f}{\partial X_{mn}} \end{bmatrix}. \tag{2.68}$$

The partial derivatives of a composite function $f = g \circ h$, where $g : \mathbb{R}^p \to \mathbb{R}^m$ and $h : \mathbb{R}^n \to \mathbb{R}^p$ are differentiable functions, can be expressed in terms of the chain rule as

$$\frac{\partial f}{\partial x^T} = \frac{\partial g(h(x))}{\partial x^T} = \frac{\partial g(y)}{\partial y^T}\bigg|_{y=h(x)} \frac{\partial h(x)}{\partial x^T}. \tag{2.69}$$

For differentiable functions $f : \mathbb{R}^n \to \mathbb{R}$ and $g : \mathbb{R}^n \to \mathbb{R}$, the product rule takes the form

$$\frac{\partial f(x)g(x)}{\partial x} = \frac{\partial f}{\partial x}g(x) + f(x)\frac{\partial g}{\partial x} = \nabla f(x)g(x) + f(x)\nabla g(x), \tag{2.70}$$

whereas for $f : \mathbb{R}^n \to \mathbb{R}^m$ and $g : \mathbb{R}^n \to \mathbb{R}$, we have that

$$\frac{\partial f(x)g(x)}{\partial x^T} = \frac{\partial f}{\partial x^T}g(x) + f(x)\frac{\partial g}{\partial x^T}. \tag{2.71}$$

Example 2.3 We now illustrate the use of the chain rule for the special case where $f = g \circ h$ is a composition of a differentiable function $g : \mathbb{R}^p \to \mathbb{R}^m$ and a linear function $h(x) = Ax$, where $A \in \mathbb{R}^{p \times n}$. We have that $\partial h / \partial x^T = A$, and hence, the chain rule yields

$$\frac{\partial f}{\partial x^T} = \frac{\partial g(y)}{\partial y^T}\bigg|_{y=Ax} A.$$

If f is real-valued, i.e. $m = 1$, then this is equivalent to the identity $\nabla f(x) = A^T \nabla g(Ax)$, and if g is also twice differentiable, then the Hessian of f is

$$\nabla^2 f(x) = A^T \nabla^2 g(Ax)A.$$

Example 2.4 Consider the function $f : \mathbb{R}^n \to \mathbb{R}$ defined as

$$f(x) = \ln\left(\sum_{i=1}^{n} e^{x_i}\right),$$

which is known as the log-sum-exp function. To derive the partial derivatives of f using the chain rule, we start by expressing f as $f = g \circ h$ where $g(y) = \ln(1^T y)$ and $h(x) = (e^{x_1}, \ldots, e^{x_n})$. We have that $\partial g / \partial y^T = \frac{1}{1^T y}1^T$ and $\partial h / \partial x^T = \mathrm{diag}(h(x))$, and hence, it follows that

$$\frac{\partial f}{\partial x^T} = \frac{1}{1^T h(x)}1^T \mathrm{diag}(h(x)) = \frac{1}{1^T h(x)}h(x)^T.$$

The Hessian of f now follows from the product rule (2.71), i.e.

$$\nabla^2 f(x) = \frac{1}{\mathbb{1}^T h(x)} \operatorname{diag}(h(x)) - \frac{1}{(\mathbb{1}^T h(x))^2} h(x) h(x)^T,$$
$$= \operatorname{diag}(\nabla f(x)) - \nabla f(x) \nabla f(x)^T.$$

Example 2.5 Recall the Laplace expansion (2.8) of the determinant of a square matrix $A \in \mathbb{R}^{n \times n}$ along the jth column, i.e.

$$\det (A) = \sum_{i=1}^{n} c_{ij} a_{ij},$$

where $c_{ij} = (-1)^{i+j} M_{ij}$ is the cofactor of the (i, j)th entry of A. None of the cofactors $c_{1j}, \dots, c_{nj}$ are a function of a_{ij}, and hence, it follows that

$$\frac{\partial}{\partial a_{ij}} \det (A) = \frac{\partial}{\partial a_{ij}} \left(\sum_{k=1}^{n} c_{kj} a_{kj} \right) = c_{ij}.$$

This result can be used to derive an expression for the derivative of $\det (A(t))$, where $A : \mathbb{R} \to \mathbb{R}^{n \times n}$, i.e.

$$\frac{d}{dt} \det (A(t)) = \sum_{i=1}^{n} \sum_{j=1}^{n} \frac{\partial \det (A(t))}{\partial a_{ij}(t)} \frac{d a_{ij}(t)}{dt} = \sum_{i=1}^{n} \sum_{j=1}^{n} c_{ij} \frac{d a_{ij}(t)}{dt}.$$

This can be expressed more compactly as

$$\frac{d}{dt} \det (A(t)) = \operatorname{tr} \left(\operatorname{adj}(A(t)) \frac{dA(t)}{dt} \right), \tag{2.72}$$

which is known as *Jacobi's formula*.

Exercises

2.1 Show that $\operatorname{tr}(B^T A) = \operatorname{tr}(AB^T)$ for all $A, B \in \mathbb{R}^{m \times n}$.

2.2 Show that the operator norm and the Frobenius norm on $\mathbb{R}^{m \times n}$ are orthogonally invariant, i.e. if $Q_1 \in \mathbb{R}^{m \times m}$ and $Q_2 \in \mathbb{R}^{n \times n}$ are orthogonal matrices, then it holds that

$$\|A\|_2 = \|Q_1 A Q_2\|_2, \quad \|A\|_F = \|Q_1 A Q_2\|_F.$$

2.3 Show that the infinity norm of a matrix $A \in \mathbb{R}^{m \times n}$ may be expressed as

$$\|A\|_\infty = \max_{i \in \mathbb{N}_m} \|a_i\|_1,$$

if $a_1^T, \dots, a_m^T \in \mathbb{R}^n$ are the m rows of A.

2.4 Let $x \in \mathbb{R}^n$ be a nonzero vector. Find a vector $v \in \mathbb{R}^n$ such that the Householder matrix $H = I - 2 \frac{vv^T}{v^T v}$ maps x to $\|x\|_2 e_1$, i.e.

$$Hx = x - 2 \frac{v^T x}{v^T v} v = \|x\|_2 e_1.$$

2.5 Recall that the Schatten p-norm of a matrix $A \in \mathbb{R}^{m \times n}$ is given by

$$\|A\|_{(p)} = \left(\sum_{i=1}^{\min(m,n)} \sigma_i^p \right)^{1/p},$$

where σ_i denotes the ith singular value of A.

(a) Show that the Schatten p-norm is orthogonally invariant, i.e. $\|A\|_{(p)} = \|Q_1 A Q_2\|_{(p)}$ if $Q_1 \in \mathbb{R}^{m \times m}$ and $Q_2 \in \mathbb{R}^{n \times n}$ are orthogonal matrices.

(b) Show that the Schatten 2-norm on $\mathbb{R}^{m \times n}$ is equivalent to the Frobenius norm, i.e. $\|A\|_{(2)} = \|A\|_F$.

(c) Show that the Schatten infinity norm on $\mathbb{R}^{m \times n}$ is equivalent to the operator norm, i.e. $\|A\|_{(\infty)} = \|A\|_2$.

2.6 Show that the inverse of a nonsingular, lower-triangular Toeplitz matrix whose first column is given by $a_0, \dots, a_{n-1}$ is another lower-triangular Toeplitz matrix.

2.7 Show that a lower-triangular Toeplitz matrix T of order n commutes with the lower shift matrix S of order n, i.e. $ST = TS$.

2.8 Show that the trace of a symmetric matrix $A \in \mathbb{S}^n$ is equal to the sum of its eigenvalues, i.e.

$$\text{tr}(A) = \sum_{i=1}^{n} \lambda_i,$$

if $A = Q\text{diag}(\lambda_1, \dots, \lambda_n)Q^T$ is a spectral decomposition of A.

2.9 Show that the smallest eigenvalue of $X \in \mathbb{S}^n$ is greater than or equal to $t \in \mathbb{R}$ if and only if $X - tI \in \mathbb{S}_+^n$.

2.10 Let $H = R + XPX^T$, where $R \in \mathbb{S}_{++}^m$, $P \in \mathbb{S}_{++}^n$, and $X \in \mathbb{R}^{m \times n}$, and define $Y = PX^T H^{-1}$. Show that the inverse of the matrix

$$M = \begin{bmatrix} I & 0 \\ Y & I \end{bmatrix}^T \begin{bmatrix} I & -X \\ 0 & I \end{bmatrix}^T \begin{bmatrix} R & 0 \\ 0 & P \end{bmatrix}^{-1} \begin{bmatrix} I & -X \\ 0 & I \end{bmatrix} \begin{bmatrix} I & 0 \\ Y & I \end{bmatrix}$$

is given by

$$M^{-1} = \begin{bmatrix} H & 0 \\ 0 & P - YHY^T \end{bmatrix}.$$

Hint: Use the identity in (2.49).

2.11 Show that a matrix $Z \in \mathbb{R}^{m \times n}$ has rank at most r if and only if there exist matrices $X \in \mathbb{S}^m$ and $Y \in \mathbb{S}^n$ such that

$$\text{rank } X + \text{rank } Y \leq 2r, \quad \begin{bmatrix} X & Z \\ Z^T & Y \end{bmatrix} \geq 0.$$

2.12 Consider a linear system

$$x_{k+1} = Ax_k + Bu_k,$$
$$y_k = Cx_k + Du_k,$$

where $x_k \in \mathbb{R}^n$, $u_k \in \mathbb{R}^m$, and $y_k \in \mathbb{R}^p$ for $0 \leq k \leq N$, and where (A, B, C, D) are real-valued matrices of compatible dimensions. Suppose $T \in \mathbb{R}^{n \times n}$ is invertible and define the transformed states $\bar{x}_k$ via $T\bar{x}_k = x_k$. Show that

$$\bar{x}_{k+1} = \bar{A}\bar{x}_k + \bar{B}u_k,$$
$$y_k = \bar{C}\bar{x}_k + Du_k,$$

where $\bar{A} = T^{-1}AT$, $\bar{B} = T^{-1}B$, and $\bar{C} = CT$.

2.13 Consider a dynamical system defined by

$$x_{k+1} = Ax_k + Bu_k,$$
$$y_k = Cx_k,$$

with $x_k \in \mathbb{R}^n$, $u_k \in \mathbb{R}$, and $y_k \in \mathbb{R}$ for $k \in \mathbb{Z}_+$, and where $A \in \mathbb{R}^{n \times n}$, $B \in \mathbb{R}^{n \times 1}$, and $C \in \mathbb{R}^{1 \times n}$ are given. For a given $N \geq n$, we introduce the so-called *observability* matrix $\mathcal{O}$ and *controllability* matrix C, which are defined as

$$\mathcal{O} = \begin{bmatrix} C \\ CA \\ \vdots \\ CA^{N-1} \end{bmatrix}, \quad C = \begin{bmatrix} B & AB & \cdots & A^{N-1}B \end{bmatrix}.$$

(a) Show that the product $\mathcal{O}C$ can be expressed as the Hankel matrix

$$\mathcal{O}C = \begin{bmatrix} h_1 & h_2 & h_3 & \cdots & h_{N-1} & h_N \\ h_2 & h_3 & h_4 & \cdots & h_N & h_{N+1} \\ h_3 & h_4 & h_5 & \cdots & h_{N+1} & h_{N+2} \\ \vdots & \vdots & \vdots & \ddots & \vdots & \vdots \\ h_{N-1} & h_N & h_{N+1} & \cdots & h_{2N-3} & h_{2N-2} \\ h_N & h_{N+1} & h_{N+2} & \cdots & h_{2N-2} & h_{2N-1} \end{bmatrix},$$

where $h_k = CA^{k-1}B$ for $k \in \mathbb{N}_{2N-1}$ are the first $2N - 1$ so-called *impulse response coefficients* or *Markov-parameters* of the dynamical system.

(b) Show that the Markov parameters are invariant under a state-transformation as in Exercise 2.12

(c) Suppose $\mathcal{O}$ has full column rank, in which case, the dynamical system is said to be *observable*. Moreover, assume that C has full row rank, which means that the dynamical system is *controllable*. What is then the rank of $\mathcal{O}C$?

(d) Suppose that instead of the matrices (A, B, C), we are given the Markov parameters $(h_1, h_2, \ldots, h_{2N-1})$ for some $N > n$. Describe a numerical procedure for finding matrices (A, B, C) such that $h_k = CA^{k-1}B$ for $k \in \mathbb{N}_{2N-1}$. How can you find (A, B, C) such that the corresponding dynamical system is both observable and controllable?

2.14 Consider the equation

$$Y = \mathcal{O}X + \mathcal{T}U,$$

where $Y \in \mathbb{R}^{p \times m}$, $\mathcal{O} \in \mathbb{R}^{p \times n}$, $X \in \mathbb{R}^{n \times m}$, $\mathcal{T} \in \mathbb{R}^{p \times q}$, and $U \in \mathbb{R}^{q \times m}$.

(a) Let $\Pi^\perp = I - U^T(UU^T)^{-1}U$ and show that

$$Y\Pi^\perp = \mathcal{O}X\Pi^\perp.$$

(b) Assume that $\begin{bmatrix} X \\ U \end{bmatrix}$ has full row rank and that $\mathcal{O}$ has full column rank. Show that

$$\text{rank}\ \left(Y\Pi^{\perp}\right) = n,$$
$$\mathcal{R}\left(Y\Pi^{\perp}\right) = \mathcal{RO}.$$

(c) Consider the factorization

$$\begin{bmatrix} U \\ Y \end{bmatrix} = \begin{bmatrix} L_{11} & 0 & 0 \\ L_{21} & L_{22} & 0 \end{bmatrix} \begin{bmatrix} Q_1 \\ Q_2 \\ Q_3 \end{bmatrix},$$

where $L_{11} \in \mathbb{R}^{q \times q}$, $L_{22} \in \mathbb{R}^{p \times p}$, and $Q_1 Q_2{}^T = 0$. Such a factorization can be obtained from a QR-factorization of $[U^T \ Y^T]$. Show that $Y\Pi^{\perp} = L_{22}Q_2$.

2.15 Let $X : \mathbb{R} \to \mathbb{R}^{m \times n}$ and $Y : \mathbb{R} \to \mathbb{R}^{n \times p}$. Show that

$$\frac{d}{dt}\left(X(t)Y(t)\right) = \frac{dX(t)}{dt}Y(t) + X(t)\frac{dY(t)}{dt}.$$

2.16 Show that

$$\frac{\partial}{\partial X}\text{tr}(AX) = A^T,$$

where $X \in \mathbb{R}^{m \times n}$ and $A \in \mathbb{R}^{n \times m}$.

2.17 Let $X : \mathbb{R} \to \mathbb{R}^{m \times m}$. Show that

$$\frac{d}{dt}(X(t)^{-1}) = -X(t)^{-1}\frac{dX(t)}{dt}X(t)^{-1}, \quad \det X(t) \neq 0.$$

2.18 Let $X \in \mathbb{R}^{m \times m}$ and assume that $\det X > 0$. Show that

$$\frac{\partial}{\partial X}\ln \det X = X^{-T}.$$

Hint: Use the chain rule and Jacobi's formula.

2.19 Let $X \in \mathbb{R}^{m \times m}$ be nonsingular and $a, b \in \mathbb{R}^m$. Show that

$$\frac{\partial}{\partial X}a^T X^{-1}b = -X^{-T}ab^T X^{-T}.$$

Hint: First use the identity $XX^{-1} = I$ to show that

$$\frac{\partial}{\partial X_{ij}}X^{-1} = -X^{-1}\frac{\partial X}{\partial X_{ij}}X^{-1}, \quad i,j \in \mathbb{N}_m.$$

2.20 Show that

$$\frac{\partial}{\partial X}\text{tr}(X^T AX) = (A + A^T)X,$$

where $X \in \mathbb{R}^{m \times n}$ and $A \in \mathbb{R}^{m \times m}$.

2.21 Let $A : \mathbb{R} \to \mathbb{R}^{m \times n}$ with $m \geq n$ and define $f : \mathbb{R} \to \mathbb{R}$ as

$$f(t) = \det \left(A(t)^T A(t) \right)^{\frac{1}{2k}},$$

for some $k \in \mathbb{N}$.

(a) Show that

$$\frac{df(t)}{dt} = \frac{f(t)}{k} \operatorname{tr} \left(A(t)^\dagger \frac{dA(t)}{dt} \right).$$

(b) Suppose that $A(t)$ has full column rank, and let $A(t) = Q_1 R_1$ be a reduced QR factorization of $A(t)$, i.e. $Q_1 \in \mathbb{R}^{m \times n}$ and $R_1 \in \mathbb{R}^{n \times n}$. Show that

$$\frac{df(t)}{dt} = \frac{1}{k} \left| \prod_{i=1}^{n} R_{ii} \right|^{1/k} \operatorname{tr} \left(R_1^{-1} Q_1^T \frac{dA(t)}{dt} \right).$$

2.22 Given a matrix $A \in \mathbb{R}^{m \times n}$, the projection of a point $x \in \mathbb{R}^m$ onto $\mathcal{N}(A^T)$ can be expressed as Px, where $P = I - AA^\dagger$ is a projection matrix. If we assuming that $\operatorname{rank}(A) = n$, then P can be expressed as

$$P = I - A(A^T A)^{-1} A^T.$$

Now, suppose that A is a function of a scalar parameter t.

(a) Show that

$$\frac{dP}{dt} = -P \frac{dA}{dt} A^\dagger - (A^\dagger)^T \frac{dA^T}{dt} P.$$

(b) Suppose $\operatorname{rank}(A) = n$ and let $A = Q_1 R_1$ be a reduced QR decomposition of A, i.e. $Q_1 \in \mathbb{R}^{m \times n}$ and $R_1 \in \mathbb{R}^{n \times n}$. Show that the projection Px and the derivative dP/dt can be evaluated efficiently using such a QR decomposition of A without explicitly computing $A^\dagger = (A^T A)^{-1} A^T$.

3

Probability Theory

In this chapter, we will discuss the basics of probability theory. It is a branch of mathematics where uncertain events are given a number between zero and one to describe how likely they are. Loosely speaking this number should be close to the relative frequency with which the event occurs when it is repeated many times. As an example we may consider throwing a fair dices one hundred times and recording how many times a one occurs. If it occurs 18 times, the relative frequency is $18/100 = 0.18$. This is close to the theoretical value of the probability which is 1/6. The reason we know it should be 1/6 is that all the possible six outcomes of the experiment should have the same probability if the dice is fair. In case a probability is one, we are almost sure the event will occur, and if it is zero, we are almost sure it will not occur.

The roots of probability theory go back to the Arab mathematician Al-Khalil who studied cryptography. Initially, probability theory only considered combinatorial problems. The theory is much easier for this case as compared to the case when the number of events is not countable. Mathematicians struggled for many years to provide a solid foundation, and it was not until in 1933 when Andrey Nikolaevich Kolmogorov made an axiomatic definition of probabilities that the problem was resolved, and modern probability theory was born. We are however not going to provide the details of the measure theoretic foundations of probability theory in this chapter; the interested reader is referred to, e.g. [98]. The presentation given here is more in line with [48]. Probability theory is the foundation for statistics and learning and used in many other branches of science.

3.1 Probability Spaces

A probability space is defined by a triplet $(\Omega, \mathcal{F}, \mathbb{P})$. Here Ω is called the *sample space*, and it is a set that contains all possible outcomes of an experiment. When throwing a dice, we can take $\Omega = \{1, 2, 3, 4, 5, 6\}$. However, if we are only interested in if the number is odd or even, we could instead take the sample space to be $\Omega = \{\text{odd}, \text{even}\}$. For other experiments, it may be more appropriate to have $\Omega = \mathbb{R}$. An example of this is when the experiment is the error with which we measure something. The sample space could also contain vectors. If we throw two dices, it is appropriate to consider $\Omega = \{1, 2, 3, 4, 5, 6\} \times \{1, 2, 3, 4, 5, 6\}$. We may sometimes have infinite-dimensional vectors, e.g. $\Omega = \mathbb{R}^{\mathbb{Z}}$, i.e. the set of real-valued functions defined on the integers, *cf.* the notation section. Here the dimensions are countable, and $\mathbb{Z}$ might be interpreted as a set of discrete time indices. It could also be the case that $\Omega = \mathbb{R}^{\mathbb{R}}$, in which case, the sample space contains all real-valued functions of a real variable. Sample spaces containing functions have applications in control and signal processing. We also encounter examples, where $\Omega = \mathbb{R}^{\mathbb{R}^n}$, i.e. the sample space contains all real-valued functions defined on $\mathbb{R}^n$, which has applications in so-called "Gaussian processes" which we will discuss in more detail in Chapter 9.

The second element $\mathcal{F}$ of a probability space should contain all events that we are interested in assigning probabilities to. This is a set of subsets of Ω, and it should be a so-called σ-*algebra*, i.e. the following properties must hold

1. $\Omega \in \mathcal{F}$
2. $\Omega \backslash A \in \mathcal{F}$, $\forall A \in \mathcal{F}$
3. $\forall A_1, A_2, \ldots \in \mathcal{F} \Rightarrow \bigcup_{i=1}^{\infty} A_i \in \mathcal{F}$.

The latter two conditions say that $\mathcal{F}$ is closed under complement and under countable unions. The difference between algebra and σ-algebra is that for an algebra, only finite unions are considered in the last condition. For finite sample spaces, there is no difference since $\mathcal{F}$ is then also a finite set. It can then at most contain all subsets of Ω. The smallest possible σ-algebra is $\mathcal{F} = \{\Omega, \emptyset\}$.

Example 3.1 Let us again consider the example of throwing a fair dice, and we take $\Omega = \{1, 2, 3, 4, 5, 6\}$. We are then interested in the events odd and even. Hence, we define the sets $A_{\text{odd}} = \{1, 3, 5\}$, and $A_{\text{even}} = \{2, 4, 6\}$, which are the sets containing the odd and even numbers, respectively. We may then take the σ-algebra to be $\mathcal{F} = \{A_{\text{odd}}, A_{\text{even}}, \Omega, \emptyset\}$. It is straightforward to verify that this is a σ-algebra.

When we carry out an experiment, like throwing a fair dice, we say that we observe an outcome of the experiment, and this will be an element of the σ-algebra, e.g. the number is either odd or even as in the example above. We then say that this is the *outcome of the experiment* or the *observation of the experiment*.

3.1.1 Probability Measure

The *probability measure* $\mathbb{P}$ is a function $\mathbb{P} : \mathcal{F} \to [0, 1]$ such that

1. $\mathbb{P}[\Omega] = 1$
2. If $A_1, A_2, \ldots \in \mathcal{F}$ is a collection of pairwise disjoint sets, then $\mathbb{P}\left[\cup_{i=1}^{\infty} A_i\right] = \sum_{i=1}^{\infty} \mathbb{P}\left[A_i\right]$.

The first condition is that $\mathbb{P}$ is normalized, and the second condition is that $\mathbb{P}$ is what is called σ-additive. Well-known properties such as $\mathbb{P}[A \cup B] = \mathbb{P}[A] + \mathbb{P}[B] - \mathbb{P}[A \cap B]$ all follow from the above axioms, see Exercise 3.1. Notice that when Ω contains uncountably many elements, say $\Omega = \mathbb{R}$, it is not possible to take $\mathcal{F}$ to contain all subsets of Ω and define $\mathbb{P}$ to satisfy the second condition above. It means that it is not possible to consider any subsets of Ω as the events of the experiment in a meaningful way. This puzzled the mathematicians for many years. One usually restricts oneself to the so-called *Borel σ-algebra*, which is the smallest σ-algebra that contains all intervals of $\mathbb{R}$. All sets in the Borel σ-algebra can be formed from open sets, or equivalently, from closed sets, through the operations of countable union, countable intersection, and relative complement. It then follows that it is possible to satisfy the second condition in the definition of the probability measure. We will not discuss how to generalize this to more complicated sample spaces like $\Omega = \mathbb{R}^{\mathbb{R}}$.

3.1.2 Probability Function

When the sample space Ω contains a countable number of elements, we may define a *probability function* (pf) $p : \Omega \to [0, 1]$ such that

$$\sum_{\omega \in \Omega} p(\omega) = 1.$$

Then the probability measure is

$$\mathbb{P}[A] = \sum_{\omega \in A} p(\omega).$$

for any $A \in \mathcal{F}$. Notice that the summations above may be infinite.

We now give an example of a probability function.

Example 3.2 We consider the case of $\Omega = \mathbb{N}_n$, and we define the probability function $p : \mathbb{N}_n \rightarrow [0, 1]$ as

$$p(k) = \begin{cases} \frac{e^{z_k}}{1 + \sum_{l=1}^{n-1} e^{z_l}}, & k \in \mathbb{N}_{n-1}, \\ \frac{1}{1 + \sum_{l=1}^{n-1} e^{z_l}}, & k = n, \end{cases}$$

for given $z_k \in \mathbb{R}, k \in \mathbb{N}_{n-1}$, which is called the *categorical* probability function. We will see that this is used in what is called logistic regression analysis. It is straightforward to verify that this is a valid probability function for any values of z_k.

The categorical probability function is an example of a *finite probability function* since the set $\mathbb{N}_n$ is finite. When p is a function of an integer, we often write p_k instead of $p(k)$. Then we may also use a vector $p = (p_1, \dots, p_n) \in [0, 1]^n$ instead of a function p to describe the probability function.

3.1.3 Probability Density Function

When the sample space is not countable, we must be more careful as mentioned above. We will here consider $\Omega = \mathbb{R}$ and define the so-called *distribution function* (df) $F : \mathbb{R} \rightarrow [0, 1]$ as

$$F(\omega) = \mathbb{P}[(-\infty, \omega]].$$

For any set $A \subset \Omega$, we have that

$$\mathbb{P}[A] = \int_A dF(\omega).$$

When F is differentiable[1] the derivative $f : \mathbb{R} \rightarrow \mathbb{R}_+$ of F exists, and we may write

$$\mathbb{P}[A] = \int_A f(\omega)d\omega.$$

The function f is called the *probability density function* (pdf). It is straightforward to generalize the results to $\Omega = \mathbb{R}^n$.

We defer giving examples of pdfs until we have introduced random variables.

3.2 Conditional Probability

We are now interested in the probability of an event A if we know that another event B has happened. As an example, we may be interested in the probability that we obtain a one when we throw a fair dice if we already know that the outcome of the experiment was an odd number, i.e. either one, three, or five. In this example, $A = \{1\}$ and $B = \{1, 3, 5\}$. The whole sample space is $\Omega = \{1, 2, 3, 4, 5, 6\}$. Clearly, we can look at a smaller sample space defined by B, i.e. what we know

---

1 It is actually enough to assume that F is absolutely continuous, and then the derivative exists almost everywhere.

has happened, and then we investigate how frequent A is in this sample space, and we obtain the probability $1/3$, i.e. A contains one of the three elements in B, which all have equal likely outcome. Another way of computing this value is to go back to the original sample space Ω and compute

$$\frac{\mathbb{P}[A \cap B]}{\mathbb{P}[B]} = \frac{1/6}{3/6} = \frac{1/6}{1/2} = \frac{1}{3},$$

i.e. we normalize the probability of both events occurring with the probability of the event that we know have occurred. If $P(B) > 0$, we define the *conditional probability* that A occurs given that B has occurred as

$$\mathbb{P}[A \mid B] = \frac{\mathbb{P}[A \cap B]}{\mathbb{P}[B]}.$$

From this it immediately follows that

$$\mathbb{P}[A \cap B] = \mathbb{P}[B]\,\mathbb{P}[A \mid B], \tag{3.1}$$

and by induction that for $A_i \in \mathcal{F}$, it holds that

$$\mathbb{P}\left[\cap_{i=1}^n A_i\right] = \mathbb{P}\left[A_1\right] \mathbb{P}\left[A_2 \mid A_1\right] \mathbb{P}\left[A_3 \mid A_1 \cap A_2\right] \cdots \mathbb{P}\left[A_n \mid A_1 \cap A_2 \cap \cdots \cap A_{n-1}\right],$$

which is called the *multiplication theorem*.

Clearly, $\mathbb{P}[A \mid B] \geq 0$, $\mathbb{P}[\Omega \mid B] = 1$, and $\mathbb{P}[B \mid B] = 1$ for any $A, B \in \mathcal{F}$. If $A_1, A_2, \ldots \in \mathcal{F}$ is a collection of pairwise disjoint sets, then $\mathbb{P}\left[\cup_{i=1}^\infty A_i \mid B\right] = \sum_{i=1}^\infty \mathbb{P}\left[A_i \mid B\right]$. Hence, $(\Omega, \mathcal{F}, \mathbb{P}[\cdot \mid B])$ is also a probability space for a fixed B such that $\mathbb{P}[B] > 0$, i.e. $\mathbb{P}[\cdot \mid B] : \mathcal{F} \to [0,1]$ is also a probability measure. This formalizes the intuition we obtained with the dice in the beginning of this section.

From (3.1) and its symmetric counterpart

$$\mathbb{P}[B \cap A] = \mathbb{P}[A]\,\mathbb{P}[B \mid A],$$

it follows that

$$\mathbb{P}[A \mid B] = \frac{\mathbb{P}[A]\,\mathbb{P}[B \mid A]}{\mathbb{P}[B]}, \tag{3.2}$$

for any $A, B \in \mathcal{F}$ such that $\mathbb{P}[A] > 0$ and $\mathbb{P}[B] > 0$. This is called *Bayes' theorem*.

Let $A_i \in \mathcal{F}$, $i \in \mathbb{N}_n$ be pairwise disjoint sets such that $\Omega = \cup_{i=1}^n A_i$. Then for any $X \in \mathcal{F}$, it holds that

$$\mathbb{P}[X] = \sum_{i=1}^n \mathbb{P}\left[A_i\right] \mathbb{P}\left[X | A_i\right], \tag{3.3}$$

which is called the *formula of total probability*. This is proven in Exercise 3.3.

Moreover, by (3.1), we have $\mathbb{P}\left[A_i \cap X\right] = \mathbb{P}\left[A_i\right] \mathbb{P}\left[X \mid A_i\right]$. From what has been said above and from (3.2) it follows that

$$\mathbb{P}\left[A_i \mid X\right] = \frac{\mathbb{P}\left[A_i\right] \mathbb{P}\left[X \mid A_i\right]}{\sum_{j=1}^n \mathbb{P}\left[A_j\right] \mathbb{P}\left[X \mid A_j\right]}, \quad i \in \mathbb{N}_n.$$

Here we have tacitly assumed that all involved events have nonzero probability.

Example 3.3 In a factory, the same items are manufactured at three different machines with a proportion given by 15% for machine 1, 45% for machine 2 and 40% for machine 3. The different machines produce defect items with probabilities 0.05, 0.04, and 0.03, respectively. Customers obtain a perfect mix of items from the different machines. We denote by A_1, A_2, and A_3 the events that an item is manufactured by machine 1, 2, and 3, respectively, and we denote by X the event that an item in the mix sent to customers is defect. Then by the formula of total probability, we have that

$$P[X] = \sum_{i=1}^{3} P[A_i] \, P[X \mid A_i] = 0.15 \times 0.05 + 0.45 \times 0.04 + 0.40 \times 0.03 = 0.0375$$

3.3 Independence

If it had not been for the concept of *independence*, probability theory would not have been a separate branch of mathematics, but instead just an example of measure theory. We have that the occurrence of the event $B \in F$ changes the probability of the event $A \in F$ to occur from $P[A]$ to $P[A \mid B]$. However, if $P[A \mid B] = P[A]$, then this is not the case. An equivalent condition to this is by (3.1) that

$$P[A \cap B] = P[A] \, P[B],$$

and this is what we take as definition of *independence* of A and B. Notice that this also implies that $P[B \mid A] = P[B]$. The definition of independence is valid also when $P[A]$ and/or $P[B]$ are zero. The relation to conditional probabilities requires that they are positive.

We can also consider a family of events, i.e. $A = \{A_1, \ldots, A_n\} \subseteq F$. We say that this family is independent if

$$P\left[\cap_{i \in J} A_i\right] = \Pi_{i \in J} P\left[A_i\right],$$

for all $J \subseteq \mathbb{N}_n$. If the family A has the property that

$$P\left[A_i \cap A_j\right] = P\left[A_i\right] P\left[A_j\right], \quad \forall i \neq j,$$

we say that the family is pairwise independent, or that the events in the family are *pairwise independent*. Independence of the family implies pairwise independence, but the converse is not necessarily true.

For a family of independent events, the probability that at least one of them happens is given by

$$P\left[\cup_{i=1}^{n} A_i\right] = 1 - \prod_{i=1}^{n} \left(1 - P\left[A_i\right]\right), \tag{3.4}$$

which is proven in Exercise 3.5.

3.4 Random Variables

A *random variable* or *stochastic variable* $X(\omega)$ is a function from the sample space Ω to some set D. If this set is countable, like $D = \mathbb{Z}$, we say that the random variable is *discrete*, and if it is uncountable, like $D = \mathbb{R}$, we say that it is *continuous*. We assume that the set D is partially ordered, see the subsection on generalized inequalities in Section 4.2 for a definition of partial ordering. This implies that $A = \{\omega \in \Omega : X(\omega) \leq x\}$ is well-defined. We also assume that the random variable $X : \Omega \to D$ is F- *measurable*, i.e. the set $A \in F$ for all $x \in D$. This means that the distribution function of the random variable $F : D \to [0, 1]$ given by

$$F(x) = P[\omega \in \Omega : X(\omega) \leq x]$$

is well defined. We will often use the shorter notation $F(x) = P[X \leq x]$. An important special case is when $D = \Omega$ and $X(\omega) = \omega$, and then the distribution function of the random variable is the same as the distribution function associated with $(\Omega, \mathcal{F}, \mathbb{P})$. Since the sample space can be countable, we realize that we have now actually defined distribution functions for countable sample spaces as well. Notice that we never did this explicitly when we talked about countable sample spaces above in this chapter. The reason was that we had not made any assumption on partial ordering of Ω,[2] *cf.* (4.23a).

For the special case, where $D = \Omega$ and $X(\omega) = \omega$, it is in some sense unnecessary to consider X to be a function, and then we will often just write $X \in D$. We will then most often not make any specific reference to the σ-algebra $\mathcal{F}$ or the probability measure $\mathbb{P}$ either. Instead, we just specify the distribution function F, or the probability function p or probability density function f defined on D. We tacitly assume that there is a well defined underlying probability measure and σ-algebra. This will in most cases be sufficient for our purposes.

3.4.1 Vector-Valued Random Variable

Just like we can have $\Omega = \mathbb{R}^n$, it is possible to define vector-valued random variables. We then have

$$F(x) = \mathbb{P}\left[\{\omega \in \Omega : X_1(\omega) \leq x_1, \dots, X_n(\omega) \leq x_n\}\right],$$

for the case of an n-dimensional random variable X. Often, we write $F(x) = \mathbb{P}[X \leq x]$ as before, but where the inequality now is to be interpreted as componentwise inequalities. The dimensions of the elements in Ω and of the random variable, i.e. the dimension of the elements in D, do not have to be the same, but sometimes they are. We will now give an important example of a pdf for an n-dimensional continuous random variable.

Example 3.4 We say that a random variable $X : \mathbb{R}^n \to \mathbb{R}^n$ defined as $X(\omega) = \omega$ with pdf $f : \mathbb{R}^n \to \mathbb{R}_+$ given by

$$f(x) = \frac{1}{\sqrt{(2\pi)^n \det(\Sigma)}} \exp\left(-\frac{1}{2}(x - \mu)^T \Sigma^{-1}(x - \mu)\right), \tag{3.5}$$

with $\mu \in \mathbb{R}^n$ and $\Sigma \in \mathbb{S}^n_{++}$ has a *Gaussian* or *normal* distribution.[3] When we want to emphasize the dependence on the parameters μ and Σ we use $\mathcal{N} : \mathbb{R}^n \times \mathbb{R}^n \times \mathbb{S}^n_{++} \to \mathbb{R}_+$ defined as $\mathcal{N}(x, \mu, \Sigma) = f(x)$.

3.4.2 Marginal Distribution

For a two-dimensional random variable $X : \Omega \to \mathbb{R}^2$, we define the *marginal distribution functions* as

$$F_1(x_1) = \mathbb{P}\left[X_1 \leq x_1\right] = \lim_{x_2 \to \infty} F(x_1, x_2), \quad F_2(x_2) = \mathbb{P}\left[X_2 \leq x_2\right] = \lim_{x_1 \to \infty} F(x_1, x_2),$$

where F is the distribution function for $X = (X_1, X_2)$, sometimes called the *joint distribution function*. This trivially generalizes to higher dimensions. When F is differentiable, it can be shown that the *marginal probability density functions* satisfy

$$f_1(x_1) = \int_\mathbb{R} f(x_1, x_2) dx_2, \quad f_2(x_2) = \int_\mathbb{R} f(x_1, x_2) dx_1.$$

---

2 For any countable sample space, we can always introduce a partial ordering by associating each element of Ω with an element in $\mathbb{Z}^n_+$ for some n.

3 We remark that it is the random variable that has a Gaussian distribution and that f is not a Gaussian distribution but a Gaussian pdf.

where f is the *joint probability density function*. For discrete-valued random variables, similar formulas hold, but then involving summations instead of integrals. For n-dimensional random variables, it is possible to look at marginal pdfs of dimension $n_1 < n$ by integrating or summing over the remaining $n_2 = n - n_1$ variables.

Example 3.5 Consider a Gaussian random variable $Z = (X, Y)$ with pdf $\mathcal{N}(z, \mu, \Sigma)$ for which

$$\mu = \begin{bmatrix} \mu_x \\ \mu_y \end{bmatrix} \in \mathbb{R}^{m+n}, \quad \Sigma = \begin{bmatrix} \Sigma_x & \Sigma_{xy} \\ \Sigma_{xy}^T & \Sigma_y \end{bmatrix} \in \mathbb{S}_{++}^{m+n}.$$

It is straightforward to show by integration that X and Y are also Gaussian random variables with marginal pdfs $\mathcal{N}(x, \mu_x, \Sigma_x)$ and $\mathcal{N}(y, \mu_y, \Sigma_y)$, respectively.

3.4.3 Independence of Random Variables

We say that two random variables $X : \Omega \to \mathbb{R}$ and $Y : \Omega \to \mathbb{R}$ are independent if the events $A = \{\omega : X(\omega) \leq x\}$ and $B = \{\omega : Y(\omega) \leq y\}$ are independent for all $x, y \in \mathbb{R}$. The generalization to vector-valued random variables is immediate, and the same definition also holds for discrete random variables with obvious modifications. The independence of the random variables X and Y is equivalent to

$$F(x, y) = F_X(x)F_Y(y),$$

and in case F is differentiable,

$$f(x, y) = f_X(x)f_Y(y).$$

Here F is the joint distribution function for (X, Y), and F_X and F_Y the marginal distribution functions. Similarly, f is the joint pdf, and f_X and f_Y are the marginal pdfs. This result also trivially generalizes to discrete random variables. When we discuss independence of $n > 2$ random variables, we realize that we define this as independence of n events, and that the criteria in terms of distribution functions and probability density functions are that we can factorize them in n factors, where these factors are the marginals.

3.4.4 Function of Random Variable

It is possible to define a *function of a random variable* $X : \Omega \to \mathcal{D}$. If $g : \mathcal{D} \to \mathcal{E}$ is such that $g(X(\omega))$ is $\mathcal{F}$-measurable, then $g(X)$ is a well-defined random variable that is a function of the random variable X. This usually holds for all "nice" functions g. If g is invertible, and the distribution function for X is $F : \mathcal{D} \to [0, 1]$, then the distribution function $F_Y : \mathcal{E} \to [0, 1]$ for $Y = g(X)$ can be obtained as

$$F_Y(y) = \mathbb{P}\left[g(X) \leq y\right] = \mathbb{P}\left[(X \leq g^{-1}(y)\right] = F\left(g^{-1}(y)\right). \tag{3.6}$$

When g is not invertible, obtaining the distribution function for $g(X)$ is more cumbersome. For continuous random variables, e.g. when $\mathcal{D} = \mathcal{E} = \mathbb{R}$, it holds that

$$F_Y(y) = \mathbb{P}\left[g(X) \leq y\right] = \int_{\{x | g(x) \leq y\}} f(x)dx,$$

where $f : \mathbb{R} \to \mathbb{R}_+$ is the pdf of X.

3.5 Conditional Distributions

We have already discussed conditional probabilities, and we have shown that we can define a conditional probability measure when the event we condition on has a positive probability. We will now define conditional pfs and conditional pdfs.

3.5.1 Conditional Probability Function

If we have random variables $X : \Omega \to \mathcal{D}$ and $Y : \Omega \to \mathcal{E}$, where $\mathcal{D}$ and $\mathcal{E}$ are countable, then we can define the *conditional distribution function* $F_{Y|X} : \mathcal{E} \to [0, 1]$ as

$$F_{Y|X}(y) = \mathbb{P}[Y \leq y \mid X = x],$$

for any x such that $\mathbb{P}[X = x] > 0$. We can also define the *conditional probability function* $p_{Y|X}(y) : \mathcal{E} \to [0, 1]$ as

$$p_{Y|X}(y) = \mathbb{P}[Y = y \mid X = x],$$

for any x such that $\mathbb{P}[X = x] > 0$. Sometimes we write $F_{Y|X}(y|x)$ and $p_{Y|X}(y|x)$ to emphasize the dependence on x. However, strictly speaking, we have just defined functions of the variable y for all values of x, i.e. a family of functions. We may, of course, consider them to be functions of x as well if we so desire. It should be clear that

$$\mathbb{P}[Y = y \mid X = x] = \frac{\mathbb{P}[Y = y, X = x]}{\mathbb{P}[X = x]},$$

which can be computed from the joint and marginal probability functions for (X, Y) and X, respectively, i.e.

$$p_{Y|X}(y|x) = \frac{p_{X,Y}(x, y)}{p_X(x)},$$

where

$$p_{X,Y}(x, y) = P(Y = y, X = x),$$

and

$$p_X(x) = \sum_{y \in \mathcal{D}} p_{X,Y}(x, y).$$

It is straightforward to verify that

$$F_{Y|X}(y|x) = \sum_{z \leq y} p_{Y|X}(z|x).$$

3.5.2 Conditional Probability Density Function

It is also possible to define conditional distribution functions for continuous random variables, but we have to be more careful, since then $\mathbb{P}[X = x] = 0$. We consider random variables $X : \Omega \to \mathcal{D}$ and $Y : \Omega \to \mathcal{E}$, where $\mathcal{D}$ and $\mathcal{E}$ are uncountable, e.g. $\mathbb{R}$. We investigate

$$\mathbb{P}\left[Y \leq y \mid x \leq X \leq x + dx\right] = \frac{\mathbb{P}\left[Y \leq y, x \leq X \leq x + dx\right]}{\mathbb{P}\left[x \leq X \leq x + dx\right]},$$

$$\approx \frac{\int_{-\infty}^{y} f_{X,Y}(x, v) dx \, dv}{f_X(x) dx},$$

$$= \int_{-\infty}^{y} \frac{f_{X,Y}(x, v)}{f_X(x)} dv,$$

for a small $dx > 0$, where $f_{X,Y} : D \times \mathcal{E} \to \mathbb{R}_+$ is the joint pdf for (X, Y) and where $f_X : D \to \mathbb{R}_+$ is the marginal pdf for X. As dx goes to zero, we obtain $\mathbb{P}[Y \le y \mid X = x]$, and hence, we define the *conditional distribution function* $F_{Y|X} : \mathcal{E} \to [0, 1]$ as

$$F_{Y|X}(y|x) = \int_{-\infty}^{y} \frac{f_{X,Y}(x, v)}{f_X(x)} dv.$$

The *conditional probability density function* $f_{Y|X} : \mathcal{E} \to \mathbb{R}_+$ is given by

$$f_{Y|X}(y|x) = \frac{f_{X,Y}(x, y)}{f_X(x)}. \tag{3.7}$$

Hence, we also obtain the same formula for continuous random variables.

Example 3.6 Consider the case when (X, Y) is jointly normal, i.e. the pdf is given by

$$f_{X,Y}(z) = \frac{1}{\sqrt{(2\pi)^{m+n} \det(\Sigma)}} \exp^{-\frac{1}{2}(z-\mu)^T \Sigma^{-1}(z-\mu)},$$

where $z = (x, y)$, $\mu = (\mu_x, \mu_y)$, and where

$$\Sigma = \begin{bmatrix} \Sigma_x & \Sigma_{xy} \\ \Sigma_{xy}^T & \Sigma_y \end{bmatrix}.$$

From (2.49) we have that

$$\begin{bmatrix} \Sigma_x & \Sigma_{xy} \\ \Sigma_{xy}^T & \Sigma_y \end{bmatrix} = \begin{bmatrix} I & 0 \\ \Sigma_{xy}^T \Sigma_x^{-1} & I \end{bmatrix} \begin{bmatrix} \Sigma_x & 0 \\ 0 & \Sigma_y - \Sigma_{xy}^T \Sigma_x^{-1} \Sigma_{xy} \end{bmatrix} \begin{bmatrix} I & \Sigma_x^{-1} \Sigma_{xy} \\ 0 & I \end{bmatrix}.$$

Notice that

$$\begin{bmatrix} I & \Sigma_x^{-1} \Sigma_{xy} \\ 0 & I \end{bmatrix}^{-1} = \begin{bmatrix} I & -\Sigma_x^{-1} \Sigma_{xy} \\ 0 & I \end{bmatrix}.$$

This factorizes the above pdf as $f_{X,Y}(z) = f_X(x) f_{Y|X}(y|x)$, where

$$f_X(x) = \frac{1}{\sqrt{(2\pi)^m \det \Sigma_x}} e^{-\frac{1}{2}(x-\mu_x)^T \Sigma_x^{-1}(x-\mu_x)},$$

and where

$$f_{Y|X}(y|x) = \frac{1}{\sqrt{(2\pi)^n \det \Sigma_{y|x}}} e^{-\frac{1}{2}(y-\mu_{y|x})^T \Sigma_{y|x}^{-1}(y-\mu_{y|x})},$$

where

$$\mu_{y|x} = \mu_y + \Sigma_{xy}^T \Sigma_x^{-1}(x - \mu_x), \quad \Sigma_{y|x} = \Sigma_y - \Sigma_{xy}^T \Sigma_x^{-1} \Sigma_{xy}.$$

From Example 3.5, we see that f_x is the marginal pdf for X. Hence, it holds by (3.7) that $f_{Y|X}(y|x)$ is the conditional pdf for Y, given $X = x$. Notice that $\Sigma_{y|x}$ is the Schur complement of Σ_x in Σ.

3.6 Expectations

Let us assume that we are interested in estimating how frequent a certain event $A \subseteq \Omega$ is when the experiment is repeated. Hence, it is enough to define $\mathcal{F} = \{\Omega, \emptyset, A, A^c\}$, where $A^c = \Omega \backslash A$. Let us assume that $\mathbb{P}[A] = p$ and that $\mathbb{P}[A^c] = 1 - p$. We then define the random variable $X : \Omega \to \{0, 1\}$ as $X(\omega) = 1$ when $\omega \in A$ and $X(\omega) = 0$ when $\omega \notin A$. If we repeat the experiment N times, it is reasonable to estimate the relative frequency of A with the *sample average approximation* (SAA)

$$\frac{1}{N}\sum_{i=1}^{N}X(\omega_i),\tag{3.8}$$

where ω_i is the outcome of the ith experiment. We realize that this quantity is very close to

$$0\times\mathbb{P}[\{\omega:X(\omega)=0\}]+1\times\mathbb{P}[\{\omega:X(\omega)=1\}]=0\times\mathbb{P}\big[A^c\big]+1\times\mathbb{P}[A]=p.$$

Inspired by this, we define the *expected value* of any discrete random variable $X:\Omega\to D\subseteq\mathbb{R}$ as

$$\mathbb{E}[X]=\sum_{x\in D}xp(x),$$

where $p(x)=\mathbb{P}[\{\omega:X(\omega)=x\}]$.[4] The expected value is a *functional* from D^Ω to the real numbers. We understand that the expected value is close to the sample average of the random variable for large values of N. For this reason, we sometimes call the expected value of a random variable the *mean* of the random variable. Sometimes, we write $\mathbb{E}X$ instead of $\mathbb{E}[X]$ to ease the notation.

For continuous random variables $X:\Omega\to\mathbb{R}$, we define the expected value as

$$\mathbb{E}[X]=\int_{\mathbb{R}}xf(x)dx,$$

where f is the pdf of the random variable. The generalization to vector-valued random variables is straightforward. We remark that expected values might be infinite.

3.6.1 Moments

For any scalar-valued random variable X, we define the kth *moment* of X as

$$m_k=\mathbb{E}\big[X^k\big],$$

and the kth *central moment* of X as

$$\mu_k=\mathbb{E}\Big[\big(X-m_1\big)^k\Big].$$

The moment m_1 is the expectation, also called the *mean*, of X and μ_2 is called the *variance* of X. The variance measures the amount by which X tends to deviate from its average. It is often also denoted by σ^2 or $\mathrm{Var}[X]$. We sometimes write $\mathrm{Var}X$ to ease the notation. The positive square root of the variance, σ, is called the *standard deviation*. It is straightforward to show that $\mu_2=m_2-m_1^2$.

3.6.2 Expected Value of Function of Random Variable

We can also define the expected value of a function g of a continuous random variable X as

$$\mathbb{E}\big[g(X)\big]=\int_{\mathbb{R}}g(x)f(x)dx,$$

where f is the pdf for X. Notice that we do not need to compute the pdf of $g(X)$ in order to compute the expectation of $g(X)$. In the special case when $g(x)=Ax$, with $A\in\mathbb{R}^{m\times n}$, it holds that $\mathbb{E}[AX]=A\mathbb{E}[X]$, and we realize that expectation is a *linear functional* on the set of random variables.

---

4 We need that the sum in the definition of the expectation is absolutely convergent, since we do not want the result to depend on the order in which we carry out the summation.

For two scalar-valued random variables X and Y with joint pdf $f_{X,Y}$, we may consider the function $g : \mathbb{R}^2 \to \mathbb{R}$ defined as $g(x, y) = x$. It then holds that

$$\mathbb{E}[g(X, Y)] = \int_{\mathbb{R}^2} x f_{X,Y}(x, y) dx\, dy = \int_{\mathbb{R}} x f_X(x) dx = \mathbb{E}[X],$$

with f_X being the marginal pdf for X. The result trivially generalizes to several random variables. This shows that we do not need to define the marginal pdf to carry out the expectation computation. This also means that we never have to distinguish between different definitions of the expectation functional. It is sufficient to consider the joint pdf for all relevant random variables involved when defining the expectation functional, independent of how many random variables there are.

3.6.3 Covariance

For two scalar-valued random variables X and Y, the product XY is a special case of a function g of the two-dimensional random variable (X, Y). Hence, the expected value of XY is given by

$$\mathbb{E}[XY] = \int_{\mathbb{R}^2} xy f_{X,Y}(x, y) dx\, dy,$$

where $f_{X,Y}$ is the pdf of (X, Y). If X and Y are independent, it is easy to see that $\mathbb{E}[XY] = \mathbb{E}[X]\mathbb{E}[Y]$. The converse is not true in general. We say that X and Y are *uncorrelated*, if $\mathbb{E}[XY] = \mathbb{E}[X]\mathbb{E}[Y]$, and otherwise, they are said to be *correlated*. The generalization to vector-valued random variables is obtained by considering the outer product. Specifically, we define the *covariance* between two random variables $X : \Omega \to \mathbb{R}^m$ and $Y : \Omega \to \mathbb{R}^n$ as

$$\text{Cov}[X, Y] = \mathbb{E}\left[(X - m_X)(Y - m_Y)^T\right] = \mathbb{E}\left[XY^T\right] - m_X m_Y^T,$$

where m_X and m_Y are the means of X and Y, respectively. The variance of X is $\text{Var}[X] = \text{Cov}[X, X]$. The random variables X and Y are uncorrelated if $\mathbb{E}\left[XY^T\right] = \mathbb{E}[X](\mathbb{E}[Y])^T = m_X m_Y^T$. This is equivalent to $\text{Cov}[X, Y] = 0$. As before, independence implies being uncorrelated. Notice that X and Y may have different dimensions.

Example 3.7 For a random vector with a Gaussian pdf as in Example 3.4, it holds that the expected value is μ and that the covariance is Σ. This result is shown in Exercise 3.8.

3.7 Conditional Expectations

Given two random variables $X : \Omega \to D$ and $Y : \Omega \to \mathcal{E}$ the *conditional expectation* of Y given $X = x$ is defined as

$$\mathbb{E}[Y \mid X = x] = \sum_{y \in \mathcal{E}} y p_{Y|X}(y|x),$$

for discrete random variables and as

$$\mathbb{E}[Y \mid X = x] = \int_{\mathcal{E}} y f_{Y|X}(y|x) dy,$$

for continuous random variables, where $p_{Y|X} : \mathcal{E} \to [0, 1]$ is the conditional probability function and where $f_{Y|X} : \mathcal{E} \to \mathbb{R}_+$ is the conditional pdf, respectively. For a given x, the conditional expectation is just a number, but it is possible to consider all values of $x \in D$, and then we can define the function $\Psi : D \to \mathbb{R}$ as $\Psi(x) = \mathbb{E}[Y \mid X = x]$. Let us now investigate the random variable $\Psi(X)$. For discrete random variables, its expectation is given by

$$\mathbb{E}[\Psi(X)] = \sum_{x \in D} \left(\sum_{y \in \mathcal{E}} y p_{Y|X}(y|x) \right) p_X(x) = \sum_{y \in \mathcal{E}} y \sum_{x \in D} p_{X,Y}(x, y),$$

$$= \sum_{y \in \mathcal{E}} y p_Y(y) = \mathbb{E}[Y],$$

where $p_{X,Y} : D \times \mathcal{E} \to [0, 1]$ is the joint probability function for (X, Y), and where $p_X : D \to [0, 1]$ is the marginal probability function for X. The same result holds for continuous random variables. We often write $\Psi(X) = \mathbb{E}[Y \mid X]$, and hence, the result can be summarized as

$$\mathbb{E}[\mathbb{E}[Y \mid X]] = \mathbb{E}[Y].$$

This formula is sometimes useful, when $\mathbb{E}[Y]$ is difficult to compute directly.

3.8 Convergence of Random Variables

Given a probability space $(\Omega, \mathcal{F}, \mathbb{P})$, we consider random variables $X_1, X_2, \ldots$ and X defined on this probability space and are interested in investigating convergence of X_k to X as $k \to \infty$. To this end, we will define four different modes of convergence:

(a) $X_k \to X$ *almost surely* $(X_k \xrightarrow{\text{a.s.}} X)$ if $\mathbb{P}\left[\omega \in \Omega \mid X_k(\omega) \to X(\omega), \ k \to \infty\right] = 1.$[5]

(b) $X_k \to X$ *in rth mean* $(X_k \xrightarrow{r} X)$ if $\mathbb{E}\left[|X_k - X|^r\right] \to 0$ as $k \to \infty.$[6]

(c) $X_k \to X$ *in probability* $(X_k \xrightarrow{P} X)$ if $\mathbb{P}\left[|X_k - K| > \epsilon)\right] \to 0$ as $k \to \infty$ for all $\epsilon > 0.$

(d) $X_k \to X$ *in distribution* $(X_k \xrightarrow{D} X)$ if $\mathbb{P}\left[X_k \leq x\right] \to \mathbb{P}[X \leq x]$ as $k \to \infty$ for all points x at which $F_X(x) = \mathbb{P}[X \leq x]$ is continuous.

The following implications hold for the different modes of convergence:

1. (a) $\Rightarrow$ (c)
2. (b) $\Rightarrow$ (c)
3. (c) $\Rightarrow$ (d)
4. If $r > s \geq 1$, then $X_k \xrightarrow{r} X \Rightarrow X_k \xrightarrow{s} X$

Without any further assumptions, no other implications hold.

Let $X_1, X_2, \ldots$ be independent identically distributed (i.i.d.) random variables with mean m. Then

$$\frac{1}{n} \sum_{i=1}^{n} X_i \xrightarrow{\text{a.s.}} m, \quad n \to \infty,$$

if and only if $\mathbb{E}\left[|X_1|\right] < \infty$. This result is known as the *strong law of large numbers*. If we assume that $\mathbb{E}\left[X_1^2\right] < \infty$, then convergence holds both almost surely and in mean square. This assumption is a sufficient condition for the strong law of large numbers. There is also a weak law of large numbers, which is related to convergence in probability; we refer the reader to [48] for details.

3.9 Random Processes

We will now discuss random variables $X : \Omega \to D^{\mathbb{Z}_+}$. Hence, the random variable is a function of $k \in \mathbb{Z}_+$, i.e. $(\omega, k) \mapsto X(\omega, k)$. We will often instead interpret it as an infinite-dimensional vector

---

5 This mode of convergence has two other notations which are $X_k \to X$ *almost everywhere* $(X_k \xrightarrow{\text{a.e.}} X)$ and $X_k \to X$ *with probability one* (w.p. 1.).

6 When $r = 1$ we say that $X_k \to X$ *in mean* and when $r = 2$ we say that $X_k \to X$ *in mean square* $(X_k \xrightarrow{\text{m.s.}} X)$.

$X = (X(\omega, 0), X(\omega, 1), \ldots)$ and use the notation $X_k(\omega) = X(k, \omega)$ to ease the notation. We may interpret the random variable X as an infinite sequence of random variables $X_k : \Omega \to D, k \in \mathbb{Z}_+$, and we may interpret $k \in \mathbb{Z}_+$ as a discrete time-index. Such a random variable X is often called a *discrete-time random process*. Sometimes one says *stochastic process* instead of random process. The set D could be a finite set like $\mathbb{N}_n$, a countable set like $\mathbb{Z}$ or an uncountable set like $\mathbb{R}$. It may also be vector-valued.

We may observe the evolution of a random process in two different ways. For each fixed outcome $\omega \in \Omega$, we obtain a *realization* or *sample path* $X(\omega)$ of X at ω. We can study the properties of this sample path. Another way of viewing the random process is to investigate a finite subset of components of the infinite-dimensional vector X, say $K = \{k_1, k_2, \ldots, k_n\} \subset \mathbb{Z}_+$. We then look at the joint distribution function $F_K : D^n \to [0, 1]$ defined as

$$F_K(x) = \mathbb{P}\left[X_{k_1} \leq x_1, \ldots, X_{k_n} \leq x_n\right].$$

The collection $\{F_K\}$ where K ranges over all finite-dimensional $K \subset \mathbb{Z}_+$ is called the collection of *finite-dimensional distributions* (fdds) of X or the *name* of X. This contains all the information that is available about X from finitely many components X_k. We mention that knowing the fdds does not in general provide a complete information about the sample paths. However, we will be content by studying only properties of the sample path that can be deduced from the fdds.

If we define $X : \Omega \to D^{\mathbb{R}_+}$, we obtain a *continuous-time random process*. We then often write $X(t)$ when it is convenient to make the dependence on $t \in \mathbb{R}_+$ explicit. The fdds are often denoted F_T, where T is now a finite subset of $\mathbb{R}_+$. An even more general concept is a *random field*, which is obtained when $X : \Omega \to D^{\mathbb{R}^n}$.

We say that a discrete-time random process is *strongly stationary* if $\{X_{k_1}, \ldots, X_{k_n}\}$ and $\{X_{k_1+l}, \ldots, X_{k_n+l}\}$ have the same joint distribution for all $k_1, \ldots, k_n$ and $l > 0$. We say that a discrete-time random process is *weakly stationary* if $\mathbb{E}\left[X_{k_1}\right] = \mathbb{E}\left[X_{k_2}\right]$ and $\mathrm{Cov}[X_{k_1}, X_{k_2}] = \mathrm{Cov}[X_{k_1+l}, X_{k_2+l}]$ for all k_1, k_2 and $l > 0$. In other words, a random process is weakly stationary if and only if it has a constant mean and the *autocovariance function* $c : \mathbb{Z}_+^2 \to \mathbb{R}$ given by

$$c(k, k + l) = \mathrm{Cov}[X_k, X_{k+l}],$$

satisfies

$$c(k, k + l) = c(0, l),$$

for all k and $l \geq 0$. Thus, for weakly stationary processes, we may define the autocovariance function as a function of only l and write $c : \mathbb{Z}_+ \to \mathbb{R}$.

Strong stationarity implies weak stationarity, but the converse is not true in general. One example where strong stationarity is equivalent to weak stationarity is when the fdds are all Gaussian. The definitions of weak and strong stationarity for a continuous-time random process are similar as for a discrete-time random process, and this also goes for a random field.

We will now discuss a generalization of the law of large numbers where the sequence of random variables X_k are a stationary process, not necessarily i.i.d. If $X_k, k \geq 1$, is a strongly stationary process with $\mathbb{E}|X_1| < \infty$, then there exists a random variable Y with the same mean as X_1 such that

$$\frac{1}{n}\sum_{k=1}^{n} X_k \to Y \text{ a.s. and in mean.}$$

If instead $X_k, k \geq 1$, is a weakly stationary process with $\mathbb{E}|X_1| < \infty$, then there exists a random variable Y with the same mean as X_1 such that

$$\frac{1}{n}\sum_{k=1}^{n} X_k \to Y \text{ in mean square.}$$

These results are called the *strong ergodic theorem* and the *weak ergodic theorem*, respectively.

3.10   Markov Processes

Discrete-time random processes $X : \Omega \to D^{\mathbb{Z}_+}$ that satisfies the so-called *Markov property*

$$\mathbb{P}\left[X_{k_n} \leq x_n \mid X_{k_{n-1}} \leq x_{n-1}\right] = \mathbb{P}\left[X_{k_n} \leq x_n \mid X_{k_{n-1}} \leq x_{n-1}, \ldots, X_{k_1} \leq x_1\right], \tag{3.9}$$

for all $k_1 \leq k_2 \leq \cdots \leq k_n$, are called *Markov processes*.

If D is a finite set, like $\mathbb{N}_N$, the process is often called a *Markov chain*. We say that a Markov process is *homogeneous* if

$$\mathbb{P}\left[X_n \leq x \mid X_{n-1} \leq y\right] = \mathbb{P}\left[X_1 \leq x \mid X_0 \leq y\right],$$

for all x, y and n. For a homogeneous Markov chain, we define the *transition matrix* $P \in \mathbb{R}^{N \times N}$ with elements p_{ij}, called the *transition probabilities*, as

$$p_{ij} = \mathbb{P}\left[X_n = j \mid X_{n-1} = i\right].$$

The *n-step transition matrix* $P_n \in \mathbb{R}^{N \times N}$ has elements $p_{ij}(n)$ called the *n-step transition probabilities* defined as

$$p_{ij}(n) = \mathbb{P}\left[X_{m+n} = j \mid X_m = i\right].$$

It holds that $P_{m+n} = P_m P_n$, which is called the *Chapman–Kolmogorov equation*, and hence, $P_n = P^n$. If we let π^k be the row vector of

$$\pi_i^k = \mathbb{P}\left[X_k = i\right],$$

it follows that $\pi^k = \pi^0 P^k$ for $k \geq 0$. Hence, the transition matrix fully characterizes a homogeneous Markov chain.

If $\pi \in \mathbb{R}^{1 \times N}$ is such that $\pi_i \geq 0$, $\sum_i \pi_i = 1$ and $\pi = \pi P$, then π is called a *stationary distribution* for the Markov chain. The reason for this name is that if $\pi^0 = \pi$, then $\pi^k = \pi$ for all $k \geq 0$, i.e. the distribution remains the same for all times, and hence, the Markov chain is for this initial distribution a stationary random process, both weakly and strongly. For other initial values, the Markov chain may or may not converge to the stationary distribution as k goes to infinity. We refer the reader to [48] for further details.

Example 3.8   An important example of a Markov process for $D = \mathbb{R}^n$ is obtained by considering the random process defined by the recursion

$$X_{k+1} = AX_k + E_k, \quad k \in \mathbb{Z}_+,$$

where $A \in \mathbb{R}^{n \times n}$, E_k are *independent, identically distributed* (i.i.d.) n-dimensional random vectors, and where X_0 is a random vector with a known distribution. When E_k has zero mean we say that E is *white noise*.

3.11   Hidden Markov Models

Consider two random processes $X : \Omega \to D^{\mathbb{Z}_+}$ and $Y : \Omega \to \mathcal{E}^{\mathbb{Z}_+}$ that are correlated. The sets D and $\mathcal{E}$ will be defined later. We will assume that X is a Markov process satisfying (3.10), and that Y_j given X_j are independent of Y_k given X_k for $j \neq k$. We are interested in computing the conditional probability function or pdf for X_k given $\bar{Y}_k = (Y_0, \ldots, Y_k)$. This can be interpreted as we have an observation of $\bar{Y}_k$. The pair (X, Y) is called a *hidden Markov Model* (HMM), since X is not observed but hidden to the observer. The process X is called the *state* and sometimes Y is called the *measurement* or *output*. The problem of computing the above conditional probability function or pdf is often called a *filtration problem* or *state estimation problem*. We will do the derivation for $\mathcal{D} = \mathbb{R}^n$ and $\mathcal{E} = \mathbb{R}^p$. The derivation for finite sets is similar and obtained by considering probability functions instead of pdfs and by replacing integrals with summations.

Let $\bar{X}_k = (X_0, \ldots, X_k)$ and let $p_{\bar{X}_k, \bar{Y}_k} : \mathcal{D}^{k+1} \times \mathcal{E}^{k+1} \to \mathbb{R}_+$ be the joint pdf for $(\bar{X}_k, \bar{Y}_k)$. We also need the conditional pdf for $\bar{Y}_k$ given $\bar{X}_k$: $p_{\bar{Y}_k|\bar{X}_k} : \mathcal{E}^{k+1} \times \mathcal{D}^{k+1} \to \mathbb{R}_+$. Using the conditional independence assumption, this can be expressed as

$$p_{\bar{Y}_k|\bar{X}_k}(\bar{y}|\bar{x}) = \prod_{i=0}^{k} p_{Y_i|X_i}(y_i|x_i),$$

where $p_{Y_i|X_i} : \mathcal{E} \times \mathcal{D} \to \mathbb{R}_+$ are the conditional pdfs for Y_k given X_k. We also define the marginal pdf for $\bar{X}$ as $p_{\bar{X}} : \mathcal{D}^{N+1} \to \mathbb{R}_+$.

We start by obtaining an expression for $p_{X_0|Y_0}(x_0|y_0)$, i.e.

$$p_{X_0|Y_0}(x_0|y_0) = \frac{p_{Y_0|X_0}(y_0|x_0)p_{X_0}(x_0)}{p_{Y_0}(y_0)},$$

where

$$p_{Y_0}(y_0) = \int p_{X_0, Y_0}(x_0, y_0) dx_0 = \int p_{Y_0|X_0}(y_0|x_0)p_{X_0}(x_0)dx_0.$$

We do in this section not write out the set over which we integrate when it is the whole domain of the functions involved. Now, assume that we know $p_{X_{k-1}|\bar{Y}_{k-1}}(x_{k-1}|\bar{y}_{k-1})$, where $\bar{y}_k = (y_0, \ldots, y_k)$. This assumption is true for $k = 1$. It then follows that

$$p_{X_k, X_{k-1}|\bar{Y}_{k-1}}(x_k, x_{k-1}|\bar{y}_{k-1}) = p_{X_k|X_{k-1}, \bar{Y}_{k-1}}(x_k|x_{k-1}, \bar{y}_{k-1})p_{X_{k-1}|\bar{Y}_{k-1}}(x_{k-1}|\bar{y}_{k-1}),$$

$$= p_{X_k|X_{k-1}}(x_k|x_{k-1})p_{X_{k-1}|\bar{Y}_{k-1}}(x_{k-1}|\bar{y}_{k-1}),$$

where we have made use of the conditional independence property

$$p_{X_k|X_{k-1}, \bar{Y}_{k-1}}(x_k|x_{k-1}, \bar{y}_{k-1}) = p_{X_k|X_{k-1}}(x_k|x_{k-1}).$$

This is proven in Exercise 3.16. Integrating over x_{k-1} results in the following Chapman–Kolmogorov equation:

$$p_{X_k|\bar{Y}_{k-1}}(x_k|\bar{y}_{k-1}) = \int p_{X_k|X_{k-1}}(x_k|x_{k-1})p_{X_{k-1}|\bar{Y}_{k-1}}(x_{k-1}|\bar{y}_{k-1})dx_{k-1}.$$

Furthermore, it holds that

$$p_{X_k|\bar{Y}_k}(x_k|\bar{y}_k) = \frac{p_{X_k, \bar{Y}_k}(x_k, \bar{y}_k)}{p_{\bar{Y}_k}(\bar{y}_k)},$$

$$= \frac{p_{Y_k|X_k, \bar{Y}_{k-1}}(y_k|x_k, \bar{y}_{k-1})p_{X_k, \bar{Y}_{k-1}}(x_k, \bar{y}_{k-1})}{p_{Y_k|\bar{Y}_{k-1}}(y_k|\bar{y}_{k-1})p_{\bar{Y}_{k-1}}(\bar{y}_{k-1})},$$

$$= \frac{p_{Y_k|X_k, \bar{Y}_{k-1}}(y_k|x_k, \bar{y}_{k-1})p_{X_k|\bar{Y}_{k-1}}(x_k|\bar{y}_{k-1})}{p_{Y_k|\bar{Y}_{k-1}}(y_k|\bar{y}_{k-1})},$$

$$= \frac{p_{Y_k|X_k}(y_k|x_k)p_{X_k|\bar{Y}_{k-1}}(x_k|\bar{y}_{k-1})}{p_{Y_k|\bar{Y}_{k-1}}(y_k|\bar{y}_{k-1})}, \tag{3.10}$$

where the last equality follows from the conditional independence property

$$p_{Y_k|X_k, \bar{Y}_{k-1}}(y_k|x_k, \bar{y}_{k-1}) = p_{Y_k|X_k}(y_k|x_k).$$

This is also proven in Exercise 3.16. We will now see how to compute the denominator in (3.10):

$$p_{Y_k|\bar{Y}_{k-1}}(y_k|\bar{y}_{k-1}) = \int p_{Y_k, X_k|\bar{Y}_{k-1}}(y_k, x_k|\bar{y}_{k-1})dx_k,$$

$$= \int \frac{p_{\bar{Y}_k, X_k}(\bar{y}_k, x_k)}{p_{\bar{Y}_{k-1}}(\bar{y}_{k-1})}dx_k,$$

$$= \int \frac{p_{\bar{Y}_k, X_k}(\bar{y}_k, x_k) p_{X_k, \bar{Y}_{k-1}}(x_k, \bar{y}_{k-1})}{p_{X_k, \bar{Y}_{k-1}}(x_k, \bar{y}_{k-1}) p_{\bar{Y}_k}(\bar{y}_{k-1})} dx_k,$$

$$= \int p_{Y_k | X_k, \bar{Y}_{k-1}}(y_k | x_k, \bar{y}_{k-1}) p_{X_k | \bar{Y}_{k-1}}(x_k | \bar{y}_{k-1}) dx_k,$$

$$= \int p_{Y_k | X_k}(y_k | x_k) p_{X_k | \bar{Y}_{k-1}}(x_k | \bar{y}_{k-1}) dx_k.$$

The last inequality follows from the conditional independence property. Thus, we may summarize the optimal filtering equations as

$$p_{X_k | \bar{Y}_{k-1}}(x_k | \bar{y}_{k-1}) = \int p_{X_k | X_{k-1}}(x_k | \xi) p_{X_{k-1} | \bar{Y}_{k-1}}(\xi | \bar{y}_{k-1}) d\xi, \tag{3.11}$$

$$p_{X_k | \bar{Y}_k}(x_k | \bar{y}_k) = \frac{p_{Y_k | X_k}(y_k | x_k) p_{X_k | \bar{Y}_{k-1}}(x_k | \bar{y}_{k-1})}{\int p_{Y_k | X_k}(y_k | \xi) p_{X_k | \bar{Y}_{k-1}}(\xi | \bar{y}_{k-1}) d\xi}, \tag{3.12}$$

for $k \geq 0$ with initial value given by

$$p_{X_0 | Y_0}(x_0 | y_0) = \frac{p_{Y_0 | X_0}(y_0 | x_0) p_{X_0}(x_0)}{\int p_{Y_0 | X_0}(y_0 | \xi) p_{X_0}(\xi) d\xi}.$$

In the following example, we derive the so-called *Kalman filter* from the filtering equations for the special case when the HMM is defined by a so-called "linear state-space equation."

Example 3.9   Consider the HMM (X, Y) defined as

$$X_{k+1} = AX_k + V_k,$$
$$Y_k = CX_k + E_k,$$

where V_k and E_k, $k \geq 0$, are Gaussian i.i.d. random variables with zero mean and covariances $R_1 \in \mathbb{S}^n_{++}$ and $R_2 \in \mathbb{S}^p_{++}$, respectively. We assume that X_0 also has a Gaussian distribution with mean $\bar{x}_0 \in \mathbb{R}^n$ and covariance $R_0 \in \mathbb{S}^n_{++}$. This is assumed to be independent of V_k and E_k for all $k \geq 0$. It is straightforward to verify that the conditional independence condition holds. We have that

$$p_{X_0}(x_0) = \mathcal{N}(x_0, \bar{x}_0, R_0),$$
$$p_{X_k | X_{k-1}}(x_k | x_{k-1}) = \mathcal{N}(x_k, Ax_{k-1}, R_1),$$
$$p_{Y_k | X_k}(y_k | x_k) = \mathcal{N}(y_k, Cx_k, R_2),$$

where $\mathcal{N}$ is defined as in Example 3.4, i.e. all the involved pdfs are Gaussian. Now suppose that

$$p_{X_k | \bar{Y}_{k-1}}(x_k | \bar{y}_{k-1}) = \mathcal{N}(x_k, x_k^f, \Sigma_k^f), \tag{3.13}$$

for some $x_k^f \in \mathbb{R}^n$ and $\Sigma_k^f \in \mathbb{S}^n_{++}$, where $k \geq 1$. We have from Exercise 3.15 that

$$p_{Y_k | X_k}(y_k | x_k) p_{X_k | \bar{Y}_{k-1}}(x_k | \bar{y}_{k-1}) = \mathcal{N}(y_k, Cx_k, R_2) \mathcal{N}(x_k, x_k^f, \Sigma_k^f),$$
$$= \mathcal{N}(y_k, Cx_k^f, H_k) \mathcal{N}(x_k, x_k^f + K_k(y_k - Cx_k^f), \Sigma_k^f - K_k H_k K_k^T),$$

where $H_k = R_2 + C\Sigma_k^f C^T$, and $K_k = \Sigma_k^f C^T H_k^{-1}$. From the second equality above, we obtain that

$$\int p_{Y_k | X_k}(y_k | x_k) p_{X_k | \bar{Y}_{k-1}}(x_k | \bar{y}_{k-1}) dx_k = \mathcal{N}(y_k, Cx_k^f, H_k),$$

and hence, from (3.12) that

$$p_{X_k | \bar{Y}_k}(x_k | \bar{y}_k) = \mathcal{N}\left(x_k, x_k^f + K_k\left(y_k - Cx_k^f\right), \Sigma_k^f - K_k H_k K_k^T\right) = \mathcal{N}(x_k, x_k^a, \Sigma_k),$$

Algorithm 3.1: Standard form Kalman filter

Input: System matrices A and C, Mean $\bar{x}_0$, Covariances R_0, R_1, and R_2, Measurement data
$(y_0, y_1, \ldots)$

Output: x_k^a for $k \in \mathbb{Z}$

$x_0^f \leftarrow \bar{x}_0$

$\Sigma_0^f \leftarrow R_0$

for $k \leftarrow 0$ **to** ∞ **do**

$\qquad x_k^a = x_k^f + \Sigma_k^f C^T \left(C\Sigma_k^f C^T + R_2 \right)^{-1} \left(y_k - Cx_k^f \right)$

$\qquad \Sigma_k = \Sigma_k^f - \Sigma_k^f C^T \left(C\Sigma_k^f C^T + R_2 \right)^{-1} C\Sigma_k^f$

$\qquad x_{k+1}^f = Ax_k^a$

$\qquad \Sigma_{k+1}^f = R_1 + A\Sigma_k A^T$

end

---

where we have defined

$$
\begin{aligned}
x_k^a &= x_k^f + K_k(y_k - Cx_k^f), \\
\Sigma_k &= \Sigma_k^f - K_k H_k K_k^T.
\end{aligned}
\tag{3.14}
$$

To make the calculations valid also for $k = 0$, we just need to define $x_0^f = \bar{x}_0$ and $\Sigma_0^f = R_0$ in view of the definition of $p_{X_0|\bar{Y}_0}(x_0|\bar{y}_0)$. We have from Exercise 3.15 that

$$
\begin{aligned}
p_{X_k|X_{k-1}}(x_k|x_{k-1})p_{X_{k-1}|\bar{Y}_{k-1}}(x_{k-1}|\bar{y}_{k-1}) &= \mathcal{N}(x_k, Ax_{k-1}, R_1)\mathcal{N}(x_{k-1}, x_{k-1}^a, \Sigma_{k-1}), \\
&= \mathcal{N}(x_k, Ax_{k-1}^a, \bar{H}_{k-1})\mathcal{N}(x_{k-1}, x_{k-1}^a + \bar{K}_{k-1}(x_k - Ax_{k-1}^a), \Sigma_{k-1} - \bar{K}_{k-1}\bar{H}_{k-1}\bar{K}_{k-1}^T),
\end{aligned}
$$

where $\bar{H}_k = R_1 + A\Sigma_k A^T$ and $\bar{K}_k = \Sigma_k A^T \bar{H}_k^{-1}$. From the last equality, we obtain

$$
\int p_{X_k|X_{k-1}}(x_k|x_{k-1})p_{X_{k-1}|\bar{Y}_{k-1}}(x_{k-1}|\bar{y}_{k-1})dx_{k-1} = \mathcal{N}(x_k, Ax_{k-1}^a, \bar{H}_{k-1}),
$$

and we obtain from (3.13) and (3.11), the update formula

$$
\begin{aligned}
x_k^f &= Ax_{k-1}^a, \\
\Sigma_k^f &= \bar{H}_{k-1} = R_1 + A\Sigma_{k-1}A^T.
\end{aligned}
\tag{3.15}
$$

We can if we like eliminate x_k^f and Σ_k from (3.14) and (3.15) to obtain

$$
\begin{aligned}
x_{k+1}^f &= (A - \tilde{K}_k C)x_k^f + \tilde{K}_k y_k, \\
\Sigma_{k+1}^f &= A\Sigma_k^f A^T + R_1 - A\Sigma_k^f C^T(R_2 + C\Sigma_k^f C^T)^{-1}C\Sigma_k^f A^T,
\end{aligned}
$$

where $\tilde{K}_k = A\Sigma_k^f C^T(R_2 + C\Sigma_k^f C^T)^{-1}$. The initial values are $x_0^f = \bar{x}_0$ and $\Sigma_0^f = R_0$. The Kalman filter is summarized in Algorithm 3.1. We see that the algorithm consists of two main steps. In the first one the measurement y_k, the matrix C, and the covariance R_2 are used to update the old predicted estimate x_k^f and its covariance. In the second step, the matrix A and the covariance matrix R_1 are used to predict the state and its covariance for the next value of k.

3.12 Gaussian Processes

A continuous-time real-valued random processes $X : \Omega \to \mathbb{R}^{\mathbb{R}}$ is called a *Gaussian process* if its fdds are all Gaussian with mean $m(T) \in \mathbb{R}^{|T|}$ and covariance $\Sigma(T) \in \mathbb{S}_{++}^{|T|}$, where $T = \{t_1, \ldots, t_n\} \subset \mathbb{R}$.

A prime example of a Gaussian process is the *Wiener process* $W : \Omega \to \mathbb{R}^{\mathbb{R}_+}$ with $W(0) = 0$, $m(T) = 0$, and $\Sigma(T)$ having entries $\Sigma_{i,j}(T) = \sigma^2 \min(t_i, t_j)$, for some $\sigma^2 > 0$.

A Gaussian process is weakly and strongly stationary if and only if $\mathbb{E}[X(t)]$ is constant for all t and $\Sigma(T + h) = \Sigma(T)$, where $T + h = \{t_1 + h, \dots, t_n + h\}$, for all T and $h > 0$. A Gaussian process is a Markov process if and only if

$$\mathbb{E}\big[X(t_n) \,|\, X(t_1), \dots, X(t_{n-1})\big] = \mathbb{E}\big[X(t_n) \,|\, X(t_{n-1})\big],$$

for all $t_1 < \cdots < t_n$. An example of a stationary Gaussian Markov process is the *Ornstein–Uhlenbeck process* which has zero mean and autocovariance function $c(t) = c(0)e^{-\alpha|t|}$ for some $\alpha > 0$ and $c(0) > 0$. It is also possible to define Gaussian processes $X : \Omega \to \mathbb{R}^{\mathbb{R}^n}$, and we will return to them and their usage in Section 10.3.

Exercises

3.1 We are given a probability space $(\Omega, \mathcal{F}, \mathbb{P})$.
   (a) Show that $\mathbb{P}[A^c] = 1 - \mathbb{P}[A]$ for any $A \in \mathcal{F}$, where $A^c = \Omega \backslash A$.
   (b) Show that $\mathbb{P}\big[\emptyset\big] = 0$.
   (c) Show that $\mathbb{P}[A \cup B] = \mathbb{P}[A] + \mathbb{P}[B] - \mathbb{P}[A \cap B]$ for any $A, B \in \mathcal{F}$.

3.2 In a collection of 100 items delivered by a company, there are five defect items. We pick randomly first one item and then out of the remaining 99, we pick another item randomly. What is the probability that both items are defect?

3.3 Prove the formula for total probability in (3.3).

3.4 Consider Example 3.3. What is the probability that, if a customer has found a defect item, that it has been manufactured at machine 1?

3.5 Prove the formula in (3.4).

3.6 A person is crossing the street 300 000 times in his/her lifetime. The probability of being hit by a car is $1/300\,000$. We consider the different crossings to be independent events. What is the probability of being hit by a car at least once in a lifetime?

3.7 Consider throwing a fair dice repeatedly many times. We define a random variable $X_i : \mathbb{N}_6 \to \mathbb{N}_6$ with value equal to the value of the dice for the ith experiment. Since the dice is fair, we have that the probability function $p_{X_i} : \mathbb{N}_6 \to [0, 1]$ is defined as $p_{X_i}(k) = 1/6$. Define the random variable $X : \mathbb{N}_6^N \to \mathcal{D}$, where $\mathcal{D} = \{1, 1 + 1/N, 1 + 2/N, \dots, 6\}$ via

$$X = \frac{1}{N}\sum_{i=1}^{N} X_i.$$

   (a) Compute the expected value and the variance of the random variable X_i.
   (b) Compute the probability function for (X_1, X_2).
   (c) Compute the probability function for X when $N = 2$.
   (d) Compute the expected value of the random variable X for $N = 2$ directly using the probability function above and indirectly by using the probability function for (X_1, X_2) and the formula for expected values of functions of random variables in Section 3.6.
   (e) Compute the variance of X when $N = 2$.

3.8 Show that the for a random vector $X : \mathbb{R}^n \to \mathbb{R}^n$ with Gaussian distribution given by the pdf

$$f(x) = \frac{1}{\sqrt{(2\pi)^n \det \Sigma}} e^{-\frac{1}{2}(x-\mu)^T \Sigma^{-1}(x-\mu)},$$

it holds that the expected value is μ and that the variance is Σ.

3.9 Consider a random variable X with Gaussian distribution with zero mean and variance $I \in \mathbb{S}^n_{++}$. Let the random variable Y be defined as

$$Y = AX + b.$$

Show that this ransom variable is also Gaussian with mean b and variance AA^T.

3.10 Consider a scalar-valued random variable X with a Gaussian distribution with zero mean and variance σ^2. Show that

$$\mathbb{E}\left[\exp\left(\frac{p}{2}(X + m)^2\right)\right] = \frac{1}{\sigma\sqrt{\beta}} \exp\left(\frac{\gamma m^2}{2}\right),$$

where $\beta = 1/\sigma^2 - p$ and $\gamma = 1 + 1/\beta$.

3.11 Consider two random variables $X : \Omega \to \mathbb{R}$ and $Y : \Omega \to \mathbb{R}$. Let $Z = (X, Y)$. We know the expected value and the variance of Z, i.e.

$$\mathbb{E}[Z] = \begin{bmatrix} \mu_x \\ \mu_y \end{bmatrix}, \quad \text{Var}[Z] = \begin{bmatrix} \sigma_x^2 & \sigma_{xy} \\ \sigma_{xy} & \sigma_y^2 \end{bmatrix}$$

are known. Consider the random variable defined as $M = X + c(Y - \mu_y)$ for an arbitrary $c \in \mathbb{R}$.
(a) Show that $\mathbb{E}[M] = \mu_x$.
(b) Show that $\text{Var}[M] = \text{Var}[X] + c^2\text{Var}[Y] + 2c\text{Cov}[X, Y] = \sigma_x^2 + c^2\sigma_y^2 + 2c\sigma_{xy}$.
(c) Show that $\text{Var}[M]$ is minimized for $c^\star = -\sigma_{xy}/\sigma_y^2$ and that the minimal value is given by $(1 - \rho^2)\sigma_x^2$, where

$$\rho = \sqrt{\frac{\sigma_{xy}}{\sigma_x\sigma_y}}.$$

3.12 Consider two scalar-valued random variables (X, Y) with joint pdf

$$f_{X,Y}(x, y) = \begin{cases} 2, & \text{if } 0 < y \leq x < 1, \\ 0, & \text{otherwise.} \end{cases}$$

Compute the conditional expectation for X given $Y = y$.

3.13 Let X be a random variable taking the values 0 and 1 with equal probability $1/2$. Also, define the random variables $X_k = X$ for all $k \in \mathbb{Z}_+$. Clearly, $X_k \overset{D}{\to} X$, since all random variables defined have the same distribution. Also define the random variable $Y = 1 - X$.
(a) Show that $X_k \overset{D}{\to} Y$.
(b) Show that X_k cannot converge to Y in any other mode.

3.14 Consider the Markov process defined in Example 3.8. Assume that E_k are zero mean with $\text{Var}(X_k) = \Sigma$. Furthermore, assume that X_0 has mean m_0 and variance P_0. Show that $m_k = \mathbb{E}X_x$ and $P_k = \text{Var}X_k$ for $k \geq 1$ satisfies the recursions

$$m_{k+1} = Am_k, \quad P_{k+1} = A^T P_k A + \Sigma.$$

3.15 Let $H = R + XPX^T$, where $R \in \mathbb{S}^m_{++}$, $P \in \mathbb{S}^n_{++}$, and $X \in \mathbb{R}^{m \times n}$, and let $Y = PX^T H^{-1}$. Furthermore, let $y \in \mathbb{R}^m$ and $x, \mu \in \mathbb{R}^n$. Show that the following relationship holds:

$$\mathcal{N}(y, Xx, R)\mathcal{N}(x, \mu, P) = \mathcal{N}(y, X\mu, H)\mathcal{N}(x, \mu + Y(y - X\mu), P - XHX^T).$$

where $\mathcal{N}$ is defined as in Example 3.4.

Hint: The formula in Exercise 2.10 is useful.

3.16 Consider a HMM (X, Y) as defined in Section 3.11. Show that the conditional independence assumption

$$p_{\bar{Y}_k | \bar{X}_k}(\bar{y} | \bar{x}) = \prod_{i=0}^{k} p_{Y_i | X_i}(y_i | x_i),$$

implies that

(a)

$$p_{X_k | X_{k-1}, \bar{Y}_{k-1}}(x_k | x_{k-1}, \bar{y}_{k-1}) = p_{X_k | X_{k-1}}(x_k | x_{k-1}),$$

(b)

$$p_{Y_k | X_k, \bar{Y}_{k-1}}(y_k | x_k, \bar{y}_{k-1}) = p_{Y_k | X_k}(y_k | x_k).$$

3.17 Consider two stationary random processes $X : \Omega \to \mathcal{D}^{\mathbb{Z}}$ and $Y : \Omega \to \mathcal{E}^{\mathbb{Z}}$, where $\mathcal{D} = \mathbb{R}^n$ and $\mathcal{E} = \mathbb{R}^p$, which are assumed to satisfy the following equations:

$$X_{k+1} = AX_k + V_k,$$
$$Y_k = CX_k + E_k,$$

where $A \in \mathbb{R}^{n \times n}$, $C \in \mathbb{R}^{p \times n}$, and where (V_k, E_k) are random variables with zero mean and variance

$$R = \begin{bmatrix} R_1 & R_{12} \\ R_{12}^T & R_2 \end{bmatrix} \in \mathbb{S}^{n+p}_{++}.$$

We assume that $R_2 \in \mathbb{S}^p_{++}$, and we also assume that (V_j, E_j) and (V_k, E_k) are independent for $j \neq k$. Let $P \in \mathbb{S}^n_+$ be the unique solution to the algebraic Riccati equation

$$P = APA^T + R_1 - (R_{12} + APC^T)(R_2 + CPC^T)^{-1}(R_{12} + APC^T)^T,$$

such that $H = R_2 + CPC^T \in \mathbb{S}^p_{++}$. Such a solution always exists when $R_2 \in \mathbb{S}^p_{++}$ and can in some cases be shown to exist also when only $R_2 \in \mathbb{S}^p_+$. Define K as the solution of

$$KH = R_{12} + APC^T.$$

Define the random process $\hat{X} : \Omega \to D^{\mathbb{Z}}$ as

$$\hat{X}_{k+1} = A\hat{X}_k + K(Y_k - C\hat{X}_k).$$

Show that

$$\hat{X}_{k+1} = A\hat{X}_k + K\tilde{Y}_k,$$
$$Y_k = C\hat{X}_k + \tilde{Y}_k,$$

where $\tilde{Y}_k$ defines a random process that has variance H and is such that $\tilde{Y}_j$ is independent of $\tilde{Y}_k$ for $j \neq k$. This is called the *innovation form* of the random process Y.

Part II

Optimization

4

Optimization Theory

Mathematical optimization is an indispensable tool in learning and control. Its history goes back to the early seventeenth century with the work by Pierre de Fermat who obtained calculus-based formulae for identifying optima. This work was later developed further by Joseph-Louis Lagrange. In this chapter, we will present the foundations of optimization theory. We will start by defining what constitutes an optimization problem and introduce some basic concepts and terminology. We will also introduce the notion of convexity, which allows us to distinguish between convex and general nonlinear optimization problems. The motivation behind this distinction is that convex problems are, roughly speaking, easier to solve than general nonlinear ones. We will pay special attention to properties that are useful for recognizing convexity. Finally, we will also discuss the concept of duality and see how it is used to derive optimality conditions for optimization problems. In Chapter 6, we will see how duality also plays an important role in some optimization methods.

4.1 Basic Concepts and Terminology

Let f be a function $f : \mathcal{X} \to \mathcal{Y}$, where $\mathcal{X}$ is the *domain* of f and $\mathcal{Y}$ is its *codomain*. The set $f(\mathcal{X}) = \{f(x) \mid x \in \mathcal{X}\}$ is the *image* of f, which is a subset of $\mathcal{Y}$. We will henceforth assume that the codomain is the extended real line $\mathbb{R} \cup \{-\infty, +\infty\} = \bar{\mathbb{R}}$ or a subset thereof unless otherwise noted. For such functions, we define the *effective domain* of f as

$$\operatorname{dom} f = \{x \in \mathcal{X} \mid f(x) < \infty\}, \tag{4.1}$$

which is clearly a subset of $\mathcal{X}$. Moreover, f is said to be *proper* if $\operatorname{dom} f \neq \emptyset$ and $f(x) > -\infty$ for all $x \in \operatorname{dom} f$, and otherwise, f is *improper*. The α-*sublevel set* of f is the set

$$S_\alpha = \{x \mid f(x) \leq \alpha\}, \tag{4.2}$$

and the *epigraph* of f is the set

$$\operatorname{epi} f = \{(x, t) \in \mathcal{X} \times \mathbb{R} \mid f(x) \leq t\}. \tag{4.3}$$

Notice that S_α is a subset of $\mathcal{X}$, whereas $\operatorname{epi} f$ is a subset of $\mathcal{X} \times \mathbb{R}$, as illustrated for a function of one variable in Figure 4.1. The function f is said to be *closed* if and only if its epigraph is a closed set. The *closure* of a set $C \subseteq \mathbb{R}^n$, which we denote $\operatorname{cl} C$, is the smallest closed set that contains C.

Optimization for Learning and Control, First Edition. Anders Hansson and Martin Andersen.
© 2023 John Wiley & Sons, Inc. Published 2023 by John Wiley & Sons, Inc.
Companion Website: www.wiley.com/go/opt4lc

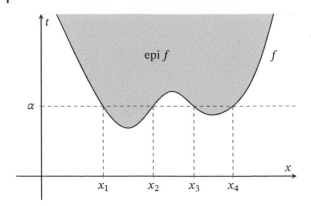

Figure 4.1 The epigraph of a function $f : \mathbb{R} \to \mathbb{R}$ and an example of a sublevel set $S_\alpha = [x_1, x_2] \cup [x_3, x_4]$.

4.1.1 Optimization Problems

We now consider an *optimization problem* of the form

$$
\begin{aligned}
&\text{minimize } f_0(x) \\
&\text{subject to } f_i(x) \le 0, \quad i \in \mathbb{N}_m, \\
&\hphantom{\text{subject to }} h_i(x) = 0, \quad i \in \mathbb{N}_p,
\end{aligned}
\tag{4.4}
$$

where $x \in \mathbb{R}^n$ is the *optimization variable*. The function $f_0 : \mathbb{R}^n \to \bar{\mathbb{R}}$ is the *objective function*, the function $f_i : \mathbb{R}^n \to \bar{\mathbb{R}}$ is the ith of m *inequality constraint functions*, and $h_i : \mathbb{R}^n \to \bar{\mathbb{R}}$ is the ith of p *equality constraint functions*. The inequality $f_i(x) \le 0$ is referred to as an *inequality constraint*, and the equation $h_i(x) = 0$ is an *equality constraint*. If the problem does not have any constraints, i.e. if $m = p = 0$, then we say that the problem is *unconstrained*, and otherwise, it is a *constrained* problem. We say that the ith inequality constraint is *active* at x if $f_i(x) = 0$, the constraint is *inactive* at x if $f_i(x) < 0$, and it is *violated* at x if $f_i(x) > 0$. Similarly, the equality constraint $h_i(x) = 0$ is violated at x if $h_i(x) \ne 0$. We say that a constraint is *satisfied* if it is not violated. For notational convenience, we define

$$
f(x) = \left(f_1(x), \dots, f_m(x) \right), \qquad h(x) = \left(h_1(x), \dots, h_p(x) \right),
$$

and we will frequently use the notation $f(x) \preceq 0$ as shorthand for $f_i(x) \le 0$ for all $i \in \mathbb{N}_m$. We return to such generalized inequalities in Section 4.2.

The *domain* of the optimization problem is the set

$$
\mathcal{D} = \left(\bigcap_{i=0}^{m} \text{dom } f_i \right) \cap \left(\bigcap_{i=1}^{p} \text{dom } h_i \right),
\tag{4.5}
$$

and the *feasible set* or *feasible region* is the subset of points in $\mathcal{D}$ that satisfy all constraints, i.e.

$$
\mathcal{F} = \{ x \in \mathcal{D} \mid f(x) \preceq 0, h(x) = 0 \}.
\tag{4.6}
$$

A point x is called *feasible* if it belongs to the feasible set $\mathcal{F}$, and x is *strictly feasible* if it is both feasible and all inequality constraints are inactive at x, i.e. if $f_i(x) < 0$, $i \in \mathbb{N}_m$. The optimization problem is said to be *feasible* if $\mathcal{F} \ne \emptyset$, and otherwise, the problem is *infeasible*.

The *optimal value* or the *minimum value* of the optimization problem (4.4) is defined as

$$
p^\star = \inf_{x \in \mathcal{F}} f_0(x).
\tag{4.7}
$$

We let $p^\star = \infty$ if the problem is infeasible, and if $x_1, x_2, \dots$ is a sequence of feasible points such that $\lim_{k \to \infty} f(x_k) = -\infty$, then the optimization problem is *unbounded below*, in which case we let $p^\star = -\infty$. The optimal value is *attained* if there exists a point $x^\star \in \mathcal{F}$ such that $p^\star = f_0(x^\star)$, and

Figure 4.2 Stationary points of a continuously differentiable function $f_0 : \mathbb{R} \to \mathbb{R}$.

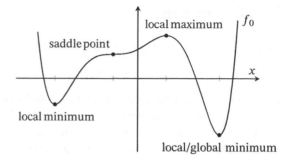

otherwise, the optimal value is *unattained*. A point $x^\star$ is called an *optimal point*, a *minimizer*, or a *solution* if it is feasible and $p^\star = f_0(x^\star)$, and the set of all optimal points is called the *optimal set*. A feasible point x that satisfies $f_0(x) \le p^\star + \epsilon$ for some $\epsilon > 0$ is called ϵ-*suboptimal*.

A feasible point x is said to be *locally optimal* if there exists a constant $r > 0$ such that

$$f_0(x) = \inf_z \left\{ f_0(z) \mid x \in \mathcal{F} \cap B_2(x, r) \right\}, \tag{4.8}$$

where

$$B_2(c, r) = \{ x \in \mathbb{R}^n \mid \|x - c\|_2 \le r \},$$

is the Euclidean ball with center c and radius r. Thus, an optimal point is also locally optimal, but the converse is not true in general. To emphasize the difference between optimal points and locally optimal points, we sometimes say that an optimal point is *globally optimal*. Similarly, we refer to $f_0(x)$ as a *local minimum value* if x is locally optimal. The notion of local and global optimality is illustrated in Figure 4.2 for an unconstrained problem with a continuously differentiable objective function $f_0 : \mathbb{R} \to \mathbb{R}$. The local extrema of such a function are stationary points, but not all stationary points are local extrema.

The special case of the optimization problem (4.4), where $f_0(x) = 0$ for all $x \in \mathcal{D}$ is called a *feasibility problem* since solving such a problem amounts to finding any feasible point. The optimal value is $p^\star = 0$ if the problem is feasible, and otherwise, $p^\star = \infty$.

4.1.2 Equivalent Problems

Two optimization problems are said to be *equivalent* if a solution to one problem can readily be obtained from a solution to the other problem and vice versa. For example, the optimization problem (4.4) is equivalent to the problem

$$
\begin{aligned}
&\text{minimize } t \\
&\text{subject to } f_0(x) - t \le 0, \\
&\qquad\qquad f_i(x) \le 0, \quad i \in \mathbb{N}_m, \\
&\qquad\qquad h_i(x) \le 0, \quad i \in \mathbb{N}_p,
\end{aligned}
\tag{4.9}
$$

with variable $z = (x, t) \in \mathbb{R}^n \times \mathbb{R}$. Note that this is a problem with $n + 1$ variables and $m + 1$ inequality constraints, and it is sometimes referred to as the *epigraph reformulation* of (4.4) since the inequality $f_0(x) \le t$ is equivalent to the constraint $z \in \text{epi } f_0$. To see that the two problems are equivalent, we now show that the solution to one is readily obtained from the other. Indeed, if $x^\star$ is a solution to (4.4) and $p^\star = f_0(x^\star)$, then $z^\star = (x^\star, p^\star)$ is a solution to (4.9) since $z = (x, t)$ is infeasible in (4.9) if $t < p^\star$. Conversely, if $z^\star = (x^\star, t^\star)$ is a solution to (4.9), then the epigraph constraint $f_0(x) \le t$ must be active at $z^\star$, and hence, $x^\star$ is a solution to (4.4).

4.2 Convex Sets

A set $C \subseteq \mathbb{R}^n$ is said to be *convex* if and only if for every $x, y \in C$,

$$\theta x + (1 - \theta)y \in C, \quad \forall\, \theta \in [0, 1]. \tag{4.10}$$

A point $\theta x + (1 + \theta)y$ for some $\theta \in [0, 1]$ is called a *convex combination* of $x \in \mathbb{R}^n$ and $y \in \mathbb{R}^n$, and the set of all convex combinations of x and y is the line segment between x and y. In other words, the definition of convexity requires that the line segment between every $x, y \in C$ is contained in C, as illustrated in Figure 4.3. We note that a linear combination of the form

$$\theta_1 x_1 + \cdots + \theta_k x_k, \quad k \geq 2, \tag{4.11}$$

with $x_1, \ldots, x_k \in \mathbb{R}^n$ and $\theta = (\theta_1, \ldots, \theta_k) \in \mathbb{R}^k$, is called

- an *affine combination* if $\mathbf{1}^T \theta = 1$,
- a *conic combination* if $\theta \in \mathbb{R}^k_+$,
- a *convex combination* if $\theta \in \mathbb{R}^k_+$ and $\mathbf{1}^T \theta = 1$.

To simplify notation, we introduce the set

$$\Delta^k = \mathbb{R}^k_+ \cap \{\theta \in \mathbb{R}^k \mid \mathbf{1}^T \theta = 1\}, \tag{4.12}$$

which is the *standard simplex* in $\mathbb{R}^k$. Figure 4.4 illustrates the set of conic and convex combinations of two points in $\mathbb{R}^2$. A direct consequence of (4.10) is that all convex combinations of any k points

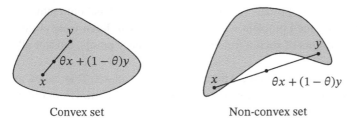

Convex set · Non-convex set

Figure 4.3 A set is convex if the line segment between two points x and y is contained in the set for every x and y in the set.

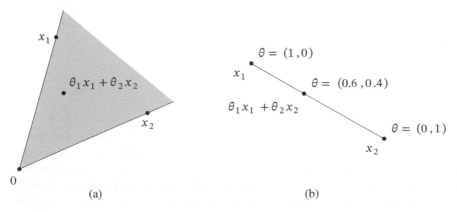

(a) · (b)

Figure 4.4 Convex and conic combinations of two points x_1 and x_2. (a) Conic combinations: $\theta \in \mathbb{R}^2_+$. (b) Convex combinations: $\theta \in \Delta^2$.

in a convex set C are contained in C, i.e., we have that

$$x_1, \ldots, x_k \in C \implies \{\theta_1 x_1 + \cdots + \theta_k x_k \mid \theta \in \Delta^k\} \subseteq C.$$

To see this, first note that this is trivially true for $k = 2$ since C is convex. The result then follows by induction: assuming that it is true for k points, we will show that it is true for $k + 1$ points $x_1, \ldots, x_{k+1} \in C$. Indeed, for all $\bar{\theta} \in \Delta^{k+1}$ such that $\bar{\theta}_{k+1} < 1$, we have that

$$y = \sum_{i=1}^{k+1} \bar{\theta}_i x_i = \sum_{i=1}^{k} \bar{\theta}_i x_i + \bar{\theta}_{k+1} x_{k+1} = (1 - \bar{\theta}_{k+1}) z + \bar{\theta}_{k+1} x_{k+1},$$

where we define

$$z = \sum_{i=1}^{k} \tilde{\theta}_i x_i, \quad \tilde{\theta}_i = \frac{\bar{\theta}_i}{\sum_{i=1}^{k} \bar{\theta}_i} = \frac{\bar{\theta}_i}{1 - \bar{\theta}_{k+1}}, \quad i \in \mathbb{N}_k.$$

This shows that z is a convex combination of $x_1, \ldots, x_k$, and hence, z belongs to C by assumption. It follows that $y \in C$ since it is a convex combination of z and x_{k+1}.

The *dimension* of a convex set $C \subseteq \mathbb{R}^n$ is the dimension of its affine hull, i.e.

$$\dim C = \dim (\text{aff } C). \tag{4.13}$$

The *relative interior* of C is the interior of C within the affine hull of C,

$$\text{relint } C = \{x \in C \mid \exists\, \epsilon > 0 \text{ such that } B_2(x, \epsilon) \cap \text{aff } C \subseteq C\}, \tag{4.14}$$

where $B_2(x, \epsilon)$ is the Euclidean ball centered at x and with radius ϵ.

The *convex hull* of a set $C \subseteq \mathbb{R}^n$ is the smallest convex set that contains C. Equivalently, it is the intersection of all convex sets that contain C, i.e.

$$\text{conv } C = \cap\{D \subseteq \mathbb{R}^n \mid D \text{ is convex and } C \subseteq D\}. \tag{4.15}$$

We note that it can be difficult to identify the convex hull of a set by using this definition. *Carathéodory's theorem* provides a characterization that is often more useful in practice. In states that every point in conv C can be expressed as a convex combination of at most $n + 1$ points in C, i.e.

$$\text{conv } C = \left\{ \sum_{i=1}^{n+1} \theta_i x_i \mid x_1, \ldots, x_{n+1} \in C,\ \theta \in \Delta^{n+1} \right\}. \tag{4.16}$$

4.2.1 Convexity-Preserving Operations

We will now consider a number of operations that preserve convexity. As we will see, knowledge of such operations is often very useful when we wish to establish convexity of a given set.

4.2.1.1 Intersection

A fundamental property of convex sets is that the intersection of any number of convex sets is itself a convex set, i.e. if C_τ is a convex set for every $\tau \in T$, then

$$C = \bigcap_{\tau \in T} C_\tau,$$

is convex. This follows directly from the definition of a convex set by noting that any two points x and y in C also belong to C_τ for all $\tau \in T$, and moreover,

$$\{\theta x + (1 - \theta) y \mid \theta \in [0, 1]\} \in C_\tau,$$

since C_τ is convex.

4.2.1.2 Affine Transformation

The image and the preimage of a convex set under an affine transformation is itself a convex set. To see this, let $f : \mathbb{R}^n \to \mathbb{R}^m$ be an affine function, i.e. f can be expressed as $f(x) = Ax + b$ for some $A \in \mathbb{R}^{m \times n}$ and $b \in \mathbb{R}^m$, and hence, f satisfies

$$\theta f(x) + (1 - \theta)f(y) = f(\theta x + (1 - \theta)y).$$

The image of a set $C \in \mathbb{R}^n$ under f, which is defined as

$$f(C) = \{f(x) \mid x \in C\},$$

is therefore convex if C is convex since for every $x, y \in C$ and $\theta \in [0, 1]$,

$$\theta x + (1 - \theta)y \in C \implies \theta f(x) + (1 - \theta)f(y) = f(\theta x + (1 - \theta)y) \in f(C).$$

Similarly, the preimage of a set $C \subseteq \mathbb{R}^m$ under f, which is defined as

$$f^{-1}(C) = \{x \in \mathbb{R}^n \mid f(x) \in C\},$$

is also convex if C is convex since for every $x, y \in f^{-1}(C)$ and $\theta \in [0, 1]$, it holds that

$$\theta f(x) + (1 - \theta)f(y) = f(\theta x + (1 - \theta)y) \in C \implies \theta x + (1 - \theta)y \in f^{-1}(C).$$

4.2.1.3 Perspective Transformation

The *perspective function* $P : \mathbb{R}^n \times \mathbb{R}_{++} \to \mathbb{R}^n$ is the function $P((u, s)) = u/s$. The perspective of a set $C \subseteq \operatorname{dom} P$ is the image of C under P, i.e.

$$P(C) = \{u/s \mid (u, s) \in C\},$$

and $P(C)$ is convex if C is a convex set, as we will now show. Suppose $x = (u, s)$ and $y = (v, t)$ are two points in $C \subseteq \operatorname{dom} P$. Using the definition of P, we find that

$$P(\theta x + (1 - \theta)y) = \frac{\theta u + (1 - \theta)v}{\theta s + (1 - \theta)t} = \frac{\theta s}{\theta s + (1 - \theta)t} u/s + \frac{(1 - \theta)t}{\theta s + (1 - \theta)t} v/t$$
$$= \bar{\theta} P(x) + (1 - \bar{\theta})P(y)$$

where $\bar{\theta} = \theta s/(\theta s + (1 - \theta)t) \in [0, 1]$. This implies that P maps the line segment between x and y to the line segment between $P(x)$ and $P(y)$, and hence, $P(C)$ is convex if C is convex.

4.2.2 Examples of Convex Sets

Using the definition of convex sets (4.10) and our knowledge of convexity-preserving operations, we will now provide some basic examples of convex sets and verify their convexity.

4.2.2.1 Hyperplanes and Halfspaces

A *hyperplane* in $\mathbb{R}^n$ is a set

$$\mathcal{H} = \{x \in \mathbb{R}^n \mid a^T x = b\}, \tag{4.17}$$

where $a \in \mathbb{R}^n$ is a nonzero vector and $b \in \mathbb{R}$, i.e. it is the solution set to a nontrivial equation with n unknowns. Given any vector $x_0 \in \mathcal{H}$, we have that

$$\mathcal{H} = \{x_0 + u \mid a^T u = 0\} = x_0 + \mathcal{R}(a)^{\perp},$$

which shows that $\mathcal{H}$ is a translation of the subspace $\mathcal{R}(a)^{\perp}$, and hence, a is normal to $\mathcal{H}$. It is straightforward to verify that $\mathcal{H}$ is a convex set: for every $x, y \in \mathcal{H}$,

$$a^T(\theta x + (1 - \theta)y) = \theta a^T x + (1 - \theta)a^T y = b, \quad \forall \theta \in \mathbb{R}.$$

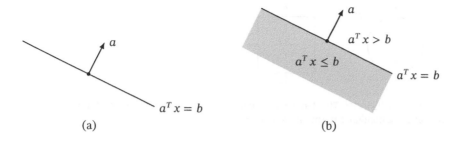

$$a^T x > b$$
$$a^T x \leq b$$
$$a^T x = b$$
$$a^T x = b$$

(a)   (b)

Figure 4.5 A hyperplane and a halfspace in $\mathbb{R}^2$: (a) hyperplane and (b) halfspace.

The set of solutions to a nontrivial linear inequality $a^T x \leq b$ with $a \in \mathbb{R}^n$, $a \neq 0$, and $b \in \mathbb{R}$ is a closed *halfspace* in $\mathbb{R}^n$,

$$\{x \in \mathbb{R}^n \mid a^T x \leq b\}. \tag{4.18}$$

The boundary of the set is the hyperplane $\{x \in \mathbb{R}^n \mid a^T x = b\}$. A halfspace is a convex set, which follows by noting that for every x and y in the halfspace, it holds that

$$a^T(\theta x + (1 - \theta)y) = \theta a^T x + (1 - \theta)a^T y \leq \theta b + (1 - \theta)b = b, \quad \forall\, \theta \in [0, 1].$$

Figure 4.5 shows a hyperplane and a halfspace in $\mathbb{R}^2$.

4.2.2.2 Polyhedral Sets

A *polyhedral set* is the intersection of a finite number of hyperplanes and halfspaces, and hence, it is a convex set. Since a hyperplane can be expressed as the intersection of two closed halfspaces, i.e.

$$\{x \in \mathbb{R}^n \mid a^T x = b\} = \{x \in \mathbb{R}^n \mid a^T x \leq b\} \cap \{x \in \mathbb{R}^n \mid a^T x \geq b\},$$

an equivalent definition is that a polyhedral set is the intersection of finitely many halfspaces. Examples of polyhedral sets include line segments, affine sets, the nonnegative orthant $\mathbb{R}^n_+$, and the standard simplex Δ^n. Figure 4.6 shows examples of polyhedral sets in $\mathbb{R}^2$ and $\mathbb{R}^3$.

4.2.2.3 Norm Balls and Ellipsoids

A *norm ball* is a set of the form

$$B(c, r) = \{x \in \mathbb{R}^n \mid \|x - c\| \leq r\}, \tag{4.19}$$

where $\| \cdot \|$ is a norm on $\mathbb{R}^n$, $c \in \mathbb{R}^n$ is the center of the ball, and $r > 0$ is the radius. It follows from the properties of a norm that $B(c, r)$ is convex. Specifically, if x and y are in $B(c, r)$, then for all $\theta \in [0, 1]$,

$$\|\theta(x - c) + (1 - \theta)(y - c)\| \leq \theta\|x - c\| + (1 - \theta)\|y - c\| \leq r,$$

and hence, $\theta x + (1 - \theta)y \in B(c, r)$ for all $\theta \in [0, 1]$.

A norm ball that is induced by a quadratic norm of the form $\|x\|_A = \sqrt{x^T A x}$ for some $A \in \mathbb{S}^n_{++}$ is an *ellipsoid*. More generally, an ellipsoid is an affine transformation of the unit Euclidean norm ball

Figure 4.6 Examples of a polyhedral sets: the set to the left is the intersection of five halfspaces in $\mathbb{R}^2$, and the set to the right is the standard simplex in $\mathbb{R}^3$, which is the intersection of a hyperplane and three halfspaces.

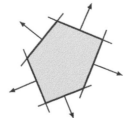

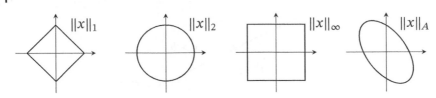

Figure 4.7 Examples of norm balls in $\mathbb{R}^2$. The 1-norm ball and the ∞-norm ball are polyhedral sets, whereas the 2-norm ball and the quadratic norm ball are ellipsoids.

$B_2(0, 1) = \{x \in \mathbb{R}^n \mid \|x\|_2 \leq 1\}$, but it is only a norm ball if it is an injective linear transformation of $B_2(0, 1)$. To see this, suppose $f : \mathbb{R}^n \rightarrow \mathbb{R}^n$ is an injective affine transformation, i.e. $f(x) = Cx + d$ for some nonsingular $C \in \mathbb{R}^{n \times n}$ and $d \in \mathbb{R}^n$. We then have that

$$f(B_2(0, 1)) = \{Cx + d \mid \|x\|_2 \leq 1\} = \{y \mid \|C^{-1}(y - d)\|_2 \leq 1\}$$
$$= \{y \mid \|y - d\|_A \leq 1\},$$

which shows that $f(B_2(0, 1))$ is a norm ball induced by the quadratic norm $\|\cdot\|_A$ with $A = C^{-T}C^{-1}$. Figure 4.7 shows some examples of norm balls in $\mathbb{R}^2$.

4.2.2.4 Convex Cones

A set $K \in \mathbb{R}^n$ is a *cone* if

$$x \in K \implies tx \in K, \quad \forall t \in \mathbb{R}_+,$$

and K is a *convex cone* if it is also convex. We say that a cone K is *pointed* if $K \cap (-K) = \{0\}$. Pointedness implies that the largest linear subspace included in K is $\{0\}$. Moreover, we say that a convex cone K is *proper* if it is pointed, closed, and full dimensional, i.e.

$$K \cap (-K) \subseteq \{0\}, \quad K = \text{cl}\, K, \quad \text{int}\, K \neq \emptyset.$$

The *conic hull* of a set $C \in \mathbb{R}^n$, which we denote cone C, is the smallest convex cone that contains C. A cone is a *polyhedral cone* if it is also a polyhedral set. One example is the nonnegative orthant in $\mathbb{R}^n$, i.e.

$$\mathbb{R}_+^n = \{x \in \mathbb{R}^n \mid x_i \geq 0, \ i \in \mathbb{N}_n\},$$

which is the intersection of n halfspaces. Figure 4.8 illustrates the nonnegative orthant in $\mathbb{R}^3$.

The *norm cone* associated with a norm on $\mathbb{R}^{n-1}$ is the n-dimensional set

$$K = \{(x, t) \in \mathbb{R}^{n-1} \times \mathbb{R} \mid \|x\| \leq t\}. \tag{4.20}$$

The case where the norm is the Euclidean norm is an important special case, which we will refer to as the *second-order cone* and denote by $\mathbb{Q}^n$. We note that the second-order cone has several other

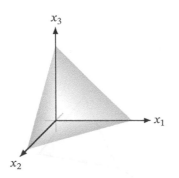

Figure 4.8 The nonnegative orthant in $\mathbb{R}^3$.

Figure 4.9 The second-order cone in $\mathbb{R}^3$.

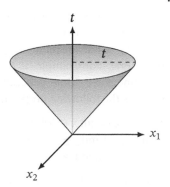

names in the literature, e.g. the *Lorentz cone*, the *quadratic cone*, and the more casual name, the *ice-cream cone*. Figure 4.9 shows the second-order cone in $\mathbb{R}^3$.

The cone of symmetric, positive semidefinite matrices of order n is the set

$$S_+^n = \{X \in \mathbb{S}^n \mid u^T X u \geq 0, \ \forall \, u \in \mathbb{R}^n\}. \tag{4.21}$$

It is easy to verify that it is indeed a cone, and convexity follows by noting that for a given $u \in \mathbb{R}^n$, the set

$$\{X \in \mathbb{S}^n \mid u^T X u \geq 0\}$$

is a closed halfspace, and hence, S_+^n can be expressed as the intersection of infinitely many halfspaces.

The *dual cone* of a convex cone $K \subseteq \mathbb{R}^n$ is the set

$$K^* = \{y \in \mathbb{R}^n \mid x^T y \geq 0, \ \forall \, x \in K\}. \tag{4.22}$$

In other words, K^* is the set of vectors that form a nonnegative inner product with all vectors in K. This is illustrated in Figure 4.10. We note that it can be shown that the dual of the dual cone K^* is the closure of K, i.e. $K^{**} = \operatorname{cl} K$. Thus, we have that $K^{**} = K$ if K is a proper convex cone. A convex cone K is called *self-dual* if $K = K^*$. The nonnegative orthant, the second-order cone, and the cone of positive semidefinite matrices are examples of self-dual cones. Finally, we note that $-(K^*)$ is called the *polar cone* of K.

4.2.3 Generalized Inequalities

Given a proper convex cone $K \subseteq \mathbb{R}^n$ and vectors $x, y \in \mathbb{R}^n$, we define the *generalized inequality* $x \succeq_K y$ and its strict counterpart $x \succ_K y$ as

$$x \succeq_K y \quad \Longleftrightarrow \quad x - y \in K, \tag{4.23a}$$

$$x \succ_K y \quad \Longleftrightarrow \quad x - y \in \operatorname{int} K. \tag{4.23b}$$

Figure 4.10 An example of a convex cone K and its dual cone K^* in $\mathbb{R}^2$. Note that K^* contains K in this example.

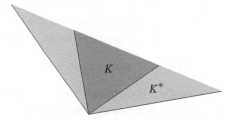

The relation $\geq_K$ defines a nonstrict *total order* on $\mathbb{R}^n$ if $x \geq_K y$ or $y \geq_K x$ for every $x, y \in \mathbb{R}^n$. For example, this is the case if $K = R_+$, in which case the generalized inequality reduces to the ordinary scalar inequality $x \geq y$. However, the relation $\geq_K$ does not define a total order on $\mathbb{R}^n$ in general. For example, if $K = \mathbb{R}_+^2$ and $x = (1, -1)$ and $y = (-1, 1)$, then neither $x - y$ nor $y - x$ is nonnegative. More generally, the relation $\geq_K$ defines a nonstrict *partial order* on $\mathbb{R}^n$, which satisfies the following properties:

1. $x \geq_K x$ for all $x \in \mathbb{R}^n$ (reflexivity),
2. if $x \geq_K y$ and $y \geq_K x$, then $y = x$ (antisymmetry),
3. if $x \geq_K y$ and $y \geq_K z$, then $x \geq_K z$ (transitivity).

Furthermore, the relation $>_K$ defines a strict partial order on $\mathbb{R}^n$, which satisfies the following properties:

1. there is no x such that $x >_K x$ (irreflexivity),
2. $x >_K y \implies y \not>_K x$ (asymmetry),
3. if $x >_K y$ and $y >_K z$, then $x >_K z$ (transitivity).

These properties of $\geq_K$ and $>_K$ follow directly from (4.23) and the assumption that K is a proper convex cone. We note that a direct consequence of the transitivity property is that

$$a \geq_K b, \ c \geq_K d \implies a + c \geq_K c + d,$$

i.e. adding two valid generalized inequalities yields another valid generalized inequality.

We will frequently encounter two types of generalized inequalities. The first type is defined with respect to the nonnegative orthant $\mathbb{R}_+^n$ and corresponds to componentwise inequality

$$x \geq_{\mathbb{R}_+^n} y \iff x_i \geq y_i, \ i \in \mathbb{N}_n.$$

The second type is defined with respect to the cone of positive semidefinite matrices or order n and corresponds to eigenvalue inequalities. Specifically, if $X, Y \in \mathbb{S}^n$ and $K = \mathbb{S}_+^n$, then

$$X \geq_{\mathbb{S}_+^n} Y \iff X - Y \in \mathbb{S}_+^n,$$

or, equivalently, $X \geq_{\mathbb{S}_+^n} Y$ if and only if the eigenvalues of $X - Y$ are nonnegative.

To simplify our notation, we will write $x \geq y$ and $x > y$ instead of $x \geq_{\mathbb{R}_+^n} y$ and $x >_{\mathbb{R}_+^n} y$ whenever x and y are vectors, and we will write $X \geq Y$ and $X > Y$ instead of $X \geq_{\mathbb{S}_+^n} Y$ and $X >_{\mathbb{S}_+^n} Y$ whenever X and Y are symmetric matrices.

4.3 Convex Functions

A function $f : \mathbb{R}^n \to \bar{\mathbb{R}}$ is *convex* if dom f is a convex set and for all $x, y \in$ dom f, it holds that

$$f(\theta x + (1 - \theta)y) \leq \theta f(x) + (1 - \theta)f(y), \quad \forall \ \theta \in [0, 1]. \tag{4.24}$$

In other words, every line segment joining two points $(x, f(x))$ and $(y, f(y))$ on the graph of f lies above or on the graph, as illustrated in Figure 4.11. The function f is *strictly convex* if (4.24) holds with strict inequality whenever $x \neq y$ and $\theta \in (0, 1)$. Moreover, f is *strongly convex* with modulus $\mu > 0$, or simply μ-strongly convex, if for all $x, y \in$ dom f and $\theta \in [0, 1]$,

$$f(\theta x + (1 - \theta)y) \leq \theta f(x) + (1 - \theta)f(y) - \frac{\theta(1 - \theta)\mu}{2} \|x - y\|_2^2. \tag{4.25}$$

Equivalently, f is μ-strongly convex if and only if $g(x) = f(x) - (\mu/2)\|x\|_2^2$ is convex. We say that f is *concave* if $-f$ is convex, and similarly, f is strictly/strongly concave if $-f$ is strictly/strongly convex.

Figure 4.11 A function f is convex if the line segment connecting the two points $(x, f(x))$ and $(y, f(y))$ is on or above the graph of the function for every x and y.

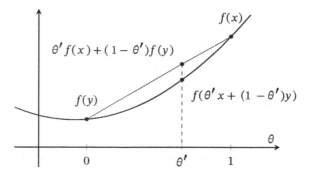

We note that the notion of convexity can be generalized to vector-valued function by replacing the scalar inequality in (4.24) by a generalized inequality. Specifically, we say that $f : \mathbb{R}^n \to \mathbb{R}^m$ is convex with respect to a proper convex cone $K \subset \mathbb{R}^m$, or simply K-convex, if dom f is a convex set and for all $x, y \in$ dom f,

$$f(\theta x + (1 - \theta)y) \preceq_K \theta f(x) + (1 - \theta)f(y), \quad \forall\, \theta \in [0, 1]. \tag{4.26}$$

The inequality (4.24) can also be expressed in terms of the epigraph of f. Indeed, using the definition of the epigraph, we may express (4.24) as

$$\theta \begin{bmatrix} x \\ f(x) \end{bmatrix} + (1 - \theta) \begin{bmatrix} y \\ f(y) \end{bmatrix} \in \text{epi}\, f, \quad \forall\, \theta \in [0, 1].$$

A direct consequence of this is that the function f is convex if and only if its epigraph is a convex set, as illustrated in Figure 4.12. This implies that all sublevel sets of a convex function are convex, but the converse is not true in general.

4.3.1 First- and Second-Order Conditions for Convexity

Let $f : \mathbb{R}^n \to (-\infty, +\infty]$ be continuously differentiable on dom f, which we assume is open. The function f is then convex if and only if dom f is a convex set and

$$f(y) \geq f(x) + \nabla f(x)^T(y - x), \quad \forall\, x, y \in \text{dom}\, f. \tag{4.27}$$

In other words, the first-order Taylor approximation of f at x is an affine lower bound of f, as illustrated in Figure 4.13. Similarly, f is strictly convex if (4.27) holds with strict inequality whenever $x \neq y$. A direct consequence of (4.27) is that all stationary points of f are global minima because

Figure 4.12 The epigraph of a function f is a convex set if and only if f is a convex function.

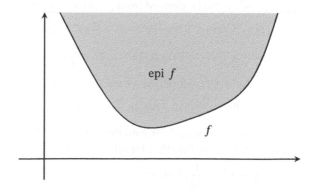

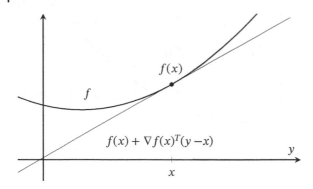

Figure 4.13 A continuously differentiable function f is convex if and only if all tangent planes lie below or on the graph of f.

$f(y) \geq f(x)$ for all $y \in \text{dom } f$ whenever $\nabla f(x) = 0$. Moreover, if f is strictly convex, then f has at most one stationary point since $\nabla f(x) = 0$ implies that $f(y) > f(x)$ for all $y \neq x$.

To prove (4.27), we start by rewriting (4.24) as

$$\theta f(y) \geq \theta f(x) + f(x + \theta(y - x)) - f(x).$$

The inequality (4.27) then follows by dividing both sides by $\theta \neq 0$, i.e.

$$f(y) \geq f(x) + \frac{f(x + \theta(y - x)) - f(x)}{\theta},$$

and taking the limit as θ goes to 0, which yields the directional derivative

$$\lim_{\theta \to 0} \frac{f(x + \theta(y - x)) - f(x)}{\theta} = \nabla f(x)^T (y - x).$$

The function f is μ-strongly convex if and only if

$$f(y) \geq f(x) + \nabla f(x)^T (y - x) + \frac{\mu}{2} \|y - x\|_2^2, \quad \forall\, x, y \in \text{dom } f, \tag{4.28}$$

for some $\mu > 0$. To see this, recall that f is μ-strongly convex if and only if $g(x) = f(x) - (\mu/2)\|x\|_2^2$ is convex, and note that (4.28) is equivalent to $g(y) \geq g(x) + \nabla g(x)^T (y - x)$ for all $x, y \in \text{dom } f$, which is the convexity condition (4.27). The condition (4.28) implies that the sublevel sets of f are bounded, which follows by noting that

$$S_t = \{y \,|\, f(y) \leq t\} \subseteq \{y \,|\, f(x) + \nabla f(x)^T (y - x) + (\mu/2)\|y - x\|_2^2 \leq t\}, \quad \forall\, x \in \text{dom } f,$$

where the set on the right-hand side of the inclusion operator is either a Euclidean ball or the empty set. Strong convexity also allows us to bound the distance from any x to the optimal point $x^\star$, if it exists, in terms of $\nabla f(x)$,

$$\|x - x^\star\|_2 \leq \frac{2}{\mu} \|\nabla f(x)\|_2. \tag{4.29}$$

This inequality can be derived from (4.28) by substituting $x^\star$ for y, i.e.

$$f(x^\star) \geq f(x) + \nabla f(x)^T (x^\star - x) + \frac{\mu}{2} \|x^\star - x\|_2^2$$

$$\geq f(x) - \|\nabla f(x)\|_2 \|x^\star - x\|_2 + \frac{\mu}{2} \|x^\star - x\|_2^2,$$

and since $f(x^\star) \leq f(x)$ for all x, we see that

$$0 \geq -\|\nabla f(x)\|_2 \|x^\star - x\|_2 + \frac{\mu}{2} \|x^\star - x\|_2^2.$$

Moreover, substituting $x^\star$ for y on the left-hand side of (4.28) and minimizing the right-hand side with respect to y, we find that for all $x \in \text{dom } f$,

$$f(x) - p^\star \leq \frac{1}{2\mu} \|\nabla f(x)\|_2^2. \tag{4.30}$$

The first-order condition for convexity can also be expressed as

$$(\nabla f(y) - \nabla f(x))^T (y - x) \geq 0, \quad \forall\, x, y \in \text{dom}\, f. \tag{4.31}$$

This follows from (4.27) by adding the inequality to itself with x and y interchanged. This means that ∇f is a *monotone* operator if and only if f is continuously differentiable and convex. Similarly, if f is strictly convex, then (4.31) holds with strict inequality whenever $x \neq y$, in which case ∇f is said to be *strictly monotone*. Finally, if f is strongly convex, then

$$(\nabla f(y) - \nabla f(x))^T (y - x) \geq \mu \|y - x\|_2^2, \quad \forall\, x, y \in \text{dom}\, f, \tag{4.32}$$

which means that ∇f is *strongly monotone* with parameter μ.

If f is twice continuously differentiable on dom f, then f is convex if and only if dom f is convex and

$$\nabla^2 f(x) \succeq 0, \quad \forall\, x \in \text{dom}\, f. \tag{4.33}$$

In other words, f is convex if and only if the Hessian matrix is positive semidefinite on dom f. To show that (4.33) follows from convexity, we first substitute $x + tz$ for y in (4.31), where $t > 0$ is chosen such that $x + tz \in \text{dom}\, f$, i.e.

$$(\nabla f(x + tz) - \nabla f(x))^T (z + tz - x) \geq 0.$$

Dividing by t and taking the limit as $t \to 0$, we find that $z^T \nabla^2 f(x) z \geq 0$. This must hold for all z since dom f is open, and hence, $\nabla^2 f(x)$ is positive semidefinite. Conversely, suppose $\nabla^2 f(x)$ is positive semidefinite on dom f, and let $\phi(t) = f(x + t(y - x))$ be the restriction of f to the line through some $x, y \in \text{dom}\, f$. It follows that

$$\phi''(t) = (y - x)^T \nabla^2 f(x)(y - x) \geq 0,$$

which implies that $\phi'(t)$ is monotone, and hence, $\phi(t)$ is convex.

The function f is strictly convex if $\nabla^2 f(x) \succ 0$ for all $x \in \text{dom}\, f$, but the converse is not true in general: the Hessian of a twice continuously differentiable strictly convex function is not necessarily positive definite on dom f. Finally, f is μ-strongly convex if and only if

$$\nabla^2 f(x) \succeq \mu I, \quad \forall\, x \in \text{dom}\, f, \tag{4.34}$$

which implies that the eigenvalues of $\nabla^2 f(x)$ are greater than or equal to μ.

4.3.2 Convexity-Preserving Operations

In Section 4.2, we saw that the knowledge of convexity-preserving operations can be useful in establishing convexity of sets. The same is true for convex function, which is why we now consider operations that preserve convexity of functions.

4.3.2.1 Scaling, Sums, and Integrals

The set of convex functions $f : \mathbb{R}^n \to (-\infty, +\infty]$ is closed under addition and scaling by positive real numbers. Thus, the sum of two proper convex functions f and g is a convex function, and αf is convex if f is convex and $\alpha \in \mathbb{R}_{++}$. This is an immediate consequence of (4.24). More generally, if $f : \mathbb{R}^n \times \mathbb{R}^p \to (-\infty, +\infty]$ and $f(x, y)$ is convex in x for every $y \in Y \subseteq \mathbb{R}^p$, then

$$h(x) = \int_Y f(x, y)\, dy \tag{4.35}$$

is also convex.

4.3.2.2 Pointwise Maximum and Supremum

The pointwise maximum of $k \geq 2$ convex functions $f_i : \mathbb{R}^n \to (-\infty, +\infty]$, $i \in \mathbb{N}_k$, is defined as

$$f(x) = \max_{i \in \mathbb{N}_k} f_i(x).$$

The epigraph of f is the intersection

$$\text{epi } f = \bigcap_{i \in \mathbb{N}_k} \text{epi } f_i,$$

which shows that epi f is a convex set, and hence, f is convex. This is illustrated in Figure 4.14, which shows the pointwise maximum of four affine functions and its epigraph. The result can be extended to the pointwise supremum of uncountably many convex functions. Specifically, if $f : \mathbb{R}^n \times \mathbb{R}^p \to (-\infty, +\infty]$ and $f(x, y)$ is convex in x for every $y \in Y \subseteq \mathbb{R}^p$, then $h : \mathbb{R}^n \to (-\infty, +\infty]$ defined as

$$h(x) = \sup_{y \in Y} f(x, y) \tag{4.36}$$

is a convex function.

4.3.2.3 Affine Transformation

Recall from Section 4.2 that the image and the preimage of a convex set under an affine transformation is a convex set. A similar result holds for convex functions, as we will now show. Suppose $f : \mathbb{R}^m \to (-\infty, +\infty]$ is a convex function and let $g : \mathbb{R}^n \to (-\infty, +\infty]$ be a composition of f and an affine function, i.e.

$$g(x) = f(Ax + b)$$

for some $A \in \mathbb{R}^{m \times n}$ and $b \in \mathbb{R}^m$. The function g is then convex, which follows immediately from (4.24). Note also that

$$\text{epi } g = \{(x, t) \mid f(Ax + b) \leq t\} = \{(x, t) \mid (Ax + b, t) \in \text{epi } f\},$$

which shows that epi g is the preimage of epi f under an affine transformation.

4.3.2.4 Perspective Transformation

The *perspective function* of a function $f : \mathbb{R}^n \to (-\infty, +\infty]$ is the function $P_f : \mathbb{R}^n \times \mathbb{R} \to (-\infty, +\infty]$ defined as

$$P_f(x, t) = \begin{cases} tf(x/t), & t > 0, \\ \infty, & \text{otherwise.} \end{cases} \tag{4.37}$$

The perspective function of a proper convex function is itself a convex function; see Exercise 4.6. We note that the epigraph of f is related to P_f in the sense that for $t > 0$, it holds that

$$t \text{ epi } f = \{(tx, ts) \mid f(x) \leq s\} = \{(x, s) \mid tf(x/t) \leq s\} = \{(x, s) \mid P_f(x, t) \leq s\}.$$

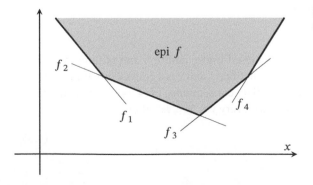

Figure 4.14 The pointwise maximum of convex functions is itself convex.

Moreover, noting that

$$\text{epi } P_f = \{(x, t, s) \mid tf(x/t) \le s\},$$

we see that for all $\alpha \ge 0$,

$$(x, t, s) \in \{0\} \cup \text{epi } P_f \quad \implies \quad (\alpha x, \alpha t, \alpha s) \in \{0\} \cup \text{epi } P_f,$$

which shows that $\{0\} \cup \text{epi } P_f(x, t)$ is a convex cone if f is a proper convex function.

4.3.2.5 Partial Infimum

We define the *partial infimum* $h : \mathbb{R}^n \to \bar{\mathbb{R}}$ of a function $f : \mathbb{R}^n \times \mathbb{R}^p \to (-\infty, +\infty]$ as

$$h(x) = \inf_y f(x, y). \tag{4.38}$$

This is a convex function if f is a convex function. This result follows by noting that the strict epigraph of h, defined as

$$\{(x, t) \mid h(x) < t\},$$

is the image of the strict epigraph of f

$$\{(x, y, t) \mid f(x, y) < t\}$$

under the linear transformation $(x, y, t) \mapsto (x, t)$.

Example 4.1 As an example of a function that is obtained as a partial infimum, consider $f : \mathbb{R}^n \to \mathbb{R}$ defined as $f(z) = z^T X z$ for some $X \in \mathbb{S}_{++}^n$, which implies that f is convex. If we partition X and z conformably as

$$X = \begin{bmatrix} A & B \\ B^T & C \end{bmatrix}, \quad z = \begin{bmatrix} x \\ y \end{bmatrix},$$

with $x \in \mathbb{R}^{n_1}$ and $y \in \mathbb{R}^{n_2}$, then the partial infimum of f with respect to y is

$$h(x) = \inf_y f(x, y) = x^T (A - BC^{-1}B^T)x,$$

where $A - BC^{-1}B^T$ is the Schur complement of C in X. Convexity of h implies that this Schur complement is positive semidefinite.

4.3.2.6 Square of Nonnegative Convex Functions

If $f : \mathbb{R}^n \to [0, \infty)$ is a nonnegative convex function, then

$$g(x) = f(x)^2 \tag{4.39}$$

is a convex function. Indeed, convexity of f implies that for all $x, y \in \mathbb{R}^n$,

$$f(\theta x + (1 - \theta)y) \le \theta f(x) + (1 - \theta)f(y), \quad \forall \, \theta \in [0, 1],$$

and by squaring both sides of the inequality, we find that

$$\begin{aligned} g(\theta x + (1 - \theta)y) &\le \theta^2 f(x)^2 + (1 - \theta)^2 f(y)^2 + 2\theta(1 - \theta)f(x)f(y) \\ &= -\theta(1 - \theta)(f(x) - f(y))^2 + \theta f(x)^2 + (1 - \theta)f(y)^2 \\ &\le \theta f(x)^2 + (1 - \theta)f(y)^2 \\ &= \theta g(x) + (1 - \theta)g(y) \end{aligned}$$

for all $\theta \in [0, 1]$.

4.3.3 Examples of Convex Functions

We start by listing some basic examples of convex and concave functions.

- Linear and affine functions are both convex and concave.
- Absolute value: $f(x) = |x|$ is convex on $\mathbb{R}$.
- Powers: $f(x) = x^\alpha$ with dom $f = \mathbb{R}_{++}$ is concave if $\alpha \in [0, 1]$ and convex if $\alpha \notin (0, 1)$.
- Powers of absolute value: $f(x) = |x|^\alpha$ is convex on $\mathbb{R}$ if $\alpha \geq 1$.
- Exponential function: $f(x) = \exp(x)$ is convex on $\mathbb{R}$.
- Logarithm: $f(x) = \ln(x)$ with dom $f = \mathbb{R}_{++}$ is concave.
- Negative entropy: $f(x) = x \ln(x)$ with dom $f = \mathbb{R}_+$ and $f(0) = 0$ is convex.
- Quadratic-over-linear: $f(x, y) = x^2/y$ with dom $f = \mathbb{R} \times \mathbb{R}_{++}$ is convex.
- One-sided square: $f(x) = \max\ (0, x)^2$ is convex on $\mathbb{R}$.

Next, we give some examples of special classes of convex functions.

4.3.3.1 Norms
All norms are convex function, which is an immediate consequence of the triangle inequality and (4.24). Moreover, the epigraph of a norm is a norm cone, which is a convex set.

4.3.3.2 Indicator and Support Functions
The *indicator function* of a set $C \subseteq \mathbb{R}^n$ is the function $I_C : \mathbb{R}^n \to \{0, \infty\}$ defined as

$$I_C(x) = \begin{cases} 0, & x \in C, \\ \infty, & x \notin C. \end{cases} \tag{4.40}$$

This is clearly a convex function if and only if the set C is convex, and it is proper if $C \neq \emptyset$.

The *support function* of a nonempty set $C \subseteq \mathbb{R}^n$ is the function $S_C : \mathbb{R}^n \to \bar{\mathbb{R}}$ defined as

$$S_C(x) = \sup_{y \in C}\ x^T y. \tag{4.41}$$

We note that this is always a convex function since it is the supremum of affine functions.

4.3.4 Conjugation

The *Legendre–Fenchel transformation* or *conjugate* of a function $f : \mathbb{R}^n \to \bar{\mathbb{R}}$ is the function $f^* : \mathbb{R}^n \to \bar{\mathbb{R}}$ defined as

$$f^*(y) = \sup_x\ \{y^T x - f(x)\}. \tag{4.42}$$

The conjugate function f^* is always a convex function, which follows by noting that it is the pointwise supremum of affine functions. Furthermore, epi f^* is the intersection of closed halfspaces, and hence, it is a closed set, i.e. f^* is closed. Evaluating f^* at 0 yields

$$f^*(0) = \sup_x\ \{-f(x)\} = -\inf_x\ f(x),$$

and, generally speaking, this means that evaluating the conjugate function is as hard as finding a global minimum of f. For a fixed y, the conjugate function $f^*(y)$ may be interpreted graphically as the largest signed vertical distance from f to the linear function $y^T x$. This is illustrated in Figure 4.15 for a univariate function.

The definition of the conjugate function implies that if $f : \mathbb{R}^n \to (-\infty, +\infty]$ is a proper function, then for every $x, y \in \mathbb{R}^n$,

$$f(x) + f^*(y) \geq x^T y.$$

Figure 4.15 The conjugate function $f^*(y)$ is the largest signed vertical distance from f to the linear function $y^T x$, as illustrated here for a univariate function.

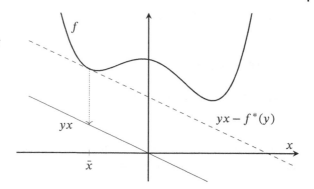

This is known as the *Fenchel–Young inequality*. Equality holds if the supremum of $y^T x - f(x)$ is attained at x.

The conjugate of f^* is called the *biconjugate* of f and is denoted $f^{**} = (f^*)^*$. An immediate consequence of the Fenchel–Young inequality is that $f^{**} \le f$, i.e.

$$f(x) \ge \sup_y \ \{x^T y - f^*(y)\} = f^{**}(x).$$

Moreover, the *Fenchel–Moreau theorem* states that if f is proper, then $f^{**} = f$ if and only if f is convex and closed. More generally, the biconjugate f^{**} is the lower convex envelope of f, which is the supremum of all closed, convex functions that lie below f. In other words, epi f^{**} is the closed convex hull of epi f. To see this, we first note that if f and g are functions from $\mathbb{R}^n$ to $\bar{\mathbb{R}}$ and $f \le g$, then

$$f(x) \le g(x) \quad \Longleftrightarrow \quad y^T x - g(x) \le y^T x - f(x),$$

which implies that $f^* \ge g^*$. Using the same argument, we also see that $f^{**} \le g^{**}$. Thus, if f is closed and convex, then $f = f^{**} \le g^{**} \le g$. Taking the supremum of all closed, convex functions $f \le g$, we conclude that g^{**} is the lower convex envelope of g.

Example 4.2 The conjugate of the indicator function of a set $C \subseteq \mathbb{R}^n$ is given by

$$I_C^*(y) = \sup_x \ \{y^T x - I_C(x)\} = \sup_{x \in C} y^T x = S_C(y),$$

i.e. it is the support function of the set C. In the special case, where C is a nonempty convex cone $K \subset \mathbb{R}^n$, the conjugate function is

$$S_K(y) = \sup_{x \in K} y^T x = I_{-(K^*)}(y),$$

which is the indicator function of the polar cone $-(K^*)$.

4.3.5 Dual Norms

The *dual norm* of a norm $\| \cdot \|$ on $\mathbb{R}^n$ is defined as

$$\|y\|_* = \sup_{\|x\| \le 1} y^T x = \sup_x \ \{y^T x - I_B(x)\} = S_B(y) \tag{4.43}$$

where $B = \{x \in \mathbb{R}^n \mid \|x\| \le 1\}$ denotes the unit norm ball for the norm $\| \cdot \|$. Thus, the dual norm $\| \cdot \|_*$ is the conjugate function of the indicator function of the norm ball B. We invite the reader to verify that the dual norm is indeed a norm, see Exercise 4.11. Since I_B is convex and closed, we conclude that the conjugate of the dual norm $\| \cdot \|_*$ is $I_B^{**} = I_B$. The definition of the dual norm can also be expressed as

$$\|y\|_* = \sup_{x \neq 0} \frac{x^T y}{\|x\|}.$$

This readily implies that for all $x, y \in \mathbb{R}^n$,

$$\|x\|\|y\|_* \geq |x^T y|,$$

which may be viewed as a generalization of the Cauchy–Schwartz inequality. We note that the dual norm of $\| \cdot \|_*$ is the norm $\| \cdot \|_{**} = \| \cdot \|$.

Example 4.3 The dual norm of the Euclidean norm on $\mathbb{R}^n$ is

$$\|y\|_* = \sup_x \{y^T x \mid \|x\|_2 \leq 1\} = \|y\|_2,$$

which follows from the fact that $x = y/\|y\|_2$ achieves the supremum if $y \neq 0$.

Example 4.4 The dual norm of $\| \cdot \|_\infty$ on $\mathbb{R}^n$ is given by

$$\|y\|_* = \sup_x \{y^T x \mid \|x\|_\infty \leq 1\} = \sum_{i=1}^n |y_i| = \|y\|_1.$$

Similarly, the dual norm of $\| \cdot \|_1$ is $\| \cdot \|_\infty$ since $\| \cdot \|_{**} = \| \cdot \|$. More generally, the dual norm of $\| \cdot \|_p$ with $p \geq 1$ is $\| \cdot \|_q$, where $1/p + 1/q = 1$.

Example 4.5 The dual norm of the matrix 2-norm on $\mathbb{R}^{m \times n}$ may be expressed as

$$\|Y\|_* = \sup_X \{\text{tr}(Y^T X) \mid \|X\|_2 \leq 1\}.$$

Now, suppose $Y = U_1 S V_1^T$ is a reduced singular value decomposition (SVD) of Y, where $S = \text{diag}(\sigma_1, \ldots, \sigma_r)$ is the matrix with the nonzero singular values of Y. The condition $\|X\|_2 \leq 1$ implies that the singular values of X are between 0 and 1, and using von Neumann's trace inequality (2.33), we find that for all $\|X\|_2 \leq 1$,

$$\text{tr}(Y^T X) \leq \sum_{i=1}^k \sigma_i = \text{tr}(S).$$

This upper bound is attained at $X = U_1 V_1^T$ if $Y \neq 0$, and hence, the dual norm of $\| \cdot \|_2$ is the nuclear norm.

4.4 Subdifferentiability

A proper function $f : \mathbb{R}^n \to \mathbb{R} \cup \{+\infty\}$ is said to be *subdifferentiable* at x if there exists a vector $g \in \mathbb{R}^n$ such that

$$f(y) \geq f(x) + g^T(y - x), \quad \forall\, y \in \text{dom}\, f.$$

Such a vector g is called a *subgradient* of f at x, and the set of all subgradients at x, i.e.

$$\partial f(x) = \{g \in \mathbb{R}^n \mid f(y) \geq f(x) + g^T(y - x) \,\forall\, y \in \text{dom}\, f\}, \tag{4.44}$$

is the *subdifferential* of f at x. Note that the subdifferential is a set-valued map, and it may be an empty set. We use the convention that $\partial f(x) = \emptyset$ if $x \notin \text{dom } f$, and hence, its effective domain is the set

$$\text{dom } \partial f = \{x \in \mathbb{R}^n \mid \partial f(x) \neq \emptyset\}.$$

The subdifferential $\partial f(x)$ can also be expressed as

$$\partial f(x) = \bigcap_{y \in \text{dom } f} \{g \mid f(y) \geq f(x) + g^T(y - x)\},$$

which shows that $\partial f(x)$ is the intersection of closed halfspaces, and hence, $\partial f(x)$ is closed and convex. Figure 4.16 illustrates the definition of the subdifferential. We note that global information is necessary to determine if a nonconvex function is subdifferentiable at a point x. In other words, the plot in Figure 4.16b alone is insufficient to determine where f is subdifferentiable. For example, if the function approaches some constant as $|x|$ tends to infinity, then f is only subdifferentiable at its minima.

The set of global minimizers of a proper function $f : \mathcal{X} \to (-\infty, +\infty]$ may be characterized in terms of the subdifferential of f using *Fermat's rule*, which states that

$$\text{argmin}_{u}\{f(u)\} = \{x \in \text{dom } f \mid 0 \in \partial f(x)\}. \tag{4.45}$$

This property is easily verified using the definition of the subdifferential (4.44), i.e.

$$x \in \text{argmin}_{u}\{f(u)\} \iff f(y) \geq f(x) + 0^T(y - x) \; \forall \, y \in \text{dom } f \iff 0 \in \partial f(x).$$

However, this property is primarily useful when f is a convex function, because it is generally difficult to characterize the subdifferential of a nonconvex function. We note that if f is convex

Figure 4.16 Illustration of subderivatives and subdifferential of nonsmooth and smooth functions. The bold parts of the graphs correspond to the intervals on which f is subdifferentiable. (a) Nonsmooth convex function: the subdifferential of f at a and b are $\partial f(a) = \{g_1\}$ and $\partial f(b) = [g_2, g_3]$, respectively. (b) Smooth nonconvex function: f is not subdifferentiable on $(a, b) \cup (c, d)$.

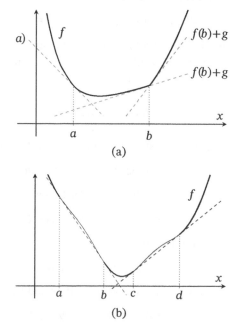

and continuously differentiable at $x \in \text{dom } f$, then $\partial f(x) = \{\nabla f(x)\}$. Conversely, if f is convex and $\partial f(x)$ is the singleton $\{g\}$, then f is differentiable at x and $\nabla f(x) = g$. This is not true for nonconvex functions, as is evident from the example in Figure 4.16b.

4.4.1 Subdifferential Calculus

We will now establish some rules that can be used to find a subgradient or the subdifferential of convex functions that are constructed via one or more convexity-preserving operations.

4.4.1.1 Nonnegative Scaling

The subdifferential of αf, where $f : \mathbb{R}^n \to (-\infty, +\infty]$ and $\alpha > 0$ may be expressed as

$$\partial(\alpha f)(x) = \alpha \partial f(x). \tag{4.46}$$

This result follows directly from the definition (4.44) by multiplying both sides of the inequality by α.

4.4.1.2 Summation

Given two closed proper convex functions $f_1 : \mathbb{R}^n \to (-\infty, +\infty]$ and $f_2 : \mathbb{R}^n \to (-\infty, +\infty]$, it holds that

$$\partial f(x) \supseteq \partial f_1(x) + \partial f_2(x) = \{u + v \mid u \in \partial f_1(x), v \in \partial f_2(x)\}. \tag{4.47}$$

The right-hand side of the inclusion is the *Minkowski sum* of $\partial f_1(x)$ and $\partial f_2(x)$. It is easy to verify (4.48) by noting that if $u \in \partial f_1(x)$ and $v \in \partial f_2(x)$, then

$$f_1(y) + f_2(y) \geq f_1(x) + f_2(x) + (u + v)^T(y - x), \quad \forall y \in \text{dom } f,$$

which implies that $u + v \in \partial f(x)$. Moreover, if f_1 and f_2 satisfy

$$\text{relint dom } f_1 \cap \text{relint dom } f_2 \neq \emptyset,$$

then it can be shown that

$$\partial f(x) = \partial f_1(x) + \partial f_2(x), \quad \forall x \in \mathbb{R}^n. \tag{4.48}$$

The result is readily extended to sums of more than two functions by means of induction. We note that the Minkowski sum of a set and the empty set is itself empty, which implies that dom $\partial f = \text{dom } \partial f_1 \cap \text{dom } \partial f_2$.

4.4.1.3 Affine Transformation

Let $f(x) = g(Ax + b)$, where $g \in \mathbb{R}^m \to (-\infty, +\infty]$ is closed and convex, and $A \in \mathbb{R}^{m \times n}$ and $b \in \mathbb{R}^m$ are given. The subdifferential of f at x then satisfies

$$\partial f(x) \supseteq A^T \partial g(Ax + b) = \{A^T u \mid u \in \partial g(Ax + b)\}.$$

To see this, note that if $u \in \partial g(Ax + b)$, then

$$g(y) \geq g(Ax + b) + u^T(y - Ax - b), \quad \forall y \in \text{dom } g,$$

which implies that

$$g(Az + b) \geq g(Ax + b) + u^T(Az - Ax), \quad \forall z : Az + b \in \text{dom } g,$$

or, equivalently, $A^T u \in \partial f(x)$. If, in addition, we assume that $\mathcal{R}(A) \cap \text{relint } (\text{dom } g) \neq \emptyset$, then

$$\partial f(x) = A^T \partial g(Ax + b), \quad \forall x \in \mathbb{R}^n. \tag{4.49}$$

4.4.1.4 Pointwise Maximum

Consider the pointwise maximum of two closed convex functions $f_1 : \mathbb{R}^n \to \mathbb{R}$ and $f_2 : \mathbb{R}^n \to \mathbb{R}$, i.e.

$$f(x) = \max \ \{f_1(x), f_2(x)\}.$$

The subdifferential of f is then

$$\partial f(x) = \begin{cases} \partial f_1(x), & f_1(x) > f_2(x), \\ \partial f_2(x), & f_1(x) < f_2(x), \\ \mathrm{conv}\ (\partial f_1(x) \cup \partial f_2(x)), & f_1(x) = f_2(x). \end{cases} \tag{4.50}$$

From the definition of the subdifferential, we find that if $u \in \partial f_1(x)$ and $f_1(x) > f_2(x)$, then for all $y \in \mathrm{dom}\ f$,

$$f(y) = \max \ \{f_1(y), f_2(y)\} \geq f_1(y) \geq f_1(x) + u^T(y-x) \geq f(x) + u^T(y-x).$$

This implies that $u \in \partial f(x) \iff u \in \partial f_1(x)$ whenever $f_1(x) > f_2(x)$. In the case where $f_1(x) = f_2(x)$, we note that if $u \in \partial f_1(x)$ and $v \in \partial f_2(x)$, then for all $\theta \in [0,1]$,

$$f(y) \geq \theta f_1(x) + (1-\theta)f_2(x) + (\theta u + (1-\theta)v)^T(y-x), \quad \forall\ y \in \mathrm{dom}\ f.$$

This shows that the line segment $\mathrm{conv}\ (\{u,v\})$ belongs to $\partial f(x)$, and since this is true for all $u \in \partial f_1(x)$ and $v \in \partial f_2(x)$, we may conclude that the right-hand side of (4.50) is a subset of the left-hand side. Showing that equality holds is more tedious, and we refer the reader to [55] for a rigorous proof.

More generally, the pointwise maximum of $k \geq 2$ closed convex functions $f_1, \dots, f_k$ from $\mathbb{R}^n$ to $\mathbb{R}$ may be expressed as

$$\partial f(x) = \mathrm{conv}\ \left(\bigcup_{i \in \mathcal{I}(x)} \partial f_i(x) \right)$$

where $\mathcal{I}(x) = \{i \in \mathbb{N}_k \mid f_i(x) = f(x)\}$ is the set of active indices. We note that this can also be generalized to the supremum of uncountably many proper convex functions.

4.4.1.5 Subgradients of Conjugate Functions

Let $f : \mathbb{R}^n \to (-\infty, +\infty]$ be a proper convex function, and suppose $y \in \partial f(x)$. The definition of the subdifferential then implies that

$$f(z) \geq f(x) + y^T(z-x), \quad \forall\ z \in \mathrm{dom}\ f,$$

or, equivalently,

$$y^T x - f(x) \geq y^T z - f(z), \quad \forall\ z \in \mathrm{dom}\ f.$$

The right-hand sides achieves its supremum at $z = x$, and hence,

$$y \in \partial f(x) \quad \implies \quad y^T x - f(x) = \sup_z \ \{y^T z - f(z)\} = f^*(y).$$

Now, suppose that f is also closed. This means that $f^{**} = f$, and hence,

$$f^{**}(x) + f^*(y) = x^T y, \quad x \in \partial f^*(y).$$

Thus, we may conclude that

$$x \in \partial f^*(y) \quad \iff \quad y \in \partial f(x) \tag{4.51}$$

if f is proper convex and closed.

Example 4.6 The subdifferential of the indicator function of a nonempty closed convex set $C \subseteq \mathbb{R}^n$ is a closed proper convex function. This means that

$$y \in \partial I_C(x) \quad \Longleftrightarrow \quad x \in \partial S_C(y),$$

where $S_C = I_C^*$ is the support function of C. We note that the subdifferential $\partial I_C(x)$ is also referred to as the *normal cone* of C at x, and it may also be expressed as

$$N_C(x) = \partial I_C(x) = \{g \in \mathbb{R}^n \mid 0 \geq g^T(y - x), \ \forall \, y \in C\}. \tag{4.52}$$

It is easy to verify that $N_C(x) = \{0\}$ if $x \in \text{int } C$, and furthermore, if $g \in N_C(x)$ and $g \neq 0$, then C is contained in the halfspace $\{y \mid g^T y \leq g^T x\}$.

4.5 Convex Optimization Problems

We say that an optimization problem of the form (4.4) is a *convex optimization problem* if $f_0, f_1, \ldots, f_m$ are convex functions and $h_1, \ldots, h_p$ are affine functions, i.e. $h(x) = Ax - b$ for some $A \in \mathbb{R}^{p \times n}$ and $b \in \mathbb{R}^p$. Thus, a convex problem can be expressed as

$$\begin{aligned} \text{minimize} \quad & f_0(x) \\ \text{subject to} \quad & f_i(x) \leq 0, \quad i \in \mathbb{N}_m, \\ & Ax = b. \end{aligned} \tag{4.53}$$

The feasible set $\mathcal{F}$ is the intersection of 0-sublevel sets of m convex functions and p hyperplanes, and hence, $\mathcal{F}$ is a convex set. We note that the feasible set may be convex even if some or all f_i are nonconvex functions, in which case the optimization problem is not a convex problem according to our definition, but it is equivalent to a convex problem if $\mathcal{F}$ is a convex set and the objective function is convex on $\mathcal{F}$.

4.5.1 Optimality Condition

We will now show that for convex optimization problems of the form (4.53) with a continuously differentiable objective function, $x \in \mathcal{F}$ is an optimal point if and only if

$$\nabla f_0(x)^T(y - x) \geq 0, \quad \forall \, y \in \mathcal{F}. \tag{4.54}$$

To prove this result, we first assume that x is feasible and satisfies (4.54). The first-order condition for convexity (4.27) then implies that $f_0(y) \geq f_0(x)$ for all $y \in \mathcal{F}$, and hence, x is optimal. Conversely, if x is optimal and (4.54) does not hold, then there exists some $y \in \mathcal{F}$ such that

$$\nabla f_0(x)^T(y - x) < 0.$$

Noting that $z(\theta) = \theta y + (1 - \theta)x \in \mathcal{F}$ for all $\theta \in [0, 1]$ and substituting $z(\theta)$ for y in (4.27), we see that

$$f_0(z(\theta)) \geq f_0(x) + \theta \nabla f_0(x)^T(y - x).$$

This means that $f_0(z(\theta)) < f_0(x)$ for some sufficiently small $\theta > 0$, which contradicts that x is optimal. Another way to see this is that the directional derivative of f_0 at x in the direction $y - x$ is negative, i.e.

$$\frac{d}{d\theta} f_0(z(\theta))|_{\theta=0} = \nabla f_0(x)^T(y - x) < 0.$$

The optimality condition (4.54) is illustrated in Figure 4.17, and as the figure shows, the feasible set $\mathcal{F}$ is a subset of the halfspace $\{x \mid \nabla f_0(x^\star)^T(x - x^\star) \geq 0\}$ if $x^\star$ is an optimal point and $\nabla f_0(x^\star) \neq 0$.

Figure 4.17 The optimality condition (4.54) implies that an optimal point $x^\star$ is either a stationary point of f_0 or $\nabla f_0(x^\star)$ defines a halfspace that contains the feasible set $\mathcal{F}$.

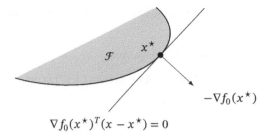

$$\nabla f_0(x^\star)^T(x - x^\star) = 0$$

If x is an the optimal point in the interior of the feasible set, there exists a ball $B_2(x, \epsilon) \subseteq \mathcal{F}$ with radius $\epsilon > 0$. The optimality condition (4.54) requires that

$$\nabla f_0(x)^T(y - x) \geq 0, \quad \forall\, y \in B_2(x, r),$$

which can also be expressed as $\nabla f_0(x)^T u \geq 0$ for all u such that $\|u\|_2 \leq \epsilon$. Clearly, this is only possible if $\nabla f_0(x) = 0$, which means that x is a stationary point of f_0.

Next, we consider the more general case, where f_0 is convex but not necessarily continuously differentiable. First, we note that (4.53) is equivalent to the problem

$$\text{minimize } f_0(x) + I_{\mathcal{F}}(x), \tag{4.55}$$

where $I_{\mathcal{F}}$ is the indicator function for the feasible set $\mathcal{F}$. Fermat's rule then implies that x is optimal if and only if

$$0 \in \partial f_0(x) + N_{\mathcal{F}}(x),$$

where $N_{\mathcal{F}} = \partial I_{\mathcal{F}}$ is the normal cone of $\mathcal{F}$. In other words, x is optimal if and only if there exists a subgradient $g \in \partial f_0(x)$ such that

$$-g \in N_{\mathcal{F}}(x) \quad \Longleftrightarrow \quad g^T(y - x) \geq 0, \quad \forall\, y \in \mathcal{F}. \tag{4.56}$$

Note that if f_0 is continuously differentiable at x, then $\partial f_0(x) = \{\nabla f_0(x)\}$ and the optimality condition (4.56) reduces to (4.54).

4.5.2 Equality Constrained Convex Problems

We now consider the special convex case of (4.53), where $f_0 : \mathbb{R}^n \to \mathbb{R}$ is convex and continuously differentiable and $m = 0$, i.e. there are no inequality constraints. The feasible set is then the affine set

$$\mathcal{F} = \{x \in \mathbb{R}^n \mid Ax = b\}.$$

Recall from Section 2.10 that this set may be expressed as $x + \mathcal{N}(A)$ for any x that satisfies $Ax = b$. The optimality condition (4.54) therefore reduces to

$$Ax = b, \qquad \nabla f_0(x)^T u \geq 0, \quad \forall\, u \in \mathcal{N}(A).$$

Noting that $u \in \mathcal{N}(A) \Longleftrightarrow -u \in \mathcal{N}(A)$, we conclude that $\nabla f_0(x)^T u = 0$ for all $u \in \mathcal{N}(A)$, or, equivalently,

$$\nabla f_0(x) \in \mathcal{N}(A)^\perp = \mathcal{R}(A^T).$$

In other words, x is optimal if and only if there exists a vector $\mu \in \mathbb{R}^p$ such that

$$Ax = b, \quad \nabla f_0(x) + A^T \mu = 0. \tag{4.57}$$

This optimality condition can also be expressed in terms to the so-called *Lagrangian* $L : \mathbb{R}^n \times \mathbb{R}^p \to \mathbb{R}$ defined as

$$L(x, \mu) = f_0(x) + \mu^T h(x).$$

Specifically, if $h(x) = Ax - b$, then (4.57) is equivalent to the conditions

$$\nabla_x L(x, \mu) = 0 \tag{4.58a}$$

$$h(x) = 0. \tag{4.58b}$$

As we will see in Section 4.7, it turns out that these conditions are necessary conditions for optimality even when h is not an affine function.

4.6 Duality

We now return to the general optimization problem (4.4) without making any assumptions about convexity, and we are interested in computing nontrivial lower bounds on the optimal value $p^\star$. We will do this by constructing a so-called *dual problem*, and such a problem can be constructed in several ways. We start out with *Lagrangian duality*, which emerged in the 1940s but is based on techniques pioneered by Lagrange in the 1780s. We then look at *Fenchel duality*, which is closely related to Lagrangian duality but offers a different perspective.

4.6.1 Lagrangian Duality

The *Lagrangian* function $L : \mathbb{R}^n \times \mathbb{R}^m \times \mathbb{R}^p \to (-\infty, +\infty]$ associated with (4.4) is defined as

$$
\begin{aligned}
L(x, \lambda, \mu) &= f_0(x) + \sum_{i=1}^{m} \lambda_i f_i(x) + \sum_{i=1}^{p} \mu_i h_i(x) \\
&= f_0(x) + \lambda^T f(x) + \mu^T h(x)
\end{aligned}
\tag{4.59}
$$

with dom $L = \mathcal{D} \times \mathbb{R}^m \times \mathbb{R}^p$. The auxiliary variables λ and μ are called *Lagrange multipliers* or *dual variables*. The variable λ_i is associated with the inequality constraint $f_i(x) \leq 0$, and μ_i is associated with the equality constraint $h_i(x) = 0$.

The Lagrangian function can be used to construct lower bounds on $p^\star$ by noting that if x is feasible and $\lambda \in \mathbb{R}_+^m$, then

$$f(x) \leq 0, \quad h(x) = 0, \quad \lambda^T f(x) \leq 0.$$

This readily implies that for all $(x, \lambda, \mu) \in \mathcal{F} \times \mathbb{R}_+^m \times \mathbb{R}^p$, it holds that

$$L(x, \lambda, \mu) \leq f_0(x).$$

Now, taking the infimum over $x \in \mathcal{D}$ on the left-hand side of the inequality and the infimum over $x \in \mathcal{F} \subseteq \mathcal{D}$ on the right-hand side, we see that

$$\inf_{x \in \mathcal{D}} L(x, \lambda, \mu) \leq p^\star, \quad \forall \lambda \geq 0.$$

The left-hand side is called the *Lagrange dual function* or simply the *dual function*. It is a function $g : \mathbb{R}^m \times \mathbb{R}^p \to \bar{\mathbb{R}}$ of λ and μ, i.e.

$$g(\lambda, \mu) = \inf_{x \in \mathcal{D}} L(x, \lambda, \mu). \tag{4.60}$$

Now, noting that L is an affine function of λ and μ, we see that the dual function g is the pointwise infimum of a family of affine functions, and hence, g is a concave function. This is true regardless of whether or not (4.4) is a convex optimization problem. We remark that if L is unbounded from below for some (λ, μ), then $g(\lambda, \mu) = -\infty$. We will say that (λ, μ) is a *dual feasible* point if $\lambda \geq 0$ and $(\lambda, \mu) \in \text{dom}(-g)$, where

$$\text{dom}(-g) = \{(\lambda, \mu) \mid g(\lambda, \mu) > -\infty\}$$

is the effective domain of $-g$.

Example 4.7 We now derive the Lagrange dual function for the special case of (4.4), where both the inequality and equality constraints are affine, i.e. the problem takes the form

$$
\begin{aligned}
& \text{minimize } f_0(x) \\
& \text{subject to } Ax \leq b \\
& \qquad\qquad\; Cx = d
\end{aligned}
\tag{4.61}
$$

where $A \in \mathbb{R}^{m \times n}, b \in \mathbb{R}^m, C \in \mathbb{R}^{p \times n}$, and $d \in \mathbb{R}^p$ are given. Using the definition of the Lagrangian, we find that

$$L(x, \lambda, \mu) = f_0(x) + \lambda^T(Ax - b) + \mu^T(Cx - d),$$

and the dual function may be expressed as

$$
\begin{aligned}
g(\lambda, \mu) &= \inf_{x \in D} L(x, \lambda, \mu) \\
&= -\lambda^T b - \mu^T d + \inf_{x \in D} \left(f_0(x) + \left(A^T \lambda + C^T \mu \right)^T x \right) \\
&= -\lambda^T b - \mu^T d - \sup_{x \in D} \left(-\left(A^T \lambda + C^T \mu \right)^T x - f_0(x) \right) \\
&= -\lambda^T b - \mu^T d - f_0^*(-A^T \lambda - C^T \mu),
\end{aligned}
$$

where the last step follows from the definition of the conjugate function. Recalling that conjugate functions are convex by construction, we immediately see that g is indeed a concave function, and moreover, $\text{dom}(-g) = \{(\lambda, \mu) \mid -A^T \lambda - C^T \mu \in \text{dom } f_0^*\}$.

4.6.2 Lagrange Dual Problem

We have seen that the Lagrange dual function can be used to construct lower bounds on the optimal value $p^\star$ of (4.4), and hence, an obvious question is what is the best bound that can be obtained from the dual function in this way? A natural idea would be to maximize the dual function, which gives rise to the *Lagrange dual problem* associated with (4.4) and defined as

$$
\begin{aligned}
& \text{maximize } g(\lambda, \mu) \\
& \text{subject to } \lambda \geq 0.
\end{aligned}
\tag{4.62}
$$

We note that in the context of duality, the problem (4.4) is referred to as the *primal problem*. The optimal value of the dual problem is

$$d^\star = \sup \{g(\lambda, \mu) \mid \lambda \geq 0\},\tag{4.63}$$

and $(\lambda^\star, \mu^\star)$ is a *dual optimal* point if it is dual feasible and $g(\lambda^\star, \mu^\star) = d^\star$. We will say that the Lagrange dual problem is a convex optimization problem since it is trivially equivalent to minimizing $-g$ subject to $\lambda \geq 0$.

The lower bound property of the dual function directly implies that $d^\star \leq p^\star$, and this property is referred to as *weak duality*. We note that weak duality holds even if $p^\star$ and/or $d^\star$ are infinite, and as a consequence, the primal problem is infeasible if the dual problem is unbounded and vice versa.

Moreover, for any dual feasible (λ, μ) and primal feasible x, it holds that $g(\lambda, \mu) \leq d^\star \leq p^\star \leq f_0(x)$, and as a consequence,

$$p^\star - d^\star \leq f_0(x) - g(\lambda, \mu).$$

The difference $f_0(x) - g(\lambda, \mu)$ is called the *duality gap* at (x, λ, μ), whereas $p^\star - d^\star$ is the *optimal duality gap*.

We say that *strong duality* holds if $d^\star = p^\star$, i.e. the optimal duality gap is zero. Strong duality does not hold in general, but it can be shown to hold for a convex optimization problem of the form (4.53) if it satisfies certain *constraint qualifications*. One example is *Slater's constraint qualification* or *Slater's condition*, which states that strong duality holds if there exists a point $x \in \text{relint } D$ that is strictly feasible in the sense that

$$f_i(x) < 0, \quad i \in \mathbb{N}_m, \quad Ax = b.$$

Slater's condition also implies that the dual optimal value is attained whenever it is finite, i.e. there exists a dual point $(\lambda^\star, \mu^\star)$ such that $g(\lambda^\star, \mu^\star) = d^\star$. We note that the duality gap can be used as a stopping criteria for optimization methods when strong duality holds. Indeed, if (x, λ, μ) is primal and dual feasible, then

$$f_0(x) - p^\star \leq f_0(x) - g(\lambda, \mu), \quad d^\star - g(\lambda, \mu) \leq f_0(x) - g(\lambda, \mu). \tag{4.64}$$

Thus, if the duality gap is $\epsilon = f_0(x) - g(\lambda, \mu)$, then x is ϵ-suboptimal for the primal problem and (λ, μ) is ϵ-suboptimal for the dual problem.

4.6.3 Fenchel Duality

We will now consider a conceptually different approach to constructing lower bounds on the optimal value associated with (4.4). To this end, we will consider problems of the form

$$\text{minimize } \phi(x) \tag{4.65}$$

where $\phi : \mathbb{R}^n \to (-\infty, +\infty]$. This problem is equivalent to (4.4) if we let $\phi(x) = f_0(x) + I_F(x)$. Now, suppose that we define a *perturbation* function $h : \mathbb{R}^n \times \mathbb{R}^q$ such that $h(x, 0) = \phi(x)$. The optimal value of the perturbed problem of minimizing $h(x, y)$ is then a function of y, i.e.

$$v(y) = \inf_x \; h(x, y),$$

where we note that $v(0) = \inf_x \phi(x)$ is the optimal value of the unperturbed problem. Recall that the biconjugate of v satisfies $v^{**} \leq v$, and hence, $v^{**}(0)$ provides a lower bound on $v(0)$. Using the definition of the conjugate function, we have that

$$v^{**}(0) = \sup_z \; \{-v^*(z)\},$$

which naturally leads to the dual problem

$$\text{maximize } -v^*(z) \tag{4.66}$$

with variable $z \in \mathbb{R}^n$. The dual objective is a concave function, since $v^*(z)$ is convex. Furthermore, we immediately see that strong duality holds if v is a closed convex function since this implies that $v^{**} = v$. The conjugate of $v(y) = \inf_x h(x, y)$ can also be expressed in terms of h^* by noting that

$$v^*(z) = \sup_y \{z^T y - \inf_x h(x, y)\}$$

$$= \sup_y \{z^T y + \sup_x \{-h(x, y)\}\}$$

$$= \sup_{x,y} \{z^T y - h(x, y)\}$$

$$= h^*(0, z).$$

Thus, the dual problem (4.66) depends on the choice of perturbation function.

Example 4.8 Let ϕ in (4.65) be defined as $\phi(x) = f_0(x) + I_{\mathbb{R}^n_+}(b - Ax)$ with $A \in \mathbb{R}^{m \times n}$ and $b \in \mathbb{R}^m$. This corresponds to an optimization problem of the form (4.4) with the objective function $f_0(x)$ and affine inequality constraints $Ax \preceq b$. Now, suppose that we define the perturbation function $h : \mathbb{R}^n \times \mathbb{R}^n \to \mathbb{R}$ as

$$h(x, y) = f_0(x) + I_{\mathbb{R}^n_+}(b - Ax - y). \tag{4.67}$$

The conjugate of h is then

$$
\begin{aligned}
h^*(v, z) &= \sup_{x,y} \{v^T x + z^T y - f_0(x) - I_{\mathbb{R}^n_+}(b - Ax - y)\} \\
&= \sup_{x,\tilde{y}} \{v^T x + z^T(b - Ax - \tilde{y}) - f_0(x) - I_{\mathbb{R}^n_+}(\tilde{y})\} \\
&= b^T z + \sup_x \{(v - A^T z)^T x - f_0(x)\} + \sup_{\tilde{y}} \{-z^T \tilde{y} - I_{\mathbb{R}^n_+}(\tilde{y})\} \\
&= b^T z + f_0^*(v - A^T z) + S_{\mathbb{R}^n_+}(-z)
\end{aligned}
$$

where the second step follows by letting $\tilde{y} = b - Ax - y$, and in the last step, we used the fact that $I_{\mathbb{R}^n_+}^*$ is the support function $S_{\mathbb{R}^n_+}$. Recall from Example 4.6 that the support function of a nonempty convex cone is the indicator function of the polar cone, so $S_{\mathbb{R}^n_+}(-z) = I_{\mathbb{R}^n_+}(z)$. This implies that we can express the dual problem as

$$
\begin{aligned}
& \text{maximize} \quad -b^T z - f_0^*(-A^T z) \\
& \text{subject to} \quad z \geq 0.
\end{aligned}
\tag{4.68}
$$

We end this section by outlining *Fenchel's duality theorem*. Specifically, we will consider the optimization problem

$$\text{minimize} \quad f(x) + g(Ax) \tag{4.69}$$

where $f : \mathbb{R}^n \to \bar{\mathbb{R}}$ and $g : \mathbb{R}^m \to \bar{\mathbb{R}}$ are proper convex functions, and $A \in \mathbb{R}^{m \times n}$ is given. The conjugate of the perturbation function $h(x, y) = f(x) + g(Ax + y)$ can be expressed as

$$
\begin{aligned}
h^*(v, z) &= \sup_{x,y} \{v^T x + z^T y - f(x) - g(Ax + y)\} \\
&= \sup_{x,\tilde{y}} \{v^T x + z^T(\tilde{y} - Ax) - f(x) - g(\tilde{y})\} \\
&= \sup_x \{(v - A^T z)^T x - f(x)\} + \sup_{\tilde{y}} \{z^T \tilde{y} - g(\tilde{y})\} \\
&= f^*(v - A^T z) + g^*(z),
\end{aligned}
$$

and hence, the dual problem is

$$\text{maximize} \quad -f^*(-A^T z) - g^*(z) \tag{4.70}$$

with variable $z \in \mathbb{R}^m$. We note that this is equivalent to the Lagrange dual problem associated with the problem

$$
\begin{aligned}
& \text{minimize} \quad f(x) + g(y) \\
& \text{subject to} \quad Ax = y
\end{aligned}
$$

with variables $x \in \mathbb{R}^n$ and $y \in \mathbb{R}^m$ and which is equivalent to (4.69). From weak duality, we have that

$$p^\star = \inf_x \{f(x) + g(Ax)\} \geq \sup_z \{-f^*(-A^T z) - g^*(z)\} = d^\star,$$

and it can be shown that strong duality holds if f and g satisfy certain conditions. For example, the condition relint $(\text{dom } f) \cap \text{relint} (\text{dom } g) \neq \emptyset$ is a sufficient condition for strong duality.

4.7 Optimality Conditions

We will now study the behavior of the primal problem (4.4) and its associated dual problem (4.62) at a point at which the optimal values are attained and equal. In other words, we assume that $(x^\star, \lambda^\star, \mu^\star)$ satisfies

$$f_0(x^\star) = g(\lambda^\star, \mu^\star) = L(x^\star, \lambda^\star, \mu^\star) = f_0(x^\star) + \sum_{i=1}^{m} \lambda_i^\star f_i(x^\star) + \sum_{i=1}^{p} \mu_i^\star h_i(x^\star).$$

An immediate consequence is that

$$h_i(x^\star) = 0, \ i \in \mathbb{N}_p, \qquad \sum_{i=1}^{m} \lambda_i^\star f_i(x^\star) = 0,$$

and since $\lambda_i^\star f_i(x^\star) \leq 0$, it follows that

$$\lambda_i^\star f_i(x^\star) = 0, \quad i \in \mathbb{N}_m. \tag{4.71}$$

This property is known as *complementary slackness*: $\lambda_i^\star$ and $f_i(x^\star)$ cannot be nonzero at the same time. Moreover, if the objective function and the constraint functions are continuously differentiable, then it must also hold that

$$\left. \frac{\partial}{\partial x} L(x, \lambda^\star, \mu^\star) \right|_{x=x^\star} = \nabla_x L(x^\star, \lambda^\star, \mu^\star) = 0,$$

i.e. the gradient of L with respect to x vanishes at $(x^\star, \lambda^\star, \mu^\star)$. These observations may be summarized as the following necessary conditions for optimality, which are known as the *Karush–Kuhn–Tucker* (KKT) conditions:

$$\nabla_x L(x^\star, \lambda^\star, \mu^\star) = 0 \tag{4.72a}$$

$$f(x^\star) \leq 0, \ h(x^\star) = 0 \tag{4.72b}$$

$$\lambda^\star \geq 0 \tag{4.72c}$$

$$\lambda_i^\star f_i(x^\star) = 0, \quad i \in \mathbb{N}_m. \tag{4.72d}$$

The condition (4.72b) corresponds to primal feasibility, (4.72c) corresponds to dual feasibility, and (4.72d) is the complementary slackness condition.

4.7.1 Convex Optimization Problems

In the special case, where the primal problem is a convex optimization problem of the form (4.53), the KKT conditions (4.72) are not only necessary conditions but also sufficient conditions for optimality. To see this, suppose that $(\bar{x}, \bar{\lambda}, \bar{\mu})$ satisfies the KKT conditions for the convex optimization problem (4.53), i.e.

$$\nabla_x L(\bar{x}, \bar{\lambda}, \bar{\mu}) = 0$$

$$f(\bar{x}) \leq 0, \ A\bar{x} = b$$

$$\bar{\lambda} \geq 0$$

$$\bar{\lambda}_i f_i(\bar{x}) = 0, \quad i \in \mathbb{N}_m.$$

The dual feasibility condition $\lambda \geq 0$ implies that $L(x, \bar{\lambda}, \bar{\mu})$ is convex in x. Moreover, the stationarity condition $\nabla_x L(\bar{x}, \bar{\lambda}, \bar{\mu}) = 0$ implies that $\bar{x}$ is a minimizer of $L(x, \bar{\lambda}, \bar{\mu})$, and hence,

$$g(\bar{\lambda}, \bar{\mu}) = \inf_{x \in D} L(x, \bar{\lambda}, \bar{\mu}) = L(\bar{x}, \bar{\lambda}, \bar{\mu}).$$

Using the complementary slackness condition, we find that $L(\bar{x}, \bar{\lambda}, \bar{\mu}) = f_0(\bar{x})$, and hence, $\bar{x}$ and $(\bar{\lambda}, \bar{\mu})$ result in a zero duality gap, which shows that they are primal and dual optimal points, respectively.

We note that for convex optimization problems, Slater's condition implies the existence of primal and dual optimal points $x^\star$ and $(\lambda^\star, \mu^\star)$ with zero duality gap.

4.7.2 Nonconvex Optimization Problems

Recall that in general, the KKT conditions are only necessary conditions for optimality. Although strong duality does hold for some nonconvex optimization problems, it is important to note that it is not a generic property. Thus, for nonconvex optimization problems, the situation is more complicated if we do not know if there exist primal and dual optimal points $x^\star$ and $(\lambda^\star, \mu^\star)$ with a zero duality gap. Moreover, we may be interested in characterizing not only global optima but also local optima. To this end, we once again return to the optimization problem (4.4), where we assume that f_0, f, and h are continuously differentiable. Given a feasible point x, we say that the *linear independence constraint qualification* (LICQ) holds at x if the gradients of the equality constraint functions and the active inequality constraint functions at x are linearly independent, i.e. the vectors

$$\{\nabla h_1(x), \dots, \nabla h_p(x)\} \cup \{\nabla f_i(x) \mid f_i(x) = 0, i \in \mathbb{N}_m\}$$

are linearly independent. Now, suppose that $\bar{x}$ is a locally optimal point and that the LICQ holds at $\bar{x}$. Then there exist $\lambda \in \mathbb{R}^m$ and $\mu \in \mathbb{R}^p$ such that

$$\nabla_x L(\bar{x}, \lambda, \mu) = 0$$
$$f(\bar{x}) \leq 0, \ h(\bar{x}) = 0$$
$$\lambda \geq 0$$
$$\lambda_i f_i(\bar{x}) = 0, \quad i \in \mathbb{N}_m.$$

A proof of this result can be found in, e.g. [74, p. 314]. This has the following important practical consequence: if we do not know $\bar{x}$ but instead compute a solution to the KKT conditions, then it is a candidate for a local minimizer if the LICQ holds at $\bar{x}$. However, further information is necessary to determine if $\bar{x}$ is indeed a local minimizer, since it can also be a local maximizer or a saddle point. If the functions f_0, f, and h are also twice continuously differentiable, then there are second-order sufficiency conditions that that guarantee that a point $\bar{x}$ is a local minimizer. For example, $\bar{x}$ is a local minimizer if there exist $\lambda \in \mathbb{R}^m$ and $\mu \in \mathbb{R}^p$ such that

$$\nabla_x L(\bar{x}, \lambda, \mu) = 0$$
$$f(\bar{x}) \leq 0, \ h(\bar{x}) = 0$$
$$\lambda \geq 0$$
$$\lambda_i f_i(\bar{x}) = 0, \quad i \in \mathbb{N}_m,$$

and the Hessian of L at $\bar{x}$ is positive definite on the subspace

$$M = \left\{ y \in \mathbb{R}^n \mid y^T \nabla h_i(\bar{x}) = 0, \ y^T \nabla f_j(\bar{x}) = 0, \ \forall i \in \mathbb{N}_p, \ \forall j \in J \right\},$$

where $J = \left\{ j \in \mathbb{N}_m \mid f_j(\bar{x}) = 0, \ \lambda_j > 0 \right\}$; see, e.g. [74, p. 316].

Exercises

4.1 Consider the optimization problem

$$\text{minimize } f(x, y)$$

where $f : \mathbb{R}^m \times \mathbb{R}^n \to \mathbb{R}$. Show that the minimal value of f can be obtained by first minimizing over x and then minimizing over y. Specifically, let $g : \mathbb{R}^n \to \mathbb{R}$ be defined as

$$g(y) = \min_x \ f(x, y)$$

and let $\bar{x}(y) \in \text{argmin}_x f(x, y)$ denote a minimizer of $f(x, y)$ for a given y. Similarly, we define $\bar{y} \in \text{argmin}_y g(y)$. Show that $(\bar{x}(\bar{y}), \bar{y})$ is a minimizer of f.

4.2   Consider the optimization problem

$$\text{minimize} \sum_{i=1}^{m} f_i(x)$$

with variable $x \in \mathbb{R}^n$, and where $f_i : \mathbb{R}^n \to \mathbb{R}$, $i \in \mathbb{R}^m$. Show that this problem is equivalent to the problem

$$\text{minimize} \ \mathbf{1}^T t$$
$$\text{subject to} \ f_i(x) - t_i \leq 0, \quad i \in \mathbb{N}_m$$

with variables $x \in \mathbb{R}^n$ and $t \in \mathbb{R}^m$.

4.3   Let $f : \mathbb{R}_+ \to \mathbb{R}$ be defined as $f(x) = x \ln x$, where we use the convention that $0 \ln 0 = 0$.
(a) Show that f is a convex function.
(b) Show that $g : \mathbb{R}_+^n \to \mathbb{R}$ defined as $g(x) = \sum_{i=1}^n f(x_i)$ is a convex function.
(c) Derive the conjugate function of f.

4.4   Show that the function $f : \mathbb{R}^n \to \mathbb{R}$ defined as

$$f(x) = \ln(e^{x_1} + \cdots + e^{x_n})$$

is convex.

4.5   Let $g : \mathbb{R} \to \mathbb{R}$ be defined as $g(t) = f(x + vt)$, where $f : \mathbb{R}^n \to \mathbb{R}$ and $v \in \mathbb{R}^n$ are given, and $\text{dom} \ g = \{t \mid x + tv \in \text{dom} \ f\}$.
(a) Show that f is a convex function if and only if g is a convex function for all $x \in \text{dom} \ f$ and $v \in \mathbb{R}^n$.
(b) Show that $f : \mathbb{S}^n \to \mathbb{R}$ defined as $f(X) = -\ln \det X$ with $\text{dom} \ f = \mathbb{S}_{++}^n$ is a convex function.
Hint: Show that $g(t) = -\ln \det (X + Vt)$ is convex for all $X \in \mathbb{S}_{++}^n$ and $V \in \mathbb{S}^n$.

4.6   Let $f : \mathbb{R}^n \to (-\infty, +\infty]$ be a convex function. Show that the perspective function $P_f : \mathbb{R}^n \times \mathbb{R} \to (-\infty, +\infty]$ defined as

$$P_f(x, t) = \begin{cases} tf(x/t), & t > 0, \\ \infty, & \text{otherwise.} \end{cases}$$

is a convex function.

4.7   Show that $f : \mathbb{S}_{++}^n \to \mathbb{R}_+$ defined as $f(X) = y^T X^{-1} y$ is a convex function.
Hint: Use the Schur complement formula (2.58) to show that epi f is a convex set.

4.8   Let $\varphi : \mathbb{R} \to \mathbb{R}$ be a convex function and let $f : X \to \mathbb{R}$ be a function with domain $X \subset \mathbb{R}^n$ that satisfies $\int_X f(x) \, dx = 1$. Show the following result, which is known as *Jensen's inequality*:

$$\varphi \left(\int_X f(x) dx \right) \leq \int_X \varphi \left(f(x) \right) dx.$$

4.9 Let $f : \mathbb{R}^m \times \mathbb{R}^n \to \mathbb{R}$ be a convex function, and let $C \subseteq \mathbb{R}^n$ be a convex set. Show that the function $g : \mathbb{R}^m \to \mathbb{R}$ defined as

$$g(x) = \inf_{y \in C} f(x, y)$$

is a convex function.

4.10 Let $x_i \in [0, 1/2]$, $i \in \mathbb{N}_n$, and $\theta \in \Delta^n$.

(a) Show that

$$\frac{\prod_{i=1}^n x_i^{\theta_i}}{\prod_{i=1}^n (1 - x_i)^{\theta_i}} \leq \frac{\sum_{i=1}^n \theta_i x_i}{\sum_{i=1}^n \theta_i (1 - x_i)}$$

which is known as the *Ky Fan* inequality.

(b) Show that the Ky Fan inequality holds with equality if and only if either $\theta^T x = 0$ or $x > 0$ satisfies $x_i = y \in (0, 1/2]$ for all $i \in \mathbb{N}_n$ such that $\theta_i > 0$.

Hint: Apply Jensen's inequality using the function $\varphi(x) = -\ln \frac{x}{1-x}$.

4.11 Let $\| \cdot \|$ be a norm on $\mathbb{R}^n$. Show that the dual norm, which is defined as

$$\|y\|_* = \sup_x \{y^T x \mid \|x\| \leq 1\},$$

is indeed a norm.

4.12 Derive the dual norm of the quadratic norm $\| \cdot \|_P : \mathbb{R}^n \to \mathbb{R}_+$ defined as $\|x\|_P = \sqrt{x^T P x}$ for some $P \in \mathbb{S}_{++}^n$.

4.13 For each of the following functions $f : \mathbb{R} \to \mathbb{R}$, derive the subdifferential.

(a) The absolute value function $f(t) = |t|$.

(b) The function $f(x) = \max(2x + 1, -x + 2)$.

(c) The function $f(x) = \max(2x + 1, -x + 2)^2$.

4.14 [22, Exercise 2.31] Let C^* be the dual cone of a convex cone $C \subseteq \mathbb{R}^n$. Prove the following statements:

(a) C^* is a convex cone.

(b) C^{**} is closed.

(c) $C_1 \subseteq C_2$ implies $C_2^* \subseteq C_1^*$.

(d) The interior of C^* is given by $\operatorname{int} C^* = \{y \mid y^T x > 0 \ \forall \, x \in C\}$.

(e) If C has a nonempty interior, then C^* is pointed.

(f) C^{**} is the closure of C.

(g) If the closure of C is pointed, then C^* has a nonempty interior.

5

Optimization Problems

Applications in learning and control give rise to a wide range of optimization problems, and we will now discuss different classes of such optimization problems. Our starting point will be the classes of linear and nonlinear least-squares problems, instances of which occur frequently in, e.g. supervised learning. It was studied already by Carl Friedrich Gauss in the eighteenth century in order to calculate the orbits of celestial bodies. We then discuss quadratic programs, which are often encountered as local surrogate models of general optimization problems and are a component of many optimization methods. Another important class of problems is the class of conic optimization problems, and we will see that any convex optimization problem can, in principle, be cast as a conic optimization problem.

Problems that involve the rank of some matrix variable as part of the objective function or a constraint function are called rank optimization problems. We will see that some special cases of these can be solved to global optimality using techniques from linear algebra. However, in general, rank optimization problems are difficult nonlinear optimization problems, and we will introduce some heuristics that often produce good approximate solutions. We will also discuss partially separable optimization problems, which are problems with a special kind of structure that can be exploited computationally. Several examples of such problems appear later in the book, e.g. optimal control problems as well as hidden Markov processes. We also consider multiparametric optimization, which is about finding a parametric solution to a family of parameterized optimization problems, and finally, we discuss stochastic optimization problems, which arise, e.g. when the objective function involves a random variable in addition to the optimization variable.

5.1 Least-Squares Problems

A nonlinear *least-squares* (LS) problem is a problem of the form

$$\text{minimize} \quad \frac{1}{2}\sum_{k=1}^{m} f_k(x)^2, \tag{5.1}$$

with variable $x \in \mathbb{R}^n$ and where $f_k : \mathbb{R}^n \to \mathbb{R}, k \in \mathbb{N}_m$. It may also be written more compactly as

$$\text{minimize} \quad \frac{1}{2}\|f(x)\|_2^2,$$

Optimization for Learning and Control, First Edition. Anders Hansson and Martin Andersen.
© 2023 John Wiley & Sons, Inc. Published 2023 by John Wiley & Sons, Inc.
Companion Website: www.wiley.com/go/opt4lc

where we define $f(x) = (f_1(x), \ldots, f_m(x))$. If f is continuously differentiable, then the necessary optimality condition associated with the nonlinear LS problem may be expressed as

$$\frac{1}{2}\sum_{k=1}^{m}\frac{\partial}{\partial x}f_k(x)^2 = \sum_{k=1}^{m}f_k(x)\nabla f_k(x) = 0. \tag{5.2}$$

This is a set of n nonlinear equations in x, which are not easy to solve in general. We discuss optimization methods for this type of problem in Chapter 6.

A linear LS problem is the special case where f is an affine function, i.e. $f(x) = Ax - b$ for some $A \in \mathbb{R}^{m \times n}$ and $b \in \mathbb{R}^m$. The resulting LS problem is a convex optimization problem. This follows by noting that the Hessian of $(1/2)\|Ax - b\|_2^2$ is $A^T A$, which is positive semidefinite. The necessary optimality condition is therefore also sufficient, and it can be expressed as

$$A^T A x = A^T b. \tag{5.3}$$

This system of equations is called the *normal equations*. We will encounter several LS problems, e.g. linear regression, see Section 10.1.

It is sometimes useful to augment the LS problem by adding constraints. For example, if we add affine equality constraints to a linear LS problem, we obtain a problem of the form

$$\begin{aligned} \text{minimize} \quad & \frac{1}{2}\|Ax - b\|_2^2 \\ \text{subject to} \quad & Cx = d, \end{aligned} \tag{5.4}$$

with $C \in \mathbb{R}^{p \times n}$ and $d \in \mathbb{R}^p$. Like the linear LS problem, this is clearly also a convex optimization problem. The Karush–Kuhn–Tucker (KKT) conditions associated with (5.4) may be expressed as

$$\begin{bmatrix} A^T A & C^T \\ C & 0 \end{bmatrix}\begin{bmatrix} x \\ \mu \end{bmatrix} = \begin{bmatrix} A^T b \\ d \end{bmatrix}, \tag{5.5}$$

where $\mu \in \mathbb{R}^p$ is the vector of Lagrange multipliers associated with the equality constraints. This is an indefinite system of equations, and it is sometimes referred to as the KKT equations since it represents the KKT conditions. We note that the linear independence constraint qualification (LICQ) is independent of x and holds if $\text{rank}(C) = p$. Moreover, as we saw in Section 2.11, the system has a unique solution if and only if $\text{rank}(C) = p$ and $A^T A$ is positive definite on the nullspace of C, i.e.

$$\text{rank}(C) = p, \quad \mathcal{N}(A) \cap \mathcal{N}(C) = \{0\}.$$

Interested readers may find more information about LS problems in [21], which is a comprehensive reference on the topic.

Example 5.1 As an example of a nonlinear least-squares problem, we consider a so-called localization problem. Suppose that $a_1, \ldots, a_m \in \mathbb{R}^D$ are m so-called anchor positions, which we assume are known positions in a D-dimensional space. Furthermore, suppose that we are interested in estimating n unknown positions $x_1, \ldots, x_n \in \mathbb{R}^D$ based on partial information about pairwise distances. Specifically, we will assume that r_{ij} is a noisy measurement of $\|x_i - x_j\|_2$ for $(i,j) \in \mathcal{E} \subset \mathbb{N}_n \times \mathbb{N}_n$, and v_{ik} is a noisy measurement of $\|x_i - a_k\|_2$ for $(i,k) \in \mathcal{E}_a \subset \mathbb{N}_n \times \mathbb{N}_m$. Figure 5.1 illustrates a two-dimensional localization problem.

The problem of estimating the n unknown positions may be posed as a LS problem, i.e.

$$\text{minimize} \sum_{(i,j)\in\mathcal{E}}\left(\|x_i - x_j\|_2 - r_{ij}\right)^2 + \sum_{(i,k)\in\mathcal{E}_a}\left(\|x_i - a_k\|_2 - v_{ik}\right)^2,$$

with variables $x_i \in \mathbb{R}^D, i \in \mathbb{N}_n$. If the measurement errors are independent and normally distributed with zero mean and the same variance, then the above problem is equivalent to a maximum likelihood problem for estimating the positions; *cf.* Section 10.1.

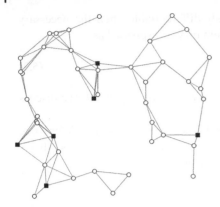

Figure 5.1 Example of a two-dimensional localization problem. The m anchors are marked with black squares, and the circles are the n unknown positions. The edges represent the noisy distance measurements.

5.2 Quadratic Programs

The linear LS problem and its linearly constrained variant are instances of a more general class of optimization problems, namely *quadratic programs* (QP). We define a QP as a problem of the form

$$
\begin{aligned}
\text{minimize} \quad & \tfrac{1}{2}x^T P x + q^T x \\
\text{subject to} \quad & Ax \geq b \\
& Cx = d,
\end{aligned}
\tag{5.6}
$$

with variable $x \in \mathbb{R}^n$ and problem data $P \in \mathbb{S}^n$, $q \in \mathbb{R}^n$, $A \in \mathbb{R}^{m \times n}$, $b \in \mathbb{R}^m$, $C \in \mathbb{R}^{p \times n}$, and $d \in \mathbb{R}^p$. This is a convex optimization problem according to our definition if and only if P is positive semidefinite. We note that the problem is equivalent to a convex optimization problem in the event that P is indefinite, but positive semidefinite on the nullspace of C. The special case of (5.6) where $P = 0$ is called a *linear program* (LP). We will see an example of an LP in Section 10.5.

QPs of the form (5.6) generally do not have a closed-form solution. The KKT conditions may be expressed as

$$
Px + A^T \lambda + C^T \mu = -q
$$

$$
Ax \leq b, \quad Cx = d
$$

$$
\lambda \geq 0
$$

$$
\text{diag}(\lambda)(Ax - b) = 0,
$$

where $\lambda \in \mathbb{R}^m$ and $\mu \in \mathbb{R}^p$ and the Lagrange multiplies associated with the inequality and equality constraints, respectively.

To derive the Lagrange dual of (5.6), we introduce the Lagrangian

$$
L(x, \lambda, \mu) = \tfrac{1}{2}x^T P x + q^T x + \lambda^T (Ax - b) + \mu^T (Cx - d).
$$

We immediately see that L is unbounded below if $P \not\succeq 0$, in which case, the dual function is $g(\lambda, \mu) = -\infty$, and hence, $d^\star = -\infty$. A more useful lower bound can be obtained if $P \succeq 0$, in which case, we find that

$$
g(\lambda, \mu) = \begin{cases} -b^T \lambda - d^T \mu - \psi(\lambda, \mu), & q + A^T \lambda + C^T \mu \in \mathcal{R}(P), \\ -\infty, & \text{otherwise,} \end{cases}
\tag{5.7}
$$

where

$$
\psi(\lambda, \mu) = \tfrac{1}{2}(q + A^T \lambda + C^T \mu)^T P^\dagger (q + A^T \lambda + C^T \mu).
$$

The range condition $q + A^T\lambda + C^T\mu \in \mathcal{R}(P)$ can be expressed as the equality constraint $Py = q + A^T\lambda + C^T\mu$ where $y \in \mathbb{R}^n$ is an auxiliary variable. Thus, using the fact that $PP^\dagger P = P$, we can express the dual of a convex QP as

$$\begin{aligned}
\text{maximize} \quad & -b^T\lambda - d^T\mu - \frac{1}{2}y^T Py \\
\text{subject to} \quad & A^T\lambda + C^T\mu - Py = -q \\
& \lambda \geq 0,
\end{aligned} \tag{5.8}$$

with variables $\lambda \in \mathbb{R}^m$, $\mu \in \mathbb{R}^p$, and $y \in \mathbb{R}^n$. Note that the dual problem is itself a QP.

QPs arise in many different applications. An important example is the subproblems that arise in some optimization methods for more general optimization problems. We will also encounter a QP in Section 10.5 in the context of support vector machines.

Example 5.2 Suppose $(\Omega, \mathcal{F}, P)$ is a probability space with $\Omega = \mathbb{N}_n$, and let $X : \Omega \to \mathbb{R}$ be a random variable. Moreover, we define $\mathcal{D} = \{x_1, \ldots, x_n\}$, where $x_k = X(k)$, $k \in \mathbb{N}_n$, and let $p = (p_1, \ldots, p_n)$ be the values of the probability function, i.e.

$$p_k = P(X = x_k), \quad k \in \mathbb{N}_n.$$

We will assume that p is unknown and consider an optimization problem of the form

$$\begin{aligned}
\text{maximize} \quad & \text{Var} f_0(X) \\
\text{subject to} \quad & \mathbb{E} f_i(X) \leq b_i, \quad i \in \mathbb{N}_m \\
& p \in \Delta^n,
\end{aligned} \tag{5.9}$$

with variable p and where $f_0 : \mathcal{D} \to \mathbb{R}$ and $f_i : \mathcal{D} \to \mathbb{R}$, $i \in \mathbb{N}_m$, and the upper bounds $b = (b_1, \ldots, b_m)$ are given. Letting $a_i = (f_i(x_1), \ldots, f_i(x_n))$, we see that the ith inequality constraint can be expressed as the affine inequality $a_i^T p \leq b_i$. Moreover, the objective function can be expressed as

$$\text{Var} f_0(X) = \mathbb{E}[f_0(X)^2] - \mathbb{E}[f_0(X)]^2 = r^T p - (c^T p)^2$$

where $r = (f_0(x_1)^2, \ldots, f_0(x_n)^2)$ and $c = (f_0(x_1), \ldots, f_0(x_n))$. It is easy to verify that $\text{Var} f_0(X)$ is a concave function of p, and hence, (5.9) can be expressed as a convex QP

$$\begin{aligned}
\text{minimize} \quad & p^T(cc^T)p - r^T p \\
\text{subject to} \quad & Ap \leq b \\
& x \in \Delta^n,
\end{aligned} \tag{5.10}$$

with variable $p \in \mathbb{R}^n$ and where $A \in \mathbb{R}^{m \times n}$ is the matrix with rows $a_1^T, \ldots, a_m^T$.

We end this section by noting that the QP in (5.6) can be generalized by allowing quadratic constraints. The resulting optimization problem is a so-called *quadratically constrained quadratic program* (QCQP), which can be expressed as

$$\begin{aligned}
\text{minimize} \quad & (1/2)x^T Px + q^T x \\
\text{subject to} \quad & (1/2)x^T A_i x + b_i^T x + c_i \leq 0, \quad i \in \mathbb{N}_m,
\end{aligned} \tag{5.11}$$

with variable $x \in \mathbb{R}^n$ and where $P \in \mathbb{S}^n$, $q \in \mathbb{R}^n$, and $(A_i, b_i, c_i) \in \mathbb{S}^n \times \mathbb{R}^n \times \mathbb{R}$, $i \in \mathbb{N}_m$. This is a convex problem if and only if P and $A_1, \ldots, A_m$ are positive semidefinite.

5.3 Conic Optimization

A *conic linear program* or *cone LP* is an optimization problem of the form

$$\begin{aligned}
\text{minimize} \quad & c^T x \\
\text{subject to} \quad & Ax = b \\
& x \succeq_K 0,
\end{aligned} \tag{5.12}$$

where $K \subset \mathbb{R}^n$ is a proper convex cone, $x \in \mathbb{R}^n$ is the optimization variable, and $c \in \mathbb{R}^n$, $A \in \mathbb{R}^{p \times n}$, and $b \in \mathbb{R}^p$ are the problem data. Recall that $x \succeq_K 0 \iff x \in K$, and hence, the special case where $K = \mathbb{R}^n_+$ corresponds to an LP. This observation allows us to view the cone LP as a conceptually simple extension of an LP. The problem (5.12) is called a *second-order cone program* (SOCP) if K is the Cartesian product of second-order cones, i.e.

$$K = \mathbb{Q}^{n_1} \times \cdots \times \mathbb{Q}^{n_m}, \qquad \sum_{i=1}^{m} n_i = n.$$

Furthermore, if we consider $\mathbb{S}^n$ instead of $\mathbb{R}^n$ and let $K = \mathbb{S}^n_+$, we obtain a *semidefinite program* (SDP)

$$
\begin{aligned}
&\text{minimize} && \text{tr}(CX) \\
&\text{subject to} && \text{tr}(A_i X) = b_i, \quad i \in \mathbb{N}_p, \\
& && X \in \mathbb{S}^n_+.
\end{aligned}
\tag{5.13}
$$

LPs, SOCPs, and SDPs are special cases that form a hierarchy of increasing complexity in the sense that LPs are SOCP representable, and SOCPs are SDP representable, i.e.

$$\text{LP} \subset \text{SOCP} \subset \text{SDP}.$$

Indeed, we have that

$$x \in \mathbb{R}^n_+ \iff x \in \mathbb{Q}^1 \times \cdots \times \mathbb{Q}^1 \iff \text{diag}(x) \in \mathbb{S}^n_+,$$

which shows that an LP can be formulated as both an SOCP and an SDP. Moreover, an SOCP can be formulated as an SDP by using the fact that

$$(y, t) \in \mathbb{Q}^n \iff \begin{bmatrix} tI & y \\ y^T & t \end{bmatrix} \in \mathbb{S}^n_+.$$

To verify this equivalence, first note that

$$(y, t) \in \mathbb{Q}^n \iff \|y\|_2 \leq t \iff y^T y \leq t^2, \; 0 \leq t.$$

Now, using (2.58), we see that

$$\begin{bmatrix} tI & y \\ y^T & t \end{bmatrix} \geq 0 \iff t \geq 0, \; t - y^T t^\dagger y \geq 0, \; y = t t^\dagger y,$$

and multiplying both sides of $t - y^T t^\dagger y \geq 0$ by t, we find that

$$\begin{bmatrix} tI & y \\ y^T & t \end{bmatrix} \geq 0 \iff t^2 - y^T y \geq 0, \; t \geq 0 \iff (y, t) \in \mathbb{Q}^n.$$

It is also possible to represent convex QPs and QCQPs as SOCPs, which is an immediate consequence of the following example.

Example 5.3 Consider the convex quadratic constraint

$$x^T P x + q^T x + s \leq 0, \tag{5.14}$$

where $x \in \mathbb{R}^n$, $P \in \mathbb{S}^n_+$, $q \in \mathbb{R}^n$, and $s \in \mathbb{R}$. Such a constraint can always be expressed as a conic constraint of the form

$$\|Ax + b\|_2 \leq c^T x + d, \tag{5.15}$$

where $A \in \mathbb{R}^{r \times n}$, $b \in \mathbb{R}^r$, $c \in \mathbb{R}^n$, and $d \in \mathbb{R}_+$. To see this, first note that (5.15) is equivalent to

$$x^T (A^T A - c c^T) x + 2(A^T b - cd)^T x + b^T b - d^2 \leq 0, \quad c^T x + d \geq 0,$$

which follows by squaring both sides of (5.15). Thus, to express (5.14) as (5.15), we need to find A, b, c, and d such that

$$P = A^T A - c c^T \geq 0, \quad q = 2(A^T b - cd), \quad s = b^T b - d^2.$$

We start by considering the case where $q \in \mathcal{R}(P)$. This implies that there exists a full-rank matrix A such that $P = A^T A$ and $q \in \mathcal{R}(A^T)$, and hence, we can choose $c = 0$ and solve for b and d, i.e.

$$b = \frac{1}{2}(A^\dagger)^T q, \quad d = |(1/4)b^T b - s|^{1/2}.$$

Now, suppose $q \neq \mathcal{R}(P)$ and let $P = B^T B$, where $B \in \mathbb{R}^{r \times n}$ has rank $r < n$. We may then decompose q as $q = \bar{q} + c$, where $\bar{q} = B^\dagger B q \in \mathcal{R}(B^T)$ and $c = (I - B^\dagger B)q \in \mathcal{N}(B)$, and hence, we have that $P = A^T A - c c^T \geq 0$ if we let $A^T = \begin{bmatrix} B^T & c \end{bmatrix}$. Moreover, $q = 2(A^T b - cd)$ is satisfied if we take

$$b = \frac{1}{2} \begin{bmatrix} (B^\dagger)^T \bar{q} \\ 1 + 2d \end{bmatrix},$$

and finally, we find d by solving

$$s = b^T b - d^2 \quad \Longleftrightarrow \quad d = s - \frac{1}{4}(\|(B^\dagger)^T \bar{q}\|_2^2 - 1).$$

5.3.1 Conic Duality

Lagrangian duality can be extended to cone LPs of the form (5.12). We will define the Lagrangian $L : \mathbb{R}^n \times \mathbb{R}^n \times \mathbb{R}^p \to \mathbb{R}$ associated with (5.12) as

$$L(x, \lambda, \mu) = c^T x - \lambda^T x + \mu^T(Ax - b).$$

Here $\mu \in \mathbb{R}^p$ is the Lagrange multiplier associated with the equality constraints $Ax = b$, and $\lambda \in \mathbb{R}^n$ is the Lagrange multiplier associated with the conic constraint $x \in K$, which can also be expressed as $-x \preceq_K 0$. Using the definition of the dual cone, i.e.

$$K^* = \{y \in \mathbb{R}^n \mid y^T x \geq 0, \ \forall \, x \in K\},$$

we immediately see that $-\lambda^T x \leq 0$ for all $x \in K$ if and only if $\lambda \in K^*$. As a consequence, we have the lower bound property $L(x, \lambda, \mu) \leq c^T x$ if x is feasible and $\lambda \in K^*$. The Lagrange dual function $g : \mathbb{R}^n \times \mathbb{R}^p \to \bar{\mathbb{R}}$ is given by

$$g(\lambda, \mu) = \inf_x \ L(x, \lambda, \mu),$$

and since L is an affine function of x, we find that

$$g(\lambda, \mu) = \begin{cases} -b^T \mu, & c - \lambda - A^T \mu = 0, \\ -\infty, & \text{otherwise.} \end{cases}$$

This leads to the following Lagrange dual problem:

$$\begin{aligned}
&\text{maximize} && -b^T \mu \\
&\text{subject to} && A^T \mu + \lambda = c \\
& && \lambda \in K^*,
\end{aligned} \tag{5.16}$$

with variables $\lambda \in \mathbb{R}^n$ and $\mu \in \mathbb{R}^p$. Equivalently, we may replace the two constraints by $c - A^T \lambda \in K^*$ if we eliminate λ. We note that (5.12) and (5.16) are each other's dual, and hence, either of the two problems can be designated as the primal problem. For example, in some applications, it is more natural to specify an optimization problem on the form (5.16), which is then referred to as the primal problem.

Following the approach in Section 4.7, we may derive the KKT conditions for optimality, which can be expressed as

$$Ax = b, \; x \in K,$$
$$A^T \mu + \lambda = c, \; \lambda \in K^*,$$
$$\mathrm{diag}(\lambda)x = 0.$$

These conditions are both necessary and sufficient for optimality if either the primal problem or the dual problem is strictly feasible, i.e. if there exists a vector $x \in \mathrm{int}\, K$ such that $Ax = b$, or if there exists a dual point (λ, μ) such that $\lambda \in \mathrm{int}\, K^*$ and $A^T \mu + \lambda = c$.

5.3.2 Epigraphical Cones

The cone K in the cone LP (5.12) can, in principle, be any proper convex cone in $\mathbb{R}^n$, and many such cones can be expressed as the epigraph of some function. The epigraph of a function f is a cone if $(u, t) \in \mathrm{epi}\, f \implies (\alpha u, \alpha t) \in \mathrm{epi}\, f$ for all $\alpha \geq 0$. Using the definition of the epigraph of $f : \mathbb{R}^{n-1} \to \bar{\mathbb{R}}$, i.e.

$$\mathrm{epi}\, f = \{(x, t) \in \mathbb{R}^{n-1} \times \mathbb{R} \mid f(x) \leq t\}, \tag{5.17}$$

we see that f must satisfy $\alpha f(x) = f(\alpha x)$ for all $\alpha \geq 0$ in order for $\mathrm{epi}\, f$ to be a cone. Such a function is said to be *positively homogeneous*. Thus, $K = \mathrm{epi}\, f$ is clearly a convex cone if f is positively homogeneous and convex, and we will call such a cone an *epigraphical cone*. An important class of such cones are norm cones, i.e. if $f(x) = \|x\|$ is a norm on $\mathbb{R}^{n-1}$, then $\mathrm{epi}\, f$ is a norm cone in $\mathbb{R}^n$.

We will now consider a class of epigraphical cones that are generated by the perspective of a proper convex function $h : \mathbb{R}^{n-2} \to (-\infty, +\infty]$. Recall that the perspective of h is the function $P_h : \mathbb{R}^{n-2} \times \mathbb{R} \to (-\infty, +\infty]$ defined as

$$P_h(u, s) = \begin{cases} sh(u/s), & s > 0, \\ \infty, & \text{otherwise.} \end{cases}$$

The set $\{0\} \cup \mathrm{epi}\, P_h$ is then a convex cone, as shown in Section 4.3. Taking the closure of $\mathrm{epi}\, P_h$, we obtained a closed convex cone

$$K = \mathrm{cl}(\mathrm{epi}\, P_h). \tag{5.18}$$

It can be shown that this cone is pointed if $\mathrm{epi}\, h$ does not contain a line. Moreover, it follows from the definition that a constraint of the form $h(x) \leq t$ can be expressed $(x, 1, t) \in K$.

An example of a cone of the form (5.18) is the *exponential cone*, which is obtained by taking $h(x) = \exp(x)$, i.e.

$$K_{\exp} = \mathrm{cl}\{(x, y, z) \in \mathbb{R}^3 \mid y > 0, \; y\exp(x/y) \leq z\}$$
$$= \{(x, y, z) \in \mathbb{R}^3 \mid y > 0, \; y\exp(x/y) \leq z\} \cup \{(x, 0, z) \in \mathbb{R}^3 \mid, x \leq 0, \; z \geq 0\}. \tag{5.19}$$

It is easy to verify that this is a proper convex cone, i.e. it is closed, pointed, and has a nonempty interior. We note that the dual cone can be expressed as

$$K_{\exp}^* = \mathrm{cl}\{(u, v, w) \in \mathbb{R}^3 \mid u < 0, \; -u\exp(v/u - 1) \leq w\}, \tag{5.20}$$

which shows that $K_{\exp}$ is not a self-dual cone. Equivalently, the dual cone $K_{\exp}^*$ can also be expressed as

$$K_{\exp}^* = \{(u, v, w) \in \mathbb{R}^3 \mid -u\ln(-u/w) \leq v - u, \; u \leq 0, \; w \geq 0\},$$

if we define $0\ln(-0/w) = 0$ for $w \geq 0$. This is sometimes referred to as the *relative entropy cone*.

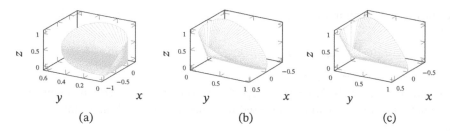

Figure 5.2 The exponential cone $K_{\exp}$ and examples of the power cone K_α: (a) exponential cone, (b) power cone: $\alpha = 2/3$, and (c) power cone: $\alpha = 4/5$.

Yet another example of a cone of the form (5.18) is a so-called *power cone*. Letting $h(x) = |x|^{1/\alpha}$ for some $\alpha \in (0,1)$, we obtain the three-dimensional cone

$$K_\alpha = \mathrm{cl}\{(x,y,z) \in \mathbb{R}^3 \mid y > 0,\ y|x/y|^{1/\alpha} \le z\}$$
$$= \{(x,y,z) \in \mathbb{R}^3 \mid y \ge 0,\ z \ge 0,\ |x| \le y^{1-\alpha}z^\alpha\}. \tag{5.21}$$

The reader may verify that this is a proper cone. The dual cone may be expressed as

$$K_\alpha^* = \left\{ (u,v,w) \in \mathbb{R}^3 \mid v \ge 0,\ w \ge 0,\ |u| \le \left(\frac{v}{1-\alpha}\right)^{1-\alpha}\left(\frac{w}{\alpha}\right)^\alpha \right\}, \tag{5.22}$$

which is the image of K_α under the linear transformation $(x,y,z) \mapsto (x,(1-\alpha)y,\alpha z)$. It is also possible to define a power cone in $\mathbb{R}^{n+1}$. Specifically, given $\alpha \in \mathrm{int}\,\Delta^n$, we define the $(n+1)$-dimensional power cone as

$$K_\alpha = \{(x,y) \in \mathbb{R} \times \mathbb{R}_+^n \mid |x| \le y_1^{\alpha_1} \cdots y_n^{\alpha_n}\},$$

and its dual may be expressed as

$$K_\alpha^* = \left\{(u,v) \in \mathbb{R} \times \mathbb{R}_+^n \mid |u| \le (v_1/\alpha_1)_1^\alpha \cdots (v_n/\alpha_n)_n^\alpha\right\}.$$

Figure 5.2 shows the exponential cone and examples of the power cone in $\mathbb{R}^3$.

We end this section with some examples of constraints that can be expressed as conic constraints.

Example 5.4 From the definition of the exponential cone $K_{\exp}$ and the power cone K_α, we immediately see that

$$e^x \le t \iff x \le \ln(t) \iff (x,1,t) \in K_{\exp},$$

and for $p \ge 1$,

$$|x|^p \le t \iff |x| \le t^{1/p} \iff (x,1,t) \in K_{1/p}.$$

This observation allows us to transform a number of constraints that involve exponential functions, logarithms, and/or powers into conic constraints. For example, the epigraph of the log-sup-exp function $f(x) = \ln(e^{x_1} + \cdots + e^{x_n})$ can be expressed as (x,t) such that

$$\ln(e^{x_1} + \cdots + e^{x_n}) \le t \iff e^{x_1} + \cdots + e^{x_n} \le e^t$$
$$\iff e^{x_1-t} + \cdots + e^{x_n-t} \le 1$$
$$\iff u_1 + \cdots + u_n \le 1,\ e^{x_i-t} \le u_i,\ i \in \mathbb{N}_n$$
$$\iff \mathbf{1}^T u \le 1,\ (x_i - t, 1, u_i) \in K_{\exp},\ i \in \mathbb{N}_n.$$

Another example of a function whose epigraph can be expressed in terms of the exponential cone is the negative entropy function, i.e. $f(x) = x\ln(x)$ with $\mathrm{dom}\,f = \mathbb{R}_+$, where we use the convention that $f(0) = 0$. Specifically, the constraint $f(x) \leq t$ can be expressed as $(-t, x, 1) \in K_{\exp}$. Indeed, for $x > 0$, it holds that

$$f(x) \leq t \quad \Longleftrightarrow \quad x \leq e^{t/x} \quad \Longleftrightarrow \quad (-t, x, 1) \in K_{\exp},$$

and for $x = 0$, we have that $0 \leq t \iff (-t, 0, 1) \in K_{\exp}$.

The geometric mean of $x_1, \ldots, x_n$ is $f(x) = (x_1 \cdots x_n)^{1/n}$ with $\text{dom} f = \mathbb{R}_{++}^n$. This is a concave function, and the constraint $|t| \leq f(x)$ can be expressed in terms the $(n+1)$-dimensional power cone defined by $\alpha = (1/n)\mathbb{1}$, i.e.

$$|t| \leq (x_1 x_2 \cdots x_n)^{1/n} \quad \Longleftrightarrow \quad (t, x) \in K_\alpha, \ \alpha = (1/n)\mathbb{1}.$$

It can also be expressed in terms of $n - 1$ power cones in $\mathbb{R}^3$. To see this, first note that $|t| \leq f(x)$ can be expressed as

$$|t| \leq x_n^{1/n} u_{n-1}^{(n-1)/n}, \quad |u_{n-1}| \leq (x_1 \cdots x_{n-1})^{1/(n-1)}, \quad x \geq 0, \quad u_{n-1} \geq 0.$$

Using this observation recursively, we see that the constraint $|t| \leq f(x)$ is equivalent to

$$t = u_n, \ x_1 = u_1, \ (u_i, u_{i-1}, x_i) \in K_{1/i}, \ i \in \{2, \ldots, n\}.$$

Example 5.5 Let $(\Omega, \mathcal{F}, P)$ be a probability space, and suppose that $X : \Omega \to \mathbb{R}$ is a random variable with moments $m_k = \mathbb{E}\left[X^k\right]$, $k \in \mathbb{N}_{2n}$, cf. Section 3.6. We define the zeroth moment as $m_0 = 1$ for convenience, and we let $H : \mathbb{R}^{2n+1} \to \mathbb{S}^{n+1}$ be the function that maps the moment sequence to the Hankel matrix

$$H(m_0, \ldots, m_{2n}) = \begin{bmatrix} m_0 & m_1 & m_2 & \cdots & m_{n-1} & m_n \\ m_1 & m_2 & m_3 & \cdots & m_n & m_{n+1} \\ m_2 & m_3 & m_4 & \cdots & m_{n+1} & m_{n+2} \\ \vdots & \vdots & \vdots & \ddots & \vdots & \vdots \\ m_{n-1} & m_n & m_{n+1} & \cdots & m_{2n-2} & m_{2n-1} \\ m_n & m_{n+1} & m_{n+2} & \cdots & m_{2n-1} & m_{2n} \end{bmatrix}.$$

The matrix $H(m_0, \ldots, m_{2n})$ is positive semidefinite since for all $x \in \mathbb{R}^{n+1}$,

$$x^T H(m_0, \ldots, m_{2n})x = \sum_{i=0}^n \sum_{j=0}^n x_i x_j \mathbb{E}[X^{i+j}] = \mathbb{E}\left[\sum_{k=0}^n x_k X^k\right]^2 \geq 0.$$

There is a partial converse of these results which states that if $m_0 = 1$ and $(m_1, \ldots, m_{2n})$ are such that $H(m_0, \ldots, m_{2n}) \succ 0$, then there exists a probability space and a random variable X such that $m_k = \mathbb{E}[X^k]$ for $k \in \mathbb{N}_{2n}$; see [22, Section 4.6.3 and Exercise 2.37]. More generally, if $m_0 = 1$ and $(m_1, \ldots, m_{2n})$ are such that $H(m_0, \ldots, m_{2n}) \succeq 0$, then there is a sequence of random variables that converges to a random variable with the given moments. This allows us to pose certain moment constraints as conic constraints involving the positive semidefinite cone. For example, suppose that we are given upper and lower bounds on the moments of a random variable X, i.e. we know that

$$l_k \leq \mathbb{E}[X^k] \leq u_k, \ k \in \mathbb{N}_{2n}.$$

The problem of finding a random variable X that satisfies these constraints and minimizes $\mathbb{E}[p(X)]$ where $p : \mathbb{R} \to \mathbb{R}$ is a polynomial defined as $p(x) = \sum_{k=0}^{2n} c_k x^k$ can then be expressed as the problem

$$\begin{aligned} \text{minimize} \quad & \sum_{k=0}^{2n} c_k m_k \\ \text{subject to} \quad & l_k \leq m_k \leq u_k, \ k \in \mathbb{N}_{2n} \\ & H(1, m_1, \ldots, m_{2n}) \in \mathbb{S}_+^{n+1}, \end{aligned}$$

with variables $m_1, \ldots, m_{2n}$. Note that the pdf of the random variable does not appear in this problem formulation.

5.4 Rank Optimization

We now turn our attention to optimization problems that involve the rank of a matrix as part of the objective function or one or more constraints. Such problems are generally difficult to solve since the rank function is both discontinuous and nonconvex. However, there are a number of important special cases where an analytic solution exists. One such special case is the so-called "matrix nearness problem"

$$
\begin{aligned}
&\text{minimize} \quad \|A - Z\| \\
&\text{subject to} \quad \text{rank}(Z) \leq k,
\end{aligned}
\tag{5.23}
$$

where $Z \in \mathbb{R}^{m \times n}$ is the variable, and $A \in \mathbb{R}^{m \times n}$ is given, and the norm is either the Frobenius norm or the 2-norm. It is easy to verify that $Z^\star = A$ is the unique solution if $k \geq \text{rank}(A)$. More generally, a solution to the problem can be obtained by means of a *truncated* singular value decomposition (SVD) of A. If we let $A = U\Sigma V^T$ be an SVD of A such that Σ is the matrix with the singular values $\sigma_1 \geq \cdots \geq \sigma_{\min(m,n)} \geq 0$ on its diagonal and

$$
A = \sum_{i=1}^{\min(m,n)} \sigma_i u_i v_i^T,
$$

then the matrix

$$
A_k = \sum_{i=1}^{k} \sigma_i u_i v_i^T,
$$

is a minimizer of (5.23) for both the Frobenius norm and the 2-norm. This follows from the *Eckart–Young–Mirsky theorem*, which states that if $k < \min(m, n)$, then

$$
\|A - A_k\|_F = \min_Z \{\|A - Z\|_F \mid \text{rank}(X) \leq k\} = \sqrt{\left(\sum_{i=k+1}^{\min(m,n)} \sigma_i^2\right)}
\tag{5.24a}
$$

$$
\|A - A_k\|_2 = \min_Z \{\|A - Z\|_2 \mid \text{rank}(Z) \leq k\} = \sigma_{k+1}.
\tag{5.24b}
$$

To prove (5.24a), suppose $\text{rank}(Z) \leq k$ and let $Z = P\Gamma Q^T$ be an SVD of a matrix Z, where Γ is the matrix with the singular values of Z on its diagonal and in decreasing order. Moreover, we decompose Σ as $\Sigma = \Sigma_k + \tilde{\Sigma}$, where Σ_k is the matrix with the singular values of A_k. It then follows from von Neumann's trace inequality in (2.33) that $|\text{tr}(A^T Z)| \leq \text{tr}(\Sigma_k^T \Gamma)$. As a result, we have that

$$
\begin{aligned}
\|A - X\|_F^2 &= \|A\|_F^2 + \|Z\|_F^2 - 2\text{tr}(A^T Z) \\
&\geq \|\Sigma\|_F^2 + \|\Gamma\|_F^2 - 2\text{tr}(\Sigma_k^T \Gamma) \\
&= \|\Sigma_k\|_F^2 + \|\tilde{\Sigma}\|_F^2 + \|\Gamma\|_F^2 - 2\text{tr}(\Sigma_k^T \Gamma) \\
&= \|\tilde{\Sigma}\|_F^2 + \|\Sigma_k - \Gamma\|_F^2 \\
&\geq \|\tilde{\Sigma}\|_F^2 = \sum_{i=k+1}^{\min(m,n)} \sigma_i^2
\end{aligned}
$$

which proves (5.24a). To verify (5.24b), we first assume that (5.24b) is false, i.e. there exists a matrix Z such that $\|A - Z\|_2 < \sigma_{k+1}$ and $\text{rank}(Z) \leq k$. An immediate consequence is that

$$
u \in \mathcal{N}(Z) \quad \Longrightarrow \quad \|Au\|_2 = \|(A - Z)u\|_2 \leq \|A - Z\|_2 \|u\| < \sigma_{k+1} \|u\|_2.
$$

On the other hand, if $v_1, \ldots, v_{k+1}$ are the leading $k+1$ right-singular vectors of A and $v \in \text{span}(v_1, \cdots, v_{k+1})$, then $\|Av\|_2 \geq \sigma_{k+1} \|v\|_2$. Noting that $\text{rank}(Z) \leq k \iff \dim \mathcal{N}(Z) \geq n - k$, we see that $\mathcal{N}(A) \cap \text{span}(v_1, \cdots, v_{k+1}) \neq \{0\}$, and hence, we have a contradiction.

The problem (5.23) is closely related to the rank-minimization problem

$$\text{minimize } \operatorname{rank}(Z)$$
$$\text{subject to } \|A - Z\| \leq \delta, \tag{5.25}$$

where the norm is either the Frobenius norm or the 2-norm, $Z \in \mathbb{R}^{m \times n}$ is the variable, and $A \in \mathbb{R}^{m \times n}$ and $\delta \in \mathbb{R}_+$ are given. Again, the optimal value and a minimizer can be expressed in terms of an SVD of A. For the Frobenius norm, the minimum rank is

$$r_{\mathrm{F}}(\delta) = \min \left\{ k \in \mathbb{Z}_+ \mid \sum_{i=k+1}^{\min(m,n)} \sigma_i^2 \leq \delta^2 \right\},$$

and for the 2-norm, it is

$$r_2(\delta) = \min \{k \in \mathbb{Z}_+ \mid \sigma_{k+1} \leq \delta\},$$

where we define $\sigma_{\min(m,n)+1} = 0$ for convenience. The optimal value $p^\star$ is equal to the minimum rank, and $A_k = \sum_{i=1}^k \sigma_i u_i v_i^T$ is a minimizer if $k = p^\star$. The connection between the rank-constrained problem (5.23) and the rank minimization problem (5.25) is illustrated in Figure 5.3b.

Next, we consider a more general rank minimization problem of the form

$$\text{minimize } \operatorname{rank}(Z)$$
$$\text{subject to } Z \in C, \tag{5.26}$$

with variable $Z \in \mathbb{R}^{m \times n}$ and where $C \subset \mathbb{R}^{m \times n}$. With few exceptions, such problems cannot be solved analytically by means of an SVD. However, several heuristics exist that give a "good solution" to the optimization problem but is not necessarily optimal; see, e.g. [37]. One such heuristic is the so-called nuclear norm heuristic, which uses the nuclear norm as a proxy for the rank function, i.e. we replace (5.26) by

$$\text{minimize } \|Z\|_*$$
$$\text{subject to } Z \in C, \tag{5.27}$$

which is a convex optimization problem if C is a convex set. Note that $\operatorname{rank}(Z)$ is equal to the number of nonzero singular values of Z, whereas $\|Z\|_* = \sum_{i=1}^{\min(m,n)} \sigma_i(Z)$ is the sum of the singular values. Moreover, it can be shown that $\|Z\|_*$ is the lower *convex envelope* of $\operatorname{rank}(Z)$ on the norm ball $\{Z \in \mathbb{R}^{m \times n} \mid \|Z\|_2 \leq 1\}$. In other words, $\|Z\|_*$ is the largest convex function such that $\|Z\|_* \leq \operatorname{rank}(Z)$ for all Z in the aforementioned norm ball. Using the fact that

$$\|Z\|_* \leq t \quad \Longleftrightarrow \quad \operatorname{tr}(X) + \operatorname{tr}(Y) \leq 2t, \quad \begin{bmatrix} X & Z \\ Z^T & Y \end{bmatrix} \succeq 0,$$

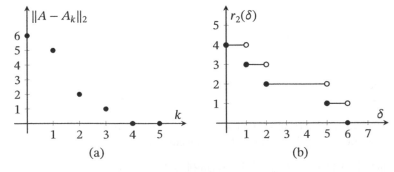

(a)   (b)

Figure 5.3 The optimal value associated with the rank-constrained problem (5.23) and the rank-minimization problem (5.25) for the 2-norm and a matrix $A \in \mathbb{R}^{10 \times 5}$ with the singular values $(6, 5, 2, 1, 0)$. (a) Rank-constrained minimization and (b) rank minimization.

cf. Exercise 5.6, we can reformulate (5.27) as

$$\text{minimize} \quad (1/2)(\text{tr}(X) + \text{tr}(Y))$$
$$\text{subject to} \quad \begin{bmatrix} X & Z \\ Z^T & Y \end{bmatrix} \succeq 0 \tag{5.28}$$
$$Z \in C,$$

with variables $X \in \mathbb{S}^m$, $Y \in \mathbb{S}^n$, and $Z \in \mathbb{R}^{m \times n}$.

Another heuristic for the problem (5.26) is the so-called "log-det heuristic" [37]. This is based on the fact that if $Z \in \mathbb{S}_+^m$ is a low-rank matrix, then $\ln \det(Z + \delta I) \ll 0$ if $\delta > 0$ is sufficiently small. This motivates the problem formulation

$$\text{minimize} \quad \ln \det(Z + \delta I)$$
$$\text{subject to} \quad Z \in C, \tag{5.29}$$

as a heuristic for solving (5.26) if $C \subset \mathbb{S}_+^m$. Note that $\ln \det(Z + \delta I)$ is a concave function of Z, and hence, it is still a difficult optimization problem in general. However, local optimization methods can be used in conjunction with a good initial guess as a heuristic solution method. The log-det heuristic can also be used in the case where $Z \in \mathbb{R}^{m \times n}$ is a rectangular matrix by embedding Z in a symmetric matrix, i.e.

$$\text{minimize} \quad \ln \det(\text{bdiag}(X, Y) + \delta I)$$
$$\text{subject to} \quad \begin{bmatrix} X & Z \\ Z^T & Y \end{bmatrix} \succeq 0 \tag{5.30}$$
$$Z \in C,$$

where $X \in \mathbb{S}^m$ and $Y \in \mathbb{S}^n$. The motivation for this problem formulation comes from the fact that $\text{rank}(Z) \leq r$ if and only if there exists matrices $X \in \mathbb{S}^m$ and $Y \in \mathbb{S}^n$ such that $\text{rank}(X) + \text{rank}(Y) \leq 2r$ and

$$\begin{bmatrix} X & Z \\ Z^T & Y \end{bmatrix} \succeq 0,$$

cf. Exercise 2.11.

Example 5.6 Given a partial impulse response $h \in \mathbb{R}^n$, the matrices $(A, B, C) \in \mathbb{R}^{r \times r} \times \mathbb{R}^{r \times 1} \times \mathbb{R}^{1 \times r}$ are said to be a minimal state-space realization of h if r is the smallest possible natural number such that $h_k = CA^{k-1}B$, $k \in \{1, \dots, n\}$. A well-known property of linear systems is that r is given by

$$r = \min_{\tilde{h}} \{\text{rank} H(\tilde{h}) \mid \tilde{h}_i = h_i, \ i \in \mathbb{N}_n\},$$

where $H : \mathbb{R}^{2n-1} \to \mathbb{S}^n$ is the Hankel matrix defined as

$$H(\tilde{h}) = \begin{bmatrix} \tilde{h}_1 & \tilde{h}_2 & \tilde{h}_3 & \cdots & \tilde{h}_{n-1} & \tilde{h}_n \\ \tilde{h}_2 & \tilde{h}_3 & \tilde{h}_4 & \cdots & \tilde{h}_n & \tilde{h}_{n+1} \\ \tilde{h}_3 & \tilde{h}_4 & \tilde{h}_5 & \cdots & \tilde{h}_{n+1} & \tilde{h}_{n+2} \\ \vdots & \vdots & \vdots & \ddots & \vdots & \vdots \\ \tilde{h}_{n-1} & \tilde{h}_n & \tilde{h}_{n+1} & \cdots & \tilde{h}_{2n-3} & \tilde{h}_{2n-2} \\ \tilde{h}_n & \tilde{h}_{n+1} & \tilde{h}_{n+2} & \cdots & \tilde{h}_{2n-2} & \tilde{h}_{2n-1} \end{bmatrix}.$$

We note that the problem of obtaining a minimal state-space realization (A, B, C) from a rank-r matrix $H(\tilde{h})$ is the topic of Exercise 2.13.

We end this section by briefly considering a more general formulation of the rank-constrained problem (5.23), namely, the problem

$$\text{minimize} \quad f(Z)$$
$$\text{subject to} \quad \text{rank}(Z) \leq k, \tag{5.31}$$

with variable $Z \in \mathbb{R}^{m \times n}$, and where $f : \mathbb{R}^{m \times n} \to \bar{\mathbb{R}}$. This problem can be reformulated by introducing two new variables $U \in \mathbb{R}^{m \times k}$ and $V \in \mathbb{R}^{n \times k}$ and eliminating the rank constraint by substituting UV^T for Z. However, the reformulated problem is generally still a nonconvex one, but it is often a convenient form for the use of local optimization methods.

5.5 Partially Separability

A function $f : \mathbb{R}^n \to \bar{\mathbb{R}}$ is called *partially separable* if it can be expressed in terms of some elementary functions $f_k : \mathbb{R}^{n_k} \to \bar{\mathbb{R}}, k \in \mathbb{N}_m$, as

$$f(x) = \sum_{k=1}^{m} f_k(A_k x) \tag{5.32}$$

where $A_k \in \mathbb{R}^{n_k \times n}, k \in \mathbb{N}_m$, are matrices with a high-dimensional nullspace. The notion of partial separability can also be extended to functions with more general domains, e.g. a discrete set. Partial separability arises frequently in both estimation and control because of the fact that descriptions of dynamical systems often involve difference equations that only introduce coupling between states that are adjacent in time.

We will focus on the special case where the range of A_k^T is spanned by a small number of standard basis vectors. This implies that each function $f_k(A_k x)$ only depends on a small subset of the entries of x. The partial separability structure can be expressed in terms of index sets $\beta_k \subset \mathbb{N}_n, k \in \mathbb{N}_m$, such that f_k depends on x_i if and only if $i \in \beta_k$. We will assume that the index sets are maximal, i.e. $\beta_i \not\subseteq \beta_j$ if $i \neq j$, and their union is $\cup_{k=1}^{m} \beta_k = \mathbb{N}_n$. Moreover, we define A_k to be a matrix with rows $\{e_i^T \mid i \in \beta_k\}$. We note that f is said to be *separable* in the special case, where $\beta_i \cap \beta_j = \emptyset$ for $i \neq j$. For example, this is the case if f is of the form $f(x) = \sum_{i=1}^{n} f_i(x_i)$.

The partial separability structure of f can also be characterized in terms of an undirected graph with vertex set $\mathbb{N}_n$ and an edge between nodes i and $j, i \neq j$, if and only if $\{i, j\} \in \beta_k$ for some $k \in \mathbb{N}_m$. We will refer to this graph as the *interaction graph* associated with (5.32). The interaction graph is closely related to the sparsity graph associated with the Hessian matrix $\nabla^2 f(x)$ provided that f is twice continuously differentiable at x. This follows by noting that

$$\nabla^2 f(x) = \sum_{k=1}^{m} A_k^T \nabla^2 f_k(A_k x) A_k,$$

and hence, $e_i^T \nabla^2 f(x) e_j = 0$ if $A_k e_i = 0$ or $A_k e_j = 0$ for all $k \in \mathbb{N}_m$. Each index set β_k is a so-called *clique* of the interaction graph, i.e. a set of pairwise adjacent vertices. The coupling between the functions $f_1, \dots, f_m$ can be described in terms of a *clique intersection graph*, which is an undirected graph with vertex set $\{\beta_1, \dots, \beta_m\}$ and an edge between vertices β_i and β_j if and only if $i \neq j$ and $\beta_i \cap \beta_j \neq \emptyset$.

5.5.1 Minimization of Partially Separable Functions

We will now look at how partial separability can be exploited in the context of optimization. To illustrate the basic principles, we start by considering a function $f : \mathbb{R}^5 \to \mathbb{R}$ defined as

$$f(x) = f_1(x_1) + f_2(x_1, x_2) + f_3(x_2, x_3) + f_4(x_1, x_4, x_5), \tag{5.33}$$

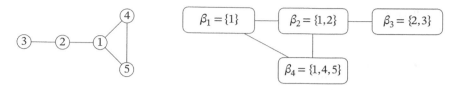

Figure 5.4 The interaction and clique intersection graphs associated with the partially separable function (5.33).

corresponding to the four index sets $\beta_1 = \{1\}$, $\beta_2 = \{1, 2\}$, $\beta_3 = \{2, 3\}$ and $\beta_4 = \{1, 4, 5\}$. Figure 5.4 shows the interaction and the clique intersection graphs associated with (5.33). Partial separability allows us to compute $\inf_x f(x)$ recursively by noting that

$$\inf_x f(x) = \inf_{x_1} \left\{ f_1(x_1) + \inf_{x_2} \left\{ f_2(x_1, x_2) + \inf_{x_3} f_3(x_2, x_3) \right\} + \inf_{x_4, x_5} f_4(x_1, x_4, x_5) \right\}.$$

In other words, we can start by computing

$$V_4(x_1) = \inf_{x_4, x_5} f_4(x_1, x_4, x_5),$$

and if we define $V_3 : \mathbb{R} \to \mathbb{R}$ in a similar manner, i.e.

$$V_3(x_2) = \inf_{x_3} f_3(x_2, x_3),$$

we may compute V_3 and V_4 in parallel. Once V_3 has been obtained, we can compute

$$V_2(x_1) = \inf_{x_2} \{f_2(x_1, x_2) + V_3(x_2)\},$$

and finally, we obtain $p^\star = \inf_x f(x)$ by computing

$$p^\star = \inf_{x_1} \{f_1(x_1) + V_2(x_1) + V_4(x_1)\}.$$

Assuming that the infimum is attained, we can also compute an optimal point $x^\star$ as follows:

$$x_1^\star \in \operatorname*{argmin}_{x_1} \{f_1(x_1) + V_2(x_1) + V_4(x_1)\}$$

$$x_2^\star = \mu_2(x_1^\star) \in \operatorname*{argmin}_{x_2} \{f_2(x_1^\star, x_2) + V_3(x_2)\}$$

$$x_3^\star = \mu_3(x_2^\star) \in \operatorname*{argmin}_{x_3} \{f_3(x_2^\star, x_3)\}$$

$$(x_4^\star, x_5^\star) = \mu_4(x_1^\star) \in \operatorname*{argmin}_{x_4, x_5} \{f_4(x_1^\star, x_4, x_5)\}.$$

The computations can be represented by a tree with four nodes as illustrated in Figure 5.5. This is a so-called *spanning tree* of the clique intersection graph, and it is often referred to as an *elimination tree* because of its connection to the order in which variables are eliminated in Gauss elimination for sparse linear systems of equations; see, e.g. [106]. The computations start at the leaves of the tree, possibly in parallel, and then proceed up the tree by adding the functions V_k to parent nodes. At each node, partial minimization is carried out with respect to the variables that are not shared with the parent node. Once the optimization problem at the root of the tree has been solved, an optimal point $x^\star$ can be computed using the functions μ_k by passing the optimal value downward from the root of the tree toward the leaves. We note that it is also possible to define any other node

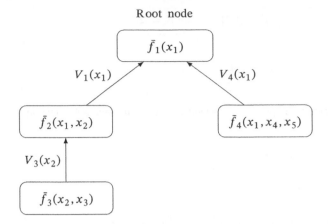

Figure 5.5 A computational tree associated with the optimization problem (5.33).

of the tree to be the root, which results in a different organization of the computations. In this example, computations cannot be carried out in parallel if one of the leaves of the tree in Figure 5.5 is used as the root.

Given a minimization problem with a partially separable objective function, it is generally difficult to find an elimination order that minimizes some notion of computational cost. However, in practice, it is possible to use heuristics like the *nested dissection algorithm* to obtain a good elimination tree; see, e.g. [60]. Optimization over a tree is sometimes called *message passing* since the functions V_k can be thought of as messages from a child node to its parent node, while the functions μ_k are used to pass information about optimizers in the opposite direction.

Another difficulty with the recursive approach to minimizing a partially separable function is that it is generally hard or even impossible to obtain analytical expressions for the functions V_k. A notable exception is when dom f is a finite set, and the values of the functions $f_1, \ldots, f_m$ can be tabulated. Another exception is when f is a quadratic function, in which case, the functions V_k are also quadratic functions. The optimality condition $\nabla f(x) = 0$ corresponds to a system of linear equations, which has a unique solution if f is strongly convex. In this special case, the elimination tree characterizes a partial order in which a sparse Cholesky factorization is computed and the solution is found.

The optimization over the tree is also often called *dynamic programming over trees*. The motivation for this name comes from the fact that dynamical systems with difference equations couple time-adjacent states. In these types of applications, it turns out that the computational tree is often a chain, i.e. it is possible to choose the root of the tree in such a way that no node of the tree has more than one child. This is also the case for the function (5.33), but it is not true in general.

5.5.2 Principle of Optimality

As a special case of optimization problems with a partially separable objective function, we now consider problems of the form

$$\text{minimize} \ \sum_{k=1}^{m-1} f_k(x_k, x_{k+1}) + f_m(x_m). \tag{5.34}$$

The associated clique intersection graph is a chain, which will also serve as our elimination tree. Assuming that the minimum is attained, we have that

$$\min_x \left\{ \sum_{i=1}^{m-1} f_i(x_i, x_{i+1}) + f_m(x_m) \right\} = \min_{x_1, \ldots, x_k} \left\{ \sum_{i=1}^{k-1} f_i(x_i, x_{i+1}) + V_k(x_k) \right\} \tag{5.35}$$

where $V_k : \mathbb{R}^n \to \bar{\mathbb{R}}$ is defined as

$$V_k(x_k) = \min_{x_{k+1},\dots,x_m} \left\{ \sum_{i=k}^{m-1} f_i(x_i, x_{i+1}) + f_m(x_m) \right\}, \quad k \in \mathbb{N}_{m-1},\tag{5.36}$$

and $V_m(x_m) = f_m(x_m)$. This is often called *the principle of optimality*, which refers to the fact that an optimal point $(x_{k+1}^\star, \dots, x_m^\star)$ in (5.36) is a subvector of an optimal point $x^\star$ in (5.34). It is a direct consequence of partially separability and the fact that for any function $F : \mathbb{R}^p \times \mathbb{R}^q \to \bar{\mathbb{R}}$, it holds that

$$\inf_{z_1, z_2} F(z_1, z_2) = \inf_{z_1} V(z_1)$$

where the function $V : \mathbb{R}^p \to \bar{\mathbb{R}}$ is given by $V(z_1) = \inf_{z_2} F(z_1, z_2)$. It now follows from (5.36) that

$$\begin{aligned} V_{k-1}(x_{k-1}) &= \min_{x_k,\dots,x_m} \left\{ \sum_{i=k-1}^{m-1} f_i(x_i, x_{i+1}) + f_m(x_m) \right\} \\ &= \min_{x_k} \left\{ f_{k-1}(x_{k-1}, x_k) + \min_{x_{k+1},\dots,x_m} \left\{ \sum_{i=k}^{m-1} f_i(x_i, x_{i+1}) + f_m(x_m) \right\} \right\} \\ &= \min_{x_k} \left\{ f_{k-1}(x_{k-1}, x_k) + V_k(x_k) \right\}. \end{aligned}$$

This recursive definition of the functions V_k is often called the *dynamic programming recursion*, and the functions V_k are called *value functions*. The recursion starts with $V_m(x_m) = f_m(x_m)$ and proceeds with $V_{m-1}(x_{m-1})$, $V_{m-2}(x_{m-2})$, and so on. We see from (5.36) that $p^\star = \min_{x_1} V_1(x_1)$ and

$$x_1^\star \in \operatorname*{argmin}_{x_1} V_1(x_1).$$

We then define functions $\mu_k : \mathbb{R} \to \mathbb{R}$ as

$$\mu_k(x_{k-1}) = \operatorname*{argmin}_{x_k} \left\{ f_{k-1}(x_{k-1}, x_k) + V_k(x_k) \right\}, \quad k \in \{2, \dots, m\},$$

such that an optimal point $x^\star$ is can be computed using the recursion $x_k^\star = \mu_k(x_{k-1}^\star)$ for $2 \leq k \leq m$. The approach can readily be extended to the more general case, where x_k is a vector instead of a scalar. Moreover, dynamic programming can also be done over general trees, and although this is a straightforward generalization, the proof is somewhat messy from a notational point of view, and hence, we omit it. We will apply the results presented in this section to optimal control problems in Chapter 8. Dynamic programming over trees also has applications to probabilistic graphical models, and we will see how it can be used for maximum likelihood estimation for hidden Markov processes in Chapter 9.

5.6 Multiparametric Optimization

A *multiparametric program* is an optimization problem on the form

$$\begin{aligned} &\text{minimize } f_0(x, \theta) \\ &\text{subject to } f_i(x, \theta) \leq 0, \quad i \in \mathbb{N}_m, \end{aligned}\tag{5.37}$$

where $x \in \mathbb{R}^n$ is the variable, $\theta \in \Theta \subseteq \mathbb{R}^r$ is a vector of parameters, and $f_i : \mathbb{R}^n \times \mathbb{R}^r \to \mathbb{R}$, $i \in \{0, 1, \dots, m\}$. We note that it is simply called a parametric program when $r = 1$. Our goal is to solve (5.37) for all values of θ, and hence, the solution $x^\star$ will be a function of θ. Notice that we already have discussed this type of problem in Section 5.5. Here, we will only consider convex multiparametric programs, i.e. multiparametric programs of the form (5.37), where f_0 and $f_1, \dots, f_m$

are convex in x for every fixed value of θ. We will use $f(x, \theta)$ as shorthand for $(f_1(x, \theta), \dots, f_m(x, \theta))$. The KKT conditions may then be expressed as follows: there exist $x(\theta) \in \mathbb{R}^n$ and $\lambda(\theta) \in \mathbb{R}^m$ such that

$$\nabla_x f_0(x(\theta), \theta) + \sum_{i=1}^{m} \lambda_i(\theta) \nabla_x f_i(x(\theta), \theta) = 0 \tag{5.38a}$$

$$f(x(\theta), \theta) \leq 0 \tag{5.38b}$$

$$\lambda(\theta) \geq 0 \tag{5.38c}$$

$$\lambda_i(\theta) f_i(\theta) = 0, \quad i \in \mathbb{N}_m. \tag{5.38d}$$

Assuming that Slater's condition is satisfied for all $\theta \in \Theta$, the KKT conditions are necessary and sufficient conditions for optimality.

As an example of a multiparametric program, we now consider a multiparametric quadratic program (mpQP), where we let $f_0(x, \theta) = \frac{1}{2} x^T H x$ with $H \in \mathbb{S}_{++}^n$ and $f(x, \theta) = Gx - w - S\theta$ with $G \in \mathbb{R}^{m \times n}$ and $S \in \mathbb{R}^{m \times r}$. The resulting KKT conditions are

$$Hx + G^T \lambda = 0 \tag{5.39a}$$

$$Gx - w - S\theta \leq 0 \tag{5.39b}$$

$$\lambda \geq 0 \tag{5.39c}$$

$$\text{diag}(\lambda)(Gx - w - S\theta) = 0. \tag{5.39d}$$

Now, suppose that $x^\star(\bar\theta)$ and $\lambda^\star(\bar\theta)$ satisfy the KKT conditions for a given parameter vector $\bar\theta \in \Theta$. We will drop the argument $\bar\theta$ when it is obvious from the context and simply write $x^\star$ and $\lambda^\star$. In order to express the KKT conditions in terms of the active and inactive constraints, we introduce the index sets

$$\mathcal{A}(\bar\theta) = \left\{ i \in \mathbb{N}_m \mid f_i(x^\star, \bar\theta) = 0 \right\},$$

$$\mathcal{N}(\bar\theta) = \mathbb{N}_m \backslash \mathcal{A}(\bar\theta),$$

and define $G_\mathcal{A}$ to be the matrix with the rows of G that correspond to the active constraints, whereas $G_\mathcal{N}$ contains the rows that correspond to inactive constraints. We use the same notation for S, w, and $\lambda^\star$. It then follows from (5.39d) that $\lambda_\mathcal{N}^\star = 0$, and from (5.39a) we find that

$$x^\star = -H^{-1} G_\mathcal{A}^T \lambda_\mathcal{A}^\star. \tag{5.40}$$

Moreover, from the definition of active constraints, we see that

$$G_\mathcal{A} x^\star - w_\mathcal{A} - S_\mathcal{A} \bar\theta = 0, \tag{5.41}$$

and combining (5.40) and (5.41), we conclude that

$$\lambda_\mathcal{A}^\star = -\left(G_\mathcal{A} H^{-1} G_\mathcal{A}^T \right)^{-1} (w_\mathcal{A} + S_\mathcal{A} \bar\theta) \tag{5.42a}$$

$$x^\star = H^{-1} G_\mathcal{A}^T \left(G_\mathcal{A} H^{-1} G_\mathcal{A}^T \right)^{-1} (w_\mathcal{A} + S_\mathcal{A} \bar\theta). \tag{5.42b}$$

Here, we have tacitly assumed that $G_\mathcal{A}$ has full row rank. If this is not the case, we may redefine $\mathcal{A}$ by removing some of its elements such that $G_\mathcal{A}$ has full row rank and $G_\mathcal{A}^T$ spans the same subspace. Notice that $x^\star$ and $\lambda_\mathcal{A}^\star$ are not only optimal for $\theta = \bar\theta$; they are optimal for all θ that satisfy the KKT conditions (5.39), i.e. all θ such that

$$GH^{-1} G_\mathcal{A}^T \left(G_\mathcal{A} H^{-1} G_\mathcal{A}^T \right)^{-1} (w_\mathcal{A} + S_\mathcal{A} \theta) - w + S\theta \leq 0$$

$$-\left(G_\mathcal{A} H^{-1} G_\mathcal{A}^T \right)^{-1} (w_\mathcal{A} + S_\mathcal{A} \theta) \geq 0.$$

An immediate consequence is that $x^\star(\theta)$ is an affine function of θ over a polyhedral subset of Θ. Moreover, the optimal value $p^\star(\theta)$ is a quadratic function of θ over the same set. Once this polyhedral subset has been determined, a new parameter vector $\theta \in \Theta$ outside the polyhedral set can be chosen, and this procedure is repeated until the whole set Θ has been explored. The result is a polyhedral decomposition of Θ that defines regions in which $x^\star(\theta)$ is an affine function and $p^\star(\theta)$ is a quadratic function of θ. Thus, on the set Θ, $x^\star(\theta)$ is a piecewise affine function and $x^\star(\theta)$ is a piecewise quadratic function. See Exercise 5.8 for an example.

Example 5.7   As a special case of an mpQP, consider the unconstrained problem

$$\text{minimize}\ \ f_0(x, \theta),$$

with variable $x \in \mathbb{R}^n$ and parameters $\theta \in \mathbb{R}^r$, and where

$$f_0(x, \theta) = \frac{1}{2}\begin{bmatrix} x \\ \theta \end{bmatrix}^T \begin{bmatrix} H_{11} & H_{21}^T \\ H_{21} & H_{22} \end{bmatrix} \begin{bmatrix} x \\ \theta \end{bmatrix} = \frac{1}{2}x^T H_{11} x + \frac{1}{2}\theta^T H_{22}\theta + \theta^T H_{21}^T x, \tag{5.43}$$

with $H_{11} \in \mathbb{S}_{++}^n$. The stationarity condition $\partial f_0/\partial x = 0$ can be expressed as

$$H_{11}x + H_{21}^T \theta = 0,$$

which implies that $x^\star(\theta) = -H_{11}^{-1}H_{21}^T\theta$ and

$$p^\star(\theta) = f(x^\star(\theta), \theta) = \frac{1}{2}\theta^T(H_{22} - H_{21}H_{11}^{-1}H_{21}^T)\theta.$$

Note that $p^\star(\theta)$ is a quadratic function of θ, and $\nabla^2 p^\star(\theta)$ is the Schur complement of H_{11} in the block matrix in (5.43).

5.7 Stochastic Optimization

A *stochastic optimization problem* is an optimization problem in which the objective function and/or one or more constraints involve some element of randomness. To illustrate the basic idea, we consider a probability space $(\Omega, \mathcal{F}, P)$ and define a random variable ξ from the sample space Ω to some set $\mathcal{D}$. The random variable could be discrete or continuous. We let $\mathbb{E}[\cdot]$ denote expectation with respect to the probability distribution associated with the random variable ξ. The general so-called *single-stage* stochastic optimization problem is a problem of the form

$$\text{minimize}\ \ \mathbb{E}[F(x, \xi)], \tag{5.44}$$

with $F : \mathbb{R}^n \times \mathcal{D} \to \bar{\mathbb{R}}$ and variable $x \in \mathbb{R}^n$.

The problem (5.44) can be reduced to a deterministic problem if we let $f : \mathbb{R}^n \to \bar{\mathbb{R}}$ be defined as $f(x) = \mathbb{E}[F(x, \xi)]$. However, the probability distribution associated with the random variable ξ is required to compute the expectation. This distribution is typically not available in practice, and even if it is known, the problem of computing the expectation may be intractable.

In statistical learning, ξ typically represents the problem data, i.e. an outcome of ξ corresponds to a random observation. One approach to avoid computing the expectation in (5.44) is to replace the expectation by a *sample average approximation* (SAA), i.e.

$$\mathbb{E}[F(x, \xi)] \approx \frac{1}{m}\sum_{i=1}^{m} F(x, a_i) \tag{5.45}$$

where $a_1, \ldots, a_m$ are m independent observations of the random variable ξ.[1] This is motivated by the close connection between expected values and averages as discussed in Section 3.6. The resulting problem is deterministic and of the form

$$\text{minimize} \ \frac{1}{m} \sum_{i=1}^{m} f_i(x), \tag{5.46}$$

where $f_i : \mathbb{R}^n \to \bar{\mathbb{R}}$ are defined as $f_i(x) = F(x, a_i)$. Problems of this form arise naturally in applications where a finite set of training examples is available. An example is *empirical risk minimization* where the functions $f_1, \ldots, f_m$ often take the form

$$f_i(x) = l(a_i^T x, b_i)$$

where $l : \mathbb{R}^2 \to \mathbb{R}$ is a loss function and $(a_i, b_i) \in \mathbb{R}^n \times \mathbb{R}$ is one of m observations. We will see a number of applications that involve optimization problems of this form in Chapter 10, including linear regression, logistic regression, support vector machines, and artificial neural networks.

Multistage stochastic optimization differs from the single-stage problem in that not all variables are included for optimization at the same time; the optimization is performed in a sequential manner in so-called *stages*, see, e.g. [34]. The partial optimization that is carried out at a stage is called a *decision*. Usually, the decision and random outcomes at the current stage affect the value of future decisions. To illustrate the basic principle, we now consider a random process $\xi : \Omega \to \mathcal{D}^{\mathbb{Z}_N}$, where we assume that the set $\mathcal{D}$ is finite,[2] and we partition ξ as $\xi = (\xi_0, \ldots, \xi_N)$, where ξ_k is a random variable associated with stage k. The decision variable x_k at stage k is actually a function of the random process and defined as $x_k : \mathcal{D}^{\mathbb{Z}_k} \to \mathcal{E}$ with $(\xi_0, \ldots, \xi_k) \mapsto x_k(\xi_0, \ldots, \xi_k)$, and where $\mathcal{E}$ is also a finite set. We let $x = (x_0, \ldots, x_N)$. We now realize that x is also a random process, and the way it is defined it is said to be *adapted* to the random process ξ. We then define $F : \mathcal{E}^{\mathbb{Z}_N} \times \mathcal{D}^{\mathbb{Z}_N} \to \mathbb{R}$ and the optimization problem

$$\text{minimize} \ \mathbb{E}[F(x, \xi)],$$

where $\mathbb{E}[\cdot]$ denotes expectation with respect to the random process ξ. This problem looks similar to the single-stage stochastic optimization problem. However, because of the constraints imposed on the decision variables, it is also possible to state the problem as

$$\mathbb{E}_{\xi_0} \left[\min_{x_0(\xi_0)} \mathbb{E}_{\xi_1} \left[\min_{x_1(\xi_0, \xi_1)} \mathbb{E}_{\xi_2} \left[\cdots \mathbb{E}_{\xi_N} \left[\min_{x_N(\xi_0, \ldots, \xi_N)} F(x, \xi) \right] \right] \right] \right] \tag{5.47}$$

where $\mathbb{E}_{\xi_k}[\cdot]$ denotes *conditional* expectation with respect to the probability distribution for ξ_k given $(\xi_0, \ldots, \xi_{k-1})$ for $k \in \mathbb{N}_N$, and where $\mathbb{E}_{\xi_0}$ is expectation with respect to ξ_0. Here, we have made use of the multiplication theorem discussed in Section 3.2. We realize that we may just as well consider ξ_0 to be known and remove the expectation with respect to ξ_0, and this is often the way the problem is stated. Also, notice that the innermost minimization has to be carried out parametrically with respect to all variables except x_N, and then expectation is taken with respect to ξ_N conditioned on given values of the random variables $(\xi_0, \ldots, \xi_{N-1})$. One technicality is that the resulting random variable after we carry out the parametric optimization might not be measurable in case we consider nonfinite sets, in which case the expectation is not well defined. Another problem can be that the parametric minimum does not exist. However, even for finite sets, the parametric optimization can be very cumbersome to carry out, but we will see that there are applications of multistage stochastic optimization to so-called "Markov decision processes," for which the computational burden is manageable; see Section 8.9

---

1 To be more precise, we let $\xi_1, \ldots, \xi_m$ be independent identically distributed random variables, and we let $a_1, \ldots, a_m$ be observations of each of these random variables, i.e. we repeat the same experiment m times.
2 The reason we restrict ourselves to a finite set is that the minima and random variables we implicitly define below are not necessarily well-defined otherwise.

Exercises

5.1 Consider the optimization problem

$$\text{minimize } \|Ax - b\|,$$

with variable $x \in \mathbb{R}^n$ and problem data $A \in \mathbb{R}^{m \times n}$ and $b \in \mathbb{R}^m$. For each of the following cases, reformulate the problem as an LP to show that it is equivalent to an LP.
(a) The objective function is the infinity norm, i.e. $\|Ax - b\|_\infty$.
(b) The objective function is the 1-norm, i.e. $\|Ax - b\|_1$.

5.2 Consider the following problem from Example 5.5:

$$\text{minimize } \sum_{k=0}^{2n} c_k m_k$$
$$\text{subject to } l_k \le m_k \le u_k, \ k \in \mathbb{N}_{2n}$$
$$H(1, m_1, \dots, m_{2n}) \in \mathbb{S}_+^{n+1},$$

with variables $m_1, \dots, m_{2n}$ and where $H : \mathbb{R}^{2n+1} \to \mathbb{S}^{n+1}$ maps a moment sequence to a Hankel matrix. Suppose that $n = 2$ and that we are interested in the possible values of m_3 given the moment constraints

$$0 \le m_1 \le 1, \qquad 0 \le m_2 \le 1, \qquad 0 \le m_4 \le 1.$$

Investigate this numerically by writing a program that solves the two problems of maximizing and minimizing m_3 subject to these moment constraints and the Hankel matrix constraint $H(1, m_1, m_2, m_3, m_4) \in \mathbb{S}_+^5$.

5.3 Write a MATLAB script that uses YALMIP to solve the realization problem in Example 5.6 using the nuclear norm heuristic in (5.27). You can directly use `norm(H,'nuclear')`, where `H` is the Hankel matrix, as your objective function in YALMIP, in which case you do not need to include any constraints.
(a) Try your code on some problems generated with `drss` using a system order of three. Neglect the direct term, and discard the system if it is close to being uncontrollable or unobservable. You can check this by computing the eigenvalues of the controllability and observability Gramians of the system, i.e., using the commands `eig(dlyap(A,B*B'))` and `eig(dlyap(A',C'*C))`. If the eigenvalues are close to zero, then discard the system.
(b) Next, suppose that you know the first five Markov parameters, which you can easily compute. You can then compute the ϵ-rank of the optimal Hankel matrix by computing its singular values. In this context, a reasonable threshold for considering a singular value to be negligible is $\epsilon = 10^{-4}$. Does the ϵ-rank agree with the true order? Check at least ten examples to get a fair statistic before you draw any conclusions.
(c) Use the result of Exercise 2.13 to obtain matrices (A, B, C). Do these matrices agree with the ones you generated randomly, and if not, why? Do the Markov parameters agree?

5.4 Consider a polynomial $p : \mathbb{R} \to \mathbb{R}$ of even degree defined as $p(x) = a_0 + a_1 x + \cdots + a_{2m} x^{2m}$ where $a_k \in \mathbb{R}, 0 \le k \le 2m$.
(a) Show that $p(x)$ is nonnegative for all $x \in \mathbb{R}$ if and only if there exist two polynomials $q : \mathbb{R} \to \mathbb{R}$ and $r : \mathbb{R} \to \mathbb{R}$ defined as $q(x) = b_0 + b_1 x + \cdots + b_m x^m$ and $r(x) = c_0 + c_1 x + \cdots + c_m x^m$ such that $p(x) = q(x)^2 + r(x)^2$.

(b) Show that $p(x)$ is nonnegative for all $x \in \mathbb{R}$ if and only if there exists a matrix $X \in \mathbb{S}_+^{m+1}$ such that

$$a_k = \sum_{0 \le i,j \le m, i+j=k} X_{i+1,j+1}, \qquad 0 \le k \le 2m.$$

(c) Compute the minimum of the polynomial $p(x) = x^4 + 3x^3 - x^2 + x - 1$.

5.5 Let $X \in \mathbb{R}^{m \times n}$ and $Y \in \mathbb{S}^n$ be given.

(a) Show that $Y = X^T X$ if and only if

$$\text{rank}\left(\begin{bmatrix} I & X \\ X^T & Y \end{bmatrix}\right) = m.$$

(b) Show that

$$\text{rank}\left(\begin{bmatrix} I & X \\ X^T & Y \end{bmatrix}\right) = m \quad \Longrightarrow \quad \begin{bmatrix} I & X \\ X^T & Y \end{bmatrix} \ge 0.$$

5.6 Recall that the nuclear norm of a matrix $Z \in \mathbb{R}^{m \times n}$ can be expressed as

$$\|Z\|_* = \sup_{W \in \mathbb{R}^{m \times n}} \{\text{tr}(W^T Z) \mid \|W\|_2 \le 1\},$$

i.e. it is the dual norm of the matrix 2-norm.

(a) Show that $\|Z\|_*$ is equal to the optimal value of the problem

$$\begin{aligned} \text{maximize} \quad & \text{tr}(W^T Z) \\ \text{subject to} \quad & \begin{bmatrix} I & W^T \\ W & I \end{bmatrix} \ge 0. \end{aligned}$$

(b) Show that $\|Z\|_* \le t$ if and only if there exists matrices $X \in \mathbb{S}^n$ and $Y \in \mathbb{S}^m$ such that

$$\text{tr}(X) + \text{tr}(Y) \le 2t, \qquad \begin{bmatrix} X & Z^T \\ Z & Y \end{bmatrix} \ge 0.$$

Hint: Derive the dual of the optimization problem in the first part of this exercise.

5.7 [14, Exercise 1.16] Suppose that we would like to compute the product of N matrices

$$M_1 M_2 \cdots M_N$$

where $M_k \in \mathbb{R}^{n_k \times n_{k+1}}$, $k \in \mathbb{N}_N$, are given. Recall that matrix multiplication is associative, and hence, there are $(N-1)!$ ways of carrying out the $N-1$ matrix–matrix multiplications. As a simple example, note that the product $M_1 M_2 M_3$ may be computed as either $(M_1 M_2) M_3$ or $M_1 (M_2 M_3)$, where the parentheses indicate the order of operations. The result is always the same in exact arithmetic, but the number of FLOPs is generally different.

Now, suppose that we would like to find a multiplication order that minimizes the total number of FLOPs, which is known as the so-called *matrix chain multiplication problem*. Derive a dynamic programming recursion using the following value functions or messages $V : \mathbb{N}_N \times \mathbb{N}_N \to \mathbb{N}$, where $V(j,k)$ is the minimum number of FLOPs required to compute the product $M_j M_{j+1} \cdots M_k$, where $j \le k$. Apply the resulting dynamic programming recursion to the case, where $N = 3$ and $(n_1, n_2, n_3, n_4) = (2, 10, 5, 1)$.

5.8 Let $A \in \mathbb{R}^{15 \times 3}$, $b \in \mathbb{R}^{15}$, and $E \in \mathbb{R}^{15 \times 2}$, and consider the following multiparametric quadratic program

$$\text{minimize } \|x - 1\|_2^2$$
$$\text{subject to } Ax \leq b + E\theta,$$

with variable $x \in \mathbb{R}^3$ and parameter $\theta \in [-1, 1] \times [-1, 1]$. Solve an instance of this problem using YALMIP and the multiparametric toolbox MPT3. The problem data (A, b, E) may be generated randomly.

5.9 Consider the optimization problem

$$\text{minimize } f_1(x_1) + f_2(x_2) + \cdots + f_m(x_m)$$
$$\text{subject to } a^T x = b,$$

with variable $x \in \mathbb{R}^n$, problem data $a \in \mathbb{R}^n$, and $b \in \mathbb{R}$, and where $f_i : \mathbb{R} \to \mathbb{R}$, $i \in \mathbb{N}_m$.

(a) Suppose that all elements of a are nonzero, i.e. $a_i \neq 0$, $i \in \mathbb{N}_m$. Show that the optimization problem is equivalent to a partially separable one if we introduce the new variables s_i, $i \in \mathbb{N}_{m-1}$, defined as

$$s_1 = a_1 x_1$$
$$s_2 = s_1 + a_2 x_2$$
$$\vdots$$
$$s_{m-1} = s_{m-2} + a_{m-1} x_{m-1}$$
$$b = s_{m-1} + a_m x_m.$$

(b) Can the problem still be reformulated as a partially separable problem if some element of a are equal to zero?

5.10 Consider the localization problem from Example 5.1, i.e.

$$\text{minimize } \sum_{(i,j) \in \mathcal{E}} \left(\|x_i - x_j\|_2 - r_{ij} \right)^2 + \sum_{(i,k) \in \mathcal{E}_a} \left(\|x_i - a_k\|_2 - v_{ik} \right)^2,$$

with variables $x_i \in \mathbb{R}^D$, $i \in \mathbb{N}_n$. Let $d_{ij} = \|x_i - x_j\|_2$ and $\delta_{ij} = d_{ij}^2$ for $(i, j) \in \mathcal{E}$ and, similarly, let $e_{ik} = \|x_i - a_k\|_2$ and $\epsilon_{ik} = e_{ik}^2$ for $(i, k) \in \mathcal{E}_a$. Moreover, let $X = \begin{bmatrix} x_1 & \cdots & x_n \end{bmatrix} \in \mathbb{R}^{D \times n}$ and $Y = X^T X \in \mathbb{S}_+^n$.

(a) Show that the localization problem is equivalent to the optimization problem

$$\text{minimize } \sum_{(i,j) \in \mathcal{E}} \left(d_{ij} - r_{ij} \right)^2 + \sum_{(i,k) \in \mathcal{E}_a} \left(e_{ik} - v_{ik} \right)^2$$
$$\text{subject to } Y_{ii} + Y_{jj} - 2Y_{ij} = \delta_{ij}$$
$$\delta_{ij} = d_{ij}^2, \ d_{ij} \geq 0, \quad (i, j) \in \mathcal{E}$$
$$Y_{ii} + \|a_k\|_2^2 - 2x_i^T a_k = \epsilon_{ik}$$
$$\epsilon_{ik} = e_{ik}^2, \ e_{ik} \geq 0, \quad (i, k) \in \mathcal{E}_a$$
$$Y = X^T X,$$

with variables X, Y, d_{ij} and δ_{ij} for $(i, j) \in \mathcal{E}$, and e_{ik} and ϵ_{ik} for $(i, k) \in \mathcal{E}_a$. Notice that the objective function is convex, but many of the constraints are not.

(b) Show that the following convex optimization problem

$$\text{minimize } \sum_{(i,j)\in\mathcal{E}} \left(d_{ij} - r_{ij}\right)^2 + \sum_{(i,k)\in\mathcal{E}_a} \left(e_{ik} - v_{ik}\right)^2$$

$$\text{subject to } Y_{ii} + Y_{jj} - 2Y_{ij} = \delta_{ij}$$

$$\begin{bmatrix} 1 & d_{ij} \\ d_{ij} & \delta_{ij} \end{bmatrix} \geq 0, \ d_{ij} \geq 0, \quad (i,j) \in \mathcal{E}$$

$$Y_{ii} + \|a_k\|_2^2 - 2x_i^T a_k = \epsilon_{ik}$$

$$\begin{bmatrix} 1 & e_{ik} \\ e_{ik} & \epsilon_{ik} \end{bmatrix} \geq 0, \ e_{ik} \geq 0, \quad (i,k) \in \mathcal{E}_a$$

$$\begin{bmatrix} I & X \\ X^T & Y \end{bmatrix}, \geq 0$$

is a *relaxation* of the localization problem, i.e. its feasible set contains the feasible set of the localization problem, and hence, its optimal value is a lower bound on the optimal value of the localization problem.

(c) Implement and apply the Levenberg–Marquardt algorithm to the nonconvex localization problem in MATLAB. The files `localization1.m`, `localization2.m`, and `localization3.m` contain data for localization problems: the first is a small problem whereas the second and third are larger problems where the latter has larger measurement errors for the range measurements.

The range measurements are stored in the matrices $R \in \mathbb{R}^{n\times n}$ and $V \in \mathbb{R}^{n\times m}$ with elements r_{ij} and v_{ik}, respectively. The zero entries in these matrices correspond to unavailable measurements, and hence nonzeros implicitly determine the sets $\mathcal{E}$ and $\mathcal{E}_a$. The columns of the matrix $A \in \mathbb{R}^{D\times m}$ are the anchor positions, and the columns of the matrix $X \in \mathbb{R}^{D\times n}$ are the true, but unknown, positions, which may be used for comparison.

(d) Use YALMIP to solve the convex relaxation. If you got bad results with the Levenberg–Marquardt algorithm, then try initializing it with the solution you obtain by solving the relaxed problem. Next, solve the relaxed problem with nuclear norm regularization, i.e., add the following terms to the objective function:

$$\gamma\left(\left\|\begin{bmatrix} I & X \\ X^T & Y \end{bmatrix}\right\|_* + \sum_{(i,j)\in\mathcal{E}} \left\|\begin{bmatrix} 1 & d_{ij} \\ d_{ij} & \delta_{ij} \end{bmatrix}\right\|_* + \sum_{(i,k)\in\mathcal{E}_a} \left\|\begin{bmatrix} 1 & e_{i,k} \\ e_{i,k} & \epsilon_{i,k} \end{bmatrix}\right\|_*\right).$$

Try different values of $\gamma > 0$ to see if you can recover the solution from the Levenberg–Marquardt algorithm. If you are not able to recover the exact solution, then try to find a value of γ that minimizes the mean square error

$$\sum_{i=1}^{n} \|x_i^\star - x_i\|_2^2$$

where $x_1^\star, \dots x_n^\star$ is the positions you obtain from the regularized relaxation and $x_1, \dots x_n$ are the positions you obtain using the Levenberg–Marquardt algorithm.

(e) Compute and report the mean square errors with respect to the true positions for each of the three investigated algorithms (Levenberg–Marquardt, convex relaxation with/without regularization). Moreover, report the value of the original objective function for all three methods. To do this you have to compute the true distances between the estimated positions, *i.e.*, $\|x_i^\star - x_j^\star\|_2$ and $\|x_i^\star - a_k\|_2$, where $(x_1^\star, \ldots x_n^\star)$ is a solution obtained with one of the three methods. Notice that these distances are not equal to d_{ij} and e_{ij} when you use the relaxed formulations. What method performs the best in terms of the mean square error criterion, and what method performs the best in terms of the original objective function?

6

Optimization Methods

We now turn our attention to numerical methods that can be used to solve different classes of optimization problems. The methods are mostly iterative: given some initial point x_0, they generate a sequence of points x_k, $k \in \mathbb{N}$, that converges to a local or global minimum. Such methods were proposed already by Isaac Newton and Carl Friedrich Gauss. We will start by reviewing some basic principles and properties that we will make use of throughout this chapter. We then discuss first-order methods for unconstrained optimization, which are methods that make use of first-order derivatives of the objective function. Second-order methods require that the Hessian of the objective function exists and is available, and we will see that the use of second-order information can dramatically reduce the number of iterations required to find a solution. However, this typically comes at the expense of more costly iterations, and we will explore the trade-off between the cost per iteration and the number of iterations through the lens of variable metric methods, which use first-order derivatives to approximate second-order derivatives. We will also consider methods for nonlinear least-squares problem and methods for optimization problems that involve nonsmooth functions and/or different types of constraints.

Many large-scale learning problems involve an objective function that is a sum of terms, and for such problems, it is if often very costly to compute the full gradient at each step. To overcome this obstacle, we will discuss stochastic optimization methods that replace full gradients with stochastic gradients that are cheap to compute. Moreover, many large-scale problems in learning are also partially separable, and we will demonstrate how this can be utilized.

6.1 Basic Principles

Many of the iterative optimization methods that we will consider in this chapter are based on some basic principles and assumptions that we will now introduce in the context of unconstrained minimization of a smooth function. We start by pointing out some important properties of smooth functions, and we then consider two conceptually different ways of constructing a so-called "descent method."

6.1.1 Smoothness

A function $f : \mathbb{R}^n \to \bar{\mathbb{R}}$ is said to be *smooth* if it is continuously differentiable on dom f, which is an open set. Moreover, f is L-smooth on dom f if ∇f is Lipschitz continuous on dom f with Lipschitz constant $L > 0$, i.e. ∇f satisfies

$$\|\nabla f(y) - \nabla f(x)\|_2 \leq L\|y - x\|_2, \quad \forall\, x, y \in \text{dom } f. \tag{6.1}$$

Optimization for Learning and Control, First Edition. Anders Hansson and Martin Andersen.
© 2023 John Wiley & Sons, Inc. Published 2023 by John Wiley & Sons, Inc.
Companion Website: www.wiley.com/go/opt4lc

This property has the important implication that

$$f(y) \leq f(x) + \nabla f(x)^T(y-x) + \frac{L}{2}\|y-x\|_2^2, \quad \forall\, x, y \in \text{dom}\, f, \tag{6.2}$$

see Exercise 6.2. For a given x, the right-hand side of (6.2) is a convex quadratic function of y and an upper bound of $f(y)$. For twice continuously differentiable functions, L-smoothness implies that $\|\nabla^2 f(x)\|_2 \leq L$, i.e. the spectral norm of $\nabla^2 f(x)$ is bounded by L for all x. We note that the inequality (6.2) is equivalent to the condition

$$|(\nabla f(y) - \nabla f(x))^T(y-x)| \leq \frac{L}{2}\|y-x\|_2^2, \quad \forall\, x, y \in \text{dom}\, f,$$

which follows by adding (6.2) to itself with x and y interchanged. If f is both L-smooth and convex, then it holds that

$$\frac{1}{L}\|\nabla f(y) - \nabla f(x)\|_2^2 \leq (\nabla f(y) - \nabla f(x))^T(y-x), \quad \forall\, x, y \in \text{dom}\, f, \tag{6.3}$$

and if f is also μ-strongly convex, then for all $x, y \in \text{dom}\, f$,

$$\frac{\mu L}{\mu + L}\|y-x\|_2^2 + \frac{1}{\mu + L}\|\nabla f(y) - \nabla f(x)\|_2^2 \leq (\nabla f(y) - \nabla f(x))^T(y-x), \tag{6.4}$$

see Exercise 6.3. We will use these inequalities to analyze the convergence behavior of gradient-based methods for unconstrained minimization of convex functions.

6.1.2 Descent Methods

Consider an unconstrained optimization problem of the form

$$\text{minimize } f(x), \tag{6.5}$$

where $f : \mathbb{R}^n \to \mathbb{R}$ is smooth. Given $x \in \mathbb{R}^n$ and $\Delta x \in \mathbb{R}^n$, we say that Δx is a *descent direction* at x if the directional derivative of f at x in the direction Δx is negative, i.e. $\nabla f(x)^T \Delta x < 0$. Given a descent direction Δx at x, it is possible to find a step size $t > 0$ such that $f(x + t\Delta x) < f(x)$. This motivates an iterative method of the form

$$x_{k+1} = x_k + t_k \Delta x_k, \quad \mathbb{Z}_+,$$

where x_0 is an initial guess, Δx_k is a descent direction at x_k, and the step size $t_k > 0$ is chosen such that $f(x_{k+1}) < f(x_k)$. This is clearly a *descent method* in the sense that the objective value decreases at each iteration so long as there exists a descent direction. However, without additional assumptions on f and restrictions on the step size and direction, such a method does not necessarily converge to a stationary point of f.

The direction $\Delta x = -\nabla f(x)$ is clearly a descent direction, and so is $\Delta x = -P\nabla f(x)$ for any $P \in \mathbb{S}_{++}^n$ as well as $\Delta x = -\text{sgn}(\partial f(x)/\partial x_i)e_i$ for any $i \in \mathbb{N}_n$ such that $\partial f(x)/\partial x_i \neq 0$. We say that Δx is a *normalized steepest descent direction* at x for some norm $\|\cdot\|$ if $\nabla f(x) \neq 0$ and

$$\Delta x \in \operatorname*{argmin}_p \{\nabla f(x)^T p \mid \|p\| \leq 1\} = \operatorname*{argmax}_p \{-\nabla f(x)^T p \mid \|p\| \leq 1\} = \partial I_B^*(-\nabla f(x)),$$

where B denotes the unit norm ball $\{p \in \mathbb{R}^n \mid \|p\| \leq 1\}$. Moreover, the directional derivative in such a direction may be expressed as

$$\inf_p \{\nabla f(x)^T p \mid \|p\| \leq 1\} = -\sup_p \{-\nabla f(x)^T p \mid \|p\| \leq 1\} = -\|\nabla f(x)\|_*.$$

For the Euclidean norm, $-\nabla f(x)/\|\nabla f(x)\|_2$ is the unique normalized steepest descent direction provided that $\nabla f(x) \neq 0$.

Before we turn our attention to any specific method, we will outline two different approaches to the problem of finding a suitable step size and/or a descent direction, namely *line search* methods and *surrogation* methods.

6.1.3   Line Search Methods

Line search methods address the problem of finding a suitable step size $t > 0$ given a descent direction Δx at x. Since we are interested in minimizing f, a natural choice is to select t as

$$t \in \underset{\tau \geq 0}{\text{argmin}}\{\phi(\tau)\}, \tag{6.6}$$

where $\phi : \mathbb{R} \to \mathbb{R}$ defined as $\phi(t) = f(x + t\Delta x)$ is the restriction of f to the line defined by x and the search direction Δx. This is a so-called *exact line search*, and although it is appealing to make the most out of a descent direction, the minimization (6.6) can be expensive if a minimizer even exists. We note that the exact line search is also called *Cauchy's rule* in the special case, where $\Delta x = -\nabla f(x)$. An alternative to Cauchy's rule is *Curry's rule*, which can be stated as

$$t = \min_{\tau \geq 0}\{\tau \mid \phi'(\tau) = 0\}, \tag{6.7}$$

i.e. the step size is the smallest nonnegative t such that $x + t\Delta x$ is stationary point of f. This is a root-finding problem, which can also be expensive to solve.

A more practical approach to the problem of finding a suitable step size is to choose t such that it satisfies a sufficient descent condition known as the *Armijo condition*. This condition requires that $t > 0$ satisfies

$$\phi(t) \leq \phi(0) + \alpha_1 t \phi'(0), \tag{6.8}$$

where $\alpha_1 \in (0, 1/2)$ is a parameter. In other words, the reduction in the objective value must be proportional to $t\phi'(0)$, which is negative since Δx is assumed to be a descent direction. The Armijo condition is illustrated in Figure 6.1. Note that it is always satisfied for some sufficiently small $t > 0$ since $\phi(t) \approx \phi(0) + \phi'(0)t$ when t is close to zero. Thus, to ensure that the step makes a reasonable amount of progress, we need to avoid short steps. One way to do this is to impose a so-called *curvature condition* of the form

$$\alpha_2 \phi'(0) \leq \phi'(t), \tag{6.9}$$

where $\alpha_2 \in (\alpha_1, 1)$ is a parameter. Roughly speaking, this means that a step size $t > 0$ is inadmissible if the slope of ϕ at t is still downward and relatively steep. A step size $t > 0$ is said to satisfy the *Wolfe conditions* if it satisfies the Armijo condition (6.8) as well as the curvature condition (6.9).

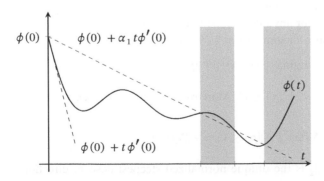

Figure 6.1   Illustration of the Armijo condition. The gray regions correspond to inadmissible step sizes.

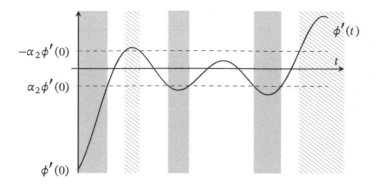

Figure 6.2 Illustration of the curvature conditions (6.9) and (6.10). The hatched regions correspond to step sizes that violate the strong curvature condition, whereas step size in the gray regions violate both the weak and strong curvature conditions.

Notice that the condition (6.9) does not rule out that $\phi'(t)$ may be large and positive. Thus, if we would like t to be close to a stationary point of ϕ, we may modify the curvature condition (6.9) to include an upper bound on $\phi'(t)$, i.e.

$$\alpha_2 \phi'(0) \leq \phi'(t) \leq -\alpha_2 \phi'(0). \tag{6.10}$$

This may also be expressed as $|\phi'(t)| \leq \alpha_2 |\phi'(0)|$ and is often referred to as the strong curvature condition. Figure 6.2 illustrates the curvature condition (6.9) and the stronger version (6.10). Collectively, the Armijo condition and the strong curvature condition (6.10) comprise the *strong Wolfe conditions*.

We end our discussion of line search methods by outlining two practical algorithms. The first one is based on the Armijo condition, whereas the second one is based on the Wolfe conditions.

6.1.3.1 Backtracking Line Search

The prototypical backtracking line search starts with some initial step size $t > 0$ that is accepted if the Armijo condition is satisfied, and otherwise, the step size is repeatedly scaled by a factor $\beta \in (0, 1)$ until the Armijo condition is satisfied. The resulting step size may be expressed as

$$t^\star = \max_{k \in \mathbb{Z}_+} \{\beta^k t \mid \phi(\beta^k t) \leq \phi(0) + \alpha_1 \beta^k t \phi'(0)\},$$

and the algorithm is summarized in Algorithm 6.1. The parameter β controls how aggressively the step size is reduced, whereas α_1 controls the sufficient descent condition. To avoid unnecessarily short steps, we require that $\alpha_1 \in (0, 1/2]$. This bound can be motivated by considering the case where $\phi(t)$ is a quadratic function of t, i.e.

$$\phi(t) = \phi(0) + t\phi'(0) + \frac{t^2}{2}\phi''(0).$$

The exact minimizer of $\phi(t)$ is then $t^\star = -\phi'(0)/\phi''(0)$, and the Armijo condition (6.8) reduces to $t \leq 2(1 - \alpha_1)t^\star$. It follows that $t = t^\star$ does not satisfy the descent condition if $\alpha_1 > 1/2$.

6.1.3.2 Bisection Method for Wolfe Conditions

To find a step size that satisfies the Wolfe conditions, it is generally insufficient to simply backtrack, and hence a more sophisticated approach is needed. An admissible step size can be found my means of bisection as shown in Algorithm 6.2. This algorithm starts with some initial guess, say, $t = 1$,

---

Algorithm 6.1: Backtracking line search

Function Backtrack $(\phi, t, \alpha_1, \beta)$:
  Input: Continuously differentiable function $\phi : \mathbb{R} \to \mathbb{R}$, initial step size $t \geq 0$,
        and parameters $\alpha_1 \in (0, 1/2]$, $\beta \in (0, 1)$.
  Output: $t > 0$ such that $\phi(t) \leq \phi(0) + \alpha_1 t \phi'(0)$.

  while $\phi(t) > \phi(0) + \alpha_1 t \phi'(0)$ **do**
  $\quad | \quad t \leftarrow \beta t$ ▷ backtracking
  end
  return t

---

and the trivial lower bound $l = 0$ and upper bound $u = +\infty$. The Armijo condition is checked at the beginning of each loop iteration. If it is violated, then the upper bound u is reduced to t and the midpoint of the interval $[l, u]$ is used as a new candidate step size. Otherwise, the curvature condition is checked, and if it is violated, then the lower bound is increased to t, and a new candidate step size is then the midpoint of $[l, u]$ if u is finite and otherwise, $2t$. Note that once the upper bound u is finite, the width of the interval $[l, u]$ is reduced by a factor of two in each loop iteration. This observation can be used to show that the algorithm either terminates after finitely many iterations, or alternatively, the upper bound u remains infinite and t doubles in every loop iterations, which implies that $\phi(t) \to \infty$ as $t \to \infty$.

Algorithm 6.2: Bisection method for Wolfe conditions.

Function BisectionWolfe $(\phi, \alpha_1, \alpha_2)$:
  Input: Continuously differentiable function $\phi : \mathbb{R} \to \mathbb{R}$ and parameters α_1 and α_2 such that
        $0 < \alpha_1 < \alpha_2 < 1$.
  Output: $t > 0$ such that $\phi(t) \leq \phi(0) + \alpha_1 t \phi'(0)$ and $\phi'(t) \geq \alpha_2 \phi'(0)$.

  $l \leftarrow 0, t \leftarrow 1, u \leftarrow \infty$
  repeat
  $\quad$ **if** $\phi(t) > \phi(0) + \alpha_1 t \phi'(0)$ **then**
  $\quad\quad | \quad u \leftarrow t$ ▷ reduce u and t
  $\quad\quad | \quad t \leftarrow (l + u)/2$
  $\quad$ **else if** $\phi'(t) < \alpha_2 \phi'(0)$ **then**
  $\quad\quad | \quad l \leftarrow t$ ▷ increase l and t
  $\quad\quad | \quad$ **if** $\beta < \infty$ **then** $t \leftarrow (l + u)/2$
  $\quad\quad | \quad$ **else** $t \leftarrow 2l$
  $\quad$ **else**
  $\quad\quad | \quad$ **return** t

6.1.4 Surrogation Methods

Another approach to the problem of finding a new point x_{k+1} that reduces the objective value is to solve a sequence of *surrogate problems*. Generally speaking, each surrogate problem should be good local approximation of the original problem, and it should be relatively easy to solve. At iteration k, we construct a surrogate model $m_k : \mathbb{R}^n \to \mathbb{R}$ that approximates the objective function f in some neighborhood of the current iterate x_k. The surrogate model m_k may incorporate information such as the previous iterates $x_0, \dots, x_k$ and the corresponding function values and partial derivatives or some order. We will introduce two kinds of surrogation methods, namely *trust-region* methods and *majorization minimization* methods.

6.1.4.1 Trust-Region Methods

Trust-region methods find a candidate for $x_{k+1} = x_k + \Delta x$ by solving a trust-region problem of the form

$$\begin{aligned}
\text{minimize} \quad & m_k(x_k + \Delta x) \\
\text{subject to} \quad & \|\Delta x\| \leq \delta_k,
\end{aligned} \tag{6.11}$$

with variable Δx. The constraint is a norm ball of radius $\delta_k > 0$, which defines a *trust-region* in which we assume that our surrogate model m_k is a good approximation of f. The surrogate model is typically a linear or quadratic function. Given a solution Δx to the trust-region subproblem, progress can be quantified in terms of the ratio of the actual reduction $f(x_k) - f(x_k + \Delta x)$ to the predicted reduction $f(x_k) - m_k(x_k + \Delta x)$, i.e.

$$\rho_k = \frac{f(x_k) - f(x_k + \Delta x)}{f(x_k) - m_k(x_k + \Delta x)}.$$

The denominator is positive unless x_k is a local minimum, and hence, $\rho_k \approx 1$ if the local quadratic model is a good approximation of f within the trust region. The step is rejected and the trust-region radius is reduced if $\rho_k < \eta_1$, where $\eta_1 \in (0, 1)$ is a parameter, and otherwise, the step is accepted, Moreover, the trust-region is expanded if $\rho \geq \eta_2$, where $\eta_2 \in [\eta_1, 1)$ is a parameter. An example of a basic trust-region is shown in Algorithm 6.3. The trust-region parameters η_1 and η_2 may be chosen as, e.g. $\eta_1 = 0.1$ and $\eta_2 = 0.9$. The interested reader may find a comprehensive treatment of trust-region methods in [27].

Algorithm 6.3: Trust-region method.

Input: Continuously differentiable function $f : \mathbb{R}^n \to \mathbb{R}$, $x_0 \in \mathbb{R}^n$, $\delta_0 > 0$, $\epsilon > 0$,
$\qquad$ and $0 < \eta_1 \leq \eta_2 < 1$.
Output: x such that $\|\nabla f(x)\| \leq \epsilon$.

$x \leftarrow x_0$
for $k = 0, 1, 2, \ldots$ **do**
$\quad$ Choose surrogate model m_k.
$\quad$ Solve (6.11) for Δx and compute ρ_k.
$\quad$ **if** $\rho_k \geq \eta_1$ **then**
$\quad\quad$ $x \leftarrow x + \Delta x$ $\qquad\qquad\qquad\qquad\qquad$ ▷ accept step
$\quad\quad$ **if** $\|\nabla f(x)\| \leq \epsilon$ **then**
$\quad\quad\quad$ stop
$\quad\quad$ **end**
$\quad$ **end**
$\quad$ **if** $\rho_k \geq \eta_2$ **then**
$\quad\quad$ $\delta_{k+1} \leftarrow 2\delta_k$ $\qquad\qquad\qquad\qquad\qquad$ ▷ expand trust region
$\quad$ **else if** $\rho_k < \eta_1$ **then**
$\quad\quad$ $\delta_{k+1} \leftarrow \delta_k/2$ $\qquad\qquad\qquad\qquad\quad$ ▷ contract trust region
$\quad$ **else**
$\quad\quad$ $\delta_{k+1} \leftarrow \delta_k$
$\quad$ **end**
end

6.1.4.2 Majorization Minimization

A surrogate model m_k is a so-called *majorization* of f at x_k if $m_k(x_k) = f(x_k)$ and $m_k(x) \geq f(x)$ for all x. This property allows us to construct a descent method without the need for a trust region. Indeed, the iteration

$$x_{k+1} \in \underset{x}{\operatorname{argmin}}\ m_k(x), \quad k \in \mathbb{Z}_+, \tag{6.12}$$

satisfies $f(x_{k+1}) \leq f(x_k)$ if m_k is a majorization of f at x_k, which follows by noting that

$$f(x_{k+1}) \leq m_k(x_{k+1}) \leq m_k(x_k) = f(x_k).$$

Moreover, we have that $f(x_{k+1}) < f(x_k)$ unless x_k is a minimizer of m_k.

6.1.5 Convergence of Sequences

A sequence of vectors $x_k \in \mathbb{R}^n$, $k \in \mathbb{N}$, is said to converge to $x^* \in \mathbb{R}^n$ if

$$\lim_{k \to \infty} x_k = x^*.$$

The sequence has a q-order convergence rate if there exists finite constants $q \geq 1$ and $\rho \geq 0$ such that

$$\lim_{k \to \infty} \frac{\|x_{k+1} - x^*\|_2}{\|x_k - x^*\|_2^q} = \rho.$$

This is also referred to as *quotient convergence* or Q-convergence because of the quotient that appears in the definition. Notable special cases include

- sublinear convergence, $q = 1$ and $\rho = 1$;
- linear convergence, $q = 1$ and $\rho \in (0, 1)$;
- superlinear convergence, $q = 1$ and $\rho = 0$;
- quadratic convergence, $q = 2$ and $\rho > 0$.

Example 6.1   The sequence $x_k = 1/k$, $k \in \mathbb{N}$, tends to $x^* = 0$ as $k \to \infty$, and we see that

$$\lim_{k \to \infty} \frac{|x_{k+1} - x^*|}{|x_k - x^*|} = \lim_{k \to \infty} \frac{k}{k+1} = 1.$$

This implies that the sequence converges sublinearly. The sequence $x_k = 2^{-k}$, $k \in \mathbb{N}$, also converges to $x^* = 0$, and we find that

$$\lim_{k \to \infty} \frac{|x_{k+1} - x^*|}{|x_k - x^*|} = \lim_{k \to \infty} \frac{2^{-k-1}}{2^{-k}} = \frac{1}{2},$$

which shows that this sequence converges linearly. Finally, the sequence defined recursively as $x_{k+1} = \alpha x_k^2$, $k \in \mathbb{N}$, with $x_1 = 1/2$ and $|\alpha| \in (0, 2)$ also converges to $x^* = 0$. It satisfies

$$\lim_{k \to \infty} \frac{|x_{k+1} - x^*|}{|x_k - x^*|^2} = \lim_{k \to \infty} \frac{|\alpha x_k^2|}{|x_k|^2} = |\alpha|,$$

and hence, the sequence converges quadratically. We note that the sequence diverges if $|\alpha| > 2$, and $x_k = 1/2$ for all $k \in \mathbb{N}$ if $|\alpha| = 2$.

6.2 Gradient Descent

We now return to the unconstrained optimization problem (6.5) where we assume that the objective function $f : \mathbb{R}^n \to \mathbb{R}$ is continuously differentiable. We will study the *gradient descent* method, which performs updates of the form

$$x_{k+1} = x_k - t_k \nabla f(x_k), \tag{6.13}$$

where $t_k > 0$ is the step size at iteration k. The step size can be chosen in several ways, e.g. using some form of line search. We will analyze the behavior of the iterates generated by (6.13) with different assumptions on f and the step size sequence.

6.2.1 *L*-Smooth Functions

We start by considering the case where f is L-smooth. Using the quadratic upper bound (6.2) as a majorization in the majorization minimization iteration (6.12), we obtain the gradient descent iteration

$$x_{k+1} = x_k - \frac{1}{L}\nabla f(x_k), \quad k \in \mathbb{Z}_+, \tag{6.14}$$

with constant step size $t_k = 1/L$. Substituting x_{k+1} for y and x_k for x in (6.2), we obtain the inequality

$$f(x_{k+1}) \leq f(x_k) - \frac{1}{2L}\|\nabla f(x_k)\|_2^2, \tag{6.15}$$

which shows that the update (6.14) always reduces the objective value unless x_k is a stationary point. Moreover, the progress $f(x_k) - f(x_{k+1})$ is at least $(2L)^{-1}\|\nabla f(x_k)\|_2^2$, so roughly speaking, gradient descent is making good progress when the norm of the gradient is large. If f is bounded below, then it can be shown that

$$\lim_{k\to\infty} \|\nabla f(x_k)\|_2^2 = 0,$$

see Exercise 6.1. However, without additional assumptions on f, there is no guarantee that the gradient descent method (6.14) converges to neither a global nor local minimum.

6.2.2 Smooth and Convex Functions

A stronger result for the gradient descent iteration (6.13) may be obtained if f is both convex and L-smooth. Recall that a continuously differentiable function f is convex if and only if

$$f(y) \geq f(x) + \nabla f(x)^T(y - x), \quad \forall\, x,y \in \mathrm{dom}\, f. \tag{6.16}$$

It immediately follows that $f(y) \geq f(x^\star)$ for all $y \in \mathrm{dom}\, f$ if $x^\star$ is a stationary point, and hence, $x^\star$ is also a global minimizer. Moreover, we have that

$$\begin{aligned}
\|x_{k+1} - x^\star\|_2^2 &= \|x_k - x^\star - t_k(\nabla f(x_k) - \nabla f(x^\star))\|_2^2 \\
&= \|x_k - x^\star\|_2^2 + t_k^2\|\nabla f(x_k) - \nabla f(x^\star)\|_2^2 \\
&\quad - 2t_k(\nabla f(x_k) - \nabla f(x^\star))^T(x_k - x^\star),
\end{aligned}$$

and applying (6.3) to the last term on the right-hand side, we obtain the inequality

$$\|x_{k+1} - x^\star\|_2^2 \leq \|x_k - x^\star\|_2^2 - t_k\left(\frac{2}{L} - t_k\right)\|\nabla f(x_k)\|_2^2.$$

This implies that $\|x_{k+1} - x^\star\|_2^2 < \|x_k - x^\star\|_2^2$ whenever $t_k \in (0, 2/L)$ and $\nabla f(x_k) \neq 0$. Similarly, substituting x_{k+1} for y and x_k for x in (6.2), we have that

$$f(x_{k+1}) \leq f(x_k) - t_k\left(1 - \frac{L}{2}t_k\right)\|\nabla f(x_k)\|_2^2, \tag{6.17}$$

and substituting $x^\star$ for y and x_k for x in (6.16) and rearranging the terms, we obtain the inequality

$$f(x_k) - p^\star \leq \nabla f(x_k)^T(x_k - x^\star)$$

where $p^\star = f(x^\star)$ is the optimal value. Applying the Cauchy–Schwarz inequality to the inner product on the right-hand side, we find that

$$f(x_k) - p^\star \leq \|\nabla f(x_k)\|_2 \|x_k - x^\star\|_2 \leq \|\nabla f(x_k)\|_2 \|x_0 - x^\star\|_2,$$

where we have used the fact that $\|x_k - x^\star\|_2$ is a nonincreasing function of k if $t_k \in (0, 2/L)$ for all k. Assuming that $x_0 \neq x^\star$, we may rewrite this inequality as

$$\frac{f(x_k) - p^\star}{\|x_0 - x^\star\|_2} \leq \|\nabla f(x_k)\|_2.$$

Combining this with (6.17), we have that

$$\delta_{k+1} \leq \delta_k - t_k \left(1 - \frac{L}{2} t_k\right) \frac{\delta_k^2}{R^2},$$

where $\delta_k = f(x_k) - p^\star$ and $R = \|x_0 - x^\star\|_2$. Dividing by $\delta_k \delta_{k+1}$ on both sides and rearranging the terms, we arrive at

$$\frac{1}{\delta_{k+1}} \geq \frac{1}{\delta_k} + t_k \left(1 - \frac{L}{2} t_k\right) \frac{\delta_k}{\delta_{k+1}} \frac{1}{R^2}$$

$$\geq \frac{1}{\delta_k} + t_k \left(1 - \frac{L}{2} t_k\right) \frac{1}{R^2}$$

$$\geq \frac{1}{\delta_0} + \frac{1}{R^2} \sum_{i=0}^{k} t_i \left(1 - \frac{L}{2} t_i\right),$$

where the last inequality follows by recursive application of the previous inequality. Choosing a constant step size $t_i = \rho/L$ with $\rho \in (0, 2)$ leads to the bound

$$\delta_{k+1} \leq \frac{\delta_0 L R^2}{LR^2 + (k+1)\delta_0 \rho (1 - \rho/2)},$$

where $\rho = 1$ minimizes the right-hand side. Thus, it follows that when f is L-smooth and convex, then the gradient descent iteration as outlined in Algorithm 6.4 satisfies

$$f(x_k) - p^\star = O(1/k),$$

which means that the worst-case rate of convergence is sublinear. However, note that this does not guarantee that x_k converges unless $p^\star$ is attained.

Algorithm 6.4: Gradient descent (constant step size)

Input: L-smooth convex function $f : \mathbb{R}^n \to \mathbb{R}$, starting point x_0, tolerance $\epsilon > 0$, and step size $t \in (0, 2/L)$.

Output: x such that $\|\nabla f(x)\| \leq \epsilon$.

$x \leftarrow x_0$
for $k = 1, 2, \ldots$ **do**
   | $x \leftarrow x - t\nabla f(x)$
   | **if** $\|\nabla f(x)\| \leq \epsilon$ **then** stop
end

Example 6.2 Consider the function $f : \mathbb{R}^2 \to \mathbb{R}$ defined as $f(x) = g(Ax + b)$, where $g : \mathbb{R}^3 \to \mathbb{R}$ is the log-sum-exp function

$$g(z) = \ln\left(e^{z_1} + e^{z_2} + e^{z_3}\right),$$

and

$$A = \begin{bmatrix} 2 & 1 \\ -1 & 1 \\ -1 & -2 \end{bmatrix}, \quad b = \begin{bmatrix} -2 \\ -1 \\ 0 \end{bmatrix}.$$

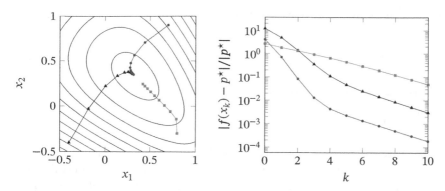

Figure 6.3 Examples of the gradient descent iteration with three different starting points. The starting points and the first ten iterations are shown.

The function f is convex and ∇f it is Lipschitz continuous since

$$\|\nabla^2 f(x)\|_2 = \|A^T \nabla^2 g(Ax + b) A\|_2 \le \|A\|_2^2 \|\nabla^2 g(Ax + b)\|_2 \le \|A\|_2^2 / 2.$$

The last inequality follows from the fact that $\|\nabla^2 g(z)\|_2 \le 1/2$, which can be shown using the so-called "Gershgorin circle theorem". Figure 6.3 shows the level curves of the function f along with the iterates of the gradient descent iterations for three different starting points and the constant step size $t_k = 2/\|A\|_2^2$. Observe that the starting point has a significant effect on the practical performance but not the asymptotic behavior.

6.2.3 Smooth and Strongly Convex Functions

Strong convexity makes it possible to derive a stronger asymptotic error bound for the gradient descent iteration (6.13). Specifically, we will assume that f is L-smooth, μ-strongly convex, and closed, which guarantees the existence of a unique minimizer $x^\star$. The error at iteration $k + 1$ can then be expressed as

$$
\begin{aligned}
\|x_{k+1} - x^\star\|_2^2 &= \|x_k - t_k \nabla f(x_k) - x^\star\|_2^2 \\
&= \|x_k - x^\star - t_k(\nabla f(x_k) - \nabla f(x^\star))\|_2^2 \\
&= \|x_k - x^\star\|_2^2 + t_k^2 \|\nabla f(x_k) - \nabla f(x^\star)\|_2^2 \\
&\quad - 2t_k(\nabla f(x_k) - \nabla f(x^\star))^T (x_k - x^\star),
\end{aligned}
\tag{6.18}
$$

where $x^\star$ is the unique minimizer, and hence, $\nabla f(x^\star) = 0$. Using (6.4), it follows that

$$
\begin{aligned}
\|x_{k+1} - x^\star\|_2^2 &\le \|x_k - x^\star\|_2^2 + t_k^2 \|\nabla f(x_k) - \nabla f(x^\star)\|_2^2 \\
&\quad - 2t_k \left(\frac{\mu L}{\mu + L} \|x_k - x^\star\|_2^2 + \frac{1}{\mu + L} \|\nabla f(x_k) - \nabla f(x^\star)\|_2^2 \right) \\
&= \left(1 - \frac{2t_k \mu L}{\mu + L} \right) \|x_k - x^\star\|_2^2 \\
&\quad - t_k \left(\frac{2}{\mu + L} - t_k \right) \|\nabla f(x_k) - \nabla f(x^\star)\|_2^2,
\end{aligned}
$$

and hence,

$$
\|x_{k+1} - x^\star\|_2^2 \le \left(1 - \frac{2t_k \mu L}{\mu + L} \right) \|x_k - x^\star\|_2^2, \quad t_k \in \left(0, \frac{2}{\mu + L} \right].
$$

Using this inequality recursively and employing a constant step size $t_k = t \in (0, 2/(\mu + L)]$, we conclude that

$$\|x_k - x^\star\|_2^2 \leq \left(1 - \frac{2t\mu L}{\mu + L}\right)^k \|x_0 - x^\star\|_2^2. \tag{6.19}$$

The best upper bound is obtained with $t = 2/(\mu + L)$, in which case

$$1 - \frac{2t\mu L}{\mu + L} = \left(\frac{L - \mu}{L + \mu}\right)^2 = \left(\frac{\kappa - 1}{\kappa + 1}\right)^2,$$

where $\kappa = L/\mu$ may be viewed as a condition number. Indeed, if f is twice continuously differentiable, μ-strongly convex, and L-smooth, then the eigenvalues of $\nabla^2 f(x)$ belong to the interval $[\mu, L]$ for all x, or equivalently, $\mu I \preceq \nabla^2 f(x) \preceq LI$. Thus, the ratio $\kappa = L/\mu$ may be viewed as an upper bound on the condition number of the Hessian.

6.3 Newton's Method

We now return to the unconstrained problem (6.5) with the additional assumption that f is twice continuously differentiable. This assumption allows us to construct a local quadratic approximation of f at x using a second-order Taylor expansion around x, i.e. $\tilde{f} : \mathbb{R}^n \to \mathbb{R}$ defined as

$$\tilde{f}(y; x) = f(x) + \nabla f(x)^T (y - x) + \frac{1}{2}(y - x)^T \nabla^2 f(x)(y - x). \tag{6.20}$$

The gradient of $\tilde{f}(x + \Delta x; x)$ with respect to a direction Δx is $\nabla f(x) + \nabla^2 f(x)\Delta x$, and setting this to 0 yields the stationarity condition

$$\nabla^2 f(x)\Delta x = -\nabla f(x). \tag{6.21}$$

This system of equations is known as the *Newton equations*. If f is a convex function, then the quadratic approximation (6.20) is convex in y since $\nabla^2 f(x) \succeq 0$ for all x. The stationarity condition (6.21) therefore characterizes the global minima of the local quadratic approximation provided that the Newton equations (6.21) are consistent. In contrast, if f is nonconvex and $\nabla^2 f(x) \nsucceq 0$, then the quadratic approximation (6.20) is unbounded below. In this case, the stationarity condition (6.21) characterizes the *saddle points* or maxima of a quadratic form. Recall that we are interested in minimizing f rather than finding a saddle point or a local maximum, so the stationarity condition (6.21) alone is of limited interest when f is nonconvex. In this case, the quadratic approximation (6.20) is typically used in combination with a trust-region method or with *cubic regularization*. The latter is based on the assumption that $\nabla^2 f$ is Lipschitz continuous with constant $L_2 > 0$, which implies that

$$f(y) \leq f(x) + \nabla f(x)^T (y - x) + \frac{1}{2}(y - x)^T \nabla^2 f(x)(y - x) + \frac{L_2}{6} \|y - x\|_2^3.$$

The right-hand side is a majorization of f at x, and hence, it can be used to construct a descent method based on the majorization minimization principle; see, e.g. [82].

In the convex case, *Newton's method* uses as search direction a vector Δx_{nt} that satisfies (6.21). Thus, if $\nabla^2 f(x)$ is nonsingular, then the search direction is the unique solution $\Delta x_{\mathrm{nt}} = -\nabla^2 f(x)^{-1} \nabla f(x)$, which naturally leads to an iteration of the form

$$x_{k+1} = x_k - t_k \nabla^2 f(x_k)^{-1} \nabla f(x_k), \quad k \in \mathbb{Z}_+, \tag{6.22}$$

where x_0 is an initial guess, and $t_k > 0$ is the step size at iteration k. A step for which $t_k = 1$ is called a *pure* Newton step, whereas a step with $0 < t_k < 1$ is referred to as a *damped* Newton step.

With additional assumptions on f, a pure Newton step can be shown to yield a descent if x_k is sufficiently close to a stationary point of f. However, a full step does not necessarily yield a descent if x_k is far away from a stationary point, in which case a damped Newton step should be used.

6.3.1 The Newton Decrement

The directional derivative of f in the Newton direction Δx_{nt} is given by

$$\frac{d}{dt}f(x + t\Delta x_{nt})\Big|_{t=0} = \nabla f(x)^T \Delta x_{nt} = -\nabla f(x)^T \nabla^2 f(x)^{-1} \nabla f(x),$$

and it is negative if $\nabla f(x) > 0$ and $\nabla f(x) \neq 0$, in which case a descent is achievable with a sufficiently small step size. When the Hessian is positive definite, the directional derivative may be expressed as $-\lambda(x)^2$, where $\lambda : \mathbb{R}^n \to \mathbb{R}$ is the so-called *Newton decrement* of f at x, defined as

$$\lambda(x) = \|\nabla f(x)\|_{\nabla^2 f(x)^{-1}} = \left(\nabla f(x)^T \nabla^2 f(x)^{-1} \nabla f(x)\right)^{1/2}. \tag{6.23}$$

Alternatively, using the Newton equations (6.21), the Newton decrement may also be expressed as

$$\lambda(x) = \|\Delta x_{nt}\|_{\nabla^2 f(x)} = \left(\Delta x_{nt}^T \nabla^2 f(x) \Delta x_{nt}\right)^{1/2}. \tag{6.24}$$

It is straightforward to verify that the Newton decrement satisfies

$$f(x) - \inf_y \tilde{f}(y; x) = \frac{1}{2}\lambda(x)^2,$$

which allows us to view $\lambda(x)^2/2$ as an estimate of $f(x) - p^\star$. This motivates the use of a stopping criterion of the form $(1/2)\lambda(x)^2 \leq \epsilon$, where $\epsilon > 0$ is a given tolerance. Newton's method with a backtracking line search and this stopping criterion is shown in Algorithm 6.5.

Algorithm 6.5: Newton's method

Input: Starting point x_0, tolerance $\epsilon > 0$, and backtracking parameters $\alpha \in (0, 1/2]$ and $\beta \in (0, 1)$.

Output: x such that $\lambda(x)^2/2 \leq \epsilon$.

$x \leftarrow x_0$
for $k = 1, 2, \ldots$ **do**
    $\Delta x_{nt} \leftarrow -\nabla^2 f(x)^{-1} \nabla f(x)$ ▷ Newton direction
    $\lambda \leftarrow \sqrt{-\nabla f(x)^T \Delta x_{nt}}$ ▷ Newton decrement
    if $\lambda^2/2 \leq \epsilon$ **then** stop
    $t \leftarrow 1$
    while $f(x + t\Delta x_{nt}) > f(x) - t\alpha\lambda^2$ **do**
        $t \leftarrow \beta t$ ▷ backtracking
    end
    $x \leftarrow x + t\Delta x_{nt}$ ▷ update x
end

6.3.2 Analysis of Newton's Method

We will now consider the convergence properties of Newton's method for minimizing a twice continuously differentiable function $f : \mathbb{R}^n \to \mathbb{R}$, as outlined in Algorithm 6.5. Given an initial guess $x_0 \in \text{dom} f$, we will assume that f is μ-strongly convex and that $\nabla^2 f$ is Lipschitz continuous on the sublevel set

$$S = \{x \mid f(x) \leq f(x_0)\},$$

i.e. there exists a constant L_2 such that

$$\|\nabla^2 f(y) - \nabla^2 f(x)\|_2 \leq L_2 \|y - x\|_2, \quad \forall \, y, x \in S. \tag{6.25}$$

We will assume that S is closed and nonempty, and strong convexity of f implies that S is also bounded; *cf.* Section 4.3. Moreover, strong convexity of f and continuity of $\nabla^2 f$ imply that there exists a constant L such that $\nabla^2 f(x) \preceq L$ for all $x \in S$. In other words, f is L-smooth on S. The backtracking line search ensures that all iterates stay within the sublevel set S.

Our analysis of algorithm 6.5 consists of three parts. In the first part, we show that Newton's method is affinely invariant. Specifically, we show that given a nonsingular matrix $A \in \mathbb{R}^{n \times n}$ and a vector $b \in \mathbb{R}^n$, there is a one-to-one correspondence between the sequences of iterates obtained when applying the method to $f(x)$ and $g(y) = f(Ay + b)$ provided that $x_0 = Ay_0 + b$. In the second part, we show that there exists a constant $\delta \in (0, \mu^2/L_2]$ such that $\|\nabla f(x_k)\|_2 \leq \delta$ is a sufficient condition for a pure Newton step to satisfy the line search condition in the current and all subsequent iterations, and moreover,

$$\|\nabla f(x_{k+l})\|_2 \leq \left(\frac{1}{2}\right)^{2^l - 1} \|\nabla f(x_k)\|_2, \quad l \in \mathbb{Z}_+. \tag{6.26}$$

This is referred to as the *pure Newton phase*. Finally, in the third part, we show that there exists a constant $\sigma > 0$ such that

$$f(x_{k+1}) - f(x_k) \leq -\sigma, \tag{6.27}$$

whenever $\|\nabla f(x_k)\|_2 > \delta$. This is referred to as the *damped Newton phase* since damped Newton steps are needed to ensure convergence.

6.3.2.1 Affine Invariance

Consider the affine transformation $x = Ay + b$, where $A \in \mathbb{R}^{n \times n}$ is a nonsingular matrix and $b \in \mathbb{R}^n$. Using the chain rule, the gradient and Hessian of $g(y) = f(Ay + b)$ can be expressed as

$$\nabla g(y) = A^T \nabla f(Ay + b), \quad \nabla^2 g(y) = A^T \nabla^2 f(Ay + b)A.$$

The Newton update for g at y is

$$\begin{aligned}
y^+ &= y + t_k \nabla^2 g(y)^{-1} \nabla g(y) \\
&= y + t_k (A^T \nabla^2 f(Ay + b)A)^{-1} A^T \nabla f(Ay + b) \\
&= y + t_k A^{-1} \nabla^2 f(Ay + b)^{-1} \nabla f(Ay + b),
\end{aligned}$$

and after left-multiplying by A and adding b on both sides, we see that

$$\begin{aligned}
Ay^+ + b &= Ay + b + t_k \nabla^2 f(Ay + b)^{-1} \nabla f(Ay + b) \\
x^+ &= x + t_k \nabla^2 f(x)^{-1} \nabla f(x),
\end{aligned}$$

which shows that there is a one-to-one correspondence between y^+ and x^+, which is the Newton update for f at x. Finally, noting that

$$\nabla g(y)^T \nabla g^2(y)^{-1} \nabla g(y) = \nabla f(x)^T \nabla^2 f(x)^{-1} \nabla f(x),$$

we see that the Newton decrement is affinely invariant, and hence, the backtracking line search does not depend on A and b.

6.3.2.2 Pure Newton Phase

We start our analysis of the pure Newton phase by showing that there exists a constant $\delta > 0$ such that a pure Newton step $x_{k+1} = x_k + \Delta x_{\mathrm{nt}}$ satisfies the backtracking line search condition

$$f(x_{k+1}) \leq f(x_k) - \alpha \lambda(x_k)^2, \tag{6.28}$$

whenever $\|\nabla f(x_k)\|_2 \leq \delta$. The Lipschitz condition (6.25) implies that

$$f(y) \leq f(x) + \nabla f(x)^T(y-x) + \frac{1}{2}(y-x)^T\nabla^2 f(x)(y-x) + \frac{L_2}{6}\|y-x\|_2^3,$$

and substituting x_k for x and $x_{k+1} = x_k + \Delta x_{nt}$ for y, we find that

$$f(x_{k+1}) \leq f(x_k) + \nabla f(x_k)^T\Delta x_{nt} + \Delta x_{nt}^T\nabla^2 f(x_k)\Delta x_{nt} + \frac{L_2}{6}\|\Delta x_{nt}\|_2^3$$

$$\leq f(x_k) - \lambda(x_k)^2 + \frac{1}{2}\lambda(x_k)^2 + \frac{L_2}{6\mu^{3/2}}\lambda(x_k)^3,$$

where we have used the result that $\mu\|\Delta x_{nt}\|_2^2 \leq \lambda(x_k)^2$, which follows from (6.24). Rewriting this inequality as

$$f(x_{k+1}) \leq f(x_k) - \left(\frac{1}{2} - \frac{L_2\lambda(x_k)}{6\mu^{3/2}}\right)\lambda(x_k)^2,$$

we see that the pure Newton step satisfies the backtracking line search condition (6.28) if

$$\frac{1}{2} - \frac{L_2\lambda(x_k)}{6\mu^{3/2}} \geq \alpha,$$

or, equivalently, if

$$\lambda(x_k) \leq \frac{3\mu^{3/2}}{L_2}(1 - 2\alpha).$$

Combining this inequality with the bound $\lambda(x_k) \leq \mu^{-1/2}\|\nabla f(x_k)\|_2$, which is readily obtained from (6.23), we conclude that a sufficient condition for the line search condition to be satisfied is that

$$\|\nabla f(x_k)\|_2 \leq \delta \leq \frac{3\mu^2}{L_2}(1 - 2\alpha). \tag{6.29}$$

Next, to derive the bound (6.26), we start by showing that the pure Newton update $x_{k+1} = x_k + \Delta x_{nt}$ satisfies

$$\|\nabla f(x_{k+1})\|_2 \leq \frac{L_2}{2\mu^2}\|\nabla f(x_k)\|_2^2. \tag{6.30}$$

This follows from the Lipschitz condition (6.25) by noting that

$$\|\nabla f(x_{k+1})\|_2 = \|\nabla f(x_k) + \int_0^1 \nabla^2 f(x_k + \tau\Delta x_{nt})\Delta x_{nt}\ d\tau)\|_2$$

$$= \|\int_0^1 \left[\nabla^2 f(x_k + \tau\Delta x_{nt}) - \nabla^2 f(x_k)\right]\Delta x_{nt}\ d\tau\|_2$$

$$\leq \frac{L_2}{2}\|\Delta x_{nt}\|_2^2,$$

and using the result that $\|\Delta x_{nt}\|_2^2 = \|\nabla^2 f(x_k)^{-1}\nabla f(x_k)\|_2^2 \leq \mu^{-2}\|\nabla f(x_k)\|_2^2$, we arrive at the inequality (6.30). Applying (6.30) recursively, we find that after l pure Newton steps,

$$\|\nabla f(x_{k+l})\|_2 \leq \left(\frac{L_2}{2\mu^2}\|\nabla f(x_k)\|_2\right)^{2^l - 1}\|\nabla f(x_k)\|_2. \tag{6.31}$$

Thus, to satisfy both (6.29) and $L_2/(2\mu^2)\|\nabla f(x_k)\|_2 \leq 1/2$, we take

$$\delta = \min(1, 3(1-2\alpha))\frac{\mu^2}{L_2}, \tag{6.32}$$

which results in the bound

$$\|\nabla f(x_{k+l})\|_2 \leq \left(\frac{1}{2}\right)^{2^l - 1}\delta, \quad l \in \mathbb{Z}_+.$$

This means that the sequence $\|\nabla f(x_{k+l})\|_2, l \in \mathbb{Z}_+$, converges at least quadratically to 0.

The strong convexity assumption can also be used to derive upper bounds on $\|x_k - x^\star\|_2$ and the suboptimality $f(x_k) - p^\star$. Indeed, using (6.26) and the fact that $\|x - x^\star\|_2 \leq (2/\mu)\|\nabla f(x)\|_2$ for all $x \in S$ when f is μ-strongly convex on S, we see that

$$\|x_{k+l} - x^\star\|_2 \leq \frac{2}{\mu}\|\nabla f(x_{k+l})\|_2 \leq \mu^{-1}\left(\frac{1}{2}\right)^{2^l-2}\|\nabla f(x_k)\|_2, \quad l \in \mathbb{Z}_+.$$

Moreover, using the fact that $f(x) - p^\star \leq 1/(2\mu)\|\nabla f(x)\|_2^2$ for all $x \in S$, *cf.* (4.30), we find that

$$f(x_{k+l}) - p^\star \leq \mu^{-1}\left(\frac{1}{2}\right)^{2^{l+1}-1}, \quad l \in \mathbb{Z}_+.$$

6.3.2.3 Damped Newton Phase

We will now show that there exists a constant $\sigma > 0$ such that $f(x_{k+1}) - f(x_k) \leq -\sigma$ whenever $\|\nabla f(x_k)\|_2 > \delta$. Since f is L-smooth on S, we have that

$$f(y) \leq f(x) + \nabla f(x)^T(y - x) + \frac{L}{2}\|y - x\|_2^2, \quad \forall\, y, x \in S.$$

Substituting $x_{k+1} = x_k - t_k \nabla^2 f(x_k)^{-1}\nabla f(x_k) = x_k + t_k \Delta x_{\mathrm{nt}}$ for y and x_k for x yields

$$f(x_{k+1}) \leq f(x_k) + t_k \nabla f(x_k)^T \Delta x_{\mathrm{nt}} + \frac{Lt_k^2}{2}\|\Delta x_{\mathrm{nt}}\|_2^2$$

$$\leq f(x_k) - \left(t_k - \frac{Lt_k^2}{2\mu}\right)\lambda(x_k)^2,$$

where we have used the lower bound $\lambda(x)^2 \geq \mu\|\Delta x_{\mathrm{nt}}\|_2^2$ and the fact that $\nabla f(x_k)^T \Delta x_{\mathrm{nt}} = -\lambda(x_k)^2$. Substituting μ/L for t_k, which minimizes the right-hand side, yields the bound

$$f(x_{k+1}) \leq f(x_k) - \frac{\mu}{2L}\lambda(x_k)^2$$

$$< f(x_k) - \alpha_1 \frac{\mu}{L}\lambda(x_k)^2,$$

since $\alpha_1 \in (0, 1/2)$. As a result, the step size μ/L satisfies the Armijo condition (6.8), which must terminate with a step size that satisfies $t_k \geq \beta \frac{\mu}{L}$. Now, using the bound $\lambda(x_k)^2 \geq L^{-1}\|\nabla f(x_k)\|_2^2$, which follows (6.23), and the assumption that $\|\nabla f(x_k)\|_2 \leq \delta$, we arrive at

$$f(x_{k+1}) - f(x_k) < -\alpha_1 \beta \frac{\mu}{L}\lambda(x_k)^2$$

$$\leq -\alpha_1 \beta \|\nabla f(x_k)\|_2^2 \frac{\mu}{L^2}$$

$$\leq -\alpha_1 \beta \delta^2 \frac{\mu}{L^2}.$$

We conclude that (6.27) is satisfied with $\sigma = \alpha_1 \beta \delta^2 \mu/L^2$.

Example 6.3 To illustrate the typical behavior of Newton's methods, we now revisit the smooth function $f : \mathbb{R}^2 \to \mathbb{R}$ from Example 6.2, i.e.

$$f(x) = \ln(\exp(2x_1 + x_2 - 2) + \exp(-x_1 + x_2 - 1) + \exp(-x_1 - 2x_2)).$$

Figure 6.4 shows the iterates obtained using Newton's method with a backtracking line search and for three different starting points. Unlike the gradient method, Newton's method clearly converges very rapidly in the vicinity of the minimizer.

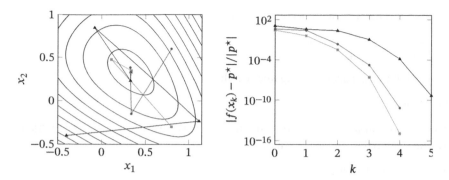

Figure 6.4 Newton's method with three different starting points.

6.3.3 Equality Constrained Minimization

Newton's method can be generalized to equality constrained problems of the form

$$\text{minimize } f(x)$$
$$\text{subject to } Ax = b, \tag{6.33}$$

with variable $x \in \mathbb{R}^n$ and where $A \in \mathbb{R}^{p \times n}$ and $b \in \mathbb{R}^p$ are given. We will restrict our attention to the case where f is convex, and we will assume that A has full rank and $p < n$. One approach to the equality constrained problem is to eliminate the constraints $Ax = b$. Specifically, suppose $F \in \mathbb{R}^{n \times (n-p)}$ is a basis for the nullspace of A and $\bar{x}$ is any vector that satisfies $A\bar{x} = b$. It follows that

$$\{x \in \mathbb{R}^n \mid Ax = b\} = \{Fz + \bar{x} \mid z \in \mathbb{R}^{n-p}\},$$

and hence, the equality constrained problem (6.33) is equivalent to

$$\text{minimize } f(Fz + \bar{x}),$$

with variable $z \in \mathbb{R}^{n-p}$. Applying Newton's method to this problem yields the Newton equation

$$F^T \nabla^2 f(Fz + \bar{x}) F \Delta z = -F^T \nabla f(Fz + \bar{x}).$$

Given a minimizer $z^\star$ of $f(Fz + \bar{x})$, a solution to the equality constrained problem may be obtained as $x^\star = Fz^\star + \bar{x}$.

Instead of eliminating the equality constraints, it is also possible to apply Newton's method to the optimality conditions associated with (6.33). The Lagrangian $L : \mathbb{R}^n \times \mathbb{R}^p \to \mathbb{R}$ is given by

$$L(x, \mu) = f(x) + \mu^T(Ax - b),$$

where $\mu \in \mathbb{R}^p$ is the Lagrange multiplier associated with $Ax = b$, and hence, the Karush–Kuhn–Tucker (KKT) conditions can be expressed as

$$\nabla f(x) + A^T \mu = 0$$
$$Ax - b = 0,$$

or equivalently, $r(z) = 0$, where $z = (x, \mu)$ and $r : \mathbb{R}^n \times \mathbb{R}^p \to \mathbb{R}^n \times \mathbb{R}^p$ is defined as

$$r(z) = \begin{bmatrix} \nabla f(x) + A^T \mu \\ Ax - b \end{bmatrix}.$$

The Newton direction $\Delta z = (\Delta x, \Delta \mu)$ for the nonlinear equation $r(z) = 0$ is the solution to the linearized equation, i.e.

$$r(z) + \frac{\partial r}{\partial z^T} \Delta z = 0.$$

Using the definition of r, this linear system of equations may be expressed as

$$\begin{bmatrix} \nabla^2 f(x) & A^T \\ A & 0 \end{bmatrix} \begin{bmatrix} \Delta x \\ \Delta \mu \end{bmatrix} = - \begin{bmatrix} \nabla f(x) + A^T \mu \\ Ax - b, \end{bmatrix}. \tag{6.34}$$

The step size can be chosen using a backtracking line search where

$$\phi(t) = \|r(z + t\Delta z)\|_2,$$

is used with the Armijo condition $\phi(t) \le \phi(0) + \alpha_1 t \phi'(0)$. If $r(z) \ne 0$ we have $\phi'(0) = -\|r(z)\|_2$ which allows us to express the Armijo descent condition as

$$\|r(z + t\Delta z)\|_2 \le (1 - t\alpha_1)\|r(z)\|_2. \tag{6.35}$$

The resulting method is often referred to as an *infeasible start* Newton method since it does not require that the initial guess x_0 satisfies the equality constraints.

6.4 Variable Metric Methods

As an alternative to the second-order Taylor approximation in (6.20) that we used to derive the Newton direction, we now consider a more general convex quadratic approximation $m_k : \mathbb{R}^n \to \mathbb{R}$ of f at x_k of the form

$$m_k(y) = f(x_k) + \nabla f(x_k)^T (y - x_k) + \frac{1}{2}(y - x_k)^T B_k(y - x_k), \tag{6.36}$$

where $B_k > 0$. The function m_k can be viewed as a local surrogate of the function f at x_k, and it is easy to check that it satisfies

$$m_k(x_k) = f(x_k), \qquad \nabla m_k(x_k) = \nabla f(x_k).$$

The approximation m_k may be used to define a search direction, i.e.

$$\Delta x = \operatorname*{argmin}_p m_k(x_k + p) = -B_k^{-1} \nabla f(x_k),$$

which is the steepest descent direction in the quadratic norm $\| \cdot \|_{B_k}$. This motivates an iteration of the form

$$x_{k+1} = x_k - t_k B_k^{-1} \nabla f(x_k), \quad k \in \mathbb{Z}_+, \tag{6.37}$$

where the step size is chosen using some form of line search. Note that the update corresponds to the gradient descent method in (6.13) if we let $B_k = I$, and the choice $B_k = \nabla^2 f(x_k)$ corresponds to Newton's method.

Generally speaking, the Newton direction is a better search direction than the negative gradient, but it is also more expensive to compute. Variable metric methods can be viewed as a compromise between the two methods in that they maintain an approximation of the Hessian or its inverse and update the approximation without computing second-order derivatives. The main condition that is used to update B_k or $H_k = B_k^{-1}$ is the so-called *secant equation*

$$B_{k+1}(x_{k+1} - x_k) = \nabla f(x_{k+1}) - \nabla f(x_k). \tag{6.38}$$

Recall that the definition of m_k implies that $\nabla m_{k+1}(x_{k+1}) = \nabla f(x_{k+1})$, and the secant equation is simply the additional condition that $\nabla m_{k+1}(x_k) = \nabla f(x_k)$. We will define $y_k = \nabla f(x_{k+1}) - \nabla f(x_k)$ and $s_k = x_{k+1} - x_k$ so that the secant equation can be expressed $B_{k+1}s_k = y_k$. We will henceforth drop the iteration index k from y_k and s_k to simplify the notation.

6.4.1 Quasi-Newton Updates

The secant equation in (6.38) is a system of n equations in $n(n+1)/2$ unknowns, and hence, it is underdetermined for $n > 1$. Quasi-Newton update formulae mostly differ in how they construct a matrix B_{k+1} that satisfies the secant equation. Taking the inner product with s on both sides of the secant equation, we see that $s^T B_{k+1} s = y^T s$, and hence, there is no positive definite solution unless $y^T s > 0$. This condition is always satisfied if f is strongly convex, which follows from the fact that ∇f is then strongly monotone; *cf.* (4.32). More generally, the condition $y^T s > 0$ can be enforced by using a line search based on the Wolfe conditions. Indeed, if $\phi(t) = f(x_k + t\Delta x)$, then $y^T s = t_k \phi'(t_k) - t_k \phi'(0)$, and hence, $y^T s > t_k(\alpha_2 - 1)\phi'(0) > 0$ if t_k satisfies the curvature condition (6.9). We will consider three different quasi-Newton updates, namely the Broyden, Fletcher, Goldfarb, and Shanno (BFGS) update, the Davidon, Fletcher, and Powell (DFP) update, and the SR1 update.

6.4.1.1 The BFGS Update

The BFGS update, which is named after Broyden, Fletcher, Goldfarb, and Shanno who independently discovered it in 1970, constructs B_{k+1} as

$$B_{k+1} = B_k - \frac{B_k s s^T B_k}{s^T B_k s} + \frac{y y^T}{y^T s}. \tag{6.39}$$

The update B_{k+1} is positive definite provided that B_k is positive definite and $y^T s > 0$, and it can be shown to be the solution to the convex optimization problem

$$\begin{aligned} \text{minimize} \quad & \operatorname{tr}(BB_k^{-1}) + \psi(B) \\ \text{subject to} \quad & Bs = y, \end{aligned} \tag{6.40}$$

with variable $B \in \mathbb{S}^n$, and where $\psi : \mathbb{S}^n \to \bar{\mathbb{R}}$ is defined as $\psi(B) = -\ln \det(B)$ with $\operatorname{dom} \psi = \mathbb{S}^n_{++}$. The BFGS update may also be expressed as

$$\begin{aligned} H_{k+1} &= H_k + \frac{(y^T s + y^T H_k y) s s^T}{(y^T s)^2} - \frac{H_k y s^T + s y^T H_k}{y^T s} \\ &= \left(I - \frac{s y^T}{y^T s} \right) H_k \left(I - \frac{y s^T}{y^T s} \right) + \frac{s s^T}{y^T s}, \end{aligned} \tag{6.41}$$

where $H_{k+1} = B_{k+1}^{-1}$ is the inverse Hessian approximation. An important consequence of this update formula is that the explicit inverse approximation H_k allows the search direction $\Delta x = -H_k \nabla f(x_k)$ to be computed by means of a matrix–vector product rather than needing to solve the system of equations $B_k \Delta x = -\nabla f(x_k)$. Algorithm 6.6 outlines the BFGS quasi-Newton algorithm using a line search based on the Wolfe conditions. We note that a backtracking line search is often used instead when the objective function is strongly convex. The initial inverse Hessian approximation H_0 can be chosen in different ways. One approach is to start with an initial gradient step $x_1 = x_0 - t_0 \nabla f(x_0)$ where t_0 is computed using a line search based on the Wolfe conditions. We then define $H_0 = \gamma I$, where

$$\gamma = \frac{y_0^T s_0}{\|y_0\|_2^2} = \frac{(\nabla f(x_1) - \nabla f(x_0))^T (x_1 - x_0)}{\|\nabla f(x_1) - \nabla f(x_0)\|_2^2},$$

before computing H_1 using the BFGS update formula. Note that γ is positive, which follows from the fact that $y_0^T s_0 > 0$ when t_0 satisfies the curvature condition. We will see the motivation behind this particular initialization later in this section when we discuss the so-called "Barzilai–Borwein" step sizes.

The BFGS update is a rank-2 update of the Hessian or inverse Hessian approximation, and B_k and H_k are typically dense for $k > 0$. As a result, the approximation requires $O(n^2)$ memory, which

---

Algorithm 6.6: BFGS algorithm

Input: Starting point x_0, initial inverse Hessian approximation $H_0 \in \mathbb{S}_{++}^n$, tolerance $\epsilon > 0$, line search parameters $0 < \alpha_1 < \alpha_2 < 1$

Output: x such that $\|\nabla f(x)\| \leq \epsilon$.

$x \leftarrow x_0, H \leftarrow H_0, g \leftarrow \nabla f(x)$

for $k = 1, 2, \dots$ **do**

    if $\|g\| \leq \epsilon$ **then** stop

    $\Delta x \leftarrow -Hg$

    $t \leftarrow \texttt{BisectionWolfe}(\phi, \alpha_1, \alpha_2)$ $\triangleright$ $\phi(t) = f(x + t\Delta x)$

    $x \leftarrow x + t\Delta x$

    $s \leftarrow t\Delta x, y \leftarrow g, g \leftarrow \nabla f(x), y \leftarrow g - y$

    $\rho \leftarrow 1/(s^T y), v \leftarrow Hy$

    $H \leftarrow H + \rho^2(y^T s + y^T v)ss^T - \rho(vs^T + sv^T)$ $\triangleright$ BFGS update

end

---

makes the method impractical for problems with large n. The *limited-memory* BFGS method, which is also known as L-BFGS, addresses this issue by storing a limited history of, say, m BFGS update pairs (s_{k-l}, y_{k-l}), $l \in \mathbb{N}_m$, that are used to implicitly define H_k for $k \geq m$. Specifically, H_k is defined as a sequence of m BFGS updates starting with an initial approximate inverse Hessian H_k^0, which is typically chosen as a diagonal matrix or a scaled identity matrix such as $H_k^0 = \gamma_k I$, where $\gamma_k = y_{k-1}^T s_{k-1} / \|y_{k-1}\|_2^2$. Note that rather than explicitly applying the m BFGS updates to form H_k, the m update pairs are used to recursively compute matrix-vector products with H_k without forming it. We note that L-BFGS method requires $O(n)$ memory for a fixed memory parameter m, which is typically between 10 and 50, and hence, L-BFGS requires significantly less memory than BFGS when n is large.

6.4.1.2 The DFP Update

Another update that is closely related to the BFGS update is the DFP update, which is named after Davidon, Fletcher, and Powell. It was first proposed in 1959, and hence, it predates the BFGS update. It can be expressed as

$$B_{k+1} = \left(I - \frac{ys^T}{y^T s}\right) B_k \left(I - \frac{sy^T}{y^T s}\right) + \frac{yy^T}{y^T s}$$

where $r = y - B_k s$, or equivalently, as an inverse Hessian approximation $H_{k+1} = B_k^{-1}$,

$$H_{k+1} = H_k - \frac{H_k yy^T H_k}{y^T H_k y} + \frac{ss^T}{y^T s}.$$

Notice the similarity with the BFGS update formulas, which can be obtained from the DFP update formulas by interchanging s and y as well as B and H. Like the BFGS update, the DFP update preserves positive definiteness if $y^T s > 0$. However, despite the similarity between the BFGS and the DFP updates, the BFGS update is known to possess some favorable self-correcting properties, and hence, it is often used in practice.

6.4.1.3 The SR1 Update

Yet another quasi-Newton update formula is the so-called "SR1 update," which is a symmetric rank-1 update of the form

$$B_{k+1} = B_k + \sigma vv^T,$$

where $\sigma \in \{-1, 1\}$. The secant equation $B_{k+1} s = y$ can then be expressed as

$$r = \sigma(v^T s)v, \tag{6.42}$$

where $r = y - B_k s$. Taking the inner product with s on both sides of this equation yields $\sigma(v^T s)^2 = r^T s$. Combining this with (6.42), we find that

$$vv^T = \frac{rr^T}{(\sigma^2 v^T s)^2} = \frac{rr^T}{\sigma r^T s},$$

and hence,

$$B_{k+1} = B_k + \frac{rr^T}{r^T s}. \tag{6.43}$$

Note that if $r = 0$ or $r^T s = 0$, we simply skip the update and take $B_{k+1} = B_k$. Unlike the BFGS and DFP updates, the SR1 update does not guarantee that B_{k+1} is positive definite, and hence, it is often used in combination with a trust-region method.

6.4.2 The Barzilai–Borwein Method

The secant condition can also be used to motivate the so-called *Barzilai–Borwein* (BB) step size rules for the gradient descent method. If we require that $B_{k+1} = \zeta_{k+1} I$, where ζ_{k+1} is a scalar, the secant condition is generally not satisfiable, but we can choose ζ_{k+1} such that it minimizes $\|B_{k+1}s - y\|_2^2$, i.e.

$$\zeta_{k+1} = \underset{\zeta}{\operatorname{argmin}} \ \|\zeta s - y\|_2^2 = \frac{y^T s}{\|s\|_2^2}.$$

Alternatively, we can define an inverse approximation $H_{k+1} = \gamma_{k+1} I$ and minimize $\|s - H_{k+1}y\|_2^2$, which yields the similar update

$$\gamma_{k+1} = \underset{\gamma}{\operatorname{argmin}} \ \|s - \gamma y\|_2^2 = \frac{y^T s}{\|y\|_2^2}.$$

We note that both ζ_{k+1} and γ_{k+1} are positive if and only if $y^T s > 0$. The so-called "Barzilai–Borwein gradient method" may be viewed as a special case of the variable metric iteration in (6.37) with $t_k = 1$ and either $B_k = \zeta_k I$ or $B_k = \gamma_k^{-1} I$, or equivalently, as the gradient iteration in (6.13) with the step size $t_k = \zeta_k^{-1}$ or $t_k = \gamma_k$. The resulting method is not a descent method: it is known to converge if f is strongly convex and quadratic, but it is not guaranteed to converge in general. Convergence can be established if the BB steps are combined with some form of safeguard such as a nonmonotone line search; see, e.g. [121].

6.5 Proximal Gradient Method

The majorization minimization approach that we used to analyze the gradient descent method for the minimization of an L-smooth function can also be applied to the more general problem

$$\text{minimize} \ \ g(x) + h(x), \tag{6.44}$$

where $g : \mathbb{R}^n \to \mathbb{R}$ is L-smooth, and $h : \mathbb{R}^n \to \bar{\mathbb{R}}$ is proper, closed, and convex but not necessarily differentiable. The assumption that g is L-smooth implies that $f : \mathbb{R}^n \to \bar{\mathbb{R}}$ defined as $f = g + h$ satisfies

$$f(y) \le g(x) + \nabla g(x)^T (y - x) + \frac{L}{2}\|y - x\|_2^2 + h(y). \tag{6.45}$$

The right-hand side is a majorization of f at x, and applying the majorization minimization algorithm yields the iteration

$$x_{k+1} = \underset{y}{\operatorname{argmin}} \left\{ h(y) + \nabla g(x_k)^T y + \frac{L}{2}\|y - x_k\|_2^2 \right\}$$

$$= \underset{y}{\operatorname{argmin}} \left\{ h(y) + \frac{L}{2}\|y - (x_k - \frac{1}{L}\nabla g(x_k))\|_2^2 \right\}.$$

The resulting method is known as the *proximal gradient* (PG) method, and it can also be expressed as the iteration

$$x_{k+1} = \text{prox}_{t_k h} \left(x_k - t_k \nabla g(x_k) \right), \quad k \in \mathbb{Z}_+, \tag{6.46}$$

where $t_k = 1/L$ and $\text{prox}_h : \mathbb{R}^n \to \mathbb{R}^n$ is the so-called *proximal operator* associated with the function h and defined as

$$\text{prox}_h(x) = \underset{y}{\text{argmin}} \left\{ h(y) + \frac{1}{2} \|y - x\|_2^2 \right\}. \tag{6.47}$$

From the definition of the proximal operator, we see that

$$u = \text{prox}_h(x) \quad \Longleftrightarrow \quad x - u \in \partial h(u),$$

and hence, the PG update in (6.46) can be expressed as

$$x_{k+1} = x_k - t_k \nabla g(x_k) - t_k v_{k+1},$$

for some $v_{k+1} \in \partial h(x_{k+1})$. Note that evaluating the proximal operator amounts to solving an optimization problem, and hence, a single PG iteration is therefore expensive unless the proximal operator is cheap to evaluate. The PG method is outlined in Algorithm 6.7. It can also be combined with a line search as an alternative to the constant step size $t_k = 1/L$.

Algorithm 6.7: Proximal gradient method (constant step size)

Input: L-smooth convex function $g : \mathbb{R}^n \to \mathbb{R}$, proper closed convex function $h : \mathbb{R}^n \to \bar{\mathbb{R}}$, starting point x_0, and tolerance $\epsilon > 0$.
Output: x such that $\|x - \text{prox}_{th} (x - t\nabla g(x))\| \leq \epsilon$.

$x \leftarrow x_0, t \leftarrow 1/L$
for $k = 1, 2, \ldots$ **do**
$\quad \tilde{x} \leftarrow \text{prox}_{th} (x - t\nabla g(x))$
$\quad$ **if** $\|\tilde{x} - x\| \leq \epsilon$ **then** stop
$\quad x \leftarrow \tilde{x}$
end

When g is nonconvex, the PG method can be shown to converge to a stationary point under some mild conditions. The first-order necessary optimality condition for (6.44) implies that a stationary point $x^\star$ must satisfy

$$-\nabla g(x^\star) \in \partial h(x^\star). \tag{6.48}$$

If we compare this condition to the stationarity condition for the majorization in (6.45) at $x^\star$, i.e.

$$-\nabla g(x^\star) - L(y - x^\star) \in \partial h(y),$$

we see that the conditions coincide when $y = x^\star$. This implies that a fixed-point of the iteration in (6.46) is a stationary point of $f = g + h$. In the convex case, i.e. when both g and h are convex, the PG iteration can be shown to satisfy

$$f(x_k) - p^\star = O(1/k).$$

An overview of other variants of the PG method, including a detailed analysis of both the convex and the nonconvex case, can be found in [10]. We end this section by mentioning a few methods that are closely related to the PG method.

6.5.1 Gradient Projection Method

The PG method can also be applied to constrained optimization problems of the form

$$
\begin{aligned}
& \text{minimize } g(x) \\
& \text{subject to } x \in C,
\end{aligned}
\tag{6.49}
$$

where C is a convex subset of $\mathbb{R}^n$. This problem is equivalent to (6.44) if we define $h(x) = I_C(x)$, where I_C denotes the indicator function of the set C. In this special case, the proximal operator $\text{prox}_h(x)$ becomes the Euclidean projection of x onto C, and the resulting method is commonly referred to as the *projected gradient* or *gradient projection* method.

6.5.2 Proximal Quasi-Newton

A generalization of the PG method is obtained if we construct a surrogate function $m_k : \mathbb{R}^n \to \bar{\mathbb{R}}$ of the form

$$
m_k(y) = h(y) + g(x_k) + \nabla g(x_k)^T(y - x_k) + \frac{1}{2}(y - x_k)^T B_k(y - x_k),
\tag{6.50}
$$

with $B_k \in \mathbb{S}^n_{++}$ and use its minimizer to define a new iterate or a search direction, e.g.

$$
\begin{aligned}
x_{k+1} &= \underset{y}{\text{argmin }} m_k(y) \\
&= \underset{y}{\text{argmin }} \left\{ h(y) + \frac{1}{2}\|y - (x_k - B_k^{-1}\nabla g(x_k))\|_{B_k}^2 \right\} \\
&= \text{prox}_h^{B_k}(x_k - B_k^{-1}\nabla g(x_k)),
\end{aligned}
\tag{6.51}
$$

where $\text{prox}_h^{B_k} : \mathbb{R}^n \to \mathbb{R}^n$ is the proximal operator of h for the norm $\| \cdot \|_{B_k}$ and defined as

$$
\text{prox}_h^{B_k}(x) = \underset{y}{\text{argmin }} \left\{ h(y) + \frac{1}{2}\|y - x\|_{B_k}^2 \right\}.
\tag{6.52}
$$

The iteration (6.51) is known as a *proximal quasi-Newton* method or *proximal Newton* method in the special case, where $B_k = \nabla^2 f(x_k)$. We refer the reader to [66] for further details.

6.5.3 Accelerated Proximal Gradient Method

The gradient method and the PG method can be combined with acceleration techniques that lead to an improved rate of convergence. One such technique is the so-called *heavy ball* method, which was proposed by Polyak in the 1960s [89]. This method may be derived from a second-order differential equation of the form

$$
\mu \frac{d^2 x}{dt^2} = -\nabla f(x) - m \frac{dx}{dt},
$$

where $x : \mathbb{R} \to \mathbb{R}^n$ is a state vector. The differential equation characterizes the motion of a body with mass, e.g. a heavy ball, under friction in a potential field. The constants μ and m are friction and mass constants, respectively. Employing a finite-difference method leads to an iteration of the form

$$
x_{k+1} = x_k - \gamma \nabla f(x_k) + \eta(x_k - x_{k-1}),
$$

where γ and η are constants and $x_k = x(t_k)$ is the state at time t_k. This is an example of a so-called "multistep method." Notice that it reduces to the gradient method if $\eta = 0$, and hence, it is the last term that introduces momentum.

Inspired by the heavy ball method, Nesterov [80] proposed an accelerated gradient method that satisfies the improved worst-case bound

$$f(x_k) - p^\star = O(1/k^2),$$

when f is convex and L-smooth. An extension of this method is the so-called *accelerated proximal gradient* (APG) method, which is due to [11] and outlined in Algorithm 6.8. Unlike the PG method, the APG method shown here is not a descent method. The next example illustrates the effect of the acceleration.

Algorithm 6.8: Accelerated proximal gradient method (constant step size)

Input: L-smooth convex function $g: \mathbb{R}^n \to \mathbb{R}$, proper closed convex function
$\qquad h: \mathbb{R}^n \to \bar{\mathbb{R}}$, starting point x_0, and tolerance $\epsilon > 0$.

$x \leftarrow x_0, y \leftarrow x, \gamma_1 \leftarrow 1, t \leftarrow 1/L$

for $k = 1, 2, \ldots$ **do**

$\qquad \tilde{x} \leftarrow \text{prox}_{th}(y - t\nabla g(y))$

$\qquad \gamma_{k+1} \leftarrow \left(1 + \sqrt{1 + 4\gamma_k^2}\right)/2$

$\qquad y \leftarrow \tilde{x} + ((\gamma_k - 1)/\gamma_{k+1})(\tilde{x} - x)$

$\qquad x \leftarrow \tilde{x}$

end

Example 6.4 Consider the 1-norm regularized least-squares problem

$$\text{minimize} \quad \frac{1}{2}\|Ax - b\|_2^2 + \gamma\|x\|_1$$

with variable $x \in \mathbb{R}^n$, problem data $A \in \mathbb{R}^{m \times n}$ and $b \in \mathbb{R}^m$, and parameter $\gamma > 0$. To apply the PG and APG methods, we define $g: \mathbb{R}^n \to \mathbb{R}$ as $g(x) = (1/2)\|Ax - b\|_2^2$ and $h: \mathbb{R}^n \to \mathbb{R}$ as $h(x) = \gamma\|x\|_1$. The function g is L-smooth with constant $L = \|A\|_2^2$, and hence, we may use the constant step size $t = 1/L$. The proximal operator $\text{prox}_{th}(x)$ can be expressed as

$$\text{prox}_{th}(x) = (S_{t\gamma}(x_1), \ldots, S_{t\gamma}(x_n)),$$

where $S_{t\gamma}: \mathbb{R} \to \mathbb{R}$ is defined as

$$S_{t\gamma}(\tau) = \text{sgn}(\tau)\max(0, |\tau| - t\gamma).$$

This known as the *soft-thresholding operator*; see Exercise 6.6.

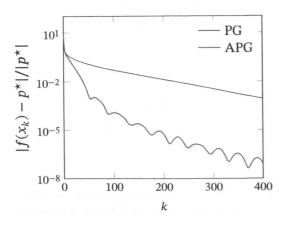

Figure 6.5 The relative suboptimality for the PG and the APG methods.

Figure 6.5 shows a numerical example for a problem instance with $m = 200$ and $n = 800$. The plot clearly shows that the PG method is a descent method, whereas the APG method is not. Both methods exhibit a sublinear convergence rate, but the effect of acceleration makes a significant difference.

6.6 Sequential Convex Optimization

We now consider optimization problems with an objective function that can be expressed as the difference of two convex functions, i.e.

$$\text{minimize } g(x) - h(x), \tag{6.53}$$

where $g : \mathbb{R}^n \to \bar{\mathbb{R}}$ and $h : \mathbb{R}^n \to \bar{\mathbb{R}}$ are both proper, closed, and convex functions. Such problems are often referred to as *difference convex* optimization problems. The difference $g - h$ is generally not a convex function, but a convex majorization of $g - h$ at a point x can easily be constructed if h is continuously differentiable or subdifferentiable at x. Indeed, since h is convex, we can use an affine lower bound on h

$$h(y) \geq h(x) + v^T(y - x), \quad v \in \partial h(x),$$

to construct a majorization of $f(x) = g(x) - h(x)$ at x, i.e.

$$f(y) \leq g(y) - h(x) - v^T(y - x), \quad \forall\, x, y \in \text{dom } f. \tag{6.54}$$

The majorization minimization iteration then amounts to the iteration

$$x_{k+1} = \underset{y}{\operatorname{argmin}}\{g(y) - v_k^T y\}, \quad k \in \mathbb{Z}_+, \tag{6.55}$$

where $v_k \in \partial h(x_k)$ is any subgradient of h at x_k. This approach is sometimes called *sequential convex optimization* since each iteration involves the solution of a convex optimization problem. The iteration (6.55) is a descent method by design, and with additional assumptions, it can be shown to converge to a stationary point or a local minimum of $f = g - h$; see, e.g. [65].

Sequential convex optimization can also be useful when the objective function is not readily decomposed into the difference of two convex functions, which we now illustrate with an example.

Example 6.5 Consider the optimization problem

$$\text{minimize } g(x) + \sum_{i=1}^{n} h(|x_i|)$$

where $g : \mathbb{R}^n \to \mathbb{R}$ is convex and $h : (-\varepsilon, \infty) \to \mathbb{R}$ is nondecreasing, concave, and continuously differentiable on $(-\varepsilon, \infty)$ with $\varepsilon > 0$. The composition $h(|x_i|)$ is not necessarily convex or concave, but concavity of h implies that $h(s) + h'(s)(t - s) \geq h(t)$, and hence,

$$\sum_{i=1}^{n} \left(h(|x_i|) + h'(|x_i|)(|y_i| - |x_i|) \right) \geq \sum_{i=1}^{n} h(|y_i|),$$

for all $y, x \in \mathbb{R}^n$. The assumption that h is nondecreasing implies that $h'(|x_i|) \geq 0$, and hence, the left-hand side is a convex function of y, and it is a majorization of the right-hand side at x. This leads to the majorization minimization iteration

$$x_{k+1} = \underset{y}{\operatorname{argmin}}\{g(y) + \|\operatorname{diag}(w_k)y\|_1\}, \quad k \in \mathbb{Z}_+, \tag{6.56}$$

where $w_k \geq 0$ is the vector with elements $w_{k,i} = h'(|x_{k,i}|)$ for $i \in \mathbb{N}_n$. For example, the function $h(t) = \ln(t + \delta)$ with $\delta > 0$ is an increasing concave function on $(-\delta, \infty)$, which leads to the weight update $w_{k,i} = (|x_{k,i}| + \delta)^{-1}$. We end this example by noting that the approach is also known as iteratively reweighted ℓ_1-regularization since each iteration involves a weighted ℓ_1-regularized optimization problem with new weights.

6.7 Methods for Nonlinear Least-Squares

Recall the general nonlinear LS problem (5.1), which is an unconstrained optimization problem of the form

$$\text{minimize} \quad \frac{1}{2}\|f(x)\|_2^2,$$

where $f : \mathbb{R}^n \to \mathbb{R}^m$ is defined as $f(x) = (f_1(x), \dots, f_m(x))$ and $f_i : \mathbb{R}^n \to \mathbb{R}, i \in \mathbb{N}_m$. This is generally a difficult nonlinear optimization problem. Several local optimization methods exist that are tailored to the specific structure of the problem. One such method is the *Gauss–Newton* (GN) method, which is applicable when the functions $f_1, \dots, f_m$ are continuously differentiable. The basic idea is to replace f by its first-order Taylor approximation around the current iterate x_k, i.e. $\tilde{f} : \mathbb{R}^n \to \mathbb{R}^m$ defined as

$$\tilde{f}(x; x_k) = f(x_k) + \frac{\partial f(x_k)}{\partial x^T}(x - x_k).$$

The GN method can then be expressed as the iteration

$$x_{k+1} \in \underset{x}{\text{argmin}} \frac{1}{2}\|\tilde{f}(x; x_k)\|_2^2, \quad k \geq 0. \tag{6.57}$$

Each iteration involves solving a linear least-squares problem since $\tilde{f}(x; x_k)$ is an affine function of x, and the update is unique if the Jacobian matrix $\partial f(x_k)/\partial x^T$ has full rank. Noting that $(1/2)\|\tilde{f}(x; x_k)\|_2^2$ is a quadratic approximation of $(1/2)\|f(x)\|_2^2$, we see that the GN method can also be viewed as a variable metric method.

If f is twice continuously differentiable, then the Hessian of $f_0(x) = (1/2)\|f(x)\|_2^2$ is given by

$$\nabla^2 f_0(x) = \sum_{i=1}^{m} \left(f_i(x)^2 \nabla^2 f_i(x) + \nabla f_i(x) \nabla f_i(x)^T \right),$$

whereas the Hessian of $\tilde{f}_0(x; x_k) = (1/2)\|\tilde{f}(x; x_k)\|_2^2$ is given by

$$\nabla^2 \tilde{f}_0(x; x_k) = \sum_{i=1}^{m} \nabla f_i(x_k) \nabla f_i(x_k)^T.$$

Comparing the two Hessians, we see that

$$\nabla^2 f_0(x_k) = \nabla^2 \tilde{f}_0(x_k; x_k) + \sum_{i=1}^{m} f_i(x_k)^2 \nabla^2 f_i(x_k).$$

This suggests that the GN update (6.57) mimics a pure Newton step if the sum on the right-hand side is negligible, in which case we can expect super-linear convergence. However, although the GN method often works well in practice, it may fail to converge. This issue can be addressed by combining the GN method with a line search provided that the sublevel set $\{x \mid f_0(x) \leq f_0(x_0)\}$ is bounded and that $\partial f(x_k)/\partial x^T$ has full rank for all k.

6.7.1 The Levenberg-Marquardt Algorithm

The *Levenberg–Marquardt* (LM) algorithm addresses the shortcomings of the GN method by introducing regularization, i.e. it uses the strongly convex quadratic function $\tilde{f}_0 : \mathbb{R}^n \to \mathbb{R}$ defined as

$$\tilde{f}_0(x; x_k) = \frac{1}{2}\|\tilde{f}(x; x_k)\|_2^2 + \frac{\mu_k}{2}\|x - x_k\|_2^2, \tag{6.58}$$

where $\mu_k \in \mathbb{R}_{++}$ is a parameter, as a local approximation of $f_0(x) = (1/2)\|f(x)\|_2^2$ at x_k. From the Hessian of $\tilde{f}_0(x; x_k)$, i.e.

$$\nabla^2 \tilde{f}_0(x; x_k) = \sum_{i=1}^{m} \nabla f_i(x_k)\nabla f_i(x_k)^T + \mu_k I,$$

we see that if μ_k is sufficiently large, then the LM method will essentially behave like gradient descent with a step size that is roughly equal to $1/\mu_k$. On the other hand, if μ_k is small, then the step is very close a Gauss–Newton step. Algorithm 6.9 adjusts μ_k in an adaptive manner and is one of many different variants of the LM algorithm. For example, there are trust-region variants of the GN method, which are also sometimes referred to as LM algorithms; see, e.g. [78].

Algorithm 6.9: Levenberg–Marquardt algorithm

Input: Starting point x_0, initial damping parameter $\mu_0 > 0$, and tolerance $\epsilon > 0$.
Output: x such that $\|\nabla f_0(x)\| \leq \epsilon$.

$x \leftarrow x_0, \mu \leftarrow \mu_0$
for $k = 1, 2, \ldots$ **do**
$\quad$ $\tilde{x} \leftarrow \operatorname{argmin}_u \{\frac{1}{2}\|\tilde{f}(u; x)\|_2^2 + \frac{\mu}{2}\|u - x\|_2^2\}$
$\quad$ **if** $f_0(\tilde{x}) < f_0(x)$ **then**
$\quad\quad$ $x \leftarrow \tilde{x}$ $\qquad\qquad\qquad\qquad$ ▷ Accept step
$\quad\quad$ $\mu \leftarrow 0.8\mu$ $\qquad\qquad\qquad\quad$ ▷ Reduce μ
$\quad$ **else**
$\quad\quad$ $\mu \leftarrow 2\mu$ $\qquad\qquad\qquad\qquad$ ▷ Increase μ
$\quad$ **end**
$\quad$ **if** $\|\nabla f_0(x)\| \leq \epsilon$ **then** stop
end

6.7.2 The Variable Projection Method

A nonlinear LS problem is called *separable* if it can be expressed as

$$\text{minimize} \quad \frac{1}{2}\|f(x, \alpha)\|_2^2 \tag{6.59}$$

with variables $x \in \mathbb{R}^n$ and $\alpha \in \mathbb{R}^p$, and where $f : \mathbb{R}^n \times \mathbb{R}^p \to \mathbb{R}^m$ is of the form

$$f(x, \alpha) = A(\alpha)x - b(\alpha)$$

with $A : \mathbb{R}^p \to \mathbb{R}^{m \times n}$ and $b : \mathbb{R}^p \to \mathbb{R}^m$. Notice that this is a linear LS problem in x if α is fixed. We can use this fact and minimize with respect to x to obtain

$$x^\star(\alpha) = A(\alpha)^\dagger b(\alpha).$$

Substituting $x^\star(\alpha)$ for x in (6.59), we obtain the nonlinear LS problem

$$\text{minimize} \quad \frac{1}{2}\|g(\alpha)\|_2^2, \tag{6.60}$$

where $g : \mathbb{R}^p \to \mathbb{R}^m$ is defined as

$$g(\alpha) = f(x^\star(\alpha), \alpha) = -(I - A(\alpha)A(\alpha)^\dagger)b(\alpha).$$

Recall that $P(\alpha) = I - A(\alpha)A(\alpha)^\dagger$ is a projection matrix, and hence $g(\alpha)$ is the projection of $-b(\alpha)$ onto the nullspace of $A(\alpha)^T$. We will henceforth assume that $\text{rank}(A(\alpha)) = n$, which implies that $A^\dagger(\alpha) = (A(\alpha)^T A(\alpha))^{-1} A(\alpha)^T$, and we will omit α from $A(\alpha)$, $b(\alpha)$, and $P(\alpha)$ for notational convenience.

The method we will derive is often called the *variable projection method*, since it is based on minimizing the variable projection functional (6.60). In principle, one could simply apply the GN method or the LM algorithm directly to this problem, but it turns out that there is a better way to proceed. The idea is to use only an approximate gradient [61]. We will follow the derivation in [86]. The gradient of $g(\alpha)$ is the transpose of $\partial g / \partial \alpha^T$, which may be expressed as

$$\frac{\partial g}{\partial \alpha^T} = -\frac{\partial P}{\partial \alpha}b - P\frac{\partial b}{\partial \alpha^T},$$

where we define $\frac{\partial P}{\partial \alpha}b$ to be the matrix with columns $\frac{\partial P}{\partial \alpha_k}b$. Now, using the fact that

$$\frac{\partial P}{\partial \alpha_k} = -\frac{\partial A}{\partial \alpha_k}\left(A^T A\right)^{-1}A^T - A\left(A^T A\right)^{-1}\frac{\partial A^T}{\partial \alpha_k}$$
$$+ A\left(A^T A\right)^{-1}\left(\frac{\partial A^T}{\partial \alpha_k}A + A^T\frac{\partial A}{\partial \alpha_k}\right)\left(A^T A\right)^{-1}A^T$$
$$= -P\frac{\partial A}{\partial \alpha_k}\left(A^T A\right)^{-1}A^T - A\left(A^T A\right)^{-1}\frac{\partial A^T}{\partial \alpha_k}P,$$

the Jacobian of g at α can be expressed as

$$\frac{\partial g}{\partial \alpha^T} = P\frac{\partial A}{\partial \alpha}\left(A^T A\right)^{-1}A^T b + A\left(A^T A\right)^{-1}\frac{\partial A^T}{\partial \alpha}Pb - P\frac{\partial b}{\partial \alpha^T}.$$

The second term, which contains the factor $Pb = -g(\alpha)$, is omitted in an approximation suggested by Kaufman [61]. This is motivated based on the observation that $g(\alpha)$ is typically small when α is close to a minimum. Using the expression for $x^\star(\alpha)$, the Kaufman approximation of the Jacobian can be expressed as

$$\frac{\partial g}{\partial \alpha^T} \approx P\frac{\partial A}{\partial \alpha}x^\star(\alpha) - P\frac{\partial b}{\partial \alpha^T}. \tag{6.61}$$

This approximation of the Jacobian does not only save about 25% in computational time, but it can also be shown to provide better convergence properties; see, e.g. [43]. The Kaufman approximation was originally proposed for the LM algorithm, but it can be used with any of the methods for nonlinear LS problems that we have discussed in this section. Applications to the training of artificial neural networks is discussed in [99], and we will discuss applications to system identification in Chapter 12.

6.8 Stochastic Optimization Methods

Recall the single-stage stochastic optimization problem

$$\text{minimize } \mathbb{E}[F(x, \xi)], \tag{6.62}$$

where $F : \mathbb{R}^n \times \mathcal{D} \to \mathbb{R}$. As mentioned in Section 5.7, the evaluation of the expectation $\mathbb{E}[F(x, \xi)]$ is often intractable, so we will focus on solution methods that only require an oracle that can return unbiased gradient estimates. One such method is the so-called *stochastic approximation* (SA) method of Robbins & Monro [96], which is a method for finding a root of a nonlinear equation.

In the context of the stochastic problem in (6.62), the nonlinear equation of interest is the stationarity condition $\nabla f(x) = 0$, where $f(x) = \mathbb{E}[\,F(x, \xi)\,]$. The resulting algorithm is an iteration of the form

$$x_{k+1} = x_k - t_k\, g_k, \quad k \in \mathbb{Z}_+, \tag{6.63}$$

where x_0 is an initial guess, $t_k > 0$ is the step size at iteration k, and $g_k \in \mathbb{R}^n$ is a realization of an estimator of $G_k \in \mathbb{R}^n$ of $\nabla f(x_k)$. It should be stressed that G_k is random variable for each k. Hence, the iteration (6.63) is a realization of a stochastic process

$$X_{k+1} = X_k - t_k\, G_k, \quad k \in \mathbb{Z}_+, \tag{6.64}$$

where the random variable X_0 is the initial state of the process. We consider an unbiased estimator G_k of $\nabla f(x_k)$ with bounded variance, i.e.

$$\mathbb{E}[\,G_k \mid X_k = x_k\,] = \nabla f(x_k), \quad \mathbb{E}[\,\|G_k - \nabla f(x_k)\|_2^2 \mid X_k = x_k\,] \leq c^2, \tag{6.65}$$

for all k and for some scalar $c \geq 0$. Note that in the special case where $c = 0$, the iteration in (6.63) is essentially the gradient descent method (6.13). We note that the gradient of $F(x, \xi)$ with respect to x, which we will denote $\nabla F(x, \xi)$, is a random variable, and it is an unbiased estimator of $\nabla f(x)$ if

$$\frac{\partial \mathbb{E}[\,F(x, \xi)\,]}{\partial x_i} = \mathbb{E}\left[\frac{\partial F(x, \xi)}{\partial x_i}\right], \quad i \in \mathbb{N}_n.$$

When this condition is satisfied, it often natural to choose $G_k = \nabla F(X_k, \xi_k)$, where ξ_k has the same distribution as ξ for all $k \geq 0$, and where ξ_k and ξ_j are independent for $j \neq k$.

The iteration (6.63) is often referred to as a *stochastic gradient* (SG) method or a *stochastic gradient descent* (SGD) method. However, it is important to note that it is not a descent method in the deterministic sense, i.e. the search direction $-g_k$ is not necessarily a descent direction, so the objective value may increase in some iteration. We note that the step size t_k is often referred to as the *learning rate* in machine learning literature.

To better understand the influence of the step size sequence t_k, we will now analyze the process under different assumptions on f.

6.8.1 Smooth Functions

We start by considering the case where f is L-smooth and bounded from below. Recall from (6.2) that L-smoothness implies that there exists a constant $L > 0$ such that

$$f(x + \Delta x) \leq f(x) + \nabla f(x)^T \Delta x + \frac{L}{2}\|\Delta x\|_2^2,$$

is satisfied for all $x, \Delta x \in \mathbb{R}^n$. Thus, we have that X_k satisfies

$$\mathbb{E}[\,f(X_k) \mid X_{k-1}\,] \leq \mathbb{E}\left[f(X_{k-1}) - t_{k-1}\nabla f(X_{k-1})^T G_{k-1} + \frac{Lt_{k-1}^2}{2}\|G_{k-1}\|_2^2 \mid X_{k-1}\right]. \tag{6.66}$$

It follows from (6.65) that

$$\mathbb{E}[\,\nabla f(X_{k-1})^T G_{k-1} \mid X_{k-1}\,] = \|\nabla f(X_{k-1})\|_2^2$$

and

$$\mathbb{E}[\,\|G_{k-1}\|_2^2 \mid X_{k-1}\,] = \mathbb{E}[\,\|G_{k-1} - \nabla f(X_{k-1}) + \nabla f(X_{k-1})\|_2^2 \mid X_{k-1}\,] \leq c^2 + \|\nabla f(X_{k-1})\|_2^2.$$

Combining these results with (6.66), we arrive at the upper bound

$$\mathbb{E}[\,f(X_k) \mid X_{k-1}\,] \leq f(X_{k-1}) - t_{k-1}\|\nabla f(X_{k-1})\|_2^2 + \frac{Lt_{k-1}^2}{2}\left(c^2 + \|\nabla f(X_{k-1})\|_2^2\right)$$

$$= f(X_{k-1}) - t_{k-1}(1 - Lt_{k-1}/2)\|\nabla f(X_{k-1})\|_2^2 + \frac{Lt_{k-1}^2 c^2}{2}.$$

Rearranging the terms leads to the inequality

$$t_{k-1}(1 - Lt_{k-1}/2)\|\nabla f(X_{k-1})\|_2^2 \leq f(X_{k-1}) - \mathbb{E}[f(X_k) \mid X_{k-1}] + t_{k-1}^2 \frac{Lc^2}{2},$$

and summing and taking expectation yields

$$\sum_{j=0}^{k-1} t_j(1 - Lt_j/2)\mathbb{E}[\|\nabla f(X_j)\|_2^2] \leq \sum_{j=0}^{k-1} [\mathbb{E}[f(X_j)] - \mathbb{E}[f(X_{j+1})]] + \sum_{j=0}^{k-1} t_j^2 \frac{Lc^2}{2}. \tag{6.67}$$

The first sum on the right-hand side satisfies

$$\sum_{j=0}^{k-1} [\mathbb{E}[f(X_j)] - \mathbb{E}[f(X_{j+1})]] = \mathbb{E}[f(X_0)] - \mathbb{E}[f(X_k)] \leq \mathbb{E}[f(X_0)] - p^{\star},$$

where $p^{\star} = \inf_x f(x)$, and combining this inequality with (6.67) yields

$$\sum_{j=0}^{k-1} t_j(1 - Lt_j/2) \min_{j=0,\ldots,k-1} \mathbb{E}[\|\nabla f(X_j)\|_2^2] \leq \mathbb{E}[f(X_0)] - p^{\star} + \sum_{j=0}^{k-1} t_j^2 \frac{Lc^2}{2}.$$

Equivalently, if we divide by the sum $\sum_{j=0}^{k-1} t_j(1 - Lt_j/2)$ on both sides, and assuming that it is positive, we see that

$$\min_{j=0,\ldots,k-1} \mathbb{E}[\|\nabla f(X_j)\|_2^2] \leq \frac{\mathbb{E}[f(X_0)] - p^{\star}}{\sum_{j=0}^{k-1} t_j(1 - Lt_j/2)} + \frac{\sum_{j=0}^{k-1} t_j^2}{\sum_{j=0}^{k-1} t_j(1 - Lt_j/2)} \frac{Lc^2}{2}. \tag{6.68}$$

It follows that a sufficient condition for the right-hand side to vanish as $k \to \infty$ is that

$$\lim_{k\to\infty} \sum_{j=0}^{k-1} t_j(1 - Lt_j/2) = \infty, \quad \lim_{k\to\infty} \frac{\sum_{j=0}^{k-1} t_j(1 - Lt_j/2)}{\sum_{j=0}^{k-1} t_j^2} = \infty,$$

or equivalently,

$$\lim_{k\to\infty} \sum_{j=0}^{k-1} t_j = \infty, \quad \lim_{k\to\infty} \frac{\sum_{j=0}^{k-1} t_j}{\sum_{j=0}^{k-1} t_j^2} = \infty. \tag{6.69}$$

Examples of step size sequences that satisfy these conditions are sequences of the form

$$t_k = \frac{t}{(k + 1 + \zeta)^{\delta}}, \quad k \in \mathbb{Z}_+, \tag{6.70}$$

where $t > 0$, $\zeta \geq 0$, and $\delta \in (0, 1]$ are fixed parameters. The parameter t scales the sequence, δ controls the asymptotic rate of decay, and ζ may be used to reduce the rate of decay in early iterations. The value of t has no effect on the asymptotic behavior, but it typically has a strong effect on the nonasymptotic behavior. To see this, first note that step-size sequences of the form in (6.70) satisfy

$$\frac{1}{\sum_{j=0}^{k-1} t_j} \propto \frac{1}{t}, \quad \frac{\sum_{j=0}^{k-1} t_j^2}{\sum_{j=0}^{k-1} t_j} \propto t.$$

Comparing with the right-hand side of (6.68), we see that the choice of t presents a trade-off between the two terms: increasing t reduces the first term, but increases the second and vice versa.

We now analyze the worst-case bound in (6.68) for different step-size sequences. We start by noting that if $c = 0$, which corresponds to the ordinary gradient method, then it suffices to choose

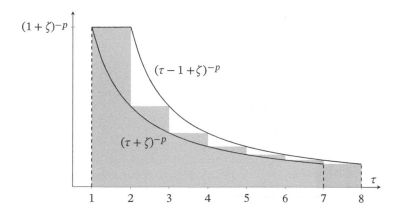

Figure 6.6 Construction of upper and lower bounds on $s_k = \sum_{j=1}^{k}(j+\zeta)^{-p}$, illustrated for $k = 7$. The gray area is equal to s_7.

a constant step-size sequence $t_k = t \in (0, 2/L)$ such that $\sum_{j=0}^{k-1} t_j(1 - Lt_j/2) = O(k)$. Indeed, this implies that the right-hand side of (6.68) decays as $O(1/k)$. However, in the stochastic setting, where $c > 0$, a constant step-size sequence does not make the right-hand side of (6.68) vanish as $k \to \infty$. In this case, we will instead consider the decreasing step-size sequence in (6.70) for different values of the decay rate parameters δ. The sum of the first k step sizes and the sum of their squares can be expressed as

$$\sum_{j=0}^{k-1} t_j = t\sum_{j=1}^{k}(j+\zeta)^{-\delta}, \qquad \sum_{j=0}^{k-1} t_j^2 = t^2\sum_{j=1}^{k}(j+\zeta)^{-2\delta},$$

which both involve sums of the form

$$s_k = \sum_{j=1}^{k}\frac{1}{(j+\zeta)^p}, \quad p \in \mathbb{R}_+.$$

To expose the asymptotic behavior of this sum, we now bound s_k from above and below in terms of the definite integral $\int_1^k (\tau + \zeta)^{-p} \, d\tau$, as illustrated in Figure 6.6. This leads to the inequalities

$$\int_1^k (\tau + \zeta)^{-p} \, d\tau \leq s_k \leq (1 + \zeta)^{-p} + \int_1^k (\tau + \zeta)^{-p} \, d\tau,$$

and using the result that

$$\lim_{k\to\infty} \int_1^k (\tau + \zeta)^{-p} \, d\tau = \begin{cases} \frac{(1+\zeta)^{1-p}}{p-1}, & p > 1, \\ \infty, & p \in [0, 1], \end{cases}$$

we can conclude that s_k converges when $p > 1$ and diverges otherwise. In the latter case, the upper and lower bounds on s_k allow us to establish the asymptotic equivalence

$$s_k \sim \begin{cases} k^{1-p}, & p \in [0, 1), \\ \ln(k), & p = 1. \end{cases}$$

This result may be used to derive upper bounds on the right-hand side of (6.68) for different values of δ, which are summarized in Table 6.1. Note that asymptotically, the upper bound decays the fastest when $\delta = 1/2$. However, this choice is not necessarily the best one in practice.

Table 6.1 Asymptotic behavior of step-size sums and the resulting upper bound on the right-hand side of (6.68) as a function of δ.

Parameter	$\sum_{j=0}^{k-1} t_j$	$\sum_{j=0}^{k-1} t_j^2$	Upper bound
$\delta > 1$	$\Theta(1)$	$\Theta(1)$	$O(1)$
$\delta = 1$	$\sim \ln(k)$	$\Theta(1)$	$O(1/\ln(k))$
$1/2 < \delta < 1$	$\sim k^{1-\delta}$	$\Theta(1)$	$O(1/k^{1-\delta})$
$\delta = 1/2$	$\sim \sqrt{k}$	$\sim \ln(k)$	$O(\ln(k)/\sqrt{k})$
$0 < \delta < 1/2$	$\sim k^{1-\delta}$	$\sim k^{1-2\delta}$	$O(1/k^{\delta})$
$\delta = 0$	$\sim k$	$\sim k$	$O(1)$

6.8.2 Smooth and Strongly Convex Functions

If the function f is μ-strongly convex in addition to being L-smooth and bounded below, we can obtain a stronger result. Akin to the analysis of the gradient descent method in Section 6.2, we have that

$$
\begin{aligned}
\mathbb{E}\big[\, \|X_{k+1} - x^\star\|_2^2 \mid X_k \big] &= \mathbb{E}\big[\, \|X_k - t_k G_k - x^\star\|_2^2 \mid X_k \big] \\
&= \mathbb{E}\big[\, \|X_k - x^\star\|_2^2 \mid X_k \big] + t_k^2 \mathbb{E}\big[\, \|G_k\|_2^2 \mid X_k \big] \\
&\quad - 2t_k \mathbb{E}\big[\, G_k^T (X_k - x^\star) \mid X_k \big] \\
&\leq \|X_k - x^\star\|_2^2 + t_k^2 (c^2 + \|\nabla f(X_k) - \nabla f(x^\star)\|_2^2) \\
&\quad - 2t_k (\nabla f(X_k) - \nabla f(x^\star))^T (X_k - x^\star),
\end{aligned}
$$

which simplifies to (6.18) in the special case, where $c = 0$. As in the deterministic case, we can apply (6.4), which yields the bound

$$
\mathbb{E}\big[\, \|X_{k+1} - x^\star\|_2^2 \mid X_k \big] \leq \left(1 - \frac{2t_k \mu L}{\mu + L}\right) \|X_k - x^\star\|_2^2 + t_k^2 c^2
$$

for $t_k \in (0, 2/(\mu + L)]$. Taking expectation and applying this inequality recursively with a fixed step size $t_k = t \in (0, 2/(\mu + L)]$, we get the bound

$$
\mathbb{E}\big[\, \|X_{k+1} - x^\star\|_2^2 \big] \leq \left(1 - \frac{2t\mu L}{\mu + L}\right)^{k+1} R^2 + c^2 t^2 \sum_{i=0}^{k} \left(1 - \frac{2t\mu L}{\mu + L}\right)^i
$$

where $R^2 = \mathbb{E}\big[\, \|X_0 - x^\star\|_2^2 \big]$. The summation in the last term on the right-hand side is the kth partial sum of a geometric series, and using the fact that

$$
\sum_{i=0}^{k} r^i = \left(\frac{1 - r^{k+1}}{1 - r}\right) \leq \frac{1}{1-r}, \quad |r| < 1,
$$

leads to the simplified bound

$$
\mathbb{E}\big[\, \|X_k - x^\star\|_2^2 \big] \leq \left(1 - \frac{2t\mu L}{\mu + L}\right)^k R^2 + tc^2 \frac{\mu + L}{2\mu L}. \tag{6.71}
$$

Thus, with a fixed step size, the sequence generated by the stochastic gradient iteration in (6.63) is only guaranteed to converge in mean square to a ball centered at $x^\star$ and with a radius that is proportional to the step size t. Clearly, a diminishing step-size sequence is needed to guarantee that $\mathbb{E}\big[\, \|X_k - x^\star\|_2^2 \big] \to 0$ as $k \to \infty$.

6.8.3 Incremental Methods

The SG method is closely related to the class of *incremental methods*, and the two terms are often used synonymously. Incremental methods deal with problems of the form (5.46), where the objective function is a finite sum of m functions $f_i(x)$, $i \in \mathbb{N}_m$, and the iterative update is of the form

$$x_{k+1} = x_k - t_k \nabla f_{i_k}(x_k), \tag{6.72}$$

where $i_k \in \mathbb{N}_m$ is chosen according to some index selection rule. Two of the most common rules are the *cyclic* rule $i_k = (k \bmod m) + 1$ and the *random* rule, where each i_k is chosen uniformly at random from $\mathbb{N}_m$. With the random index selection rule, the incremental gradient method may be viewed as a SG method applied to (5.44) if the random variable ξ is discrete with m equiprobable outcomes. We note that more general *incremental proximal gradient* (IPG) methods that can handle simple nondifferentiable functions akin to the proximal gradient method (6.46) have been proposed and analyzed in [16].

We now return to the problem in (5.46), where the objective function consists of a sum of m functions, e.g. corresponding to m observations. Applying the gradient method to this problem requires the full gradient, i.e.

$$\nabla f(x_k) = \frac{1}{m} \sum_{i=1}^{m} \nabla f_i(x_k).$$

In other words, the gradient of all the functions must be computed in order to obtain the search direction, and as a result, the gradient method is sometimes referred to as a *batch* gradient method. In contrast to the gradient method, the incremental gradient method uses as search direction the negative gradient of a single function, i.e. $-\nabla f_{i_k}(x_k)$, and as a consequence, an incremental gradient iteration can be much cheaper than a gradient iteration when m is large. However, the gradient of a single function may be viewed as a noisy approximation of the full gradient, and hence, the search direction is not necessarily a descent direction. A compromise between the gradient and incremental gradient methods may be obtained by using a subset of p functions at each iteration, i.e.

$$x_{k+1} = x_k - \frac{t_k}{p} \sum_{i \in S_k} \nabla f_i(x_k),$$

where $S_k \subset \mathbb{N}_m$ and $|S_k| = p$. This approach, which is known as a *mini-batch* method, provides a way to reduce the variance of the search direction at the expense of increased computation cost per iteration.

Example 6.6 Consider the optimization problem

$$\text{minimize} \quad \frac{\gamma}{2}\|x\|_2^2 + \frac{1}{m} \sum_{i=1}^{m} \ln(1 + \exp(-y_i a_i^T x)),$$

with variable $x \in \mathbb{R}^n$, regularization parameter $\gamma > 0$, and problem data $y \in \{-1, 1\}^m$ and $A \in \mathbb{R}^{m \times n}$, where a_i^T is the ith row of A. To apply the stochastic gradient method to this problem, we define a realization of a stochastic gradient of the objective function at x_k as

$$g_k = \gamma x_k + \frac{1}{p} \sum_{i \in S_k} a_i \frac{y_i \exp(-y_i a_i^T x_k)}{1 + \exp(-y_i a_i^T x_k)},$$

where S_k is a sample of p elements from $\mathbb{N}_m$ drawn at random without replacement. Note that $g_k = \nabla f(x_k)$ in the special case where $p = m$. Figure 6.7 illustrates the typical behavior of the stochastic gradient method based on a data set with $m = 32561$ and $n = 123$ using the mini-batch

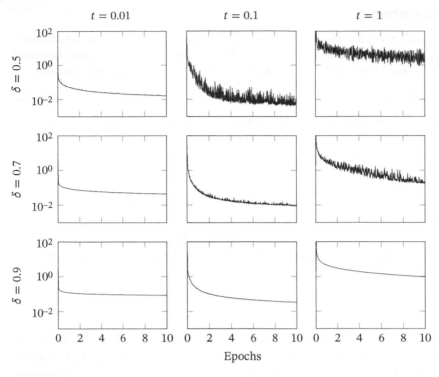

Figure 6.7 The relative suboptimality for the stochastic gradient method using step-size sequences of the form (6.70) with initial step size t and decay parameter δ.

size $p = 326$, which corresponds to roughly 1% of the data. The plots show the relative suboptimality $|f(x_k) - f(x^\star)|/|f(x^\star)|$ obtained with different step-size sequences of the form (6.70) with $\zeta = 0$ and different values of t and the decay parameter δ. The primary axis shows the number of *epochs*, which is the number of iterations scaled by p/m. The figure clearly demonstrates that progress can be very slow if the initial step size is too small or too large.

6.8.4 Adaptive Methods

One of the main challenges with incremental/stochastic gradient methods is the choice of step-size sequence. We have seen through our analysis that a diminishing step-size sequence is required to guarantee convergence, but the asymptotic convergence rate can be slow, and in practice, it is desirable to have adaptive step size rules with as few parameters to tune as possible. It is also possible to replace the scalar step size t_k by a symmetric matrix B_k such that, in the simplest case, the update is of the form

$$x_{k+1} = x_k - B_k^{-1} g_k,$$

where B_k is a symmetric scaling matrix and g_k is the gradient estimate at iteration k. This can be viewed as an incremental/stochastic variant of the variable metric method in (6.51), or equivalently, an adaptively preconditioned stochastic gradient method.

The variance of the gradient estimator also has an effect on the practical performance of stochastic gradient methods. Indeed, we have seen that a diminishing step size sequence is needed to ensure convergence even when the objective function is smooth and strongly convex. Stochastic gradient methods can be combined with variance reduction techniques that, roughly speaking, replace a simple gradient estimator with a more sophisticated unbiased estimator with lower variance. For example, in the incremental setting where the objective function is a finite sum of m functions, the so-called *stochastic variance-reduced gradient* (SVRG) method uses an update of the form

$$x_{k+1} = x_k - t_k(\nabla f_{i_k}(x_k) - \nabla f_{i_k}(\tilde{x}) + \tilde{\mu})$$

where $i_k \in \mathbb{N}_m$ is selected uniformly at random. The vector $\tilde{x}$ is an approximation of x_k and is updated only every M iterations, and $\tilde{\mu} = \nabla f(\tilde{x})$ is the full gradient at $\tilde{x}$. In the smooth and strongly convex setting, this can be shown to converge linearly, in expectation, with a suitable constant step size. We note that the technique is closely related to the method of control variates, which is illustrated in Exercise 3.11. Several other variance reduction techniques exist; see, e.g. [46] for an overview.

We end this section by outlining some examples of popular stochastic methods with adaptive step-size strategies.

6.8.4.1 AdaGrad

The *adaptive gradient method*, or AdaGrad, is a method for stochastic problems of the form (5.44), where the function $F(x, \xi)$ is assumed to be of the form $F(x, \xi) = G(x, \xi) + h(x)$ with $G(\cdot, \xi)$ and h closed and convex. The method uses either a diagonal matrix or a full matrix to adaptively scale stochastic gradients. Specifically, if $g_k \in \partial G(x_k, \xi_k)$, where ξ_k denotes a realization of ξ at iteration k and

$$\hat{G}_k = \sum_{i=0}^{k} g_k(g_k)^T,$$

then AdaGrad uses either $B_k = \gamma \mathrm{diag}((\mathrm{diag}(\hat{G}_k) + \varepsilon \mathbb{1})^{1/2}$ or $B_k = \gamma(\hat{G}_k + \varepsilon I)^{1/2}$ where $\gamma > 0$ and $\varepsilon > 0$ are parameters. The diagonal variant of AdaGrad can be expressed as the iteration

$$v_k = v_{k-1} + g_k \circ g_k$$
$$B_k = \gamma \, \mathrm{diag}(v_k + \varepsilon \mathbb{1})^{1/2}$$
$$x_{k+1} = \mathrm{prox}_h^{B_k}(x_k - B_k^{-1}g_k)$$

where the operator $\circ$ denotes the Hadamard product, i.e. elementwise product, and $v_{-1} = 0$ and x_0 are initial values. The matrix B_k can also be expressed as

$$B_k = \gamma \sqrt{k+1}\mathrm{diag}\left(\frac{1}{k+1}\mathrm{diag}(\hat{G}_k) + \frac{\varepsilon}{k+1}\mathbb{1}\right)^{1/2},$$

which may be implemented recursively as

$$\tilde{v}_k = \frac{k}{k+1}\tilde{v}_{k-1} + \frac{1}{k+1}(g_k \circ g_k)$$
$$B_k = \gamma\sqrt{k+1} \, \mathrm{diag}(\tilde{v}_k + \varepsilon/(k+1) \mathbb{1})^{1/2}.$$

The vector $\tilde{v}_k$ is the elementwise second raw sample moment of the sequence of gradient estimates $g_0, \ldots, g_k$, i.e.

$$\tilde{v}_k = \frac{1}{k+1}\sum_{i=0}^{k} g_k \circ g_k.$$

The AdaGrad update may be expressed as

$$x_{k+1} = \mathrm{prox}_h^{B_k}(x_k - t_k\tilde{g}_k),$$

with $t_k = \frac{1}{\gamma\sqrt{k+1}}$ and $\tilde{g}_k = \mathrm{diag}(\tilde{v}_k + \varepsilon/(k+1)\,\mathbb{1})^{-1/2}g_k$ which shows that AdaGrad implicitly employs a diminishing step-size sequence. In the convex setting, AdaGrad can be shown to satisfy the worst-case bound

$$\mathbb{E}[f(X_k)] - p^\star \le O(1/\sqrt{k}).$$

We refer the reader to [35] for further analysis of AdaGrad and details regarding convergence.

6.8.4.2 RMSprop

The *root mean square propagation* (RMSprop) method is in many ways similar to AdaGrad. It is a stochastic gradient iteration of the form

$$x_{k+1} = x_k - B_k^{-1}g_k,$$

where $g_k = \nabla F(x_k, \xi_k)$ is a gradient estimate, and B_k is a diagonal matrix that is chosen proportionally to the elementwise root mean square of previous gradient estimates. Unlike AdaGrad that uses all gradient estimates to compute an adaptive scaling, RMSprop emphasizes more recent stochastic gradients through the use of an exponential moving average, i.e.

$$v_k = \beta v_{k-1} + (1-\beta)(g_k \circ g_k)$$
$$B_k = \gamma\,\mathrm{diag}(v_k + \varepsilon\,\mathbb{1})^{1/2}$$
$$x_{k+1} = x_k - B_k^{-1}g_k$$

where the parameter $\beta \in (0,1)$ controls the adaptiveness of the estimate. RMSprop was proposed in a lecture note [103] along with the suggested parameter value $\beta = 0.9$. Note that unlike AdaGrad, RMSprop does not implicitly result in a diminishing step-size sequence, and in fact, the method need not converge, as shown in [95].

6.8.4.3 Adam

The *adaptive moment* estimation method, which is known as Adam, combines the adaptive scaling approach of RMSprop with *gradient aggregation* or *momentum*. It employs an exponential moving average of the form

$$\mu_k = \beta_1 \mu_{k-1} + (1-\beta_1)g_k,$$

to compute a weighted average of previous gradient estimates. The vector μ_k can be viewed as an estimate of the first raw sample moment of the weighted sequence of gradient estimates, and this is used instead of g_k to compute a search direction at iteration k, i.e.

$$\mu_k = \beta_1 \mu_{k-1} + (1-\beta_1)g_k$$
$$v_k = \beta_2 v_{k-1} + (1-\beta_2)(g_k \circ g_k)$$
$$B_k = \gamma(1-\beta_1)\left((1-\beta_2)^{-1/2}\mathrm{diag}(v_k)^{1/2} + \varepsilon\,I\right)$$
$$x_{k+1} = x_k - B_k^{-1}\mu_k.$$

The method was proposed in [63] with the recommended parameter values $\beta_1 = 0.9$ and $\beta_2 = 0.999$. Like RMSprop, the method need not converge, as shown in [95]. A variant of Adam, known as *AdaMax*, can be obtained by replacing v_k and B_k by

$$u_k = \max\left(\beta_2 u_{k-1}, |g_k|\right), \qquad B_k = \gamma(1-\beta_1)\mathrm{diag}(u_k),$$

where $|g_k|$ is the elementwise absolute value of g_k and with $u_{-1} = 0$.

6.9 Coordinate Descent Methods

We now consider a class of methods for solving unconstrained optimization problems of the form (6.5) using coordinatewise updates. The prototypical coordinate descent iteration is of the form

$$x_{k+1} = x_k - t_k [\nabla f(x_k)]_{i_k} e_{i_k}, \tag{6.73}$$

where $i_k \in \mathbb{N}_n$ is a coordinate index and $t_k > 0$ is a step size. Common index selection strategies include the cyclic order $i_k = (k \bmod n) + 1$, a randomized cyclic order where the order is reshuffled every n iterations, and a fully randomized order where the indices are selected uniformly at random. The step size can be chosen in a similar way to the gradient method, e.g. using some form of line search. With an exact line search, the method performs coordinatewise minimization, which can be expressed as

$$t_k \in \operatorname*{argmin}_t \{f(x_k + te_{i_k})\},$$

$$x_{k+1} = x_k + t_k e_{i_k}. \tag{6.74}$$

Note that only the i_kth element of x is updated at iteration k. This is a descent method by construction. However, without additional assumptions on f, the iteration in (6.74) does not necessarily converge: one can construct an example, where (6.74) with the cyclic update order will enter a cycle for some initializations, as demonstrated by Powell [90]. Moreover, even if the iteration (6.74) does converge, it may not be to a stationary point if f is nonsmooth. Figure 6.8 shows an example of a nonsmooth convex function where this can happen.

Next, we consider the case where f is smooth and convex, and we assume that the set $\{x \mid f(x) \le f(x_0)\}$ is nonempty and compact, which ensures that f attains its minimum. Using a suitable step-size sequence, the coordinate descent iteration (6.73) can then be shown to converge to a minimizer of f. The first-order condition for convexity (4.27) implies that

$$f(x + te_i) \ge f(x) + \frac{\partial f(x)}{\partial x_i} t, \quad i \in \mathbb{N}_n,$$

for all $x \in \mathbb{R}^n$, and hence, we have that

$$\nabla f(x) = 0 \iff f(x + te_i) \ge f(x) \text{ for all } t \in \mathbb{R}, \ i \in \mathbb{N}_n.$$

In other words, x is a minimizer of f if and only if x minimizes f along all n coordinate directions. Now, using the exact line search (6.74), iteration k yields a descent unless the i_kth element of $\nabla f(x_k)$ is equal to zero, and hence, a cycle through all n coordinates yields a descent unless x is a minimum. With additional assumptions on f and the step sizes, the iteration (6.74) can be shown to converge to a minimizer of f. An overview of more sophisticated variants of the coordinate descent method and detailed analyses can be found in, e.g. [12, 116].

Figure 6.8 Contour plot of the function $f(x) = \max(|2x_1 - x_2|, |2x_2 - x_1|)$, which is convex but nonsmooth. The function is nondifferentiable whenever $x_1 = x_2$ or $x_1 = -x_2$, and none of the coordinate directions is a descent direction when $x_1 = x_2$.

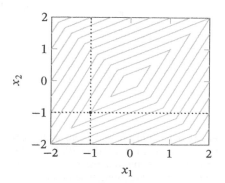

Example 6.7 Coordinate descent methods typically work well when the coupling between variables is weak. To illustrate this, we now consider convex quadratic functions of the form $f(x) = x^T Q x$ with a fixed condition number $\kappa(Q) = 20$. Specifically, we take $Q = U\text{diag}(20, 1)U^T$ for different choices of $U \in \mathbb{R}^{2\times2}$ such that $U^T U = I$. Figure 6.9 shows the coordinate descent method in action for different choices of U corresponding to different orientations of the coordinate system. The problem of minimizing $f(x)$ is separable in the special case where Q is diagonal, and in this case, the minimum point is reached in two iterations (one cycle). In contrast, progress is slow when x_1 and x_2 are maximally coupled, which is the case when

$$U = \pm\frac{1}{\sqrt{2}} \begin{bmatrix} 1 & \pm1 \\ \mp1 & 1 \end{bmatrix}.$$

Coordinate descent methods are often useful for regularized risk minimization problems of the form

$$\text{minimize} \quad \frac{1}{m}\sum_{i=1}^{m} g(a_i^T x) + \lambda\, h(x), \tag{6.75}$$

where $a_1, \ldots, a_m \in \mathbb{R}^n$ are problem data, and where the regularizer $h(x)$ is a separable function (e.g. $\|x\|_1$ or $\|x\|_2^2$). Notice that (6.75) is of the form (5.46) with $f_i(x) = \frac{1}{m}(g(a_i^T x) + \lambda\, h(x))$. The dual problem may be expressed as

$$\text{maximize} \quad -\frac{1}{m}\sum_{i=1}^{m} g^*(-mz_i) - \lambda\, h^*(\lambda^{-1}A^T z), \tag{6.76}$$

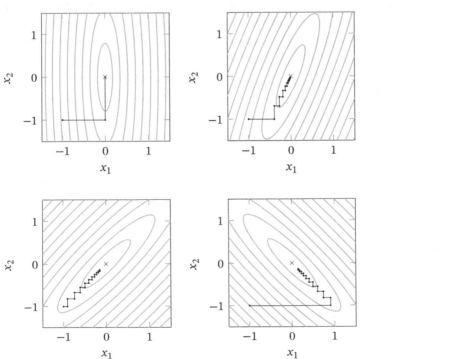

Figure 6.9 Coordinate descent with exact line search applied to convex quadratic functions of the form $f(x) = x^T Q x$ with $Q = U\,\text{diag}(20, 1)U^T$, where U is an orthogonal matrix. Each plot includes 20 iterations in addition to the initial guess $x_0 = (-1, -1)$. The minimum is reached in two iterations, i.e. one cycle through all coordinate directions, when Q is diagonal (upper left).

with variable $z \in \mathbb{R}^m$, and where $A \in \mathbb{R}^{m \times n}$ is the matrix with rows $a_1^T, \dots, a_m^T$. The sum is a separable function, whereas the last term involves all the dual variables. To apply coordinate ascent to the dual problem, we restrict the dual objective to one of its coordinate directions. Letting $z = \bar{z} + t e_i$, the dual function reduces to

$$-\frac{1}{m} \sum_{i=1}^{m} g^*(-m(\bar{z}_i + t)) - \lambda\, h^*(\lambda^{-1}(A^T \bar{z} + t a_i)) + \text{const.},$$

which can be minimized efficiently for certain choices of g and h.

6.10 Interior-Point Methods

The methods that we have discussed so far are not suitable for problems that involve all but simple inequality constraints. For example, the gradient projection method is inefficient unless projections onto the feasible set are cheap, and Newton's method cannot be applied directly to problems with inequality constraints. We will now look at a conceptually simple technique for handling inequality constraints in a general convex optimization problem on the form (4.52), i.e.

$$
\begin{aligned}
\text{minimize} \quad & f_0(x) \\
\text{subject to} \quad & f_i(x) \le 0, \quad i \in \mathbb{N}_m \\
& Ax = b
\end{aligned}
\tag{6.77}
$$

where we assume that $f_0 : \mathbb{R}^n \to \mathbb{R}$ and $f_i : \mathbb{R}^n \to \mathbb{R}$, $i \in \mathbb{N}_m$, are twice continuously differentiable convex functions and where $A \in \mathbb{R}^{p \times n}$ and $b \in \mathbb{R}^p$. We will assume that Slater's condition holds and that $\text{rank}(A) = p$. We remind the reader that the Lagrangian $L : \mathbb{R}^n \times \mathbb{R}^m \times \mathbb{R}^p \to \mathbb{R}$ is

$$L(x, \lambda, \mu) = f_0(x) + \sum_{i=1}^{m} \lambda_i f_i(x) + \mu^T(Ax - b),$$

and the Lagrange dual function is $g(\lambda, \mu) = \inf_x L(x, \lambda, \mu)$.

The problem (6.77) is equivalent to the equality constrained problem

$$
\begin{aligned}
\text{minimize} \quad & f_0(x) + \sum_{i=1}^{m} I_+(-f_i(x)) \\
\text{subject to} \quad & Ax = b
\end{aligned}
$$

where $I_+ : \mathbb{R} \to (-\infty, +\infty]$ denotes the indicator function of $\mathbb{R}_+$. Thus, the objective is finite if and only if the inequality constraints in (6.77) are satisfied. This reformulation introduces another difficulty, namely that the objective function is no longer differentiable on $\mathbb{R}^n$. To address this issue, we introduce a so-called *barrier function* $\phi : \mathbb{R}_{++} \to \mathbb{R}$ given by $\phi(\tau) = -\ln(\tau)$ such that $(1/t)\phi(\tau)$ with $t > 0$ is a smooth approximation to $I_+(\tau)$ on $\mathbb{R}_{++}$, as illustrated in Figure 6.10. This motivates the so-called *barrier problem*

$$
\begin{aligned}
\text{minimize} \quad & f_0(x) + \frac{1}{t} \sum_{i=1}^{m} \phi(-f_i(x)) \\
\text{subject to} \quad & Ax = b
\end{aligned}
\tag{6.78}
$$

as an approximation to (6.77), and where $t > 0$ controls the accuracy of the approximation. It is natural to expect that a solution to (6.78) approaches a solution to (6.77) as $t \to \infty$. We will soon formalize this intuition. The problem (6.78) is an equality constrained convex optimization problem, which follows by noting that $\phi(-f_i(x))$ is a convex function. Moreover, the domain of the barrier

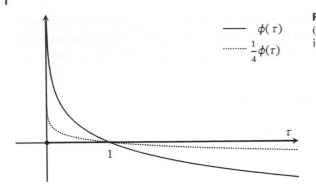

function is $\mathbb{R}_{++}$, and this implies that the domain of the barrier problem (6.78) is the relative interior of the feasible set of the original problem (6.77). This gives rise to the name *interior-point* (IP) method for a method that solves the barrier problem (6.78).

We will now compare the optimality conditions associated with the original problem (6.77) and the barrier problem (6.78). The KKT conditions for (6.77) may be expressed as

$$\nabla f_0(x) + \sum_{i=1}^{m} \lambda_i \nabla f_i(x) + A^T \mu = 0 \tag{6.79a}$$

$$f_i(x) \leq 0, \quad i \in \mathbb{N}_m \tag{6.79b}$$

$$Ax = b \tag{6.79c}$$

$$\lambda \geq 0 \tag{6.79d}$$

$$\lambda_i f_i(x) = 0, \quad i \in \mathbb{N}_m, \tag{6.79e}$$

and the KKT conditions for the barrier problem are

$$\nabla f_0(x) + \sum_{i=1}^{m} \frac{1}{-t f_i(x)} \nabla f_i(x) + A^T \mu = 0 \tag{6.80a}$$

$$Ax = b, \tag{6.80b}$$

with the additional implicit constraints that $f_i(x) < 0$, $i \in \mathbb{N}_m$, which are imposed by the domain of the barrier function. Now, suppose that $x^\star(t)$ and $\mu^\star(t)$ satisfy the optimality conditions for the barrier problem and let $\lambda^\star(t)$ be the vector with elements $\lambda_i^\star(t) = -1/(t f_i(x^\star(t)))$, $i \in \mathbb{N}_m$. It then follows that $x^\star(t)$, $\lambda^\star(t)$, and $\mu^\star(t)$ satisfy (6.79a)–(6.79d) but not the complementarity condition (6.79e) since $\lambda_i^\star(t) f_i(x^\star(t)) = -1/t$. Moreover, $x^\star(t)$ minimizes the Lagrangian $L(x, \lambda^\star(t), \mu^\star(t))$, and hence, $\lambda^\star(t)$ and $\mu^\star(t)$ are dual feasible and

$$p^\star \geq g(\lambda^\star(t), \mu^\star(t)) = L(x^\star(t), \lambda^\star(t), \mu^\star(t))$$

$$= f_0(x^\star) + \sum_{i=1}^{m} \frac{-1}{t f_i(x^\star(t))} f_i(x^\star(t)) + (\mu^\star(t))^T (Ax^\star(t) - b)$$

$$= f_0(x^\star(t)) - m/t.$$

This shows that the duality gap at $(x^\star(t), \lambda^\star(t), \mu^\star(t))$ is m/t. Thus, the solution to the barrier problem $x^\star(t)$ defines a trajectory of strictly feasible (m/t)-suboptimal points, which is known as the *central path*.

6.10.1 Path-Following Method

The barrier problem (6.78) can be solved using Newton's method for equality constrained minimization. The Newton direction $(\Delta x, \Delta \mu) \in \mathbb{R}^n \times \mathbb{R}^p$ can be expressed as the solution to the linear system of equations:

$$\begin{bmatrix} H & A^T \\ A & 0 \end{bmatrix} \begin{bmatrix} \Delta x \\ \Delta \mu \end{bmatrix} = \begin{bmatrix} -g \\ r \end{bmatrix}$$

where

$$H = \nabla^2 f_0(x) + \frac{1}{t} \sum_{i=1}^{m} \left(\frac{-1}{f_i(x)} \nabla^2 f_i(x) + \frac{1}{f_i(x)^2} \nabla f_i(x) \nabla f_i(x)^T \right)$$

$$g = \nabla f_0(x) + A^T \mu + \frac{1}{t} \sum_{i=1}^{m} \frac{-1}{f_i(x)} \nabla f_i(x)$$

$$r = b - Ax.$$

This system of equations is obtained by linearizing the optimality conditions (6.80) around (x, μ) and is referred to as the Newton equations. Rather than choosing t very large and solving a single barrier problem, it turns out that it is advantageous to solve a sequence of barrier problems with increasing values of t. Roughly speaking, the barrier problem is easier to solve when t is small, and given a point $x^\star(t)$, we can expect that Newton's methods converges quadratically to $x^\star(t^+)$ if t^+ is sufficiently close to t. The *path-following method* outlined in Algorithm 6.10 uses this approach to find an ϵ-suboptimal point. Each loop iteration is commonly referred to as an *outer iteration*, whereas the iterations involved in solving a barrier subproblem are referred to as *inner iterations*. The parameter $\gamma > 1$ determines how quickly t is increased. If the initial value of t is t_0, then the number of barrier problems to solve is given by

$$1 + \left\lceil \frac{\ln(m) - \ln(t_0 \epsilon)}{\ln \gamma} \right\rceil,$$

provided that $0 < t_0 \leq m/\epsilon$. This is a decreasing function of γ. In contrast, we may expect the number of Newton iterations required to solve a single barrier problem to increase with γ. We note that our analysis of Newton's method in Section 6.3 does not apply to the barrier problem in (6.78): the problem is not necessarily strongly convex and the barrier function $\phi(t) = -\ln(x)$ does not have a Lipschitz continuous Hessian on dom ϕ. However, convergence can still be established if f_0 is a so-called *self-concordant* function; see, e.g. [81].

Algorithm 6.10: Path-following method

Input: Strictly feasible x_0, tolerance $\epsilon > 0$, and parameters $t_0 > 0$ and $\gamma > 1$.
Output: An ϵ-suboptimal point x.

$x \leftarrow x_0, t \leftarrow t_0$
repeat
    Find $x^\star(t)$ by solving (6.78) with starting point x.
    $x \leftarrow x^\star(t)$
    if $m/t \leq \epsilon$ **then** stop
    $t \leftarrow \gamma t$

Example 6.8 To illustrate the basic principle behind the path-following method, we will now apply it to find an ϵ-suboptimal solution to the convex optimization problem

$$\begin{aligned} \text{minimize } & x_1 + x_2/5 \\ \text{subject to } & \exp(x_1) + x_2 - 3 \leq 0 \\ & x_1^2 - x_2 \leq 0 \end{aligned}$$

with variable $x \in \mathbb{R}^2$. The corresponding barrier problem can be expressed as

$$\text{minimize } x_1 + \frac{x_2}{5} + \frac{1}{t}\left(-\ln(3 - x_2 - \exp(x_1)) - \ln(x_2 - x_1^2)\right),$$

which is an unconstrained convex optimization problem. Figure 6.11 shows the iterates generated using Algorithm 6.10 for two different values of the parameter γ. In both cases, we used the point $x_0 = (0, 0.5)$ as a strictly feasible starting point and the parameters $t_0 = 0.5$ and $\epsilon = 10^{-3}$.

Algorithm 6.10 requires a strictly feasible initial point x_0, and it is not always easy to find such a point. One approach is to solve a so-called *phase I* problem, which can be expressed as

$$\begin{aligned} \text{minimize } & s \\ \text{subject to } & f_i(x) \leq s, \quad i \in \mathbb{N}_m \\ & Ax = b, \end{aligned} \tag{6.81}$$

with variables $x \in \mathbb{R}^n$ and $s \in \mathbb{R}$. If the optimal value is attained at $(x^\star, s^\star)$, then $x^\star$ is clearly a strictly feasible point for the original problem (6.77) if $s^\star < 0$. On the other hand, if $s^\star > 0$, the original problem is infeasible. Note that it is straightforward to find a strictly feasible point for the phase I problem, and hence, it can be solved using Algorithm 6.10.

6.10.2 Generalized Inequalities

The barrier approach to inequality constrained minimization can be extended to optimization problems with generalized inequalities, which we will now illustrate for a cone linear program (LP). Recall that this is a problem of the form

$$\begin{aligned} \text{minimize } & c^T x \\ \text{subject to } & Ax = b \\ & x \succeq_K 0, \end{aligned} \tag{6.82}$$

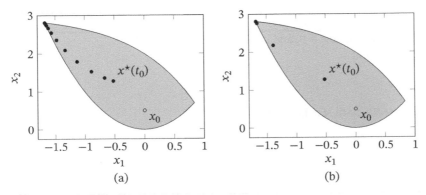

(a)   (b)

Figure 6.11 The path-following method converges to an ϵ-suboptimal point x by solving a sequence of barrier problems. The two plots show the iterates obtained with (a) $\gamma = 2$ and (b) $\gamma = 20$, respectively. With $\gamma = 2$, 13 barrier problems were solved using a total of 48 Newton iterations, whereas with $\gamma = 20$, 4 barrier problems were solved using a total of 21 Newton iterations.

with variable $x \in \mathbb{R}^n$ and problem data $A \in \mathbb{R}^{p \times n}, b \in \mathbb{R}^p$, and $c \in \mathbb{R}^n$, and where $K \subset \mathbb{R}^n$ is a proper convex cone. The Lagrangian $L : \mathbb{R}^n \times \mathbb{R}^n \times \mathbb{R}^p \to \bar{\mathbb{R}}$ is defined as

$$L(x, \lambda, \mu) = c^T x - \lambda^T x + \mu^T (Ax - b),$$

and the dual function is $g(\lambda, \mu) = \inf_x L(x, \lambda, \mu)$.

To handle the generalized inequality $x \succeq_K 0$, we introduce a barrier function $\phi : \mathbb{R}^n \to \bar{\mathbb{R}}$ with dom $\phi = \mathrm{int}\, K$ such that $(1/t)\phi(x)$ is a smooth approximation of the indicator function $I_K(x)$. In other words, we require that $\phi(x) \to \infty$ as x approaches the boundary of K. Moreover, we will require that ϕ is convex and a so-called *generalized logarithm* for K, which means that it is logarithmically homogeneous with some constant $\theta \geq 1$, i.e. for all $x \in \mathrm{dom}\, \phi$ and $t > 0$, it holds that

$$\phi(tx) = \phi(x) - \theta \ln(t).$$

Taking the derivative with respect to t on both sides, we see that $\nabla \phi(tx)^T x = -\theta/t$.

The barrier problem associated with (6.82) is given by

$$\begin{aligned} \text{minimize} \quad & c^T x + (1/t)\phi(x) \\ \text{subject to} \quad & Ax = b, \end{aligned} \tag{6.83}$$

with variable $x \in \mathbb{R}^n$, and the corresponding KKT conditions can be expressed as

$$c + (1/t)\nabla \phi(x) + A^T \mu = 0$$

$$Ax = b.$$

Now, suppose that $x^\star(t)$ and $\mu^\star(t)$ satisfy these conditions. We then see that $x^\star(t)$ minimizes $L(x, \lambda^\star(t), \mu^\star(t))$ with $\lambda^\star(t) = -(1/t)\nabla \phi(x^\star)$, and hence,

$$\begin{aligned} g(\lambda^\star(t), \mu^\star(t)) &= \inf_x L(x, \lambda^\star(t), \mu^\star(t)) \\ &= c^T x^\star(t) + (1/t)\phi(x^\star(t))^T x^\star(t) + (\mu^\star(t))^T (Ax^\star(t) - b) \\ &= c^T x^\star(t) - \theta/t. \end{aligned}$$

This shows that the duality gap is θ/t. Thus, we can solve the cone LP to ϵ-suboptimality using a modified version of Algorithm 6.10 with the barrier problem (6.83) and the stopping criterion $\theta/t \leq \epsilon$.

The cone K in (6.82) can be a composition of cones, i.e.

$$K = K_1 \times \cdots \times K_m,$$

where $K_i \subset \mathbb{R}^{n_i}, i \in \mathbb{N}_m$, are proper convex cones and $n = \sum_{i=1}^m n_i$. The conic inequality $x \succeq_K 0$ is then equivalent to

$$x_i \succeq_{K_i} 0, \quad i \in \mathbb{N}_m,$$

where $x = (x_1, \ldots, x_m)$ is a partition of x into subvectors $x_i \in \mathbb{R}^{n_i}, i \in \mathbb{N}_m$. A barrier for K can then be constructed as

$$\phi(x) = \sum_{i=1}^m \phi_i(x_i),$$

where $\phi_i : \mathbb{R}^{n_i} \to \bar{\mathbb{R}}$ with dom $\phi_i = \mathrm{int}\, K_i$ is a barrier for $K_i, i \in \mathbb{N}_m$. Moreover, it is easy to verify that ϕ is logarithmically homogeneous with constant $\theta = \sum_{i=1}^m \theta_i$ if for all $i \in \mathbb{N}_m$, ϕ_i is logarithmically homogeneous with constant θ_i. Table 6.2 lists logarithmically homogeneous barrier functions for some elementary proper convex cones.

Table 6.2 Barrier functions for elementary proper convex cones.

Conic constraint	Barrier function	θ
$x \in \mathbb{R}_+^n$	$-\sum_{i=1}^n \ln(x_i)$	n
$(x,t) \in \mathbb{Q}^n$	$-\ln(t - x^T x/t) - \ln(t)$	2
$X \in \mathbb{S}_+^n$	$-\ln \det(X)$	n
$(x,y,z) \in K_\alpha$	$-\ln(z^{2\alpha} y^{2(1-\alpha)-x^2}) - (1-\alpha)\ln(y) - \alpha \ln(z)$	3
$(x,y,z) \in K_{\exp}$	$-\ln(y \exp(z/y) - x) - \ln(z) - \ln(y)$	3

Example 6.9 Consider the semidefinite programming problem

$$
\begin{aligned}
\text{minimize} \quad & \text{tr}(CX) \\
\text{subject to} \quad & \text{tr}(A_i X) = b_i, \quad i \in \mathbb{N}_p \\
& X \succeq 0
\end{aligned}
\tag{6.84}
$$

with variable $X \in \mathbb{S}^m$ and problem data $C \in \mathbb{S}^m$, $b \in \mathbb{R}^p$, and $A_i \in \mathbb{S}^m$, $i \in \mathbb{N}_p$. This problem is equivalent to the conic problem in (6.82). Indeed, using the symmetric vectorization operator defined in (2.22), we can express the semidefinite programming problem as a conic problem of the form (6.82) by letting $n = m(m+1)/2$ and defining $c = \text{svec}(C)$, $x = \text{svec}(X)$, $K = \{\text{svec}(X) \mid X \in \mathbb{S}_+^m\}$, and $a_i = \text{svec}(A_i)$, $i \in \mathbb{N}_p$, such that $A \in \mathbb{R}^{p \times n}$ is the matrix with rows $a_1^T, \ldots, a_p^T$.

The barrier problem associated with (6.84) can be expressed as

$$
\begin{aligned}
\text{minimize} \quad & \text{tr}(CX) + \frac{1}{t}\phi(X) \\
\text{subject to} \quad & \text{tr}(A_i X) = b_i, \quad i \in \mathbb{N}_p
\end{aligned}
\tag{6.85}
$$

where $\phi(X) : \mathbb{S}^m \to \bar{\mathbb{R}}$ is defined as $\phi(X) = -\ln \det(X)$ with dom $\phi = \mathbb{S}_{++}^n$. The Newton equations may then be expressed as

$$
\frac{1}{t}\nabla^2 \phi(X)[\Delta X] + \sum_{i=1}^p \Delta \mu_i A_i = -G
$$

$$
\text{tr}(A_i \Delta X) = r_i, \quad i \in \mathbb{N}_p
$$

where $\nabla \phi(X) = -X^{-1}$, $\nabla^2 \phi(X)[\Delta X] = X^{-1} \Delta X X^{-1}$, and

$$
G = C + \sum_{i=1}^p \mu_i A_i + \frac{1}{t}\nabla \phi(X), \qquad r_i = b_i - \text{tr}(A_i X), \quad i \in \mathbb{N}_p.
$$

By eliminating ΔX from the first equation, i.e.

$$
\Delta X = -X(G + \sum_{j=1}^p \Delta \mu_j A_j)X,
$$

we obtain the reduced system

$$
H \Delta \mu = -g,
$$

where the entries of $H \in \mathbb{S}^p$ and $g \in \mathbb{R}^p$ are given by

$$
H_{ij} = \text{tr}(A_i X A_j X), \quad i,j \in \mathbb{N}_p, \qquad g_i = r_i + \text{tr}(A_i X G X), \quad i \in \mathbb{N}_p.
$$

The cost of forming H and g is often the dominating cost of a single iteration, but it is sometimes possible to reduce the cost, e.g. if some or all of the matrices $A_1, \ldots, A_p$ are sparse. We note that H

is positive definite if and only if $A_1, \dots, A_p$ are linearly independent, in which case the Cholesky factorization $H = LL^T$ can be used to solve $H\Delta\mu = -g$.

We end this section by mentioning that there are much more advance IP methods than the basic path-following scheme that we have presented in this section. These methods generally maintain both primal and dual variables, and rather than solving a sequence of barrier subproblems to high accuracy, they stay inside some neighborhood of a primal–dual central path and update the parameter t adaptively; see, e.g. [81, 115]. We also note that there are IP methods for local optimization of more general nonlinear optimization problems; see, e.g. [111].

6.11 Augmented Lagrangian Methods

We will now consider methods for equality constrained optimization problems of the form

$$\begin{aligned} &\text{minimize } f_0(x) \\ &\text{subject to } h(x) = 0, \end{aligned} \tag{6.86}$$

with variable $x \in \mathbb{R}^n$, and where $f_0 : \mathbb{R}^n \to \bar{\mathbb{R}}$ and $h : \mathbb{R}^n \to \mathbb{R}^p$. We define the Lagrangian $L : \mathbb{R}^n \times \mathbb{R}^p \to \bar{\mathbb{R}}$ as $L(x, \mu) = f_0(x) + \mu^T h(x)$ and the dual function $g : \mathbb{R}^p \to \bar{\mathbb{R}}$ as $g(\mu) = \inf_x L(x, \mu)$.

A conceptually simple approach to finding an approximate local minimizer is to consider an unconstrained problem as a proxy for the problem in (6.86). For example, we may consider a so-called *penalty problem*

$$\text{minimize } f_0(x) + \frac{\rho}{2}\|h(x)\|_2^2,$$

where $\rho > 0$ is a *penalty parameter*. Roughly speaking, we can expect that the constraint violation will be small when ρ is large, and this observation is the motivation behind *penalty methods* that solve a sequence of penalty problems with increasing values of ρ. Unfortunately, the penalty problem typically becomes very ill-conditioned when ρ is large, which makes it difficult to solve it reliably and accurately.

An alternative to the penalty approach is to consider the optimization problem

$$\begin{aligned} &\text{minimize } f_0(x) + \frac{\rho}{2}\|h(x)\|_2^2 \\ &\text{subject to } h(x) = 0, \end{aligned} \tag{6.87}$$

with penalty parameter $\rho > 0$. This problem is equivalent to (6.86), which follows immediately by noting that $\|h(x)\|_2^2 = 0$ whenever x is feasible. The Lagrangian for the problem in (6.87), which is the so-called *augmented Lagrangian* for the problem in (6.86), is the function $L_\rho : \mathbb{R}^n \times \mathbb{R}^p \to \bar{\mathbb{R}}$ defined as

$$\begin{aligned} L_\rho(x, \mu) &= f_0(x) + \mu^T h(x) + \frac{\rho}{2}\|h(x)\|_2^2 \\ &= f_0(x) + \frac{\rho}{2}\|h(x) + \rho^{-1}\mu\|_2^2 - \frac{1}{2\rho}\|\mu\|_2^2. \end{aligned}$$

The corresponding dual function $g_\rho : \mathbb{R}^p \to \bar{\mathbb{R}}$ is given by

$$g_\rho(\mu) = \inf_x L_\rho(x, \mu) = \inf_x \left\{ f_0(x) + \frac{\rho}{2}\|h(x) + \rho^{-1}\mu\|_2^2 \right\}.$$

Notice the similarity with the penalty problem: the penalty term in the definition of g_ρ includes a shift that is determined by the dual variable μ.

6.11.1 Method of Multipliers

The dual function g_ρ can be used to construct a local optimization method for the problem in (6.86). To illustrate the basic idea, we will start by assuming that $\mu_k \in \mathbb{R}^p$ is given and that $\inf_x L_\rho(x, \mu_k)$ is finite and attained at x_{k+1}. It then holds that

$$
\begin{aligned}
g_\rho(\mu) &= \inf_x L_\rho(x, \mu) \\
&= \inf_x \left(f_0(x) + \mu^T h(x) + \frac{\rho}{2} \|h(x)\|_2^2 \right) \\
&\leq f_0(x_{k+1}) + \mu^T h(x_{k+1}) + \frac{\rho}{2} \|h(x_{k+1})\|_2^2 \\
&= f_0(x_{k+1}) + \mu_k^T h(x_{k+1}) + \left(\mu - \mu_k \right)^T h(x_{k+1}) + \frac{\rho}{2} \|h(x_{k+1})\|_2^2 \\
&= g_\rho(\mu_k) + h(x_{k+1})^T (\mu - \mu_k),
\end{aligned}
$$

which implies that $-h(x_{k+1})$ is a subgradient of $-g_\rho$ at μ_k. Furthermore, if L_ρ is differentiable with respect to x at x_{k+1}, then it must also hold that

$$
0 = \nabla_x L_\rho(x_{k+1}, \mu_k) = \nabla f_0(x_{k+1}) + \frac{\partial h(x_{k+1})^T}{\partial x} (\mu_k + \rho h(x_{k+1})).
$$

This motivates the update

$$
\mu_{k+1} = \mu_k + \rho h(x_{k+1}),
$$

which implies that (x_{k+1}, μ_{k+1}) satisfies the necessary stationarity condition

$$
0 = \nabla_x L(x_{k+1}, \mu_{k+1}) = \nabla f_0(x_{k+1}) + \frac{\partial h(x_{k+1})^T}{\partial x} \mu_{k+1}.
$$

However, x_{k+1} does not necessarily satisfy the primal feasibility condition $h(x) = 0$.

The above procedure may be repeated iteratively, i.e. we fix μ and find a new x by minimizing the augmented Lagrangian, and then we update μ. The augmented Lagrangian may be a nonconvex function of x, and hence, the problem of minimizing it can be a difficult one. Thus, to make the procedure more practical, the augmented Lagrangian is typically only minimized approximately, and furthermore, the penalty parameter ρ may be increased if $\|h(x_k)\|_2$ decays too slowly as a function of k. A conceptual outline of the so-called *method of multipliers* is shown in Algorithm 6.11. The parameters β and γ determine how to adjust the penalty parameter, and typical values are $\beta = 10$ and $\gamma = 1/4$. A more detailed treatment of the method, including a convergence analysis, can be found in, e.g. [15]. We note that the method of multipliers can also be extended to handle inequality constraints, i.e. problems of the more general form in (4.4).

Algorithm 6.11: Method of Multipliers

Input: Starting point $(x_0, \mu_0) \in \mathbb{R}^n \times \mathbb{R}^p$ and parameters $\rho_0 > 0, \gamma \in (0, 1)$, and $\beta > 1$.

for $k = 0, 1, 2, \ldots$ **do**

> Compute approximate minimizer x_{k+1} of $L_{\rho_k}(\cdot, \mu_k)$.
>
> $\mu_{k+1} \leftarrow \mu_k + \rho h(x_{k+1})$ $\triangleright$ Update dual variables
>
> **if** $\|h(x_{k+1})\|_2 \leq \gamma \|h(x_k)\|_2$ **then**
> > $\rho_{k+1} = \rho_k$
>
> **else**
> > $\rho_{k+1} = \beta \rho_k$
>
> **end**

end

Example 6.10 Constrained nonlinear LS problems is an example of a class of problems that can be solved efficiently to local optimality with the augmented Lagrangian method, i.e. problems of the form

$$\text{minimize} \quad \frac{1}{2}\|f(x)\|_2^2$$
$$\text{subject to} \quad h(x) = 0$$

with variable $x \in \mathbb{R}^n$, and where $f : \mathbb{R}^n \to \mathbb{R}^m$ and $h : \mathbb{R}^n \to \mathbb{R}^p$ are nonlinear functions. The augmented Lagrangian $L_\rho : \mathbb{R}^n \times \mathbb{R}^p$ can be expressed as

$$
\begin{aligned}
L_\rho(x, \mu) &= \frac{1}{2}\|f(x)\|_2^2 + \mu^T h(x) + \frac{\rho}{2}\|h(x)\|_2^2 \\
&= \frac{1}{2}\|f(x)\|_2^2 + \frac{\rho}{2}\|h(x) + \mu/\rho\|_2^2 - \frac{1}{2\rho}\|\mu\|_2^2 \\
&= \frac{1}{2}\left\|\begin{bmatrix} f(x) \\ \sqrt{\rho}h(x) + \mu/\sqrt{\rho} \end{bmatrix}\right\|_2^2 - \frac{1}{2\rho}\|\mu\|_2^2.
\end{aligned}
$$

Thus, for fixed value of μ, the problem of minimizing L_ρ with respect to x is an unconstrained nonlinear LS problem, which can be minimized locally using, e.g. the LM algorithm.

6.11.2 Alternating Direction Method of Multipliers

We will now consider a variant of the problem in (6.86) that can be expressed as

$$\text{minimize} \quad f(x) + g(y)$$
$$\text{subject to} \quad Ax + By = c \tag{6.88}$$

with variables $x \in \mathbb{R}^n$ and $y \in \mathbb{R}^m$, and where $f : \mathbb{R}^n \to \bar{\mathbb{R}}, g : \mathbb{R}^m \to \bar{\mathbb{R}}, A \in \mathbb{R}^{p \times n}, B \in \mathbb{R}^{p \times m}$, and $c \in \mathbb{R}^p$. We will make the additional assumption that both f and g are proper, closed, and convex but not necessarily smooth. The augmented Lagrangian $L_\rho : \mathbb{R}^n \times \mathbb{R}^m \times \mathbb{R}^p \to \bar{\mathbb{R}}$ can be expressed as

$$
\begin{aligned}
L_\rho(x, y, \mu) &= f(x) + g(y) + \mu^T (Ax + By - c) + \frac{\rho}{2}\|Ax + By - c\|_2^2 \\
&= f(x) + g(y) + \frac{\rho}{2}\|Ax + By - c + \mu/\rho\|_2^2 - \frac{\mu}{\rho}\|\mu\|_2^2.
\end{aligned}
$$

To apply the method of multipliers, we need to minimize L_ρ over (x, y) for a given μ, but this is often an expensive problem to solve. The *alternating direction method of multipliers* (ADMM) approximates this step via block-coordinate minimization, which results in the iteration

$$x_{k+1} \in \underset{x}{\arg\min}\, L_\rho(x, y_k, \mu_k) \tag{6.89a}$$

$$y_{k+1} \in \underset{y}{\arg\min}\, L_\rho(x_{k+1}, y, \mu_k) \tag{6.89b}$$

$$\mu_{k+1} = \mu_k + \rho\left(Ax_{k+1} + By_{k+1} - c\right), \tag{6.89c}$$

for $k \in \mathbb{Z}_+$ and where $y_0 \in \mathbb{R}^m$ and $\mu_0 \in \mathbb{R}^p$ are initial values. If L_0 has a saddle-point, then the assumption that f and g are proper, closed, and convex ensures that the updates (6.89a) and (6.89b) are well-defined, i.e. a minimizer exists, but it is not necessarily unique. If we let $u_k = \mu_k/\rho$, then the ADMM updates can be expressed in a more convenient form as shown in Algorithm 6.12, which is a basic implementation of ADMM. Is can be advantageous to adaptively update the penalty parameter ρ and/or make use of preconditioning techniques, which is often done in more sophisticated variants.

Algorithm 6.12: Alternating Direction Method of Multipliers

Input: Initialization $y_0 \in \mathbb{R}^m$ and $u_0 \in \mathbb{R}^p$, penalty parameter $\rho > 0$, and tolerances $\epsilon_p > 0$ and $\epsilon_d > 0$.

for $k = 0, 1, 2, \ldots$ **do**

$\quad x_{k+1} \in \underset{x}{\operatorname{argmin}} \left\{ f(x) + \dfrac{\rho}{2} \|Ax + By_k - c + u_k\|_2^2 \right\}$

$\quad y_{k+1} \in \underset{y}{\operatorname{argmin}} \left\{ g(y) + \dfrac{\rho}{2} \|Ax_{k+1} + By - c + u_k\|_2^2 \right\}$

$\quad u_{k+1} = u_k + Ax_{k+1} + By_{k+1} - c$

$\quad$ **if** $\|u_{k+1} - u_k\| \le \epsilon_p$ *and* $\|\rho A^T B(y_{k+1} - y_k)\| \le \epsilon_d$ **then** stop

end

The ADMM updates are often cheaper to compute compared to the cost of the joint minimization over x and y in the method of multipliers. In the special case where $A = I$, the update of x can be expressed as

$$x_{k+1} = \operatorname{prox}_{\rho^{-1}f}(c - u_k - By_k)$$

where $\operatorname{prox}_{\rho^{-1}f}$ is the proximal operator associated with $\rho^{-1}f$. Similarly, if $B = I$, then $y_{k+1} = \operatorname{prox}_{\rho^{-1}g}(c - u_k - Ax_{k+1})$.

6.11.3 Variable Splitting

Many of the learning problems that we will discuss in Chapters 8–12 can be expressed as optimization problems of the form

$$\text{minimize } f(x) + g(x)$$

with variable $x \in \mathbb{R}^n$ and where $f : \mathbb{R}^n \to \bar{\mathbb{R}}$ and $g : \mathbb{R}^n \to \bar{\mathbb{R}}$ are proper, closed, and convex functions. For example, f may represent a data fitting term, and g may represent some form of regularization or constraints. By introducing an auxiliary variable $y \in \mathbb{R}^n$, we can equivalently consider the problem

$$\begin{aligned} \text{minimize } & f(x) + g(y) \\ \text{subject to } & x - y = 0 \end{aligned} \tag{6.90}$$

which is a special case of the problem (6.88) with $A = I$ and $B = -I$. This reformulation technique is sometimes referred to as *variable splitting*, and it allows us apply the ADMM to the problem. The scaled form of the ADMM updates in Algorithm 6.12 may be expressed as

$$x_{k+1} = \operatorname{prox}_{\rho^{-1}f}(y_k - u_k) \tag{6.91a}$$

$$y_{k+1} = \operatorname{prox}_{\rho^{-1}g}(x_{k+1} + u_k) \tag{6.91b}$$

$$u_{k+1} = u_k + x_{k+1} - y_{k+1}. \tag{6.91c}$$

We note that in the special case where $g(x) = I_C(x)$ for some closed convex set $C \subset \mathbb{R}^n$, the proximal operator $\operatorname{prox}_{\rho^{-1}g}$ reduces to the Euclidean projection onto C, which can be computed efficiently for many simple convex sets; see, e.g. [85].

Variable splitting may also be used when $f : \mathbb{R}^n \to \mathbb{R}$ is defined as $f(x) = \sum_{i=1}^{N} f_i(x)$ where the ith function $f_i : \mathbb{R}^n \to \bar{\mathbb{R}}$ is associated with, e.g. a subset of the available data. By introducing auxiliary variables $x_i \in \mathbb{R}^n$, $i \in \mathbb{N}_N$, we may instead consider the optimization problem

$$\begin{aligned} \text{minimize } & \sum_{i=1}^{N} f_i(x_i) + g(y) \\ \text{subject to } & x_i - y = 0, \quad i \in \mathbb{N}_N, \end{aligned} \tag{6.92}$$

with variables $y \in \mathbb{R}^n$ and $x = (x_1, \ldots, x_N)$ with $x_i \in \mathbb{R}^n$, $i \in \mathbb{R}^n$. Applying the ADMM to this problem, it is straightforward to see that the ADMM update of x becomes separable. In other words, $x_1, \ldots, x_N$ can be updated in parallel; see Exercise 6.14.

Exercises

6.1   Consider the gradient descent iteration (6.14), i.e.

$$x_{k+1} = x_k - \frac{1}{L}\nabla f(x_k), \quad k \in \mathbb{Z}_+,$$

and assume that f is bounded below and L-smooth. Show that $\lim_{k \to \infty} \|\nabla f(x_k)\|_2^2 = 0$.

6.2   Recall that a function $f : \mathbb{R}^n \to \mathbb{R}$ is L-smooth if there exists a finite constant $L > 0$ such that

$$\|\nabla f(y) - \nabla f(x)\|_2 \le L\|y - x\|_2, \quad \forall x, y \in \mathbb{R}^n.$$

(a) Show that if f is L-smooth, then for all $x, y \in \mathbb{R}^n$,

$$f(y) \le f(x) + \nabla f(x)^T(y - x) + \frac{L}{2}\|y - x\|_2^2.$$

Hint: Let $h(\tau) = f(x + \tau(y - x))$ and start with the Newton–Leibniz formula

$$h(1) - h(0) = \int_0^1 h'(\tau)\, d\tau.$$

(b) Suppose $f : \mathbb{R}^n \to \mathbb{R}$ is L-smooth and convex, and define $h : \mathbb{R}^n \to \mathbb{R}$ as $h(y) = f(y) - \nabla f(x)^T y$, where $x \in \mathbb{R}^n$ is given. Show that h is L-smooth and convex and that for all $y \in \mathbb{R}^n$,

$$h(x) \le h(y - (1/L)\nabla h(y)) \le h(y) - \frac{1}{2L}\|\nabla h(y)\|_2^2.$$

(c) Show that if f is L-smooth and convex, then for all $x, y \in \mathbb{R}^n$,

$$f(x) + \nabla f(x)^T(y - x) + \frac{1}{2L}\|\nabla f(y) - \nabla f(x)\|_2^2 \le f(y).$$

Hint: Use the result from the previous part of this exercise.

(d) Show that if f is L-smooth and convex, then for all $x, y \in \mathbb{R}^n$,

$$\frac{1}{L}\|\nabla f(y) - \nabla f(x)\|_2^2 \le (\nabla f(y) - \nabla f(x))^T(y - x).$$

Hint: Use the result from the previous part of this exercise.

6.3   Let $f : \mathbb{R}^n \to \mathbb{R}$ be a μ-strongly convex and L-smooth function, and define $h : \mathbb{R}^n \to \mathbb{R}$ as $h(x) = f(x) - \mu/2\|x\|_2^2$.

(a) Show that h is convex and $(L - \mu)$-smooth.

(b) Show that for all $x, y \in \mathbb{R}^n$,

$$\frac{\mu L}{\mu + L}\|y - x\|_2^2 + \frac{1}{\mu + L}\|\nabla f(y) - \nabla f(x)\|_2^2 \le (\nabla f(y) - \nabla f(x))^T(y - x).$$

Hint: Use the fact that h is $(L - \mu)$-smooth and consider two cases, namely $\mu = L$ and $\mu < L$.

6.4   Many learning problems involve objective functions of the form $f(x) = g(x) + h(x)$, where g is convex and L_g-smooth and where h is a μ-stongly convex and L_h-smooth regularization function. Show that this implies that f is $(L_g + L_h)$-smooth and μ-strongly convex.

6.5 Let $g : \mathbb{R}^n \to \mathbb{R}$ and let $f(x) = \lambda g(x/\lambda)$ for some $\lambda > 0$. Show that

$$\text{prox}_f(x) = \lambda\text{prox}_{\lambda^{-1}g}(x/\lambda).$$

6.6 (a) Show that if $f : \mathbb{R}^n \to \mathbb{R}$ is of the form $f(x) = \sum_{i=1}^n f_i(x_i)$, where $f_i : \mathbb{R} \to \mathbb{R}$, then

$$\text{prox}_f(x) = \left(\text{prox}_{f_1}(x_1), \dots, \text{prox}_{f_n}(x_n)\right).$$

(b) Show that the proximal operator associated with $f(x) = \gamma\|x\|_1$ with $\gamma > 0$ may be expressed as

$$\text{prox}_f(x) = \left(S_\gamma(x_1), \dots, S_\gamma(x_n)\right),$$

where $S_\gamma : \mathbb{R} \to \mathbb{R}$ is defined as

$$S_\gamma(t) = \begin{cases} 0, & |t| \leq \gamma, \\ t - \gamma, & t > \gamma, \\ t + \gamma, & t < -\gamma, \end{cases}$$

or, equivalently,

$$S_\gamma(t) = \text{sgn}(t)\max(0, |t| - \gamma).$$

6.7 The nuclear norm of an $m \times n$ matrix X can be expressed as

$$\|X\|_* = \sup_{Z \in \mathbb{R}^{m \times n}} \{\text{tr}(X^T Z) \mid \|Z\|_2 \leq 1\},$$

i.e. the nuclear norm is the dual norm of the operator norm $\| \cdot \|_2$ on $\mathbb{R}^{m \times n}$. Let $X = U\Sigma V^T$ be an singular value decomposition (SVD) of $X \in \mathbb{R}^{m \times n}$, where $U \in \mathbb{R}^{m \times m}$ and $V \in \mathbb{R}^{n \times n}$ are orthogonal matrices, and $\Sigma \in \mathbb{R}^{m \times n}$ has the singular values of X on its main diagonal in descending order and zeros elsewhere. Now, letting $r = \text{rank}(X)$, a thin SVD of X may be expressed as

$$X = U_1 S V_1^T,$$

where $U_1 \in \mathbb{R}^{m \times r}$ and $V_1 \in \mathbb{R}^{n \times r}$ are the first r columns U and V, respectively, and $S = \text{diag}(\sigma_1, \dots, \sigma_r)$ is a submatrix of Σ.

(a) Show that the nuclear norm of X is the sum of its singular values, i.e.

$$\|X\|_* = \sum_{i=1}^r \sigma_i(X).$$

(b) Show that the subdifferential of $h(X) = \|X\|_*$ may be expressed as

$$\partial h(X) = \{U_1 V_1^T + W \mid \|W\|_2 \leq 1, \ U_1^T W = 0, \ WV_1 = 0\}.$$

(c) Show that the proximal operator associated with $\gamma\|X\|_*$ with $\gamma > 0$ may be expressed as

$$\text{prox}_{\gamma h}(X) = U_1 \text{diag}(\tilde{\sigma}_1, \dots, \tilde{\sigma}_r)V_1^T,$$

where $h(X) = \|X\|_*$ and $\tilde{\sigma}_i = \max(0, \sigma_i - \gamma), i \in \mathbb{N}_r$.

6.8 Consider the 1-norm regularized least-squares problem

$$\text{minimize } \frac{1}{2}\|Ax - b\|_2^2 + \gamma\|x\|_1, \tag{6.93}$$

with variable $x \in \mathbb{R}^n$, problem data $A \in \mathbb{R}^{m \times n}$ and $b \in \mathbb{R}^m$, and where $\gamma > 0$ is a given regularization parameter. We will denote by $x^*(\gamma)$ a solution to (6.93) for a given γ.

(a) Show that $x^\star(\gamma) = 0$ is a solution if $\gamma \geq \|A^T b\|_\infty$.

(b) Use CVX or CVXPY to solve the problem (6.93) for different values of γ using the problem data (A, b) from the file `11regls.mat`. Create a trade-off plot with the points $(\|x^\star(\gamma)\|_1, \frac{1}{2}\|Ax^\star(\gamma) - b\|_2^2)$. A brief introduction to CVX can be found in the appendix.

Hint: Use, say, 10-20 logarithmically spaced values of γ, say, between 10^{-2} and $\|A^T b\|_\infty$.

(c) Implement the PG method and the APG method for solving (6.93). Apply both implementations to the problem data from `11regls.mat` using $\gamma = 0.1$ and 300 iterations. Compare your solution to that obtained with CVX, and plot the relative suboptimality as a function of the number of iterations for both the PG and the APG method. The relative suboptimality at x_k may be defined as

$$\frac{|f(x_k) - p^\star|}{|p^\star|},$$

where $f(x) = \frac{1}{2}\|Ax - b\|_2^2 + \gamma \|x\|_1$ is the objective function. Use a log–linear plot (i.e. a logarithmically scaled ordinate axis), and use the optimal value returned by CVX as an estimate of $p^\star$.

(d) Implement ADMM for solving the equivalent problem

$$\begin{aligned}
\text{minimize} \quad & \tfrac{1}{2}\|Ax - b\|_2^2 + \gamma \|y\|_1 \\
\text{subject to} \quad & x = y,
\end{aligned} \tag{6.94}$$

with variables $x, y \in \mathbb{R}^n$. Run 300 iterations with $\rho = 10^{-1}$, and make a plot of the relative suboptimality and another plot of $\|x_k - y_k\|_2$.

(e) Implement an interior-point method for solving the equivalent problem

$$\begin{aligned}
\text{minimize} \quad & \tfrac{1}{2}\|Ax - b\|_2^2 + \gamma\, \mathbf{1}^T u \\
\text{subject to} \quad & -u \leq x \leq u,
\end{aligned} \tag{6.95}$$

with variables $x, u \in \mathbb{R}^n$. Plot the relative suboptimality as a function of the number of outer iterations.

6.9 Consider the problem of minimizing

$$f(x) = \frac{\gamma}{2}\|x\|_2^2 + \frac{1}{m}\sum_{i=1}^{m} \ln(1 + \exp(-y_i a_i^T x)), \tag{6.96}$$

with variable $x \in \mathbb{R}^n$, problem data $y \in \{-1, 1\}^m$ and $A \in \mathbb{R}^{m \times n}$ where a_i^T denotes the ith row of A, and where $\gamma > 0$ is a given parameter.

(a) Show that f is γ-strongly convex and L-smooth with $L = \frac{1}{4m}\|A\|_2^2 + \gamma$.

(b) Implement Newton's method for minimizing (6.96), and use the condition $(1/2)\lambda(x)^2 \leq 10^{-6}$ as a stopping criterion. Test your implementation using the problem data contained in the file `classification_small.mat` and with $\gamma = 10^{-6}$.

(c) Implement a quasi-Newton method with BFGS updates of the inverse Hessian approximation. Test your implementation and compare with your implementation of Newton's method.

(d) Implement the gradient method for minimizing (6.96) with the option to use a constant step size or a backtracking line search.

(e) Implement a stochastic gradient method for minimizing (6.96). The expression

$$g_k = \gamma x_k - \frac{1}{|S_k|}\sum_{i \in S_k} \frac{y_i \exp(-y_i a_i^T x_k)}{1 + \exp(-y_i a_i^T x_k)} a_i,$$

can be used to generate a realization of a stochastic gradient at x_k by randomly choosing a subset $S_k \subseteq \mathbb{N}_m^p$ for some $|S_k| = p \in \mathbb{N}_m$. Use a step-size sequence of the form (6.70), i.e.

$$t_k = \frac{t}{(k + 1 + \zeta)^\delta}, \quad k \in \mathbb{Z}_+.$$

Test your implementation with $p/m \approx 0.05$ and compare with your implementation of Newton's method and the gradient method. Plot the objective value versus the number of epochs for realizations with different values of the parameters $\delta \in (0, 1]$ and $t > 0$.

6.10 In this exercise, we will derive a method for solving the QP in (5.6) based on ADMM. We will here state the QP slightly differently as

$$\text{minimize} \quad \frac{1}{2} x^T Q x + r^T x$$
$$\text{subject to} \quad Ax \in C$$

with variable $x \in \mathbb{R}^n$. The problem data are $Q \in \mathbb{S}_+^n$, $r \in \mathbb{R}^n$, and $A \in \mathbb{R}^{m \times n}$. We will assume that $C \subseteq \mathbb{R}^m$ is a nonempty, closed, and convex set of the form

$$C = \{z \in \mathbb{R}^m \mid l \le z \le u\}$$

with $l, u \in \mathbb{R}^m$. Notice that the constraints $Bx = c$ and $Cx \le d$ in (5.6) can be cast in the above format by defining A, l, and u appropriately, e.g. an equality constraint is obtained by defining $l_i = u_i$.

(a) Consider the equivalent optimization problem

$$\text{minimize} \quad \frac{1}{2} \bar{x}^T Q \bar{x} + r^T \bar{x} + I_{\mathcal{A}}(\bar{x}, \bar{y}) + I_C(y)$$
$$\text{subject to} \quad \bar{x} - x = 0$$
$$\bar{y} - y = 0$$

with variables $x, \bar{x} \in \mathbb{R}^n$ and $y, \bar{y} \in \mathbb{R}^m$, and where $\mathcal{A} = \{(x, y) \in \mathbb{R}^n \times \mathbb{R}^m \mid Ax = y\}$. Let (λ, μ) be Lagrange multipliers associated with the equality constraints, and let $L_\rho(x, y, \bar{x}, \bar{y}, \lambda, \mu)$ be the augmented Lagrangian. Show that the three ADMM updates are equivalent to the following problems:

1. Minimize $L_\rho(x, y, \bar{x}, \bar{y}, \lambda, \mu)$ with respect $\bar{x}$ and $\bar{y}$:

$$\text{minimize} \quad \frac{1}{2} \bar{x}^T Q \bar{x} + r^T \bar{x} + \lambda_k^T \bar{x} + \mu_k^T \bar{y} + \frac{\rho}{2} \|\bar{x} - x_k\|_2^2 + \frac{\rho}{2} \|\bar{y} - y_k\|_2^2$$
$$\text{subject to} \quad A\bar{x} = \bar{y}.$$

2. Minimize $L_\rho(x, y, \bar{x}, \bar{y}, \lambda, \mu)$ with respect x and y:

$$\text{minimize} \quad \frac{\rho}{2} \|\bar{x}_{k+1} - x\|_2^2 + \frac{\rho}{2} \|\bar{y}_{k+1} - y\|_2^2 - \lambda_k^T x - \mu_k^T y$$
$$\text{subject to} \quad y \in C.$$

3. Update Lagrange multipliers:

$$\lambda_{k+1} = \lambda_k + \rho(\bar{x}_{k+1} - x_{k+1})$$
$$\mu_{k+1} = \mu_k + \rho(\bar{y}_{k+1} - y_{k+1}).$$

(b) Write down the optimality conditions for the first update with v as Lagrange multiplier for the equality constraint, and show that the optimality conditions are equivalent to

$$\begin{bmatrix} Q + \rho I & A^T \\ A & -\rho^{-1}I \end{bmatrix} \begin{bmatrix} \bar{x}_{k+1} \\ v \end{bmatrix} = \begin{bmatrix} -r + \rho x_k - \lambda_k \\ y_k - \rho^{-1}\mu_k \end{bmatrix}$$

together with

$$\bar{y}_{k+1} = \rho^{-1}(v - \mu_k) + y_k.$$

(c) Show that the second update has the solution

$$x_{k+1} = \bar{x}_{k+1} + \rho^{-1}\lambda_k$$

and that y_{k+1} is the projection of $\bar{y}_{k+1} + \rho^{-1}\mu_k$ onto the set C. Also, show that it follows that $\lambda_k = 0$ for all k, and hence, the update for λ_k can be omitted.

(d) Define

$$r_k^p = Ax_k - y_k$$
$$r_k^d = Qx_k + r + A^T v$$

which are primal and dual residuals for the optimization problem. Let

$$\epsilon^p = \epsilon^a + \epsilon^r \max\{\|Ax_k\|_\infty, \|y_k\|_\infty\}$$
$$\epsilon^d = \epsilon^a + \epsilon^r \max\{\|Qx_k\|_\infty, \|A^T v\|_\infty, \|r\|_\infty\}$$

where $\epsilon^a > 0$ and $\epsilon^r > 0$ are some absolute and relative tolerances, respectively. The termination criteria for the algorithm are

$$\|r_k^p\|_\infty \le \epsilon^p, \quad \|r_k^d\|_\infty \le \epsilon^d.$$

Write a MATLAB code that implements the ADDM algorithm you have derived above. Make sure to use sparse linear algebra routines in MATLAB.

(e) Consider the following so-called "support vector machine" problem

$$\begin{aligned} \text{minimize} \quad & \tfrac{1}{2}x^T x + 1^T t \\ \text{subject to} \quad & \text{diag}(b)Ax \ge 1 - t \\ & t \ge 0 \end{aligned}$$

with variables $x \in \mathbb{R}^n$ and $t \in \mathbb{R}^m$ and where $b \in \{-1, 1\}^m$ is a label vector and the rows of $A \in \mathbb{R}^{m \times n}$ are m feature vectors of length n. We will return to this problem in Section 10.5. Generate problem instances for given values of m and n as

$$b_i = \begin{cases} 1, & i \le m/2, \\ -1, & i > m/2, \end{cases}$$

and let the elements A_{ij} of the matrix A come from a normal probability distribution with standard deviation $1/n$ and mean given by

$$\begin{cases} 1/n, & i \le m/2, \\ -1/n, & i > m/2, \end{cases}$$

with 15% nonzeros in each row. This means that 85% of the entries of a row of the matrix A should be zero. What entries that are zero should be chosen in a random way for each row. Notice that the matrix A of this subproblem is not the same as the matrix A we defined before. The same goes for the variable x. Solve several instances of the problem for different values of m and n. Try different values of m and n. For how large values of m and n does your code perform well? Make plots of the solution time versus the total number of nonzero elements in the matrices A and Q. You may use $\epsilon^a = 10^{-3}$ and $\epsilon^r = 10^{-3}$.

6.11 Let $\mathcal{A} : \mathbb{R}^n \to \mathbb{R}^{p \times q}$ be a linear function, and let $H \in \mathbb{S}_+^n$, $A_0 \in \mathbb{R}^{p \times q}$, and $a \in \mathbb{R}^n$. Consider the optimization problem

$$\text{minimize } \|\mathcal{A}(x) + A_0\|_* + \frac{1}{2}(x - a)^T H(x - a)$$

which is equivalent to

$$\text{minimize } \|X\|_* + \frac{1}{2}(x - a)^T H(x - a)$$
$$\text{subject to } \mathcal{A}(x) + A_0 = X$$

with variables $x \in \mathbb{R}^n$ and $X \in \mathbb{R}^{p \times q}$.

Show that an ADMM algorithm for the above optimization problem can be expressed as the following updates:

1. Compute x_{k+1} by solving

$$(H + \rho M)x_{k+1} = \mathcal{A}_{\text{adj}}\left(\rho X_k + A_0 - Z_k\right) + Ha$$

where the matrix $M \in \mathbb{S}^n$ is defined as the matrix that satisfies $\mathcal{A}_{\text{adj}}(\mathcal{A}(z)) = Mz$ for all $z \in \mathbb{R}^n$, and where $\mathcal{A}_{\text{adj}} : \mathbb{R}^{p \times q} \to \mathbb{R}^n$ is the adjoint of $\mathcal{A}$.

2. Compute

$$X_{k+1} = \underset{X}{\text{argmin}} \left(\|X\|_* + \frac{\rho}{2} \left\| \frac{1}{\rho} Z + \mathcal{A}(x_k) + A_0 - X \right\|_F^2 \right)$$

$$= \sum_{i=1}^{\min(p,q)} \max\left(0, \sigma_i - \frac{1}{\rho}\right) u_i v_i^T$$

where u_i, v_i and σ_i are given by an SVD

$$\frac{1}{\rho} Z + \mathcal{A}(x_k) + A_0 = \sum_{i=1}^{\min(p,q)} \sigma_i u_i v_i^T.$$

3. Compute

$$Z_{k+1} = Z_k + \rho\left(\mathcal{A}(x_k) + A_0 - X_k\right).$$

Hint: When defining the augmented Lagrangian, the appropriate inner product is $\langle \cdot, \cdot \rangle$: $\mathbb{R}^{p \times q} \times \mathbb{R}^{p \times q} \to \mathbb{R}$ is $\langle X, Y \rangle = \text{tr}(X^T Y)$, and the appropriate norm is the Frobenius norm $\|X\|_F = \langle X, X \rangle^{1/2} = \sqrt{\text{tr}(X^T X)}$.

6.12 Show that if $\psi : \text{int } K \to \mathbb{R}$ is a generalized logarithm for a cone $K \subseteq \mathbb{R}^n$, then it holds that

$$\nabla \psi(x)^T x = \theta, \quad x \in \text{int } K,$$

where θ is the degree of ψ.

6.13 Show that $\psi : \text{int } \mathbb{Q}^n \to \mathbb{R}$ defined as $\psi(x) = \ln\left(x_n^2 - x_1^2 - \cdots - x_{n-1}^2\right)$ is a generalized logarithm for $\mathbb{Q}^n$ with degree $\theta = 2$.

6.14 Consider the problem (6.92), i.e.

$$\text{minimize } \sum_{i=1}^N f_i(x_i) + g(y)$$
$$\text{subject to } x_i - y = 0, \quad i \in \mathbb{N}_N,$$

where $g : \mathbb{R}^n \to \bar{\mathbb{R}}$ and $f_i : \mathbb{R}^n \to \bar{\mathbb{R}}$, $i \in \mathbb{N}_N$, are proper, closed, and convex functions, and the variables are $y \in \mathbb{R}^n$ and $x = (x_1, \ldots, x_N)$ with $x_i \in \mathbb{R}^n$, $i \in \mathbb{N}_N$. Show that the scaled form of the ADMM updates for this problem can be expressed as

$$x_i^{(k+1)} = \text{prox}_{\rho^{-1}f_i}(y^{(k)} - u_i^{(k)}), \quad i \in \mathbb{N}_N$$

$$\tilde{x}^{(k+1)} = \frac{1}{N} \sum_{i=1}^{N} x_i^{(k+1)}$$

$$\tilde{u}^{(k)} = \frac{1}{N} \sum_{i=1}^{N} u_i^{(k)}$$

$$y^{(k+1)} = \text{prox}_{\rho^{-1}g}(\tilde{x}^{(k+1)} + \tilde{u}^{(k)})$$

$$u_i^{(k+1)} = u_i^{(k)} + x_i^{(k+1)} - y^{(k+1)}, \quad i \in \mathbb{N}_N$$

where we have used a superscript for the iteration index to avoid abiguity.

Part III

Optimal Control

7

Calculus of Variations

So far we have discussed optimization when the variables belong to finite-dimensional vector spaces. However, it is often of interest to also optimize over infinite-dimensional vector spaces. The most simple case is when the variable is a real-valued function of a real variable. This has applications in optimal control in continuous time, where the optimal control signal is a real-valued function of time. Another important application is the derivation of probability density functions from the principle of maximizing the entropy subject to moment constraints. In this case, the variable is the probability density function. For probability density functions, the argument is often a vector. How to solve these types of optimization problems is called *calculus of variations*. The origin of this theory goes back to Newton's minimal resistance problem. Major contributions were made by Euler and Lagrange. The generalizations to optimal control were made by Pontryagin. We will present the theory in this general form, but we will not be able to prove all the results. The interested reader is referred to the vast literature on optimal control for most of the proofs, especially for the most general results. However, we will provide the proofs for some special cases to build intuition.

7.1 Extremum of Functionals

We consider a normed linear space of continuously differentiable real-valued functions defined on $D = [a, b] \subset \mathbb{R}$, which we denote by $C(a, b)$. The norm $\| \cdot \| : C(a, b) \to \mathbb{R}$ is defined as

$$\|y\| = \max_{x \in D} |y(x)| + \max_{x \in D} |y'(x)|.$$

We then consider real-valued functionals J defined on $C(a, b)$, i.e. $J : C(a, b) \to \mathbb{R}$. A typical example is

$$J[y] = \int_a^b f\left(x, y(x), y'(x)\right) dx, \tag{7.1}$$

for some function $f : \mathbb{R} \times \mathbb{R} \times \mathbb{R} \to \mathbb{R}$. We are interested in characterizing extrema of such functionals. To this end, we define the *increment* $\Delta J : C(a, b) \to \mathbb{R}$ of a functional J for a fixed $y \in C(a, b)$ as

$$\Delta J[\delta y] = J[y + \delta y] - J[y].$$

In case

$$\Delta J[\delta y] = \delta J[\delta y] + \epsilon \|\delta y\|,$$

Optimization for Learning and Control, First Edition. Anders Hansson and Martin Andersen.
© 2023 John Wiley & Sons, Inc. Published 2023 by John Wiley & Sons, Inc.
Companion Website: www.wiley.com/go/opt4lc

where $\delta J : C(a, b) \to \mathbb{R}$ is a linear functional and $\epsilon \to 0$ as $\|\delta y\| \to 0$, we say that J is *differentiable*, and we call the functional δJ the *first variation* or *differential* of J. It can be shown that the differential of a differentiable functional is unique; see, e.g. [41].

Example 7.1 The functional in (7.1) is differentiable if f is a differentiable function in its last two arguments. This follows from a Taylor series expansion:

$$\Delta J[\delta y] = \int_a^b \left(f\left(x, y(x) + \delta y(x), y'(x) + \delta y'(x)\right) - f\left(x, y(x), y'(x)\right) \right) dx,$$

$$= \int_a^b \left(\frac{\partial f\left(x, y(x), y'(x)\right)}{\partial y} \delta y(x) + \frac{\partial f\left(x, y(x), y'(x)\right)}{\partial y'} \delta y'(x), \right.$$

$$\left. + h\left(y(x), y'(x)\right) \left\| (\delta y(x), \delta y'(x)) \right\|_2 \right) dx,$$

where $h : \mathbb{R} \times \mathbb{R} \to \mathbb{R}$ is a function that goes to zero as $(\delta y(x), \delta y'(x)) \to 0$. The latter is implied by $\|\delta y\| \to 0$. Hence, we may define ϵ as

$$\epsilon = \frac{\int_a^b h\left(y(x), y'(x)\right) \left\| (\delta y(x), \delta y'(x)) \right\|_2 dx}{\|\delta y\|},$$

$$\leq \frac{\int_a^b h\left(y(x), y'(x)\right) \|\delta y\| dx}{\|\delta y\|} = \int_a^b h\left(y(x), y'(x)\right) dx,$$

which converges to zero as h goes to zero. Hence, this functional is differentiable with first variation

$$\delta J[\delta y] = \int_a^b \left(\frac{\partial f\left(x, y(x), y'(x)\right)}{\partial y} \delta y(x) + \frac{\partial f\left(x, y(x), y'(x)\right)}{\partial y'} \delta y'(x) \right) dx. \tag{7.2}$$

It is left as an exercise to show that this functional is linear, see Exercise 7.2. Here, we understand why the norm we use in the definition of $C(a, b)$ has to include the derivative. However, in case we can assure that bounded norm of y implies bounded norm of y', we may use another definition.

7.1.1 Necessary Condition for Extremum

A central theme in calculus of variations is to characterize so-called *extrema* of functionals. We say that a functional J has a weak extremum[1] at $y^\star$ if its increment has the same sign for all y in a neighborhood of $y^\star$.[2] More formally, there should exist $\epsilon > 0$ such that $\Delta J[\delta y]$ has the same sign for all δy such that $\|\delta y\| \leq \epsilon$. If the sign is positive, we have a *weak minimum*, and if the sign is negative, we have a *weak maximum*.

It probably does not come as a surprise that in order to characterize extrema for differentiable functionals, it is sufficient to consider the first variation. Assume that a differentiable functional J has an extremum at $y^\star$. Then the first variation vanishes, i.e. $\delta J[\delta y] = 0$. The proof of this is as follows: we have

$$\Delta J[\delta y] = \delta J[\delta y] + \epsilon \|\delta y\|,$$

---

1 The reason to use the word "weak" is that there is also a definition of strong extremum that is based on another norm for the linear function space, which is defined as $\|y\| = \max_{x \in D} \|y(x)\|$. Clearly, strong extremum implies weak extremum. Necessary conditions for weak extremum are, hence, also necessary conditions for strong extremum.
2 In case there are constraints on y for $x = a$ or $x = b$, δy should be constrained to be zero at those values of x. This is often stated as $\delta y(x)$ is *admissible*. We will tacitly assume that we only consider such admissible δy.

where $\epsilon \to 0$ as $\|\delta y\| \to 0$. Hence for sufficiently small $\|\delta y\|$, the sign of $\Delta J[\delta y]$ will be the same as the sign of $\delta J[\delta y]$. Now assume that $\delta J[\delta y_0] \neq 0$ for some δy_0. Then for any $\alpha > 0$, we have

$$\delta J[-\alpha \delta y_0] = -\delta J[\alpha \delta y_0],$$

since δJ is linear. Hence, the increment can be made to have either sign for arbitrary small δy contradicting that J has an extremum.

7.1.2 Sufficient Condition for Optimality

So far we have derived necessary conditions for a differentiable functional to have a weak extremum. We are also interested in the sufficient conditions. To this end, we consider the so-called *second variation*. We say that a functional J is twice differentiable if the increment can be written as

$$\Delta J[\delta y] = \delta J[\delta y] + \delta^2 J[\delta y] + \epsilon^2 \|\delta y\|^2,$$

where δJ is the first variation, linear in δy, $\delta^2 J$ is the second variation, quadratic in δy, and $\epsilon \to 0$ as $\|\delta y\| \to 0$. It can also be shown that the second variation is unique.

Example 7.2 The functional in (7.1) is twice differentiable if f is a twice differentiable function in its last two arguments. This follows from a Taylor series expansion similar to what was done in the previous example. The second variation is given by

$$\delta^2 J[\delta y] = \frac{1}{2} \int_a^b \begin{bmatrix} \delta y(x) \\ \delta y'(x) \end{bmatrix}^T \begin{bmatrix} \dfrac{\partial^2 f(x, y(x), y'(x))}{\partial y^2} & \dfrac{\partial^2 f(x, y(x), y'(x))}{\partial y \partial y'} \\ \dfrac{\partial^2 f(x, y(x), y'(x))}{\partial y \partial y'} & \dfrac{\partial^2 f(x, y(x), y'(x))}{\partial y'^2} \end{bmatrix} \begin{bmatrix} \delta y(x) \\ \delta y'(x) \end{bmatrix} dx.$$

It can now, with similar techniques as used above, be proven that a necessary condition for $y^\star$ to be a minimum for J is that $\delta^2 J[\delta y] \geq 0$ for all δy; see, e.g. [41], This is not a sufficient condition. We say that the second variation is strongly positive if there exists a constant $k > 0$ such that $\delta^2 J[\delta y] \geq k\|\delta y\|^2$ for all y and δy. A sufficient condition for $y^\star$ to be optimal for J is that its first variation vanishes and that its second variation is strongly positive. This is also straightforward to prove; see, e.g. [41].

7.1.3 Constrained Problem

We now consider the constrained problem

$$\begin{aligned} &\text{minimize } J[y], \\ &\text{subject to } K[y] = 0, \end{aligned} \tag{7.3}$$

with variable $y \in C(a, b)$, where J and K are functionals from $C(a, b)$ to $\mathbb{R}$ and $\mathbb{R}^p$, respectively. We assume that K is affine and that J is twice differentiable with a strongly positive second variation. We define the Lagrangian functional $L : C(a, b) \times \mathbb{R}^p \to \mathbb{R}$ as

$$L[y, \mu] = J[y] + \mu^T K[y].$$

Now assume that $\bar{y} \in C(a, b)$ and $\bar{\mu} \in \mathbb{R}^p$ satisfy

$$\delta L[\delta y] = \delta J[\delta y] + \bar{\mu}^T \delta K[\delta y] = 0,$$

$$K[\bar{y}] = 0.$$

It then follows that $\bar{y}$ is optimal for (7.3). To see this we realize that $\bar{y}$ by the first condition above minimizes $L[y, \bar{\mu}]$, since L is twice differentiable with a strongly positive second variation. Now, assume that $\bar{y}$ is not optimal for the above optimization problem, but that $\tilde{y}$ is optimal. Then $J[\tilde{y}] < J[\bar{y}]$ and $K[\tilde{y}] = 0$ implies that

$$L[\tilde{y}, \bar{\mu}] = J[\tilde{y}] < J[\bar{y}],$$

which contradicts that $\bar{y}$ minimizes $L[y, \mu]$.

We will now show that strong duality holds. Let $g : \mathbb{R}^p \to \mathbb{R}$ be the *Lagrange dual function* defined via

$$g(\mu) = \min_{y \in C(a,b)} L[y, \mu].$$

We have

$$g(\mu) = \min_{y \in C(a,b)} L[y, \mu] \le \min_{y \in C(a,b),\ K[y]=0} L[y, \mu] = \min_{y \in C(a,b),\ K[y]=0} J[y] = p^{\star}.$$

Since $\bar{\mu}$ achieves equality above, we have that

$$d^{\star} = \max_{\mu} g(\mu) = g(\bar{\mu}) = p^{\star},$$

and hence strong duality holds. The extension to inequality constraints is straightforward and similar to what was done in Chapter 4. Necessary conditions for optimality of (7.3) are a bit tricky. For finite-dimensional optimization problems, we saw in Chapter 4 that constraint qualifications, such as Slater's conditions, are needed in order to guarantee strong duality, on which the proof of necessity was based.

7.1.4 Du Bois–Reymond Lemma

When the functional is described in terms of an integral, the following lemma called the fundamental lemma of calculus of variations or the *du Bois–Reymond lemma* is instrumental. If $y : [a, b] \to \mathbb{R}$ is a continuous function and

$$\int_a^b y(x)h(x)dx = 0,$$

for all $h \in C(a, b)$ such that $h(a) = h(b) = 0$, then $y(x) = 0$ for all $x \in [a, b]$. The proof is by contradiction. Suppose the function y is positive at some point $c \in [a, b]$. By continuity, it is also positive in some interval $[x_1, x_2] \subset [a, b]$. Let

$$h(x) = (x - x_1)(x_2 - x),$$

for $x \in [x_1, x_2]$ and zero, otherwise. Then it follows that

$$\int_a^b y(x)h(x)dx > 0,$$

which is a contradiction. In case the function is negative, at some point, a similar argument can be done. The lemma still holds true if we do not constrain h at a and/or b. We now consider an example.

Example 7.3 Let $f(y) = y \log y$, and consider the functionals

$$J[y] = \int_a^b f(y(x))dx,$$

$$K[y] = \int_a^b xy(x)dx - k,$$

for some constant k. We have $f'(y) = \log y + 1$ and $f''(y) = 1/y$ and hence, from (7.2) and from Example 7.2, the variations of J are

$$\delta J[\delta y] = \int_a^b \left(\log y(x) + 1\right) \delta y(x) dx,$$

$$\delta^2 J[\delta y] = \frac{1}{2} \int_a^b \frac{1}{y(x)} \delta y(x)^2 dx.$$

The first variation of K is $\delta K[\delta y] = \int_a^b x \delta y(x) dx$, and the second variation of K is zero. We define the Lagrangian as $L[y, \mu] = J[y] + \mu K[y]$, and we see that its first variation is

$$\delta L[\delta y] = \int_a^b \left(\log y(x) + 1 + \mu x\right) \delta y(x) dx.$$

By the du-Bois–Reymond lemma, it holds that if the first variation is zero, then

$$\log y(x) + 1 + \mu x = 0,$$

and hence, we have that $y(x) = \exp(-1 - \mu x)$. The constraint $K[y] = 0$ can be used to determine μ in terms of a, b, and k. We should try to verify that the second variation of L is strictly positive in order to show that y constitutes a minimum. This is however not so easy, or even possible, and we will in Section 9.2 prove optimality in another way for similar problems. Understand that strict positivity is just a sufficient condition that might be too strong in some cases.

7.1.5 Generalizations

Most of what has been discussed in this section generalizes to $D \subseteq \mathbb{R}^n$. This will be important when we revisit the example above and extensions of it in Section 9.2. We will also consider $y(x)$ to be vector valued, i.e. $y : D \to \mathbb{R}^n$. To this end, we just define the norm as

$$\|y\| = \sup_{x \in D} \|y(x)\|_2 + \sup_{x \in D} \|y'(x)\|_2,$$

where $\| \cdot \|_2$ as usual is the Euclidean vector norm. From now on, $C^n(a, b)$ is the normed linear space of differentiable functions $y : D \to \mathbb{R}^n$ with the above norm, where $D = [a, b]$. We will be less formal in our derivation of results in the remaining part of this chapter.

7.2 The Pontryagin Maximum Principle

We consider a dynamical system described by

$$\dot{x}(t) = F\left(t, x(t), u(t)\right), \tag{7.4}$$

where the *vector field* $F : \mathbb{R} \times \mathbb{R}^n \times \mathbb{R}^m \to \mathbb{R}^n$ is continuously differentiable. Here, $x \in C^n(0, T)$ is called the *state*, and $u \in C^m(0, T)$ is called the *input signal*. The signal u is also called the *control signal*. The *initial value* is given, i.e. $x(0) = x_0$ for some known $x_0 \in \mathbb{R}^n$. Let us consider the following optimal control problem:

$$\begin{aligned} &\text{minimize} \quad \phi\left(x(T)\right) + \int_0^T f\left(t, x(t), u(t)\right) dt, \\ &\text{subject to} \quad \dot{x}(t) = F\left(t, x(t), u(t)\right), \end{aligned} \tag{7.5}$$

with variables x and u, where the *incremental cost* $f : \mathbb{R} \times \mathbb{R}^n \times \mathbb{R}^m \to \mathbb{R}$ and the *terminal cost* or *final cost* $\phi : \mathbb{R}^n \to \mathbb{R}$ are continuously differentiable. We realize that this is an infinite-dimensional optimization problem, since the variables that we optimize over are the functions x and u.

We assume that the optimal u, which we denote by $u^\star$, is continuous.[3] We also assume that the corresponding solution $x^\star$ to the differential equation is unique, and that in case $u^\star$ is perturbed with a small amount, the corresponding perturbation of $x^\star$ is also small. We refrain from giving detailed conditions when this is satisfied, but just mention that it is related to what is called Lipschitz continuity of F.

We define the *Lagrangian* functional $L : C^m(0, T) \to \mathbb{R}$ as

$$L[u] = \phi(x(T)) + \int_0^T \left(f(t, x(t), u(t)) + \lambda(t)^T (F(t, x(t), u(t)) - \dot{x}(t)) \right) dt,$$

$$= \phi(x(T)) + \int_0^T \left(H(t, x(t), u(t), \lambda(t)) - \lambda(t)^T \dot{x}(t) \right) dt,$$

where $H : \mathbb{R} \times \mathbb{R}^n \times \mathbb{R}^m \times \mathbb{R}^n \to \mathbb{R}$ is the *Hamiltonian* defined as

$$H(t, x, u, \lambda) = f(t, x, u) + \lambda^T F(t, x, u).$$

We could have defined the Lagrangian functional to also depend explicitly on x and λ, but we will not need this in our derivations, and hence, we refrain from doing so. Similarly, as with the multiplier rule of Lagrange in (4.57a) and (4.57b) we expect to get a necessary condition for optimality by letting the first variation of L be zero. We make a perturbation $u = u^\star + \delta u$ of $u^\star$, where δu is small, i.e. $\|\delta u\| < \epsilon$. See Section 7.1 for the definition of the norm. The corresponding perturbed trajectory x, which is the solution of (7.4) for u differs from the original solution $x^\star$ with the quantity $\delta x = x - x^\star$, which by our assumptions is small, i.e. $\|\delta x\|$ can be made as small as we like by taking ϵ sufficiently small. We have that the increment of L is given by

$$\Delta L[\delta u] = L[u^\star + \delta u] - L[u^\star] = \phi\left(x^\star(T) + \delta x(T)\right) - \phi\left(x^\star(T)\right)$$

$$+ \int_0^T \left(H\left(t, x^\star(t) + \delta x(t), u^\star(t) + \delta u(t), \lambda(t)\right) - H\left(t, x^\star(t), u^\star(t), \lambda(t)\right) \right) dt$$

$$- \int_0^T \lambda(t)^T \frac{d}{dt} \left(x^\star(t) + \delta x(t)\right) dt + \int_0^T \lambda(t)^T \frac{d}{dt} x^\star(t)) dt.$$

We now make a Taylor series expansion and obtain the first variation

$$\delta L = \frac{\partial \phi}{\partial x^T} \delta x(T) + \int_0^T \left(\frac{\partial H}{\partial x^T} \delta x(t) + \frac{\partial H}{\partial u^T} \delta u(t) - \lambda(t)^T \frac{d \delta x(t)}{dt} \right) dt.$$

Here, we have not written out the arguments of the partial derivatives. This is clearly a linear functional. The assumption on small perturbations of δu resulting in small perturbations of δx is necessary. Otherwise, the remainder term does not converge to zero as $\|\delta u\| \to 0$. By integration of parts, it follows that

$$\delta L = \frac{\partial \phi}{\partial x^T} \delta x(T) + \int_0^T \left(\frac{\partial H}{\partial x^T} \delta x(t) + \frac{\partial H}{\partial u^T} \delta u(t) + \frac{d \lambda(t)^T}{dt} \delta x(t) \right) dt - \left[\lambda(t)^T \delta x(t) \right]_0^T.$$

Since the initial value $x(0)$ is given, it follows that $\delta x(0) = 0$, and hence, we obtain

$$\delta L = \left(\frac{\partial \phi}{\partial x^T} - \lambda(T)^T \right) \delta x(T) + \int_0^T \left(\frac{\partial H}{\partial x^T} \delta x(t) + \frac{\partial H}{\partial u^T} \delta u(t) + \frac{d \lambda(t)^T}{dt} \delta x(t) \right) dt.$$

We have so far made no assumptions on λ. We assume that it satisfies the *adjoint equations* defined as

$$\dot{\lambda}(t) = -\frac{\partial H(t, x^\star(t), u^\star(t), \lambda(t))}{\partial x}, \quad \lambda(T) = \frac{\partial \phi(x^\star(T))}{\partial x}.$$

---

3 An extension to the case when $u^\star$ is piecewise continuous is possible.

This is a linear time-varying differential equation for λ, and hence, it has a solution [97, Chapter 3], under mild conditions on H.[4] For this λ, it follows that

$$\delta L = \int_0^T \frac{\partial H}{\partial u^T} \delta u(t) dt.$$

From the du Bois–Reymond lemma, it then follows that

$$\frac{\partial H(t, x^\star(t), u^\star(t), \lambda(t))}{\partial u^T} = 0,$$

in order for the first variation to vanish. For differentiable $u^\star$, it holds that

$$\frac{dH}{dt} = \frac{\partial H}{\partial t} + \frac{\partial H}{\partial x^T} \dot{x}^\star + \frac{\partial H}{\partial u^T} \dot{u}^\star + \frac{\partial H}{\partial \lambda^T} \dot{\lambda} = \frac{\partial H}{\partial t} + \left(\frac{\partial H}{\partial x} + \dot{\lambda} \right)^T F = \frac{\partial H}{\partial t},$$

where all functions are evaluated for $(x^\star, u^\star)$. It is possible to show that the result holds also for piecewise continuous $u^\star$. In case H does not explicitly depend on t, which is called an *autonomous optimal control problem*, we realize that H is a constant independent of t.

We have now proven the following necessary conditions of Pontryagin, also called the *Pontryagin maximum principle* (PMP). Given optimal $u^\star$ and $x^\star$ for (7.5), there exists an *adjoint variable* λ such that

$$\dot{\lambda}(t) = -\frac{\partial H(t, x^\star(t), u^\star(t), \lambda(t))}{\partial x}, \quad \lambda(T) = \frac{\partial \phi(x^\star(T))}{\partial x}, \tag{7.6a}$$

$$\frac{\partial H(t, x^\star(t), u^\star(t), \lambda(t))}{\partial u^T} = 0, \tag{7.6b}$$

$$\frac{dH(t, x^\star(t), u^\star(t), \lambda(t))}{dt} = \frac{\partial H(t, x^\star(t), u^\star(t), \lambda(t))}{\partial t}. \tag{7.6c}$$

We remark that the necessary conditions do not distinguish between maximum or minimum. They hold for any extremum, and this is the reason why the conditions are called a maximum principle. They could just as well have been called a minimum principle.

We are also interested in the sufficient conditions for a locally optimal solution of (7.5). We will not provide details of the derivation. The condition is based on the second variation of the Lagrangian L. Let $(\bar{u}, \bar{x})$ and λ satisfy

$$\dot{x}(t) = F(t, \bar{x}(t), \bar{u}(t)), \quad x(0) = x_0,$$
$$\dot{\lambda}(t) = -\frac{\partial H(t, \bar{x}(t), \bar{u}(t), \lambda(t))}{\partial x}, \quad \lambda(T) = \frac{\partial \phi(\bar{x}(T))}{\partial x},$$
$$\frac{\partial H(t, \bar{x}(t), \bar{u}(t), \lambda(t))}{\partial u} = 0,$$
$$\frac{dH(t, \bar{x}(t), \bar{u}(t), \lambda(t))}{dt} = \frac{\partial H(t, \bar{x}(t), \bar{u}(t), \lambda(t))}{\partial t},$$
$$\frac{d^2 \phi(\bar{x}(T))}{dx dx^T} \geq 0, \quad \frac{\partial^2 H(t, \bar{x}(t), \bar{u}(t), \lambda(t))}{\partial z \partial z^T} \geq 0, \quad \frac{\partial^2 H(t, \bar{x}(t), \bar{u}(t), \lambda(t))}{\partial u \partial u^T} > 0,$$

where $z = (x, u)$. Then $(\bar{u}, \bar{x})$ is a local minimum of (7.5).

---

4 The system matrix and the input signal should be bounded for the existence of a solution. This holds if $\frac{\partial f(t, x^\star(t), u^\star(t))}{\partial x}$ and $\frac{\partial F(t, x^\star(t), u^\star(t))}{\partial x}$ are bounded functions of t on $[0, T]$.

7.2.1 Linear Quadratic Control

We will now look at an optimal control problem which has linear dynamics and a quadratic cost. The problem is called a *linear quadratic* (LQ) control problem and is given by

$$\text{minimize } \frac{1}{2}x(T)^T Q_0 x(T) + \frac{1}{2}\int_0^T \left(x(t)^T Q x(t) + u(t)^T R u(t)\right) dt,$$

$$\text{subject to } \dot{x}(t) = Ax(t) + Bu(t),$$

with variables x and u for given initial value $x(0) = x_0$. The Hamiltonian is given by

$$H = \frac{1}{2}\left(x^T Q x + u^T R u\right) + \lambda^T (Ax + Bu).$$

We realize that the adjoint equations are

$$\dot{\lambda} = -\frac{\partial H}{\partial x} = -Qx - A^T \lambda, \quad \lambda(T) = Q_0 x(T).$$

From the PMP, we have that

$$\frac{\partial H}{\partial u} = Ru + B^T \lambda = 0.$$

If we assume that $R > 0$, i.e. positive definite, we have that $u = -R^{-1}B^T \lambda$. If we insert this into the differential equations for x and λ, we obtain

$$\begin{bmatrix} \dot{x} \\ \dot{\lambda} \end{bmatrix} = \begin{bmatrix} A & -BR^{-1}B^T \\ -Q & -A^T \end{bmatrix}\begin{bmatrix} x \\ \lambda \end{bmatrix}, \quad \begin{bmatrix} x(0) \\ \lambda(T) \end{bmatrix} = \begin{bmatrix} x_0 \\ Q_0 x(T) \end{bmatrix}.$$

This is a two-point boundary value problem. Such problems are in general not easy to solve. The reason is that we do now know the initial value for λ. However, for the problem above, we may define

$$\mathcal{A} = \begin{bmatrix} A & -BR^{-1}B^T \\ -Q & -A^T \end{bmatrix}, \quad \Phi(t,s) = \begin{bmatrix} \Phi_{11}(t,s) & \Phi_{12}(t,s) \\ \Phi_{21}(t,s) & \Phi_{22}(t,s) \end{bmatrix} = e^{\mathcal{A}(t-s)},$$

where the partitions are conformable. The matrix Φ is called the *transition matrix*, and it can be shown that it is always invertible, and that $\Phi(t,t) = I$ for any t, see [97]. It then follows that

$$\begin{bmatrix} x_0 \\ \lambda(0) \end{bmatrix} = \begin{bmatrix} \Phi_{11}(0,T) & \Phi_{12}(0,T) \\ \Phi_{21}(0,T) & \Phi_{22}(0,T) \end{bmatrix}\begin{bmatrix} x(T) \\ Q_0 x(T) \end{bmatrix} = \begin{bmatrix} \Phi_{11}(0,T) + \Phi_{12}(0,T)Q_0 \\ \Phi_{21}(0,T) + \Phi_{22}(0,T)Q_0 \end{bmatrix} x(T).$$

We may now use these equations to express $\lambda(0)$ in terms of x_0, i.e. we obtain

$$\lambda(0) = \left(\Phi_{21}(0,T) + \Phi_{22}(0,T)Q_0\right)\left(\Phi_{11}(0,T) + \Phi_{12}(0,T)Q_0\right)^{-1} x_0,$$

assuming the inverse exists. It actually does; see, e.g. [20]. We are now in a position to solve the linear differential equation, which will then give us the optimal control signal.

7.2.2 The Riccati Equation

It turns out that we can actually be even more explicit. The above formula that we derived for $\lambda(0)$ can be generalized from 0 to any t, i.e.

$$\lambda(t) = P(t)x(t),$$

where

$$P(t) = \left(\Phi_{21}(t,T) + \Phi_{22}(t,T)Q_0\right)\left(\Phi_{11}(t,T) + \Phi_{12}(t,T)Q_0\right)^{-1}.$$

From this, it follows that

$$\dot\lambda = \dot Px + P\dot x = \dot Px + P(Ax - BR^{-1}B^T\lambda) = \dot Px + P(Ax - BR^{-1}B^T Px),$$

but we also have from the adjoint equations that

$$\dot\lambda = -Qx - A^T\lambda = -Qx - A^T Px,$$

and hence,

$$-Qx - A^T Px = \dot Px + PAx - PBR^{-1}B^T Px.$$

This is implied by

$$\dot P + A^T P + PA + Q - PBR^{-1}B^T P = 0,$$

with a boundary condition

$$P(T) = \left(\Phi_{21}(T, T) + \Phi_{22}(T, T)Q_0\right)\left(\Phi_{11}(T, T) + \Phi_{12}(T, T)Q_0\right)^{-1}$$
$$= (0 + Q_0)(I + 0)^{-1} = Q_0.$$

The above equation is called a *differential Riccati equation*. Once this equation has been solved, it holds that the optimal control signal is given by

$$u(t) = -R^{-1}B^T\lambda(t) = -R^{-1}B^T P(t)x(t).$$

We have actually been able to derive a feedback policy for the optimal solution, i.e. u depends explicitly on x, and not only on t, which is the case when we express u in terms of λ. It is not in general possible to obtain a feedback policy from the PMP. We were lucky here that the we could use the transition matrix. This is based on that all differential equations involved are linear. The result still holds for time-varying A, B, Q, and R.

7.3 The Euler–Lagrange Equations

We will now relate the above result to a classical result in the calculus of variations which is the special case when $\dot x = u$ or equivalently $F(t, x, u) = u$. It then follows that $H = f + \lambda^T u$. Hence, the conditions in (7.6a and b) become

$$\dot\lambda = -\frac{\partial f}{\partial x},$$

$$\frac{\partial f}{\partial u} + \lambda = 0,$$

from which it follows that

$$\frac{d}{dt}\left(\frac{\partial f}{\partial \dot x}\right) - \frac{\partial f}{\partial x} = 0, \tag{7.7}$$

which are called the *Euler–Lagrange* equations. They are the necessary conditions for an optimal $x^\star$ of the problem

$$\text{minimize } \int_0^T f(t, x(t), \dot x(t))\, dt,$$

with variable x. We will now look at an example from mechanical systems.

Example 7.4 In mechanical systems, the state x contains positions and angels, and $\dot x$ is called generalized velocity. We define the potential energy of the system as $V(x)$ and the kinetic energy as

$T(x, \dot{x})$, where it is assumed that $T(x, \dot{x}) = \dot{x}^T A(x)\dot{x}$ for some symmetric matrix $A(x)$. From this, we define $f(x, \dot{x}) = T(x, \dot{x}) - V(x)$, which is the difference between kinetic energy and potential energy. In mechanics, this is called the Lagrangian, and this should not be confused with the Lagrangian we have defined previously. The Lagrangian of mechanics is our incremental cost. A fundamental principle for conservative mechanical systems is the principle of least action, which says that the states should be a stationary solution to

$$\int_0^T f(x(t), \dot{x}(t))dt.$$

Hence, x should be a solution of the Euler–Lagrange equations, i.e.

$$\frac{d}{dt}\left(\frac{\partial T}{\partial \dot{x}}\right) - \frac{\partial T}{\partial x} + \frac{\partial V}{\partial x} = 0.$$

We also have that the Hamiltonian is

$$H = f + \lambda^T \dot{x} = T - V + \lambda^T \dot{x} = T - V - \frac{\partial f}{\partial \dot{x}}\dot{x} = \dot{x}^T A\dot{x} - V - 2\dot{x}A\dot{x} = -T - V.$$

Since the system is autonomous, the Hamiltonian should be a constant, and hence, we have proven that the sum of potential energy and kinetic energy is constant for conservative mechanical systems.

7.3.1 Beltrami's Identity

For the case, when f does not depend explicitly on t, we may derive a very useful identity called *Beltrami's identity* from the Euler–Lagrange equations in (7.7). We have for this case

$$\frac{df}{dt} = \frac{\partial f}{\partial x}\dot{x} + \frac{\partial f}{\partial \dot{x}}\ddot{x} = \frac{d}{dt}\left(\frac{\partial f}{\partial \dot{x}}\dot{x}\right),$$

where the first equality follows by assumption and the second equality by the chain rule. Hence,

$$\frac{d}{dt}\left(f - \frac{\partial f}{\partial \dot{x}}\dot{x}\right) = 0,$$

from which it follows that

$$f - \frac{\partial f}{\partial \dot{x}}\dot{x} = C,$$

where C is a constant. This is Beltrami's identity.

Example 7.5 We will now derive the shape of a chain hanging from two given points by minimizing the total potential energy of the chain. A differential piece of the chain, of length ds has mass $dm = \rho ds$, where ρ is the mass density of the chain. Here, $ds = \sqrt{1 + \dot{x}^2}$, where $x(t)$ is the position $x(t)$ of the differential segment. The position $x(t)$ of the segment multiplied by its mass, and the gravitational constant g, is equal to the potential energy of the segment, i.e. $g\rho x(t)ds$. The total potential energy is therefore given by

$$\int_0^T f(x, \dot{x})dt,$$

where $f(x, \dot{x}) = g\rho x\sqrt{1 + \dot{x}^2}$. This integral should be minimized subject to the constraints $x(0) = x_0$ and $x(T) = x_T$, which defines where the endpoints of the chain are located. We see that for this example we do not have a constraint on $\lambda = -\frac{\partial f}{\partial \dot{x}}$ at $t = T$. Instead, we use a constraint on $x(T)$. Beltrami's identity gives

$$g\rho x\sqrt{1 + \dot{x}^2} - g\rho x\frac{\dot{x}^2}{\sqrt{1 + \dot{x}^2}} = C.$$

We let $a = C/(g\rho)$ and obtain by multiplying with the square root

$$x = a\sqrt{1 + \dot{x}^2},$$

or equivalently

$$\left(\frac{x}{a}\right)^2 - \dot{x}^2 = 1.$$

It is straightforward to verify that

$$x(t) = a\cosh\left(\frac{t - t_0}{a}\right)$$

satisfies the equation for any t_0. The boundary conditions give

$$a\cosh\left(\frac{-t_0}{a}\right) = x_0,$$

$$a\cosh\left(\frac{T - t_0}{a}\right) = x_T,$$

from which a and t_0 can be determined.

7.4 Extensions

We will now discuss different extensions of the PMP. These are for the cases when there are constraints on the control signal, when the final time T is optimized, and for the cases when the initial value of the state $x(0)$ and/or the final value of the state $x(T)$ are constrained to belong to manifolds. We will not give the proof for these cases, but we will instead state the corresponding version of the PMP and solve the examples. We will from now on only discuss the autonomous case.

Consider the optimal control problem

$$\text{minimize} \quad \phi(x(T)) + \int_0^T f(x(t), u(t))\, dt,$$

$$\text{subject to} \quad \dot{x}(t) = F(x(t), u(t)),$$

$$x(0) \in S_0, \quad x(T) \in S_T, \tag{7.8}$$

$$u(t) \in U \subset \mathbb{R}^m,$$

$$T \geq 0,$$

with variables x, u, and T, where $f : \mathbb{R}^n \times \mathbb{R}^m \to \mathbb{R}$, $F : \mathbb{R}^n \times \mathbb{R}^m \to \mathbb{R}$, and $\phi : \mathbb{R}^n \to \mathbb{R}$ are continuously differentiable. The sets S_0 and S_T are subsets of $\mathbb{R}^n$ and *manifolds*. We assume that S_0 can be described as

$$S_0 = \left\{ x \in \mathbb{R}^n : G_0(x) = 0 \right\},$$

where $G_0 : \mathbb{R}^n \to \mathbb{R}^p$ with $p \leq n$ is differentiable with a full rank Jacobian for x in a neighborhood of the optimal solution point on S_0. We assume a similar description of S_T with a function G_T. Notice that we in this formulation also optimize the final time T.

Define the Hamiltonian $H : \mathbb{R}^n \times \mathbb{R}^m \times \mathbb{R}^{n+1} \to \mathbb{R}$ as

$$H(x, u, \tilde{\lambda}) = \lambda_0 f(x, u) + \lambda^T F(x, u),$$

where $\tilde{\lambda} = (\lambda_0, \lambda)$. Assume that $(x^\star, u^\star, T^\star)$ are optimal for the optimal control problem above. Then there exists a nonzero adjoint function $\tilde{\lambda} : [0, T] \to \mathbb{R}^{n+1}$ such that

(i) $\dot{\lambda}(t) = -\frac{\partial H(x^\star(t), u^\star(t), \tilde{\lambda}(t))}{\partial x}$, $\lambda_0 = c \geq 0$, where $c \in \mathbb{R}$ is a constant

(ii) $H(x^\star(t), u^\star(t), \tilde{\lambda}(t)) = \min_{v \in U} H(x^\star(t), v, \tilde{\lambda}(t)) = 0$, $\forall t \in [0, T^\star]$

(iii) $\lambda(0) \perp S_0$

(iv) $\lambda(T) - \frac{\partial \phi(x^\star(T^\star))}{\partial x} \perp S_T$

Above we have used the notation $\lambda_0 \perp S_0$ to mean that $\lambda(0)^T v = 0$ for all v such that $\frac{\partial G_0(x(0))}{\partial x^T} v = 0$, where G_0 is the function defining the manifold S_0. In case the final time T is not optimized, then condition (ii) is replaced with that the Hamiltonian is constant and not necessarily zero along the optimal solution. For many problems, it turns out that $\lambda_0 > 0$, and then since H is homogeneous in $\tilde{\lambda}$, there is no loss in generality to take $\lambda_0 = 1$. We then obtain the same definition of the Hamiltonian as we had before. It is typically pathological cases for which $\lambda_0 = 0$, and hence, we will often in the examples we investigate assume that $\lambda_0 = 1$.

When using the above conditions to try to find a solution to the optimal control problem, the first step is to define the Hamiltonian. Then one should try to minimize this with respect to u. This should be done parametrically with respect to $\tilde{\lambda}$ and x. This means that we obtain a function $\mu : \mathbb{R}^n \times \mathbb{R}^{n+1} \to \mathbb{R}^m$ such that $u^\star = \mu(x, \tilde{\lambda})$. This function is then substituted into the dynamical equations for x and $\tilde{\lambda}$, i.e.

$$\dot{x} = F(x, \mu(x, \tilde{\lambda})), \quad x(0) \in S_0, \ x(T) \in S_T,$$

$$\dot{\lambda} = -\frac{\partial H(x, \mu(x, \tilde{\lambda}), \tilde{\lambda})}{\partial x}, \quad \lambda(0) \perp S_0, \ \lambda(T) - \frac{\partial \phi(x(T))}{\partial x} \perp S_T, \tag{7.9}$$

This is a so-called *two-point boundary value problem* (TPBVP). We should also use (ii) when solving the above equations. The equations are by no means easy to solve in general. To carry out the parametric optimization of H can be very difficult, especially when u is constrained by the set U. We should also remember that the PMP are only necessary conditions for optimality. Hence, they may not provide enough information to uniquely determine the optimal control. Moreover, they do not guarantee optimality, but only stationarity. Further investigations are necessary to prove that a candidate solution obtained from the PMP is indeed optimal.

We will now investigate some examples that can be solved analytically. In the first example, we illustrate how to carry out the parametric optimization of H for a constrained problem.

Example 7.6 Consider the problem

$$\text{minimize} \ \frac{1}{2}(x(T))^2,$$

$$\text{subject to} \ \dot{x}(t) = u(t),$$

$$x(0) = x_0,$$

$$u(t) \in [-1, 1], \quad \forall \ t \in [0, T],$$

with variable u. We will show using the PMP that the optimal control signal is given by the feedback

$$\mu^*(t, x) = -\text{sgn}(x(t)), \tag{7.10}$$

where $\text{sgn} : \mathbb{R} \to \{-1, 0, 1\}$ is defined as

$$\text{sgn}(x) = \begin{cases} 1, & x > 0, \\ 0, & x = 0, \\ -1, & x > 0. \end{cases}$$

Such a control signal that only takes values on the boundary of the set of allowed values for the control signal is often called *bang–bang*. The reason for this is that if implemented using a mechanical actuator, a bang will often be heard when switching from one of the values to the other. The Hamiltonian is given by

$$H(t, x, u, \lambda) = \lambda u.$$

Pointwise minimization yields

$$\mu(t, x) = \operatorname*{argmin}_{|u| \le 1} \{\lambda u\} = \begin{cases} 1, & \lambda < 0, \\ -1, & \lambda > 0, \\ \tilde{u}, & \lambda = 0, \end{cases}$$

where $\tilde{u} \in [-1, 1]$ is arbitrary. The adjoint equation is given by

$$\dot{\lambda}(t) = -\frac{\partial H}{\partial x}(t, x, u, \lambda) = 0, \quad \lambda(T) = \frac{\partial \phi}{\partial x}(x(T)) = x(T),$$

which has the solution $\lambda(t) = x(T)$. We now have two cases:

- $x(T) \neq 0$: In this case, $\lambda(t) \neq 0$ for all t, and we can write

$$\mu(t, x) = -\operatorname{sgn}(\lambda) = -\operatorname{sgn}(x(T)) = -\operatorname{sgn}(x(t)).$$

  The last equality holds since x will have the same sign as $x(T)$ during the whole time interval.
- $x(T) = 0$: In this case, $\lambda = 0$ for all t and we may use any control signal $\tilde{u} \in [-1, 1]$, which obeys the constraint $x(T) = 0$. One such control signal is

$$\mu(t, x) = -\operatorname{sgn}(x(t)),$$

  since this will drive x to zero and stay there.

Consequently, one optimal control is

$$\mu^*(t, x) = -\operatorname{sgn}(x(t)).$$

The next example will discuss control to a manifold.

Example 7.7 We are interested in finding the path with the shortest distance from a given point $x_0 \in \mathbb{R}^2$ to a manifold $S_T = \{x \in \mathbb{R}^2 : G(x) = 0\}$, where $G : \mathbb{R}^2 \to \mathbb{R}$ is a differentiable function. This can be done by finding the shortest time T it takes to "drive" with constant speed of one from the given point to the manifold. This driving is described by the differential equations:

$$\dot{x}_1(t) = \cos \theta(t),$$
$$\dot{x}_2(t) = \sin \theta(t),$$

where $\theta(t) \in [0, 2\pi)$ is the heading angle. We express the time as $T = \int_0^T dt$. The corresponding optimal control problem is hence

$$\text{minimize} \quad \int_0^T dt,$$
$$\text{subject to} \quad \dot{x}_1(t) = \cos \theta(t),$$
$$\dot{x}_2(t) = \sin \theta(t),$$
$$x(0) = x_0, \quad x(T) \in S_T,$$
$$\theta(t) \in [0, 2\pi),$$

with variable θ and T, where the final time T should be optimized. The Hamiltonian is for this example given by

$$H(x, \theta, \lambda) = 1 + \lambda_1 \cos \theta + \lambda_2 \sin \theta.$$

The adjoint equations are

$$\dot{\lambda}(t) = 0,$$

since the Hamiltonian does not depend on x. Hence, the adjoint variables are constants, i.e. $\lambda(t) = \lambda^0$ for some constant $\lambda^0 \in \mathbb{R}^2$. Since the Hamiltonian has to be zero along the optimal solution, we have that

$$1 + \sqrt{\lambda_1^2 + \lambda_2^2} \sin(\theta(t) + \delta) = 0,$$

where δ satisfies $\sin \delta = \lambda_1 / \sqrt{\lambda_1^2 + \lambda_2^2}$ and $\cos \delta = \lambda_2 / \sqrt{\lambda_1^2 + \lambda_2^2}$. This follows from the formula for the sinusoidal of the sum of two angles. Since λ is a constant, this implies that θ also has to be a constant. Hence, the shortest path is a straight line. Moreover, from the fact that the optimal θ should minimize the Hamiltonian, we obtain the necessary condition that

$$\frac{dH}{d\theta} = -\lambda_1 \sin \theta + \lambda_2 \cos \theta = 0,$$

from which it follows that

$$\frac{\cos \theta}{\sin \theta} = \frac{\lambda_1}{\lambda_2},$$

if $\theta \in (0, 2\pi)$. We will now make use of the condition $\lambda(T) \perp S_T$, which is equivalent to

$$\lambda(T) = \alpha \frac{\partial G(x(T))}{\partial x},$$

for some $\alpha \in \mathbb{R}$. From the differential equations, we have that the slope of the straight line is given by

$$\frac{dx_1}{dx_2} = \frac{\cos \theta}{\sin \theta} = \frac{\lambda_1}{\lambda_2}.$$

We hence conclude that the shortest path has to be perpendicular to the manifold S_T. The case $\theta = 0$ can be taken care of by investigating dx_2/dx_1 instead.

7.5 Numerical Solutions

Except for very simple optimal control problems, it is not possible to obtain analytical solutions. Hence, we must often resort to numerical solutions. All methods use ideas from *ordinary differential equation* (ODE) solvers to either integrate or approximate differential equations. There are essentially two types of methods:

1. Indirect methods
2. Direct methods

The indirect methods aim at solving the necessary conditions of the PMP by integrating the differential equations in a recursive manner. The direct methods aim at solving a discretization of the optimal control problem in (7.8) directly.

Regarding the indirect methods, there are two main ideas that are used and which can be summarized in the two methods

1. Shooting method
2. Gradient method

In the shooting method, the TPBVP is solved by integrating the dynamical equations in (7.9) forward in time using an ODE solver. The challenge here is that normally no initial value is known for the adjoint variable λ, and one has to guess an initial value and then modify this guess based on what value is obtained for $\lambda(T)$. The method is conceptually simple, and the control constraints are easily accommodated. However, it is crucial to find a good initial guess for λ, and the method can be numerically unstable due to the fact that the adjoint equations are often unstable when integrated forward in time. A nonlinear equation solver is needed in order to find the correct initial value for λ. This method was successfully used for launching satellites in the 1950s.

In the gradient method the control signal is used as an optimization variable, and the dynamical equation relating the control signal and the state is integrated forward in time, and the adjoint equation is integrated backward in time. This is done iteratively using an ODE solver until convergence of the gradient of the objective function to zero. The gradient of the objective function is the partial derivative of the Hamiltonian with respect to the control signal. The gradient method has the advantage that both differential equations are integrated in their stable direction. Control signal constraints can be taken care of by projection on the feasible set of control signals. The method has rapid convergence for the first iterates, but then tends to be slow. It was used successfully in the 1960s for a large number of aeronautical problems. To speed up the convergence, second-order derivatives may be used, however, with the drawback of making each iteration much more computationally expensive.

Regarding the direct methods, there are three main ideas which can be summarized in the three methods:

1. Discretization method
2. Collocation method
3. Multiple shooting method

The methods are all based on approximating the control signal as, e.g. piecewise constant or as a polynomial. They basically differ in how the state trajectory is approximated. The first one uses a very simple Euler forward difference. The latter two use ideas from ODE solvers. Collocation methods explicitly make use of the polynomial approximations used in ODE solvers to approximate the state trajectory, whereas multiple shooting methods explicitly make use of an ODE solver with the advantage of using adaptive step-lengths. This ODE solver has to be able to deliver derivatives with respect to the control signal parameters. All methods rely on an efficient nonlinear programming solver at the top level to optimize the control signal parameters.

In-between the indirect methods and the direct methods is the method of consistent approximation which borrows ideas from both methods. This method approximates the control signal as, e.g. piecewise constant, as a polynomial or using orthogonal functions. It also uses a nonlinear programming solver at the top level. However, ideas from ODE solvers is not used directly. Instead, differential equations for adjoint variables are first derived. Then an ODE solver is used to integrate first the differential equation relating the state to the control signal forward in time, and then the adjoint equation backward in time. This will explicitly provide gradients for the nonlinear programming solver. Hence, the ODE solver does not have to be able to deliver derivatives with respect to the control signal parameters. More details regarding consistent approximations is given in, e.g. [75].

We will now in detail explain several of the different algorithms. We will assume that we have numerical algorithms for integrating ODEs, for finding roots of systems of nonlinear equations, and for solving finite-dimensional optimization problems. General purpose algorithms in MATLAB for this are, e.g. `ode45`, `fsolve`, and `fmincon`, respectively.

In case one does not want to write one's own code for solving optimal control problems, there are dedicated solvers for optimal control problems such as ACADO [58], and CasADi [5].

7.5.1 The Gradient Method

We will only present here the gradient method for the optimal control problem in (7.5). More general cases are treated in e.g. [24]. It can be shown that the gradient of the objective function with respect to $u(t)$ is given by

$$\frac{\partial H(t, x(t), u(t), \lambda(t))}{\partial u},$$

for any $x(t)$ as long as $\lambda(t)$ satisfies the adjoint equation in (7.6a); see, e.g. [73]. The algorithm is as follows:

1. Guess an initial value for the control signal $u(t)$, $t \in [0, T]$.
2. Solve

$$\dot{x}(t) = F(t, x(t), u(t)), \quad x(0) = x_0,$$

   using an ODE solver.
3. Solve

$$\dot{\lambda}(t) = -\frac{\partial H(t, x(t), u(t), \lambda(t))}{\partial x}, \quad \lambda(T) = \frac{\partial \phi(x(T))}{\partial x},$$

   using an ODE solver.
4. Update $u(t)$ as

$$u(t) \leftarrow u(t) - \alpha \frac{\partial H(t, x(t), u(t), \lambda(t))}{\partial u}.$$

5. Repeat steps 2–4 until

$$\int_0^T \left| \frac{\partial H(t, x(t), u(t), \lambda(t))}{\partial u} \right|^2 dt$$

   is small enough.

Here, care has to be taken to integrate the adjoint equation backward in time. In case there are constraints such that $u(t) \in U$ for all $t \in [0, T]$, this can be taken care of by projecting the new value of $u(t)$ in Step 4 onto U. The parameter $\alpha > 0$ is a step size that has to be chosen with care.

7.5.2 The Shooting Method

The shooting method solves the TPBVP in (7.9) using an ODE solver, which assumes that the initial values $x(0)$ and $\lambda(0)$ are given. As can be seen in (7.9), this is not the case. Only some of them are given, and not necessarily explicitly. However, we assume that we can summarize what we know about the initial values in a nonlinear equation:

$$G_0(x(0), \lambda(0)) = 0,$$

where $G_0 : \mathbb{R}^n \times \mathbb{R}^n \to \mathbb{R}^p$. Similarly, we assume that we can summarize what we know about the final values in

$$G_T(x(T), \lambda(T)) = 0,$$

where $G_T : \mathbb{R}^n \times \mathbb{R}^n \to \mathbb{R}^{2n-p}$. We also define a function $G : \mathbb{R}^n \times \mathbb{R}^n \to \mathbb{R}^n \times \mathbb{R}^n$ that takes as input the initial values $(x(0), \lambda(0))$, integrates the differential equations, and outputs the final values $(x(T), \lambda(T))$. This can be implemented with an ODE solver, e.g. ode45 in MATLAB. Then we need to solve the system of equations given by

$$G_0(x(0), \lambda(0)) = 0,$$

$$G_T(G(x(0), \lambda(0))) = 0.$$

In MATLAB, this can be done with fsolve. One of the challenges is as mentioned above to choose a good enough initial guess for $\lambda(0)$. Another challenge is to carry out the minimization of the Hamiltonian explicitly in order to obtain the function μ used in (7.9). In case this cannot be done analytically, we need to resort to numerical solutions. In case the minimum of the Hamiltonian is obtained when its partial derivative with respect to u is zero, then this equation can be added to the differential equations in (7.9), i.e. we consider

$$\dot{x} = F(x, u), \quad x(0) \in S_0, \ x(T) \in S_T,$$
$$\dot{\lambda} = -\frac{\partial H(x, u, \tilde{\lambda})}{\partial x}, \quad \lambda(0) \perp S_0, \ \lambda(T) - \frac{\partial \phi(x(T))}{\partial x} \perp S_T, \quad (7.11)$$
$$\frac{\partial H(x, u, \tilde{\lambda})}{\partial u} = 0,$$

instead of (7.9) when we define the function G. The equation above is not an ODE, but a *differential algebraic equation* (DAE), which in MATLAB can be solved with, e.g. ode15i for given initial conditions. In case the minimum of the Hamiltonian could be on the boundary of the domain U, then we would need to carry out the minimization of the Hamiltonian at each step of the DAE solver, and this would require us to write special purpose code.

7.5.3 The Discretization Method

The discretization method is the most straightforward approach of all numerical approaches for solving optimal control problems. We will apply this to

$$\text{minimize} \quad \phi(x(T)) + \int_0^T f(t, x(t), u(t)) \, dt,$$
$$\text{subject to} \quad \dot{x}(t) = F(t, x(t), u(t)),$$
$$x(0) \in S_0, \quad x(T) \in S_T, \quad (7.12)$$
$$u(t) \in U \subset \mathbb{R}^m,$$
$$T \geq 0,$$

with variables x and u, where $f : \mathbb{R} \times \mathbb{R}^n \times \mathbb{R}^m \to \mathbb{R}$, $F : \mathbb{R} \times \mathbb{R}^n \times \mathbb{R}^m \to \mathbb{R}$ and $\phi : \mathbb{R}^n \to \mathbb{R}$ are continuously differentiable. The sets S_0 and S_T are subsets of $\mathbb{R}^m$ and manifolds.

First, we define a partitioning of the time interval $[0, T]$ as

$$0 = t_0 \leq t_1 \leq \cdots \leq t_N = T,$$

in N intervals $[t_i, t_{i+1}]$, where $i \in \mathbb{N}_{N-1}$. We also denote the length of the intervals by $h_i = t_{i+1} - t_i$. Often, the length of the intervals are equal, i.e. $h_i = T/N$ for all i. We approximate $u(t)$ for $t \in [t_i, t_{i+1}]$

as $u(t) = u_i \in \mathbb{R}^m$, i.e. $u(t)$ is approximated as a piecewise constant function of t. We then approximate the derivative of the state with a forward Euler approximation as

$$\dot{x}(t) \approx \frac{x(t_{i+1}) - x(t_i)}{h_i}, \quad t \in [t_i, t_{i+1}].$$

This results in the following approximation of the differential equation:

$$x_{i+1} = x_i + h_i F(t_i, x_i, u_i),$$

where $x_i = x(t_i)$. We finally approximate the objective function as

$$\phi(x_N) + \sum_{i=0}^{N-1} h_i f(t_i, x_i, u_i).$$

We then realize that the whole optimal control problem in (7.12) may be approximated with the discrete time optimal control problem

$$\text{minimize} \quad \phi(x_N) + \sum_{i=0}^{N-1} h_i f(t_i, x_i, u_i),$$

$$\text{subject to} \quad x_{i+1} = x_i + h_i F(t_i, x_i, u_i), \quad i \in \mathbb{Z}_{N-1},$$

$$x_0 \in S_0, \quad x_N \in S_T,$$

$$u_i \in U \subset \mathbb{R}^m,$$

with variables $x = (x_0, \ldots, x_N)$, and $u = (u_0, \ldots, u_{N-1})$. How this optimal control problem can be solved as a finite-dimensional optimization problem is explained for a very similar problem in Section 8.1. We will anyway give the details, since they are useful later on. We define $F_0 : \mathbb{R}^{(N+1)n} \times \mathbb{R}^{Nm} \to \mathbb{R}$ as

$$F_0(x, u) = \phi(x_N) + \sum_{i=0}^{N-1} h_i f(t_i, x_i, u_i).$$

We assume that $u \in U \times U \times \cdots \times U$ can be described as $\mathcal{F}(u) \leq 0$ for some $\mathcal{F} : \mathbb{R}^{Nm} \to \mathbb{R}^{Nq}$. We assume that $x_0 \in S_0$ and $x_N \in S_T$ can be described as

$$G_0(x_0) = 0, \quad G_T(x_N) = 0,$$

for some functions $G_0 : \mathbb{R}^n \to \mathbb{R}^p$ and $G_T : \mathbb{R}^n \to \mathbb{R}^{n-p}$. We define $\mathcal{H} : \mathbb{R}^{(N+1)n} \times \mathbb{R}^{Nm} \to \mathbb{R}^{(N+1)n}$ as

$$\mathcal{H}(x, u) = \begin{bmatrix} x_1 - x_0 - h_0 F(t_0, x_0, u_0) \\ x_2 - x_1 - h_1 F(t_1, x_1, u_1) \\ \vdots \\ x_N - x_{N-1} - h_{N-1} F(t_{N-1}, x_{N-1}, u_{N-1}) \\ G_0(x_0) \\ G_T(x_N) \end{bmatrix}.$$

Then the discrete time optimal control problem can be stated as

$$\text{minimize} \quad F_0(x, u),$$

$$\text{subject to} \quad \mathcal{F}(u) \leq 0,$$

$$\mathcal{H}(x, u) = 0,$$

with variables x and u. It is straightforward to generalize to the case when there are also inequality constraints related to the states x. The above optimization problem is a finite-dimensional optimization problem as discussed already in Chapter 4. The above optimization problem is often of very high dimension, but it has a lot of structure. The objective function is what is called separable. The constraint functions are what is called partially separable, see Section 5.5. This can be utilized to solve the problem efficiently.

7.5.4 The Multiple Shooting Method

The multiple shooting method uses similar ideas as in the discretization method for solving (7.12). However, the approximations are more accurate. This is first accomplished by considering a more general description of $u(t)$ in the intervals $[t_i, t_{i+1}]$. To this end, we define functions $\varphi_i : \mathbb{R} \times \mathbb{R}^{k_i} \to \mathbb{R}^m$ and we let

$$u(t) = \varphi_i(t, a_i), \quad t \in [t_i, t_{i+1}], \quad i \in \mathbb{Z}_{N-1}.$$

The vector a_i parameterizes the function, and these vectors will be optimization variables. We collect them in the vector $a \in \mathbb{R}^k$, where $k = \sum_{i=0}^{N-1} k_i$. The functions φ_i could be constants as in the previous section, and then $a_i = u_i$, but we can also define affine functions or general polynomials. We can also let them be orthogonal basis functions.

We define initial values $s_i, i \in \mathbb{Z}_{N-1}$, and we then use an ODE solver to integrate the N differential equations

$$\dot{x}(t) = F(t, x(t), \varphi_i(t, a_i)), \quad x(t_i) = s_i, \quad i \in \mathbb{Z}_{N-1},$$

over the intervals $[t_i, t_{i+1}]$. For each s_i, we integrate over disjoint intervals. We denote with an abuse of notation the solutions by $x(t, a_i, s_i)$ for $t \in [t_i, t_{i+1}]$. In order to ensure continuity of the solutions, we introduce the matching conditions

$$h(s_i, a_i, s_{i+1}) = s_{i+1} - x(t_{i+1}, a_i, s_i) = 0,$$

where $h : \mathbb{R}^n \times \mathbb{R}^{k_i} \times \mathbb{R}^n \to \mathbb{R}^n$. We can also use an ODE solver to solve the differential equations:

$$\dot{J}(t) = f(t, x(t, a_i, s_i), \varphi_i(t, a_i)), \quad J(t_i) = 0,$$

over the intervals $[t_i, t_{i+1}]$. We denote with an abuse of notation the solutions by $J_i(t, s_i, a_i)$ for $t \in [t_i, t_{i+1}]$. Then the objective function in (7.12) is approximated as

$$F_0(s, a) = \phi(s_N) + \sum_{i=1}^{N-1} J_i(t_{i+1}, s_i, a_i),$$

where $F_0 : \mathbb{R}^{(N+1)n} \times \mathbb{R}^k \to \mathbb{R}$. To take the constraint $u(t) \in U$ for all $t \in [0, T]$ into account, one possibility is to sample the constraint at each time t_i. This means that for all i, we add the constraint $u(t_i) \in U$. We assume that this can equivalently be expressed as $F^i(a_i) \leq 0$ for some functions $F^i : \mathbb{R}^{k_i} \to \mathbb{R}^q$. All inequality constraints can then be expressed in terms of $F : \mathbb{R}^k \to \mathbb{R}^{Nq}$, where

$$F(a) = \begin{bmatrix} F^0(a_0) \\ \vdots \\ F^{N-1}(a_{N-1}) \end{bmatrix}. \tag{7.13}$$

As in the previous section, we assume that the functions G_0 and G_T can be used to describe the constraints on the initial state and final state values. All equality constraints can be described using the function $\mathcal{H} : \mathbb{R}^{(N+1)n} \times \mathbb{R}^k \to \mathbb{R}^{(N+1)n}$ as

$$\mathcal{H}(s, a) = \begin{bmatrix} h(s_0, a_0, s_1) \\ h(s_1, a_1, s_2) \\ \vdots \\ h(s_{N-1}, a_{N-1}, s_N) \\ G_0(s_0) \\ G_T(s_N) \end{bmatrix}.$$

Then the discrete time optimal control problem can be stated as

$$\text{minimize} \quad \mathcal{F}_0(s, a),$$
$$\text{subject to} \quad \mathcal{F}(a) \leq 0,$$
$$\mathcal{H}(s, a) = 0,$$

with variables s and a. It is straightforward to generalize to the case when there are also inequality constraints related to the states x. The above optimization problem is also a finite-dimensional optimization problem. Here, we are not able to compute analytical derivatives of the functions defining the optimization problem. However, often ODE solvers can also deliver derivatives of the solutions with respect to (a, s), which can be used to compute derivatives of all the involved functions. The fact that the differential equations can be solved in parallel can be used to speed up the solver. A good reference to multiple shooting methods is [32]. The ACADO toolkit which implements multiple shooting methods is described in [58].

7.5.5 The Collocation Method

We consider the optimal control problem in (7.12). The idea in the collocation method is to merge the ideas from the discretization method and from the multiple shooting method. The same type of approximation of the control signal as in the multiple shooting method is used. However, the numerical integration of the differential equation is replaced by representing the state as a cubic polynomial, which is more accurate than using the forward Euler approximation of the discretization method. The cubic polynomial is given by

$$x_a(t) = \sum_{k=0}^{3} c_k^i \left(\frac{t - t_i}{h_i} \right)^k,$$

for $t \in [t_i, t_{i+1}]$, where $c_k^i \in \mathbb{R}^n$ are the coefficients of the vector-valued polynomial. The polynomial has to agree with the true state $x(t)$ and its derivative $\dot{x}(t)$ at the endpoints of the interval. This results in the following four equations for the coefficients c_k^i:

$$x_a(t_i) = x_i,$$
$$x_a(t_i + 1) = x_{i+1},$$
$$\dot{x}_a(t_i) = F(t_i, x_i, u_i),$$
$$\dot{x}_a(t_{i+1}) = F(t_{i+1}, x_{i+1}, u_{i+1}),$$

where

$$\dot{x}_a(t) = \sum_{k=1}^{3} \frac{k c_k^i}{h_i} \left(\frac{t - t_i}{h_i} \right)^{k-1}.$$

Since the equations are linear in c_k^i, it is straightforward to obtain the solution:

$$c_0^i = x_i,$$

$$c_1^i = h_i F_i,$$

$$c_2^i = -3x_i - 2h_i F_i + 3x_{i+1} - h_i F_{i+1},$$

$$c_3^i = 2x_i + h_i F_i - 2x_{i+1} + h_i F_{i+1},$$

where $F_i = F(t_i, x_i, u_i)$, and where $x_i = x(t_i)$, $u_i = u(t_i) = \varphi_i(t_i, a_i)$ and $h_i = t_{i+1} - t_i$ as before. Notice that the differential equation is satisfied at the endpoints of the interval by construction of the coefficients of the polynomials. Depending on the dimension of a_i, we can try to enforce the differential equation to hold at one or several points inside the interval $[t_i, t_{i+1}]$. We will only use the center point $t_i^c = (t_i + t_{i+1})/2$, which results in the constraints

$$\dot{x}_a(t_i^c) = F(t_i^c, x_a(t_i^c), \varphi_i(t_i^c, a_i)).$$

This constraint involves the variables x_i, x_{i+1} and a_i and is just a nonlinear equation in these variables. In case, the polynomial instead would have been of first degree we would have obtained the forward Euler approximation in the discretization method.

The objective function is approximated using the center points as

$$\phi(x_N) + \sum_{i=0}^{N-1} h_i f(t_i^c, x_a(t_i^c), \varphi_i(t_i^c, a_i)).$$

This results in the following parametric optimization problem:

$$\text{minimize} \quad \phi(x_N) + \sum_{i=0}^{N-1} h_i f(t_i^c, x_a(t_i^c), \varphi_i(t_i^c, a_i)),$$

$$\text{subject to} \quad \dot{x}_a(t_i^c) = F(t_i^c, x_a(t_i^c), \varphi_i(t_i^c, a_i)),$$

$$G_0(x_0) = 0, \quad G_T(x_N) = 0,$$

$$\varphi(t_i, a_i) \in U \subset \mathbb{R}^m,$$

with variables $x = (x_0, \ldots, x_N)$ and $a = (a_0, \ldots, a_{N-1})$, where we as before assume that the constraint involving the control signal can be written as inequalities involving a function as in (7.13). Understand that the variables x and a are implicitly present in x_a and $\dot{x}_a$. An advantage as compared to the multiple shooting method is that analytical derivatives with respect to (x, a) of the functions defining the optimization problem are possible to compute. However, the cubical approximation of the state might be less accurate as compared to the numerical integration of the state performed in the multiple shooting method. A good reference to the collocation method presented here is [53]. More general collocation methods involving orthogonal polynomials is discussed in e.g. [93].

Exercises

7.1 Show that $\| \cdot \| : C(a, b) \to \mathbb{R}$ defined as

$$\|y\| = \max_{x \in D} |y(x)| + \max_{x \in D} |y'(x)|$$

is a norm on the linear space of continuously differentiable real-valued functions $C(a, b)$ defined on $D = [a, b] \subset \mathbb{R}$.

7.2   Show that the first variation in (7.2), i.e.

$$\delta J[\delta y] = \int_a^b \left(\frac{\partial f\left(x, y(x), y'(x)\right)}{\partial y} \delta y(x) + \frac{\partial f\left(x, y(x), y'(x)\right)}{\partial y'} \delta y'(x) \right) dx$$

is a linear functional

7.3   Solve the optimal control problem

$$\text{minimize} \quad \int_0^T \left(x(t) + u(t)^2 \right) dt,$$

$$\text{subject to} \quad \dot{x}(t) = x(t) + u(t) + 1,$$

$$x(0) = 0,$$

with variables x and u for a fixed $T > 0$, using the PMP.

7.4   Find the extremal to the functional

$$J[y] = \int_0^1 \left(y(t)^2 + \dot{y}(t)^2 \right) dt,$$

satisfying $y(0) = 0$ and $y(1) = 1$.

7.5   Consider Newton's minimal resistance problem:

$$\text{minimize} \quad \int_0^L \frac{y\dot{y}^3}{1 + \dot{y}^2} dx,$$

with variable y subject to $y(0) = H$ and $y(L) = h$, where $H > h$. Here, $\dot{y} = dy/dx$. Show that for the optimal solution it holds that x is related to $p = -\dot{y}$ as

$$x = C + D \left(\ln p + \frac{1}{p^2} + \frac{3}{4p^4} \right),$$

for some constants C and D.

Hint: It is a good idea to consider x to be a function of y.

7.6   We are interested in computing optimal transportation routes in a circular city. The cost for transportation per unit length is given by a function $g(r)$ that only depends on the radial distance r to the city center. This means that the total cost for transportation from a point P_1 to a point P_2 is given by

$$\int_{P_1}^{P_2} g(r) ds,$$

where s represents the arc length along the path of integration. In polar coordinates (θ, r), the total cost reads

$$\int_{P_1}^{P_2} g(r) \sqrt{1 + (r\dot{\theta})^2} \, dr,$$

where $\theta = \theta(r)$, and $\dot{\theta} = d\theta/dr$.

(a) Formulate the problem of computing an optimal path as an optimal control problem.

(b) For the case of $g(r) = \alpha/r$ for some positive α, show that any optimal path satisfies the equation $\theta = a \ln r + b$ for some constants a and b.

(c) Show that if the initial point and the final point are at the same distance from the origin, then the optimal path is a circle segment. You may use the claim in (b).

7.7 Consider the dynamical system

$$\dot{x}(t) = u(t) + w(t),$$

where $x(t)$ is the state, $u(t)$ is the control signal, $w(t)$ is a disturbance signal, and where $x(0) = 0$. In the so-called H_∞-control, the objective is to find a control signal that solves

$$\min_u \max_w J[u, w],$$

with

$$J[u, w] = \int_0^T \left(\rho^2 x(t)^2 + u(t)^2 - \gamma^2 w(t)^2\right) dt,$$

where ρ is a given parameter and where $\gamma < 1$.

(a) Express the solution as a TPBVP problem using the PMP. Note that you do not need to solve the equations.

   Hint: You can consider $\bar{u} = \begin{bmatrix} u \\ w \end{bmatrix}$ as a "control signal" in the PMP framework.

(b) Solve the TPBVP and express the solution in the feedback form, i.e. $u(t) = \mu(t, x(t))$ for some μ.

(c) For what values of γ does the control signal exist for all $t \in [0, T]$.

7.8 Consider the motion of a robotic manipulator with joint angles $q \in \mathbb{R}^n$, which may be described as a function of the applied joint torques $\tau \in \mathbb{R}^n$ as

$$\tau = M(q)\ddot{q} + C(q, \dot{q})\dot{q} + G(q), \tag{7.14}$$

where $M(q) \in \mathbb{R}^{n \times n}$ is a positive definite mass matrix and $C(q, \dot{q}) \in \mathbb{R}^{n \times n}$ is a matrix accounting for Coriolis and centrifugal effects, which is *linear* in the joint velocities $\dot{q} = dq/dt$, and where $G(q) \in \mathbb{R}^n$ is a vector accounting for gravity and other joint angle-dependent torques.

Consider a path $q(s)$ as a function of a scalar path coordinate s. The path coordinate determines the spatial geometry of the path, whereas the trajectory's time dependency follows from the relation $s(t)$ between the path coordinate s and time t.

(a) Show that (7.14) can be expressed in terms of s as

$$\tau(s) = m(s)\ddot{s} + c(s)\dot{s}^2 + g(s),$$

where

$$m(s) = M(q(s)) q'(s),$$
$$c(s) = M(q(s)) q''(s) + C(q(s), q'(s)) q'(s),$$
$$g(s) = G(q(s)).$$

(b) Consider the time-optimal path-tracking problem

$$\text{minimize } T,$$
$$\text{subject to } \tau(s) = m(s)\ddot{s} + c(s)\dot{s}^2 + g(s),$$
$$s(0) = 0,$$
$$s(T) = 1,$$
$$\dot{s}(0) = 0,$$
$$\dot{s}(T) = 0,$$
$$\underline{\tau}(s(t)) \leq \tau(s(t)) \leq \bar{\tau}(s(t)),$$

with variables s and T, where the torque lower bounds $\underline{\tau}$ and upper bounds $\bar{\tau}$ may depend on s. Using the fact that $dt = (dt/ds)ds$, show that

$$T = \int_0^1 \frac{1}{\dot{s}}\, ds,$$

and that for the change of variables,

$$a(s) = \ddot{s},$$
$$b(s) = \dot{s}^2,$$

it holds that $b'(s) = 2a(s)$.

(c) Show that, the optimization problem in (b) is equivalent to

$$\text{minimize} \quad \int_0^1 \frac{1}{\sqrt{b(s)}}\, ds$$

$$\text{subject to} \quad \tau(s) = m(s)a(s) + c(s)b(s) + g(s),$$
$$b(0) = 0,$$
$$b(T) = 0,$$
$$b'(s) = 0,$$
$$b'(s) = 2a(s),$$
$$b(s) \geq 0,$$
$$\underline{\tau} \leq \tau \leq \bar{\tau},$$

with variables a and b.

(d) Compute the optimal path s, from the problem posed in (c) when

$$q(s) = s,$$
$$M(q) = l^2 m = 1,$$
$$C(q, \dot{q}) = 0,$$
$$G(q) = mlg\cos(s) = \cos(s),$$
$$\underline{\tau}(s) = -2, \quad \bar{\tau}(s) = 2.$$

Hint: Consider s as a "time-variable."

7.9 Consider controlling the triple integrator

$$\dot{x}_1 = x_2,$$
$$\dot{x}_2 = x_3,$$
$$\dot{x}_3 = u, \qquad |u| \leq 1,$$

from a certain arbitrary initial condition $x(0)$ to a final value $x(T) = 0$ in minimum time. Show that the necessary conditions for optimality are satisfied by a control of the form

$$u(t) = \text{sign}\,(p(t)),$$

where $p(t)$ is a polynomial. What is the maximum degree of the polynomial? How many times can u change sign? It is not necessary to compute the values of the coefficients of $p(t)$.

7.10 A community living around a lake wants to maximize the yield of fish taken out of the lake. The amount of fish at a certain time is denoted as x. The growth rate of the fish is kx and fish is caught at a rate of ux, where u is the control variable, which is assumed to satisfy $0 \leq u \leq u_{max}$. The dynamics of the fish population is then given by

$$\dot{x} = (k - u)x, \quad x(0) = x_0.$$

Here, $k > 0$ and $x_0 > 0$. The total amount of fish obtained during a time period T is

$$J = \int_0^T ux \, dt.$$

(a) Derive the necessary conditions given by the PMP for the problem of maximizing J.

(b) Show that the necessary conditions are satisfied by a bang–bang control. How many switching times are there?

(c) Determine an equation for calculating the switching time(s).

7.11 Consider a motion model of a particle with position $z(t)$ and speed $v(t)$. Define the state vector $x = (z, v)$ and the continuous time model

$$\dot{x}(t) = F(x(t), u(t)) = \begin{bmatrix} 0 & 1 \\ 0 & 0 \end{bmatrix} x(t) + \begin{bmatrix} 0 \\ 1 \end{bmatrix} u(t). \tag{7.15}$$

The problem is to go from the state $x_i = x(0) = (1,1)$ to $x_f = x(t_f) = (0,0)$, where $t_f = 2$, and such that the control input energy $\int_0^{t_f} u^2(t)dt$ is minimized. Thus, the optimization problem is

$$\text{minimize} \quad \int_0^{t_f} f(x, u)dt,$$

$$\text{subject to} \quad \dot{x}(t) = F(x(t), u(t)),$$

$$x(0) = x_i, \tag{7.16}$$

$$x(t_f) = x_f,$$

with variables x and u, where $f(t) = u(t)^2$ and F is given in (7.15).

(a) Show that the Euler approximation with $h = t_f/N$ as described in the subsection on discretization methods in Section 7.5 results in the discrete time model

$$x_{k+1} = \bar{F}(x_k, u_k) = \begin{bmatrix} 1 & h \\ 0 & 1 \end{bmatrix} x_k + \begin{bmatrix} 0 \\ h \end{bmatrix} u_k, \tag{7.17}$$

and the following discretized version of (7.16):

$$\text{minimize} \quad h \sum_{k=0}^{N-1} u_k^2,$$

$$\text{subject to} \quad x_{k+1} = \bar{F}(x_k, u_k), \tag{7.18}$$

$$x_0 = x_i,$$

$$x_N = x_f,$$

with variables x and u.

(b) Define the optimization parameter vector as

$$y = \left(x_0, u_0, x_1, u_1, \ldots, x_{N-1}, u_{N-1}, x_N, u_N \right).$$

Note that u_N is superfluous, but it is included to make the presentation and the code more convenient. Furthermore, define

$$F_0(y) = h \sum_{k=0}^{N-1} u_k^2 \tag{7.19}$$

and

$$\mathcal{H}(y) = \begin{bmatrix} h_1(y) \\ h_2(y) \\ h_3(y) \\ h_4(y) \\ \vdots \\ h_{N+2}(y) \end{bmatrix} = \begin{bmatrix} x_1 - \bar{F}(x_0, u_0) \\ x_2 - \bar{F}(x_1, u_1) \\ \vdots \\ x_N - \bar{F}(x_{N-1}, u_{N-1})x_0 - x_i \\ x_N - x_f \end{bmatrix}. \tag{7.20}$$

Show that the optimization problem in (7.18) can be expressed as the constrained problem

minimize $F_0(y)$,

subject to $\mathcal{H}(y) = 0$,

with variable y. Complete the files `secOrderSysEqDisc.m` and `secOrderSysCostCon.m`. Note that the constraints are linear for this system model, and this is very important to exploit in the optimization routine. In fact the problem is a quadratic programming problem, and since it only contains equality constraints, it can be solved as a linear system of equations. Complete the file `mainDiscLinconSecOrderSys.m` by creating the linear constraint matrix A_{eq} and the vector B_{eq} such that the constraints defined by $\mathcal{G}(y) = 0$ is expressed as $A_{eq}y = B_{eq}$. Then run the MATLAB script.
Hint: The structure of the matrix A_{eq} is

$$A_{eq} = \begin{bmatrix} -F & E & 0 & 0 & \dots & 0 & 0 \\ 0 & -F & E & 0 & \dots & 0 & 0 \\ 0 & 0 & -F & E & \dots & 0 & 0 \\ \vdots & & & & \ddots & & \vdots \\ 0 & 0 & 0 & 0 & \dots & -F & E \\ E & 0 & 0 & 0 & \dots & 0 & 0 \\ 0 & 0 & 0 & 0 & \dots & 0 & E \end{bmatrix}, \tag{7.21}$$

where F and E are 2×3-matrices.

(c) Now, suppose that the constraints are nonlinear. This case can be handled by passing a function that computes $\mathcal{H}(y)$ and its Jacobian. Let the initial and terminal state constraints, $h_{n+1}(y) = 0$ and $h_{N-2}(y) = 0$, be handled as above and complete the function `secOrderSysNonlcon.m` by computing

$$ceq = (h_1(y), h_2(y), \dots, h_N(y)), \tag{7.22}$$

and the Jacobian

$$ceqJac = \frac{\partial ceq^T}{\partial y}. \tag{7.23}$$

Each row in (7.23) is computed in the loop in the for-statement in `secOrderSysNonlcon.m`. Note that the length of the state vector is 2. Also, note that MATLAB wants the Jacobian to be transposed, but that operation is already added last in the file.

Finally, run the MATLAB script `mainDiscNonlinconSecOrderSys.m` to solve the problem again.

Hint: You can use the function `checkCeqNonlcon` to check that the Jacobian of the equality constraints are similar to the numerical Jacobian. You can also compare the Jacobian to the matrix A_{eq} in this example.

7.12   In this exercise, we will also use the discrete time model in (7.17). However, the terminal constraints are removed by approximating them with a penalty term in the objective function. If the terminal constraints are not fulfilled, they are penalized in the objective function. Now, the optimization problem can be defined as

$$\text{minimize} \quad c\|x_N - x_f\|_2^2 + h \sum_{k=0}^{N-1} u_k^2,$$

$$\text{subject to} \quad x_{k+1} = \bar{F}(x_k, u_k), \tag{7.24}$$

$$x_0 = x_i,$$

with variables x and u, where $\bar{F}$ is given in (7.17), $N = t_f/h$ and c is a predefined constant. This problem can be rewritten as an unconstrained quadratic optimization problem:

$$\text{minimize} \quad \mathcal{F}(u_0, u_1, \ldots, u_{N-1}), \tag{7.25}$$

with variable u. The optimization problem will be solved by using the MATLAB function: `fminunc`.

(a) Write down an explicit algorithm for the evaluation of $\mathcal{F}$.

(b) Complete the file `secOrderSysCostUnc.m` with the cost function $\mathcal{F}$. Solve the problem by running the script `mainDiscUncSecOrderSys.m`.

(c) Compare the method in this exercise with the one in the previous exercise. What are the advantages and disadvantages? Which method handles constraints best? Suppose the task was also to implement the optimization algorithm, which method would be easiest to implement, assuming that the state dynamics would be nonlinear?

(d) Implement an unconstrained gradient method that can replace the `fminunc` function in `mainDiscUncSecOrderSys.m`.

7.13   We consider the same optimal control problem as in the previous exercise. We will investigate the gradient method based on the PMP method. The optimal control problem is

$$\text{minimize} \quad J = c\|x(t_f) - x_f\|_2^2 + \int_0^{t_f} u^2(t)dt,$$

$$\text{subject to} \quad \dot{x}(t) = F(x(t), u(t)),$$

$$x(0) = x_i,$$

with variables x and u.

(a) Write down the Hamiltonian and show that the Hamiltonian partial derivatives with respect to x and u are

$$\frac{\partial H}{\partial x} = \begin{bmatrix} 0 \\ \lambda_1 \end{bmatrix},$$

$$\frac{\partial H}{\partial u} = 2u(t) + \lambda_2, \tag{7.26}$$

respectively.

(b) What are the adjoint equation and its terminal constraint?

(c) Complete the files `secOrderSysEq.m` with the system model, the file `secOrder-SysAdjointEq.m` with the adjoint equations, the file `secOrderSysFinal-Lambda.m` with the terminal values of λ, and the file `secOrderSysGradient.m` with the control signal gradient. Finally, complete the script `mainGradientSec-OrderSys.m` and solve the problem.

(d) Try some different values of the penalty constant c. What happens if c is "small?" What happens if c is "large?"

7.14 In this exercise, we will solve the problem in (7.16) using a shooting method as discussed in the subsection on shooting methods in Section 7.5. Such methods are based on successive improvements of the unspecified initial conditions of the TPBVP.

Complete the files `secOrderSysEqAndAdjointEq.m` with the combined system and adjoint equation which implements the function G in the abovementioned section. Also, complete the file `theta.m` with the final constraints, which implement the function G_T. We do not need to specify the function G_0, since the initial value for the state is fully known. The main script is `mainShootingSecOrderSys.m`, and it solves the equation $G_T(G(x(0), \lambda(0))) = 0$ with respect to $\lambda(0)$ for a given value of $x(0)$.

7.15 (a) What are the advantages and disadvantages of discretization methods?

(b) Discuss the advantages and disadvantages of the problem formulation in Exercise 7.11 compared to the formulation in Exercise 7.12.

(c) What are the advantages and disadvantages of gradient methods?

(d) Compare the methods in Exercises 7.11–7.13 in terms of accuracy and complexity/speed. Also, compare the results for different algorithms used by `fmincon` in Exercises 7.11. Can all algorithms handle the optimization problem?

(e) Assume that there are constraints on the control signal, and you can use either the discretization method in Exercise 7.11 or the gradient method in Exercise 7.13. Which method would you use?

(f) What are the advantages and disadvantages of shooting methods?

7.16 Consider the problem of finding the curve with the minimum length from a point (0,0) to (x_f, y_f). The solution is of course obvious, but we will use this example as a starting point for introducing CasADi, see `https://web.casadi.org`, to solve optimal control problems using a collocation method. We will use a control signal that is constant in each discretization interval.

The problem can be formulated by using an expression of the length of the curve from (0,0) to (x_f, y_f):

$$s = \int_0^{x_f} \sqrt{1 + y'(x)^2}\, dx. \tag{7.27}$$

Note that x is the "time" variable. The optimal control problem is solved in `minCurve-LengthCol.m` by using MATLAB and CasADi. Notice that (x, y) are represented as (t, x) in the file.

(a) Derive the expression of the length of the curve in (7.27).

(b) Write the problem on standard optimal control form, i.e. determine ϕ, f, F, and so on.

(c) Use the PMP to show that the solution is a straight line.

(d) Examine and run the script `minCurveLengthCol.m` and compare the solution with what the theory says.

7.17 Consider the minimum length curve problem again. The problem can be reformulated by using a constant speed model where the control signal is the heading angle. This problem is solved in the CasADi/MATLAB file `minCurveLengthHeadingCtrlMS.m`.

(a) Examine the CasADi/MATLAB file and write down the optimal control problem that is solved in this example, i.e. what are F, f, and ϕ in the standard optimal control formulation.

(b) Run the script and compare with the result from the previous exercise.

7.18 In this exercise, we will investigate the so-called "Brachistochrone problem," which was posed by Johann Bernoulli in Acta Eruditorum in 1696. The history of this problem involves several of the greatest scientists ever, such as Galileo, Pascal, Fermat, Newton, Lagrange, and Euler. It is about finding the curve between two points, A and B, that is covered in the least time by a body that starts in A with zero speed and is constrained to move along the curve to point B, under the action of gravity only and assuming no friction, see Figure 7.1. The word "brachistochrone" comes from the Greek language: brachistos – the shortest, chronos – time.

Let the motion of the particle, under the influence of gravity g, be defined by $\dot{z} = F(z, \theta)$, where the state vector is defined as $z = (x, y, v)$ and (x, y) is the Cartesian position of the particle in a vertical plane and v is the speed, i.e.

$$\dot{x} = v\sin(\theta),$$
$$\dot{y} = -v\cos(\theta). \tag{7.28}$$

The motion of the particle is constrained by a path that is defined by the angle $\theta(t)$.

(a) Give an explicit expression for $F(z, \theta)$. Only the expression for $\dot{v}$ is missing.

(b) Define the Brachistochrone problem as an optimal control problem based on this state-space model. Assume that the initial position of the particle is at the origin and that the initial speed is zero. The final position of the particle is $(x(t_f), y(t_f)) = (x_f, y_f) = (10, 3)$.

(c) Modify the script `minCurveLengthHeadingCtrlMS.m` of the minimum curve length example in Exercise 7.17 above and solve the Brachistochrone problem with CasADi.

7.19 The time it takes for a particle to travel on a curve between the points $p_0 = (0,0)$ and $p_f = (x_f, y_f)$ is

$$t_f = \int_{p_0}^{p_f} \frac{1}{v} ds, \tag{7.29}$$

where ds is an element of arc length, and v is the speed.

Figure 7.1 The Brachistochrone problem.

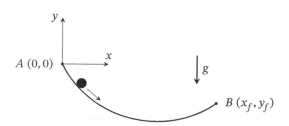

(a) Show that the travel time t_f is

$$t_f = \int_0^{x_f} \frac{\sqrt{1 + y'(x)^2}}{\sqrt{-2gy(x)}} dx, \tag{7.30}$$

for the Brachistochrone problem, where the speed of the particle is due to gravity, and its initial speed is zero c.f. Exercise 7.18.

(b) Define the Brachistochrone problem as an optimal control problem based on the expression in (7.30). Note that the time variable is eliminated and that y is a function x.

(c) Use the PMP to show that the optimal trajectory is a cycloid

$$y'(x) = \sqrt{\frac{C - y(x)}{y(x)}}, \tag{7.31}$$

with solution

$$
\begin{aligned}
x &= \frac{C}{2}(\phi - \sin(\phi)), \\
y &= \frac{C}{2}(1 - \cos(\phi)).
\end{aligned} \tag{7.32}
$$

where ϕ parameterizes the curve.

7.20 An alternative formulation of the Brachistochrone problem, c.f. Exercise 7.18, can be obtained by considering the "law of conservation of energy," which is derived from the principle of least action in Example 7.4. Consider the position (x, y) of the particle and its velocity $(\dot{x}, \dot{y})$.

(a) Write the kinetic energy T and the potential energy V as functions of $x, y, \dot{x}, \dot{y}$, the mass m, and the gravity constant g.

(b) Define the Brachistochrone problem as an optimal control problem based on the law of conservation of energy.
Hint: You should introduce $u = \dot{x}$ and $v = \dot{y}$ as control signals. Notice that the problem will contain an algebraic constraint that is not present in a standard optimal control problem as we have defined it. This means that the state evolution is not described by an ODE but by a differential algebraic equation.

(c) Solve this optimal control problem with CasADi by modifying the file `brachis-tochroneHeadingCtrlMS.m`. Assume that the mass of the particle is $m = 1$.
Hint: You need to use a value for N of at least 40.

7.21 When solving the Brachistochrone problem, we have used three different problem formulations in the three previous exercises. We will discuss here the pros and cons of the different formulations.

(a) Discuss why not only the choice of optimization algorithm is important but also the problem formulation, when deciding how to solve an optimal control problem numerically.

(b) Compare the different approaches to the Brachistochrone problem in the three previous exercises. Try to explain the advantages and disadvantages of the different formulations of the Brachistochrone problem.

7.22 We will solve here the so-called "Zermelo" problem. From the point (0,0) on the bank of a wide river, a boat starts with relative speed to the water equal to v. The stream of the river becomes faster as it departs from the bank, and the speed is $g(y)$ parallel to the bank. The movement of the boat is described by

$$\dot{x}(t) = v\cos(\phi(t)) + g(y(t)),$$
$$\dot{y}(t) = v\sin(\phi(t)),$$

where ϕ is the angle between the boat direction and bank. We want to determine the movement angle $\phi(t)$ so that $x(T)$ is maximized, where T is a fixed time. We will investigate the case when $g(y) = y$, $v = 1$, and $T = 1$. Use CasADi to solve this optimal control problem by modifying the file `minCurveLengthCol.m`.

7.23 Solve the problem in Exercise 7.11 using CasADi and both the multiple shooting method and the collocation method. How many iterations do the methods need to converge to a solution?

8

Dynamic Programming

In Chapter 7, we discussed how to solve optimal control problems over a finite time interval. We specifically considered the continuous time case, since for discrete time dynamics, it is fairly straightforward how to solve the optimal control problem. The solutions we obtained are so-called *open-loop* solutions, i.e. the value of the control signal at a certain time depends only on the initial value of the state and not on the current value of the state. For the linear quadratic (LQ) control problem, we were able to restate the solution as a *feedback policy*, i.e. to explicitly write the control signal as $u(t) = \mu(t, x(t))$ for a feedback function μ. This is very desirable since it is known that feedback solutions are more robust to unmodeled dynamics and disturbances. We will in this chapter look in more detail into the problem of obtaining feedback solutions. We will only treat the discrete time case, since it is easier from a mathematical point of view. However, many of the ideas can be extended to the continuous time case. We will first consider a finite time interval and then we will discuss the case of an infinite time interval. Optimal feedback control goes back to the work by Richard Bellman who in 1953 introduced what is known as *dynamic programming* to solve these types of problems. We will see that this is a special case of message passing as discussed in Section 5.5. The case of infinite time interval is of practical importance in many applications, since under certain conditions stability can be proven. It is unfortunately often very difficult to compute the solution, and hence, we will introduce what is known as *model predictive control* (MPC) as a remedy. This is today commonly used in industry. We will at the end of the chapter discuss how to treat uncertainty by considering a stochastic setting of the control problem. This treatment is based on the stochastic multistage decision problem introduced in Section 5.7.

8.1 Finite Horizon Optimal Control

We are given a *dynamical system* of the form

$$x_{k+1} = F_k(x_k, u_k), \tag{8.1}$$

where $F_k : D^n \times \mathcal{E}^m \to D^n, k \in \mathbb{Z}_{N-1}$ are functions describing how the *state* x_k evolves depending on the *input signal* u_k. Here, k can be interpreted as *discrete time* instances or as general *stages*. The sets D and $\mathcal{E}$ could be, e.g. $\mathbb{R}$, or they could be sets with a finite number of elements, e.g. subsets of the integers. We assume that the initial value x_0 is given. The input signal is called the *control signal*. The control signal is used to control the dynamical system, i.e. the evolution of the states. To this end, we introduce the *incremental costs* $f_k : D^n \times \mathcal{E}^m \to \mathbb{R}$ for $k \in \mathbb{Z}_{N-1}$. The incremental costs are functions of the states and the control signal. We also introduce a cost associated with the

Optimization for Learning and Control, First Edition. Anders Hansson and Martin Andersen.
© 2023 John Wiley & Sons, Inc. Published 2023 by John Wiley & Sons, Inc.
Companion Website: www.wiley.com/go/opt4lc

final state x_N called the *terminal cost* of *final cost* $\phi : \mathcal{D}^n \to \mathbb{R}$. The *optimal control problem* is then the problem of solving

$$\text{minimize} \quad \phi(x_N) + \sum_{k=0}^{N-1} f_k(x_k, u_k),$$

$$\text{subject to} \quad x_{k+1} = F_k(x_k, u_k), \quad k \in \mathbb{Z}_{N-1}, \tag{8.2}$$

$$u_k \in U_k(x_k), \quad k \in \mathbb{Z}_{N-1},$$

$$x_k \in X_k, \quad k \in \mathbb{N}_N,$$

with variables $(u_0, x_1, \dots, u_{N-1}, x_N)$, where $X_k \subset \mathcal{D}^n$, $U_k : \mathcal{D}^n \to \mathcal{V}$, and where $\mathcal{V} = \{E : E \subset \mathcal{E}^m\}$. The final stage N that we consider in the formulation of the problem is called the *time horizon* of the problem. Above, the optimal control problem is solved for a given value of the initial state x_0, and we denote the optimal value of the objective function with $J^\star : \mathcal{D}^n \to \mathbb{R}$ which is a function of x_0. For later reference, we will also introduce the *optimal cost-to-go function* $J_l^\star : \mathcal{D}^n \to \mathbb{R}$ as the optimal value of the objective function for a problem starting at time $k = l$ instead of at $k = 0$ and with initial value $x_l = x$. Clearly, $J^\star(x_0) = J_0^\star(x_0)$.

8.1.1 Standard Optimization Problem

Let $z = (u_0, x_1, \dots, u_{N-1}, x_N)$ and define $f : \mathcal{E}^m \times \mathcal{D}^n \times \dots \times \mathcal{E}^m \times \mathcal{D}^n \to \mathbb{R}$ as

$$f(z) = \phi(x_N) + \sum_{k=0}^{N-1} f_k(x_k, u_k),$$

and let $h : \mathcal{E}^m \times \mathcal{D}^n \times \dots \times \mathcal{E}^m \times \mathcal{D}^n \to \mathbb{R}^{nN}$ be given by

$$h(z) = \left(x_1 - F_0(x_0, u_0), x_2 - F_1(x_1, u_1), \dots, x_N - F_{N-1}(x_{N-1}, u_{N-1}) \right).$$

Also, assume that it is possible to equivalently express $u_k \in U_k(x_k)$ for $k \in \mathbb{Z}_{N-1}$ and $x_k \in X_k$ for $k \in \mathbb{N}_N$ as $g(z) \leq 0$ for some function $g : \mathcal{E}^m \times \mathcal{D}^n \times \dots \times \mathcal{E}^m \times \mathcal{D}^n \to \mathbb{R}^p$. Then the above optimization problem can equivalently be written

$$\text{minimize} \quad f(z),$$

$$\text{subject to} \quad g(z) \leq 0, \tag{8.3}$$

$$h(z) = 0,$$

with variable z. This can be solved with several of the different optimization methods presented in Chapter 6. It should be stressed that there is a lot of structure in the above optimization problem. Gradients and Hessians of the functions f, g, and h are sparse. This is often utilized at the linear algebra level of optimization software to increase efficiency. Unfortunately, this would not result in a feedback solution, but an *open loop* solution, i.e. the control signal does only depend on the initial value of the state and not explicitly on the current value of the state. Moreover, the formulation above also assumes that we not only know the incremental costs but also the functions F_k that describe how the states evolve. In reinforcement learning, see Chapter 11, we do not know these latter functions. Motivated by this we will discuss a different way of solving the optimal control problem that later on will enable us to generalize to reinforcement learning. The method we present is also useful for the case when the states and/or the control signals are not real-valued, but instead take values in some finite set. Even if the functions F_k were known, then the optimization methods presented so far in the book could not be used. The reason is that the optimization problem is then a combinatorial optimization problem.

8.1.2 Dynamic Programming

We will base our derivation on the fact that the optimization problem is partially separable as discussed in Section 5.5. We essentially just need to identify (x_k, u_{k-1}) in this section with x_k in that section, and let f_k in Section 5.5 be the incremental cost f_k plus an indicator function to describe our constraints. When carrying out minimization over (x_k, u_{k-1}), we realize that because of the constraint $x_k = F_{k-1}(x_{k-1}, u_{k-1})$ we can substitute x_k with $F_{k-1}(x_{k-1}, u_{k-1})$, remove the constraint, and hence, we only need to minimize with respect to u_{k-1}. The details are omitted, but it is then straightforward to show that we obtain the following *dynamic programming recursion*. Define $V_N :$ $\mathcal{D}^n \to \mathbb{R}$ as

$$V_N(x) = \begin{cases} \phi(x), & x \in X_N, \\ \infty, & x \notin X_N. \end{cases}$$

Then determine recursively starting with $k = N - 1$ and finishing with $k = 0$, the functions $V_k :$ $\mathcal{D}^n \to \mathbb{R}$ defined as

$$V_k(x) = \min_{u \in U_k(x), F_k(x,u) \in X_{k+1}} \left\{ f_k(x, u) + V_{k+1}(F_k(x, u)) \right\}. \tag{8.4}$$

Here, the minimization over u should be carried out for all possible values of x and hence, the optimal u is a function of x, i.e. we have a multiparametric optimization problem as discussed in Section 5.6. If it is possible to carry out the minimizations above,[1] then the optimal control signal for (8.2) is given by the minimizing argument in the dynamic programming recursion, i.e. $u_k^\star = \mu_k(x)$, where $\mu_k : \mathcal{D}^n \to \mathcal{E}^m$ for $k \in \mathbb{Z}_{N-1}$ is given by

$$\mu_k(x) = \operatorname*{argmin}_{u \in U_k(x), F_k(x,u) \in X_{k+1}} Q_k(x, u),$$

and where $Q_k : \mathcal{D}^n \times \mathcal{E}^m \to \mathbb{R}$ is called the Q-function and defined as

$$Q_k(x, u) = f_k(x, u) + V_{k+1}(F_k(x, u)), \tag{8.5}$$

for $k \in \mathbb{Z}_{N-1}$. If the minimizing u is not unique for each x, then μ_k should be interpreted as the set of all u that minimizes Q_k. It should be stressed that $V_k(x) = J_k^\star(x)$, i.e. the value functions agree with the optimal cost-to-go functions, and this follows from the derivations carried out in Section 5.5. The function μ_k is called the *feedback function*, and it defines a so-called *feedback policy* or *control policy*. This is for practical applications often more tractable than the so-called *open-loop policy* that is the result of solving (8.3). We summarize the dynamic programming recursion in Algorithm 8.1.

Algorithm 8.1: Dynamic programming

Input: Final state penalty ϕ, incremental costs f_k, functions F_k
Output: V_k for $k \in \mathbb{Z}_{N-1}$

$V_N(x, u) \leftarrow \phi(x)$
for $k \leftarrow N - 1$ **to** 0 **do**
$\quad \mid \quad V_k(x) \leftarrow \min_u \left(f_k(x, u) + V_{k+1}(F_k(x, u)) \right)$
end

We next consider an example.

---

1 We assume that the minimum exists.

Example 8.1 Consider the problem in (8.2) for the case when the dynamic equation is linear, i.e. $F_k(x,u) = A_k x + B_k u$, where $A_k \in \mathbb{R}^{n\times n}$ and $B_k \in \mathbb{R}^{n\times m}$ and where $D = \mathcal{E} = \mathbb{R}$. We assume that the incremental costs and the final cost are quadratic functions given by $f_k(x,u) = x^T S_k x + u^T R_k u$ and $\phi(x) = x^T S_N x$, respectively, where $R_k \in \mathbb{S}_{++}^m$ and $S_k \in \mathbb{S}_+^n$. We also assume that there are no state constraints or control signal constraints. This problem is called the linear quadratic (LQ) control problem. Application of the dynamic programming recursion in (8.4) gives $V_N(x) = x^T S_N x$ and

$$V_k(x) = \min_u \left\{ x^T S_k x + u^T R_k u + V_{k+1}\left(A_k x + B_k u\right)\right\}.$$

We will make the guess that $V_k(x) = x^T P_k x$ for some $P_k \in \mathbb{S}^n$. This is clearly true for $k = N$ with $P_N = S_N$. We now assume that it is true for $k+1$. Then the right-hand side of the above equation reads

$$Q(x,u) = x^T S_k x + u^T R_k u + (A_k x + B_k u)^T P_{k+1}(A_k x + B_k u),$$
$$= \begin{bmatrix} x \\ u \end{bmatrix}^T \begin{bmatrix} S_k + A_k^T P_{k+1} A_k & A_k^T P_{k+1} B_k \\ B_k^T P_{k+1} A_k & R_k + B_k^T P_{k+1} B_k \end{bmatrix} \begin{bmatrix} x \\ u \end{bmatrix},$$

which should be minimized with respect to u. If $R_k + B_k^T P_{k+1} B_k \in \mathbb{S}_{++}^m$, then the above optimization problem is convex in u, and the solution is obtained similarly as in Example 5.7 as

$$u = -\left(R_k + B_k^T P_{k+1} B_k\right)^{-1} B_k^T P_{k+1} A_k x,$$

which defines the feedback policy $\mu_k(x)$. Back-substitution of this expression for u results in the following expression for the right-hand side of the dynamic programming recursion:

$$x^T \left\{ S_k + A_k^T P_{k+1} A_k - A_k^T P_{k+1} B_k \left(R_k + B_k^T P_{k+1} B_k\right)^{-1} B_k^T P_{k+1} A_k \right\} x.$$

It now follows that our assumption is also true for k if we define P_k as

$$P_k = S_k + A_k^T P_{k+1} A_k - A_k^T P_{k+1} B_k \left(R_k + B_k^T P_{k+1} B_k\right)^{-1} B_k^T P_{k+1} A_k.$$

This defines a recursion for the matrices P_k which is called the discrete-time *Riccati recursion*. It remains to prove that $R_k + B_k^T P_{k+1} B_k \in \mathbb{S}_{++}^m$ for the recursion to be well defined. If $P_{k+1} \in \mathbb{S}_+^n$, then the result follows. This is true for $k = N-1$, since $P_N = S_N \in \mathbb{S}_+^n$. Assume that it is true for $k+1$, then the above minimization will result in a minimal value that is nonnegative. Hence, the result also holds for k.

Example 8.2 We will now look at the problem of finding the shortest path between two nodes in a graph, see Figure 8.1. The nodes could represent different cities, and the numbers on the edges could represent the distances between the cities. This is called a *shortest path problem*. We cast this problem as an optimal control problem as in (8.2) with the following definitions. Let $N = 5$, $x_0 = 0$, $X_k = \{-1,0,1\}$ for $k = 1, \ldots, N-1$ and $X_N = \{0\}$. We define the control signal to take the values $-1, 0, 1$, which mean to go down, stay, or go up in the graph, respectively. Hence, the control signal constraint set is

$$U_k(x) = \begin{cases} \{-1,0\}, & x = 1, \\ \{-1,0,1\}, & x = 0, \\ \{0,1\}, & x = -1, \end{cases}$$

for $k \in \mathbb{N}_{N-2}$ and

$$U_{N-1}(x) = \begin{cases} \{-1\}, & x = 1, \\ \{0\}, & x = 0, \\ \{1\}, & x = -1. \end{cases}$$

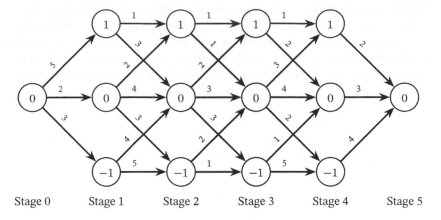

Stage 0   Stage 1   Stage 2   Stage 3   Stage 4   Stage 5

Figure 8.1 Graph for the shortest path problem in Example 8.2.

Moreover, we let $F_k(x, u) = x + u$, and hence, the next state value is the sum of the current state value and the control signal value. The final cost is $\phi(x) = 0$. Finally, the incremental cost is $f_k(x, u) = c^k_{i,j}$, where $c^k_{i,j}$ is the cost on the arrow from node i at stage k to node j at stage $k + 1$. It should be stressed that the definitions above are not unique. We could, e.g. change the control signal to take values $-2, 0, 2$ and change the function F_k to $F_k(x, u) = x + 0.5u$.

We now apply the dynamic programming recursion. At the final stage $N = 5$, we have $V_5(x) = 0$ for $x = 0$ and $V_5(x) = \infty$ for any other x, since $X_5 = \{0\}$. Then

$$V_4(x) = \min_{u \in U_4(x), F_4(x,u) \in X_5} \{f_4(x, u) + V_5(F_4(x, u))\}.$$

We realize that x and u must satisfy $F_4(x, u) = 0$ to obtain a minimum, since otherwise the second term will be infinity. Hence, for Stage 4, we get

$$V_4(x) = \begin{cases} c^4_{1,0} = 2, & x = 1, \\ c^4_{0,0} = 3, & x = 0, \\ c^4_{-1,0} = 4, & x = -1, \end{cases}$$

where the value for $x = 1$ is obtained for $u = -1$, the value for $x = 0$ is obtained for $u = 0$, and the value for $x = -1$ is obtained for $u = 1$. For Stage 3, we have

$$V_3(x) = \min_{u \in U_3(x), F_3(x,u) \in X_4} \{f_3(x, u) + V_4(F_3(x, u))\},$$

$$= \begin{cases} \min \left\{ c^3_{1,1} + V_4(1), c^3_{1,0} + V_4(0) \right\} = \min\{3, 5\} = 3, & x = 1, \\ \min \left\{ c^3_{0,1} + V_4(1), c^3_{0,0} + V_4(0), c^3_{0,-1} + V_4(-1) \right\} = \min\{5, 7, 6\} = 5, & x = 0, \\ \min \left\{ c^3_{-1,0} + V_4(0), c^3_{-1,-1} + V_4(-1) \right\} = \min\{4, 9\} = 4, & x = -1, \end{cases}$$

where the value for $x = 1$ is obtained for $u = 0$, the value for $x = 0$ is obtained for $u = 1$, and the value for $x = -1$ is obtained for $u = 1$. We can now continue for the remaining stages, and the result is depicted in Figure 8.2. The shortest path correspond to the thick arcs. Above each node, the value of $V_k(x)$ is given, and the small arrows show the optimal control signal at each node. From these arrows, we are hence able to also obtain the shortest path from any other node than the initial one to the final node. Hence, the optimal cost to go from Node 1 at Stage 1 to the final node is 5, and the optimal path is to go to Node 1 at each stage except for at Stage 4, where one has to go to node 0.

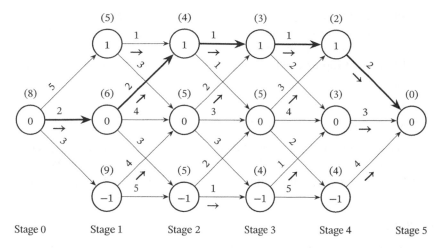

Figure 8.2 Graph for the shortest path problem in Example 8.2. The shortest path correspond to the thick arcs. Above each node, the value of $V_k(x)$ is given, and the small arrows show the optimal control signal at each node.

8.2 Parametric Approximations

It is clear that the dynamic programming recursion is difficult to solve analytically. Even if each component of the states takes values in finite sets with say p elements, the tasks are challenging for high- dimension n of the state. The total number of possible values of a state is then p^n, which also for moderate values of p and n can be a huge number. This is called the *curse of dimensionality*. For problems where the state is a real-valued vector, one may approximate it with discrete values obtaining a piecewise constant approximation, but then very course of approximations are necessary.

We will now discuss how the curse of dimensionality can be circumvented to some extent. We introduce vector-valued functions $\varphi_k : \mathcal{D}^n \to \mathbb{R}^p$ called *feature vectors* and parameters $a_k \in \mathbb{R}^q$ for $k \in \mathbb{Z}_{N-1}$. We would like to approximate the functions V_k in (8.4) with $\hat{V}_k : \mathcal{D}^n \times \mathbb{R}^q \to \mathbb{R}$, where $\tilde{V}_k(x, a_k) = \hat{V}_k(\varphi_k(x), a_k)$ for a suitable choice of $\hat{V}_k : \mathbb{R}^p \times \mathbb{R}^q \to \mathbb{R}$ and $a_k \in \mathbb{R}^q$. One possible choice of $\hat{V}_k$ is to use a linear regression model for which $p = q$, i.e.

$$\hat{V}_k(\varphi_k(x), a_k) = a_k^T \varphi_k(x).$$

There are of course many more possibilities. The definitions made above are general enough to model $\hat{V}_k$ as an artificial neural network (ANN), see Section 10.7. In case one has problem-specific insight, more clever choices can be made. In a preprocessing step, features could be extracted from data or from sufficient statistics, see [17]. For the sake of simplicity, we will in what follows only discuss the linear regression case. Notice that a piecewise constant approximation of V_k can be obtained by taking

$$\varphi_k(x) = \begin{cases} 1, & x \in D_k, \\ 0, & x \notin D_k, \end{cases}$$

where $D_k, k \in \mathbb{N}_p$ is a partition of $\mathbb{R}^n$. More general approximations are obtained by taking φ_k as basis functions. In this section, we will not consider constraints on x_k or u_k.

8.2.1 Fitted-Value Iteration

In order to find good parameters a_k, we are going to use an algorithm known as *fitted-value iteration*. It is a sequential supervised learning algorithm based on a least-squares (LS) criterion at each iteration. It is based on sampling the state-space $\mathcal{D}^n$. This can be done in many ways, and how well this is done is crucial for the success of the algorithm. A popular choice is to use some sort of Monte Carlo technique. It is important that the states that are sampled are representative for what states are typically visited by a close to optimal policy. We start by defining approximate Q-functions $\tilde{Q}_k : \mathcal{D}^n \times \mathcal{E}^m \times \mathbb{R}^q \to \mathbb{R}$ as

$$\tilde{Q}_k(x, u, a) = \begin{cases} f_k(x, u) + \tilde{V}_{k+1}(F_k(x, u), a), & k \in \mathbb{Z}_{N-2}, \\ f_k(x, u) + \phi(F_k(x, u)), & k = N - 1. \end{cases}$$

Here, we have just replaced V_{k+1} with $\tilde{V}_{k+1}$ in the expression for Q_k in (8.5). Notice that $\tilde{Q}_k$ does not depend on any parameter a for $k = N - 1$, which is the iteration index we start with. We then consider samples $x_k^s \in \mathcal{D}^n$, where $s \in \mathbb{N}_r$, and define the minimal values of the approximate Q-functions as

$$\beta_k^s = \min_u \tilde{Q}_k(x_k^s, u, a_{k+1}), \tag{8.6}$$

where a_{k+1} is a known value from the previous iterate. This is in general a nonconvex optimization problem that can be challenging to solve. It should be stressed that the quality of the approximation that is obtained depends critically on the choice of x_k^s. After this, we define the following LS problem for obtaining the next value of the parameter a_k:

$$\text{minimize } \frac{1}{2} \sum_{s=1}^r \left(\tilde{V}_k(x_k^s, a) - \beta_k^s \right)^2,$$

with variable a. The iterations start with $k = N - 1$ and proceed backward in time to $k = 0$. For the case when $\hat{V}_k$ is a linear regression model, the LS problem is a linear LS problem with closed-form solution. Otherwise, we need to use some iterative optimization method as discussed in Chapter 6. Once all the parameters a_k have been computed, the approximate feedback function is given by

$$\mu_k(x) = \operatorname*{argmin}_u \tilde{Q}_k \left(x, u, a_{k+1} \right). \tag{8.7}$$

Example 8.3   We will in this example perform fitted-value iteration for the optimal control problem in Example 8.1. We will specifically consider the case when $m = 1$ and $n = 2$, and we let A_k, B_k, R_k, S_k be independent of k, and we write them as A, B, R, S. Since we know that the value function is quadratic, we will use a feature vector that is $\varphi(x) = (x_1^2, x_2^2, 2x_1x_2)$, where $x = (x_1, x_2) \in \mathbb{R}^2$. Notice that the indices refer to components of the vector and not to time. We let

$$\hat{V}_k(\varphi(x), a) = a^T \varphi(x),$$

where $a \in \mathbb{R}^3$. With

$$\tilde{P} = \begin{bmatrix} a_1 & a_3 \\ a_3 & a_2 \end{bmatrix},$$

we may then write

$$\tilde{V}_k(x, a) = a^T \varphi(x) = x^T \tilde{P} x.$$

Hence, the true value function $V_k(x) = x^T P_k x$, and the approximate value function $\tilde{V}_k(x, a)$ agrees if $\tilde{P} = P_k$. We, moreover, have

$$\tilde{Q}_k(x, u, a) = x^T S x + u^T R u + (Ax + Bu)^T \tilde{P}(Ax + Bu),$$

$$= \begin{bmatrix} x \\ u \end{bmatrix}^T \begin{bmatrix} S + A^T \tilde{P} A & A^T \tilde{P} B \\ B^T \tilde{P} A & R + B^T \tilde{P} B \end{bmatrix} \begin{bmatrix} x \\ u \end{bmatrix}. \tag{8.8}$$

For $k = N - 1$ down to $k = 0$, we then solve the linear LS problem in (8.6) to obtain β_k^s. From Example 5.7, we realize that

$$\beta_k^s = \left(x_k^s\right)^T \left\{ S + A^T \tilde{P}_{k+1} A - A^T \tilde{P}_{k+1} B \left(R + B^T \tilde{P}_{k+1} B\right)^{-1} B^T \tilde{P}_{k+1} A \right\} x_k^s,$$

assuming that $R + B^T \tilde{P}_{k+1} B$ is positive definite. Here, $\tilde{P}_N = S$. We then obtain a_k as the solution to the linear LS problem

$$\text{minimize } \frac{1}{2} \sum_{s=1}^{r} \left(\varphi^T(x_k^s) a - \beta_k^s\right)^2,$$

with variable a. This defines $\tilde{P}_k$. The solution a_k satisfies the normal equations, cf. (5.3),

$$\Phi_k^T \Phi_k a_k = \Phi_k^T \beta_k,$$

where

$$\Phi_k = \begin{bmatrix} \varphi^T(x_k^1) \\ \vdots \\ \varphi^T(x_k^r) \end{bmatrix}; \quad \beta_k = \begin{bmatrix} \beta_k^1 \\ \vdots \\ \beta_k^r \end{bmatrix}.$$

It is crucial here to choose x_k^s such that $\Phi_k^T \Phi_k$ is invertible. We realize that we need $r \geq 3$ for this hold. Moreover, we need to choose x_k^s sufficiently different for $\Phi_k^T \Phi_k$ to be well conditioned. For a general n, we will need that $r \geq n(n+1)/2$. From (8.7), (8.8), and Example 5.7 it follows that the optimal control is given by

$$u_k = -\left(R_k + B_k^T \tilde{P}_{k+1} B_k\right)^{-1} B_k^T \tilde{P}_{k+1} A_k x_k.$$

In the example above, it should be possible to obtain the same solution as in Example 8.4. In general, it is not the case that fitted value iteration will provide the exact solution to the problem in (8.2). The reason is that in general, we cannot represent the value function exactly with the feature vectors.

8.3 Infinite Horizon Optimal Control

Let $F : \mathcal{D}^n \times \mathcal{E}^m \to \mathcal{D}^n$ and $f : \mathcal{D}^n \times \mathcal{E}^m \to \mathbb{R}$ be given functions and consider an infinite horizon optimal control problem defined as

$$\text{minimize } \sum_{k=0}^{\infty} \gamma^k f(x_k, u_k),$$
$$\text{subject to } x_{k+1} = F(x_k, u_k), \quad k \in \mathbb{Z}_+, \tag{8.9}$$
$$u_k \in U(x_k) \subseteq \mathcal{D}^m, \quad k \in \mathbb{Z}_+,$$

with variables $(u_0, x_1, u_1, x_2, \ldots)$, where x_0 is given, and where $0 < \gamma \leq 1$ is a discount factor. Here, U is a set-valued function from $\mathcal{D}^n$. We now have that the incremental costs f and the state-dynamics F are independent of the stage index k. We define the optimal value function $J^\star : \mathcal{D}^n \to \mathbb{R}$ as the optimal value of the above optimization problem for different values of the initial state x_0. Before we continue, we will make some assumptions. It is assumed that $F(0,0) = 0$ and that $0 \in U(0)$ so that zero is a stationary point of the dynamical equation.[2] If this is not the case, we can make a change

---

2 The point (x^0, u^0) is a stationary point of $x_{k+1} = F(x_k, u_k)$ if $x^0 = F(x^0, u^0)$, i.e. the state remains at x^0 for all values of k.

of coordinates to make this hold. We also assume that the incremental cost f is such that $f(0,0) = 0$. Hence, if the state reaches the stationary point, only zero value is added to the cost function for each stage we remain in stationarity. If this assumption would not be made, it would not be possible to obtain *closed-loop stability*, i.e. $(x_k, u_k) \to (0,0)$ as $k \to \infty$ with finite cost for the case when $\gamma = 1$. To simplify the presentation below, we will also restrict ourselves to the case when the incremental cost is *strictly positive definite*.[3]

8.3.1 Bellman Equation

If the above assumptions are satisfied, and there exists a strictly positive definite and *quadratically bounded*[4] solution $V : \mathcal{D}^n \to \mathbb{R}_+$ to the *Bellman equation*

$$V(x) = \min_{u \in U(x)} \{f(x,u) + \gamma V(F(x,u))\}, \tag{8.10}$$

then $V(x) = J^\star(x)$ and $u_k^\star = \mu(x_k)$ is the optimal feedback control, where $\mu : \mathcal{D}^n \to \mathcal{E}^m$ is defined as $\mu(x) = \operatorname{argmin}_{u \in U(x)} \{f(x,u) + \gamma V(F(x,u))\}$. If in addition γ is sufficiently close to one, this feedback results in closed-loop stability in the sense defined above. The proof of this result is given in Section 8.10. For later reference, we define $Q : \mathcal{D}^n \times \mathcal{E}^m \to \mathbb{R}$ as

$$Q(x,u) = f(x,u) + \gamma V(F(x,u)).$$

We next consider an example which is known as *infinite horizon LQ control*.

Example 8.4 Let us consider the case when $F(x,u) = Ax + Bu$ for matrices $A \in \mathbb{R}^{n \times n}$ and $B \in \mathbb{R}^{n \times m}$. We also assume that $f(x,u) = x^T S x + u^T R u$, where $S \in \mathbb{S}_+^n$ and $R \in \mathbb{S}_{++}^m$ and that $U(x) = \mathbb{R}^m$ for all x. Clearly, we satisfy the assumptions on the functions f and F. We will make the guess that $V(x) = x^T P x$ for some $P \in \mathbb{S}_{++}^n$. Then

$$Q(x,u) = \begin{bmatrix} x \\ u \end{bmatrix}^T \begin{bmatrix} S + \gamma A^T P A & \gamma A^T P B \\ \gamma B^T P A & R + \gamma B^T P B \end{bmatrix} \begin{bmatrix} x \\ u \end{bmatrix}.$$

Since this expression is strictly convex in u, it follows from Example 5.7 that Q is minimized for

$$u = -\gamma(R + \gamma B^T P B)^{-1} B^T P A x.$$

Back substitution of this results in

$$x^T P x = x^T \left(S + \gamma A^T P A - \gamma^2 A^T P B (R + \gamma B^T P B)^{-1} B^T P A \right) x.$$

This equation holds if P is the solution to the following *discounted algebraic Riccati equation*:

$$P = S + \gamma A^T P A - \gamma^2 A^T P B (R + \gamma B^T P B)^{-1} B^T P A. \tag{8.11}$$

It can be shown that there is a unique solution to the above equation under our assumptions and if (A, B) is controllable.[5] This solution is such that $P \in \mathbb{S}_{++}^n$, and hence, the function V is strictly positive definite and quadratically bounded. The assumptions can be relaxed, see, e.g. [51].

---

3 A function $V : \mathcal{D}^n \to \mathbb{R}$ is said to be strictly positive definite if $V(0) = 0$, and there exist $\epsilon > 0$ such that $V(x) \geq \epsilon \|x\|_2^2$.
4 A function $V : \mathcal{D}^n \to \mathbb{R}$ is said to be quadratically bounded if there exists $c > 0$ such that $V(x) \leq c\|x\|_2^2$.
5 It holds that (A, B) is controllable if and only if the matrix $\begin{bmatrix} B & AB & \cdots & A^{n-1}B \end{bmatrix}$ has full rank.

An equivalent formulation of the Bellman equation is obtained by subtracting an arbitrary function $W : \mathcal{D}^n \to \mathbb{R}$ from V to obtain the function $\hat{V} : \mathcal{D}^n \to \mathbb{R}$ as $\hat{V}(x) = V(x) - W(x)$. The function $\hat{V}$ satisfies

$$\hat{V}(x) = \min_{u \in U(x)} \{f(x, u) - W(x) + \gamma W(F(x, u)) - \gamma W(F(x, u)) + \gamma V(F(x, u))\},$$

$$= \min_{u \in U(x)} \{\hat{f}(x, u) + \gamma \hat{V}(F(x, u))\},$$

where $\hat{f} : \mathcal{D}^n \times \mathcal{E}^m \to \mathbb{R}$ is defined as

$$\hat{f}(x, u) = f(x, u) - W(x) + \gamma W(F(x, u)). \tag{8.12}$$

The equation above is called the *variational form of the Bellman equation*, and $\hat{f}$ is called the *temporal difference* corresponding to W. This equivalent formulation plays an important role when solving the Bellman equation numerically.

8.4 Value Iterations

The Bellman equation is in general a difficult equation to solve. It is possible to show that the dynamic programming iteration in (8.4) converges to the solution $V(x)$ of the Bellman equation when $k \to -\infty$.

We will for convenience restate it in a format where the iteration index proceeds forward instead of backward as

$$V_{k+1}(x) = \min_{u \in U(x)} \{f_k(x, u) + \gamma V_k(F_k(x, u))\}, \tag{8.13}$$

with initial value $V_0(x) = 0$. If one has a clever guess of an approximate solution to the Bellman equation, this can be used as initial value instead. This will make the iterates converge much faster. The iteration above is called *value iteration* (VI). The algorithm for VI is summarized in Algorithm 8.2, where T is defined in (8.14).

Algorithm 8.2: Value iteration

Input: Incremental cost f, function F, $\gamma \in (0, 1]$, tolerance $\varepsilon > 0$
Output: V solving $V = T(V)$

$V \leftarrow 0$
while $\|V - T(V)\|_\infty \geq \varepsilon$ **do**
$\quad \mid \quad V \leftarrow T(V)$
end

Example 8.5 Here, we will consider VI for infinite horizon LQ control as in Example 8.4. To this end, we let $V_k(x) = x^T P_k x$ with $P_0 = 0$. Similarly, as in Example 8.1, we realize that if P_k satisfies the recursion

$$P_{k+1} = S + \gamma A^T P_k A - \gamma^2 A^T P_k B (R + B^T P_k B)^{-1} B^T P_k A,$$

then the VI recursion is satisfied. Based on a similar argument as in Example 8.1, we also realize that the inverse in the recursion exists and that $P_k \in \mathbb{S}_+^n$ for all k.

The proof of convergence of VI is based on the contraction property of the *Bellman operator* $T : \mathbb{R}^{\mathcal{D}^n} :\to \mathbb{R}^{\mathcal{D}^n}$[6] defined as

$$T(V)(x) = \min_{u \in U(x)} \{f(x, u) + \gamma V(F(x, u))\}. \tag{8.14}$$

We first show that the Bellman operator is monotone. Assume that V_1 and V_2 are two functions from $\mathcal{D}^n$ to $\mathbb{R}$ such that $V_1(x) \leq V_2(x)$ for all $x \in \mathcal{D}^n$. Then

$$T(V_1)(x) - T(V_2)(x) \leq \gamma V_1(F(x, u_2^\star)) - \gamma V_2(F(x, u_2^\star)) \leq 0,$$

where $u_2^\star$ is the minimizing argument for $T(V_2)$. Let

$$c = \sup_{x \in \mathcal{D}^n} |V_1(x) - V_2(x)| = \|V_1 - V_2\|_\infty.$$

Then from the monotonicity

$$V_1(x) - c \leq V_2(x) \leq V_1(x) + c \Longrightarrow T(V_1)(x) - \gamma c \leq T(V_2)(x) \leq T(V_1)(x) + \gamma c,$$

and hence, the *contraction mapping property*

$$\|T(V_1) - T(V_2)\|_\infty \leq \gamma \|V_1 - V_2\|_\infty,$$

holds. From this, it follows by the contraction-mapping theorem [73, p. 272] that VI converges to a solution of the Bellman equation for the case when $\gamma < 1$. In case $\gamma = 1$, it is sometimes possible to still prove convergence for VI. See, e.g. [19] for the LQ control case.

8.5 Policy Iterations

It turns out that the convergence of VI can be slow, and there is another approach that can be pursued. This is called *policy iteration* (PI). Introduce the *Bellman policy operator*: $T_\mu : \mathbb{R}^{\mathcal{D}^n} \to \mathbb{R}^{\mathcal{D}^n}$ defined as

$$T_\mu(V)(x) = f(x, \mu(x)) + \gamma V(F(x, \mu(x))), \tag{8.15}$$

for a given function $\mu : \mathcal{D}^n \to \mathcal{E}^m$. Given a feedback policy $\mu_k : \mathcal{D}^n \to \mathcal{E}^m$ for some k, we compute $V_k : \mathcal{D}^n \to \mathbb{R}$ as the solution of the equation:

$$V_k(x) = T_{\mu_k}(V_k)(x), \tag{8.16}$$

which is the same as the Bellman equation except for that we in place of the optimal u substitute a feedback policy. This is a linear system of equations for V_k. Depending on $\mathcal{D}$ there could be finitely many equations or infinitely many equations. Solving this linear system of equations is called the *policy evaluation step*. We then obtain a new feedback policy by solving

$$\mu_{k+1}(x) = \operatorname*{argmin}_{u \in U(x)} \{f(x, u) + \gamma V_k(F(x, u))\}. \tag{8.17}$$

This is called the *policy improvement step*. We summarize the PI algorithm in Algorithm 8.3. The proof that PI converges to a solution of the Bellman equation goes as follows: it holds

---

6 The set $\mathbb{R}^{\mathcal{D}^n}$ is the set of all functions from $\mathcal{D}^n$ to $\mathbb{R}$, *cf.* the notation section.

Algorithm 8.3: Policy iteration

Input: Incremental cost f, function F, parameter $\gamma \in (0, 1]$, tolerance $\varepsilon > 0$, initial feedback
policy μ

Output: V solving $V = T(V)$

while $\|V - T(V)\|_\infty \geq \varepsilon$ **do**

   | Solve $V(x) = f(x, \mu(x)) + \gamma V(F(x, \mu(x)))$ w.r.t. V

   | $\mu(x) \leftarrow \text{argmin}_u(f(x, u) + \gamma V(F(x, u)))$

end

---

that

$$V_k(x) = T_{\mu_k}\left(V_k\right)(x) \geq T_{\mu_{k+1}}\left(V_k\right)(x),$$

where the equality is by definition of the policy evaluation step, and where the inequality is by definition of the policy improvement step. Just like the Bellman operator, the Bellman policy operator is a monotone operator, see Exercise 8.7. Hence, repeated application of $T_{\mu_{k+1}}$ results in

$$V_k(x) \geq T_{\mu_{k+1}}\left(V_k\right)(x) \geq \left(T_{\mu_{k+1}}\right)^2\left(V_k\right)(x) \geq \cdots,$$

$$\geq \lim_{n \to \infty}\left(T_{\mu_{k+1}}\right)^n\left(V_k\right)(x) = V_{k+1}(x),$$

where the equality follows from the fact that also the Bellman policy operator is a contraction mapping when $\gamma < 1$, see Exercise 8.7, and therefore, the limit satisfies $T_{\mu_{k+1}}(V_{k+1}) = V_{k+1}$. Hence, PI results in an improving sequence of value functions, and in case $V_k(x) = V_{k+1}(x)$ for all $x \in \mathcal{D}^n$, we realize that the first inequality above also holds with equality, proving that V_k satisfies the Bellman equation proving that μ_{k+1} is optimal. Hence, PI either results in a strict improvement of the value function in each iteration or termination at a solution of the Bellman equation.

We remark that neither VI nor PI are normally tractable methods. Hence, approximations are in most cases required. We now give an example where the calculations can be carried out exactly.

Example 8.6 We consider PI for the infinite horizon LQ control problem in Example 8.4. We guess that $V_k(x) = x^T P_k x$ for some $P_k \in \mathbb{S}_+^n$ and that $\mu_k(x) = -L_k x$ for some $L_k \in \mathbb{R}^{m \times n}$. The policy evaluation step is then given by finding a solution P_k of

$$x^T P_k x = x^T S x + x^T L_k^T R L_k x + \gamma x^T\left(A - B L_k\right)^T P_k\left(A - B L_k\right)x,$$

for given L_k. This can be obtained by solving the algebraic Lyapunov equation

$$P_k - \gamma\left(A - B L_k\right)^T P_k\left(A - B L_k\right) = S + L_k^T R L_k,$$

which has a positive definite solution P_k since the right-hand side is positive definite. This assumes that $\sqrt{\gamma}(A - B L_k)$ has all its eigenvalues strictly inside the unit disk. The policy improvement step is then

$$\mu_{k+1}(x) = \underset{u}{\text{argmin}}\left\{x^T S x + u^T R u + \gamma(Ax - Bu)^T P_k(Ax - Bu)\right\}.$$

The solution is given by

$$\mu_{k+1}(x) = -\gamma\left(R + \gamma B^T P_k B\right)^{-1} B^T P_k A x,$$

and hence, we obtain

$$L_{k+1} = \gamma\left(R + \gamma B^T P_k B\right)^{-1} B^T P_k A.$$

It can be shown that if we start with L_0 that is stabilizing, then so will all L_k be, see [51], where it is also shown that convergence holds for the case when $\gamma = 1$. The iterations for the LQ control problem derived above are called the *Hewer iterations*.

8.5.1 Approximation

In general, it is not possible to carry out the computations in PI exactly. One has to resort to approximations. We will use here a similar idea as in Section 8.2. It is based on defining $\tilde{V} : \mathcal{D}^n \times \mathbb{R}^p \to \mathbb{R}$ using an ANN or as a linear regression with p parameters. This function will be used to approximate V_k in (8.16). Before we do that we notice that (8.16) implies

$$
\begin{aligned}
V_k(x_0) &= f\left(x_0, \mu_k(x_0)\right) + \gamma V_k\left(F\left(x_0, \mu_k(x_0)\right)\right), \\
&= f\left(x_0, \mu_k(x_0)\right) + \gamma V_k\left(x_1\right), \\
&= f\left(x_0, \mu_k(x_0)\right) + \gamma f\left(x_1, \mu_k(x_1)\right) + \gamma^2 V_k\left(x_2\right), \\
&\;\;\vdots \\
&= \sum_{i=0}^{N-1} \gamma^i f\left(x_i, \mu_k(x_i)\right) + \gamma^N V_k\left(x_N\right),
\end{aligned}
\tag{8.18}
$$

where $x_{i+1} = F(x_i, \mu_k(x_i))$. In case N is large and μ_k is stabilizing, we have that x_N is close to zero and that also $V_k(x_N)$ is close to zero. Hence, a way to evaluate V_k for a value x_0 of the state is to just simulate the dynamical system and add up the incremental costs. In case one has an idea about what $V_k\left(x_N\right)$ might be, that can also be used. This can in particular be beneficial in case x_N is not very small.

We define these sums for different initial values x^s for $s \in \mathbb{N}_r$ as

$$
\beta_k^s = \sum_{i=0}^{N-1} \gamma^i f\left(x_i, \mu_k(x_i)\right),
$$

where $x_{i+1} = F(x_i, \mu_k(x_i))$. We then find the approximation of V_k by solving

$$
\text{minimize } \frac{1}{2} \sum_{s=1}^{r} \left(\tilde{V}(x^s, a) - \beta_k^s\right)^2,
$$

with variable a. The solution is denoted a_k. After this, we use the following exact policy improvement step

$$
\mu_{k+1}(x) = \underset{u \in U(x)}{\operatorname{argmin}} \left\{ f(x, u) + \gamma \tilde{V}\left(F(x, u), a_k\right) \right\}.
\tag{8.19}
$$

We remark that it is possible to reuse the simulated trajectory and use it to compute several costs. This follows from the simple fact that we can use also $x_1 = F(x^s, u^s)$ as an initial value, for which the simulated trajectory is obtained from the one starting at x^s by omitting the first value. This means that costs from any state on the simulated trajectory can be computed. These are just the tail-costs of the overall cost when starting at x^s. We should however stress that this might not provide enough representative initial states to obtain a good approximation of V_k.

Example 8.7 We will in this example consider the optimal control problem in Example 8.4. We will specifically consider the case when $m = 1$ and $n = 2$. Since we know that the value function is quadratic, we will use a feature vector that is $\varphi(x) = (x_1^2, x_2^2, 2x_1x_2)$, where $x = (x_1, x_2) \in \mathbb{R}^2$. Notice that the indices refer to components of the vector and not to time. We let

$$
\tilde{V}(x, a) = a^T \varphi(x),
$$

where $a \in \mathbb{R}^3$. With

$$\tilde{P} = \begin{bmatrix} a_1 & a_3 \\ a_3 & a_2 \end{bmatrix},$$

we may then write

$$\tilde{V}(x, a) = x^T \tilde{P} x.$$

Hence, the true value function $V(x) = x^T P x$ and the approximate value function $\tilde{V}(x, a)$ agree if $\tilde{P} = P$. Here, with an abuse of notation $a_k \in \mathbb{R}^3$, which defines $\tilde{P}_k$, is obtained as the solution to the linear LS problem

$$\text{minimize } \frac{1}{2} \sum_{s=1}^{r} \left(\varphi^T(x^s) a - \beta_k^s \right)^2,$$

with variable a. The solution a_k satisfies the normal equations, *cf.* (5.3),

$$\Phi_k^T \Phi_k a_k = \Phi_k^T \beta_k,$$

where

$$\Phi_k = \begin{bmatrix} \varphi^T(x^1) \\ \vdots \\ \varphi^T(x^r) \end{bmatrix}, \quad \beta_k = \begin{bmatrix} \beta_k^1 \\ \vdots \\ \beta_k^r \end{bmatrix},$$

and where

$$\beta_k^s = \sum_{i=0}^{N-1} \gamma^i \left(x_i^T S x_i + \mu_k(x_i)^T R \mu_k(x_i) \right),$$

and where $x_{i+1} = A x_i + B \mu_k(x_i)$ with initial values x^s, $s \in \mathbb{N}_r$. It is crucial here to choose x^s such that $\Phi_k^T \Phi_k$ is invertible. We realize that we need $r \geq 3$ for this hold. Moreover, we need to choose x^s sufficiently different for $\Phi_k^T \Phi_k$ to be well conditioned. For a general n, we will need $r \geq n(n+1)/2$.

We define

$$\tilde{Q}_k(x, u, a) = f(x, u) + \gamma \tilde{V}(Ax + Bu, a),$$

$$= x^T S x + u^T R u + \gamma (Ax + Bu)^T \tilde{P} (Ax + Bu),$$

$$= \begin{bmatrix} x \\ u \end{bmatrix}^T \begin{bmatrix} S + \gamma A^T \tilde{P} A & \gamma A^T \tilde{P} B \\ \gamma B^T \tilde{P} A & R + \gamma B^T \tilde{P} B \end{bmatrix} \begin{bmatrix} x \\ u \end{bmatrix}. \tag{8.20}$$

From Example 5.7, we realize that the solution to (8.19) is given by

$$\mu_{k+1}(x) = \underset{u}{\text{argmin}} \, \tilde{Q}_k(x, u, a_k) = -\gamma (R + \gamma B^T \tilde{P}_k B)^{-1} B^T \tilde{P}_k A x,$$

assuming that $R + \gamma B^T \tilde{P}_k B$ is positive definite. Here, $\tilde{P}_k$ is defined from a_k in the same way as $\tilde{P}$ is defined from a above. We may hence write

$$\mu_{k+1}(x) = -L_{k+1} x,$$

where $L_{k+1} = \gamma (R + \gamma B^T \tilde{P}_k B)^{-1} B^T \tilde{P}_k A$. It is a good idea to start with some L_0 that is stabilizing.

We remark that it may be highly beneficial to consider the variational form of the Bellman equation by replacing the incremental cost with the temporal difference in (8.12). Since the Bellman equation for the variational form looks the same as the original Bellman equation with the only difference that the incremental cost is replaced with the temporal difference, all formulas in this section remain the same when the temporal difference is used as incremental cost. The reason that the variational form may be beneficial is that $W(x)$ may be taken as an initial guess of the solution of the Bellman equation, and then we only need to parameterize the difference between the true solution and the initial guess, which might require less parameters if the initial guess is good.

8.6 Linear Programming Formulation

Our intention is now to show that the solution to the Bellman equation can also be obtained as the solution of an LP. To this end we now start the VI with a V_0 such that $V_1 = T(V_0) \geq V_0$ for all $x \in \mathcal{D}^n$. One possible choice is $V_0 = 0$ for all $x \in \mathcal{D}^n$. We obtain $T(V_1) \geq T(V_0)$ from the monotonicity property, and hence, $V_2 = T(V_1) \geq V_1 \geq V_0$. If we repeat this, we obtain $V_k = T(V_{k-1}) \geq V_{k-1} \geq V_0$. Since V_k converges to the solution V of the Bellman equation, we have shown that V is the maximum element, see [22, p. 45] of the set of functions V that satisfy the linear inequalities:

$$V(x) \leq f(x, u) + \gamma V(F(x, u)), \quad \forall (x, u) \in \mathcal{D}^n \times \mathcal{E}^m \text{ such that } u \in U(x).$$

The maximum element can be obtained by solving the LP

$$\text{maximize } \sum_{x \in \mathcal{D}^n} c(x) V(x),$$

$$\text{subject to } V(x) \leq f(x, u) + \gamma V(F(x, u)), \quad \forall (x, u) \in \mathcal{D}^n \times \mathcal{E}^m \text{ such that } u \in U(x),$$

where $c(x) > 0$ is arbitrary. This follows from the dual characterization of maximum elements, see [22, p. 54]. The optimization variable is $V(x)$ for all values of x. We remark that the LP formulation is only tractable if it does not have too many constraints or variables.

Example 8.8 We consider an example where $\mathcal{D} = \mathcal{E} = \{-1, 0, 1\}$ and where $m = n = 1$. We let $f(x, u) = x^2 + u^2$ and we let the values of the function F be defined by the following table:

x \ u	-1	0	1
-1	-1	-1	0
0	-1	0	1
1	0	1	1

We assume that $0 \leq \gamma < 1$. The LP formulation of the optimal control problem is then given by

$$\text{maximize } V(-1) + V(0) + V(1),$$

$$\text{subject to } V(-1) \leq (-1)^2 + (-1)^2 + \gamma V(-1),$$

$$V(-1) \leq (-1)^2 + 0^2 + \gamma V(-1),$$

$$V(-1) \leq (-1)^2 + 1^2 + \gamma V(0),$$

$$V(0) \leq 0^2 + (-1)^2 + \gamma V(-1),$$

$$V(0) \leq 0^2 + 0^2 + \gamma V(0),$$

$$V(0) \leq 0^2 + 1^2 + \gamma V(1),$$

$$V(1) \leq 1^2 + (-1)^2 + \gamma V(0),$$

$$V(1) \leq 1^2 + 0^2 + \gamma V(1),$$

$$V(1) \leq (1)^2 + 1^2 + \gamma V(1),$$

over the variables $(V(-1), V(0), V(1))$. We immediately realize that the first constraint is implied by the second constraint and that the last constraint is implied by the second last constraint. Moreover, the fifth constraint is equivalent to $V(0) = 0$, if $0 \leq \gamma < 1$, since $V(x) \geq 0$ from the fact that the objective function is bounded by zero from below. From this, we see that the fourth and sixth constraint cannot be active at optimality, since they only provide lower bounds on the variables. The remaining constraints can be summarized as $V(-1), V(1) \leq v$, where $v = \min \{2, 1/(1 - \gamma)\}$.

Hence, the optimal solution is $(V(-1), V(0), V(1)) = (v, 0, v)$, where $v \in (1, 2]$, since $0 < \gamma < 1$. It now remains to compute the optimal u for different values of $x \in \mathcal{D}$. We have for $x = -1$ that the right-hand side of the Bellman equation is given by

$$(-1)^2 + u^2 + \gamma \begin{cases} v, & u = -1, \\ v, & u = 0, \\ 0, & u = 1. \end{cases}$$

We see that whatever value v has it can never be that $u = -1$ is optimal. If $\gamma v < 1$, it is optimal to have $u = 0$, which can be seen to be the case when $0 < \gamma < 1/2$, and hence, for $1/2 \leq \gamma < 1$, it is optimal to take $u = 1$. A similar argument shows that when $x = 1$, $u = 0$ is optimal when $0 < \gamma < 1/2$ and $u = -1$ optimal when $1/2 \leq \gamma < 1$. When $x = 0$ the right-hand side of the Bellman equation is given by

$$0^2 + u^2 + \gamma \begin{cases} v, & u = -1, \\ 0, & u = 0, \\ v, & u = 1, \end{cases}$$

and hence, $u = 0$ is optimal. We summarize our findings by noting that when $0 < \gamma < 1/2$ the optimal solution is always zero. The future costs are discounted so much that it is not worth the effort to take any action with the control signal. Otherwise, it pays off to steer away from a nonzero state value to make it zero.

8.6.1 Approximations

The LP formulation is often not tractable in general, since there might be many variables. A remedy to this is to approximate $V(x)$ with, e.g. a feature-based linear regression $\tilde{V} : \mathcal{D}^n \times \mathbb{R}^p \to \mathbb{R}$ as

$$\tilde{V}(x, a) = a^T \varphi(x),$$

where $\varphi : \mathcal{D}^n \to \mathbb{R}^p$. The approximate optimization problem is

$$\text{maximize} \quad \sum_{x \in \mathcal{D}^n} c(x) \tilde{V}(x, a),$$

$$\text{subject to} \quad \tilde{V}(x, a) \leq f(x, u) + \gamma \tilde{V}(F(x, u), a), \quad \forall (x, u) \in \mathcal{D}^n \times \mathcal{E}^m \text{ such that } u \in U(x),$$

with variable a. This might still be an intractable problem because of the many constraints. One way to overcome this obstacle is to sample the constraints, i.e. omit some of them by just considering (x^s, u^s) for $s \in \mathbb{N}_r$ as above. Notice that the approach presented here can also be used to approximately evaluate a fixed policy μ. Just replace u above with $\mu(x)$. This might reduce the number of constraints significantly and can then be used together with PI.

8.7 Model Predictive Control

It is clear that the Bellman equation can be very difficult to solve in many practical applications. However, it is the case that the resulting feedback law has the very attractive property that it is insensitive to disturbances and unmodeled dynamics. This is a property that an open-loop solution of (8.3) does not have. Here, we will discuss *MPC*, which is a way to obtain nearly optimal feedback control laws by computing a sequence of open-loop controls as in (8.3). We may suspect that such an approach should be very appealing, since it is a feedback control law, and at the same time, it is obtained not by solving the Bellman equation, but by solving open-loop control problems, which are known to be tractable from a computational point of view.

8.7.1 Infinite Horizon Problem

We consider a nondiscounted infinite time horizon optimal control problem defined as

$$
\begin{aligned}
\text{minimize} \quad & \sum_{k=0}^{\infty} f(x_k, u_k), \\
\text{subject to} \quad & x_{k+1} = F(x_k, u_k), \quad k \in \mathbb{Z}_+, \\
& u_k \in U \subseteq \mathcal{E}^m, \quad k \in \mathbb{Z}_+, \\
& x_k \in X \subseteq \mathcal{D}^n, \quad k \in \mathbb{N},
\end{aligned}
\tag{8.21}
$$

with variables $(u_0, x_1, \ldots)$, where x_0 is given. The only difference compared to our previous setup is that the set U does not depend on x and that we have a constraint set X for the state. We define the optimal value function $J^\star : \mathbb{R}^n \to \mathbb{R}$ as the optimal value of the above optimization problem for different values of the initial state x_0. We make the same technical assumptions as before on the functions and sets defining the problem.

If there exists a strictly positive definite solution $V : \mathbb{R}^n \to \mathbb{R}_+$ to the Bellman equation

$$
V(x) = \min_{u \in U, F(x,u) \in X} \{ f(x, u) + V(F(x, u)) \},
\tag{8.22}
$$

then $V(x) = J^\star(x)$ and $u_k^\star = \mu(x_k)$, where $\mu : \mathcal{D}^n \to \mathcal{E}^m$ defined as

$$
\mu(x) = \operatorname*{argmin}_{u \in U, F(x,u) \in X} \{ f(x, u) + V(F(x, u)) \}
$$

is an optimal feedback that results in closed-loop stability. The only difference as compared to our previous presentation of the Bellman equation is that we have added a constraint on $F(x, u)$.

8.7.2 Guessing the Value Function

It is clear that if one for some reason is able to guess an approximate solution $\hat{V}$ to the Bellman equation, then

$$
\mu(x) = \operatorname*{argmin}_{u \in U, F(x,u) \in X} \{ f(x, u) + \hat{V}(F(x, u)) \}
\tag{8.23}
$$

will be an approximation of the optimal feedback function.

This approximation can also be interpreted as the optimal solution to the one time-step horizon optimal control problem

$$
\begin{aligned}
\text{minimize} \quad & \hat{V}(x_{k+1}) + f(x_k, u_k), \\
\text{subject to} \quad & x_{k+1} = F(x_k, u_k), \\
& u_k \in U, \quad x_{k+1} \in X,
\end{aligned}
$$

with variables (u_k, x_{k+1}), where $x_k \in X$ is given. This is a special case of (8.2). In case (8.23) still is difficult to solve, then the one time-step horizon optimal control problem can be solved using some of the methods presented in Chapter 6 on-line, i.e. the problem can be solved repeatedly for each time-instant k as soon as a new measurement of x_k becomes available. In case the resulting optimization problem is of moderate size and the sampling time is not too short, this can be carried out in real-time. Notice that the on-line solution only computes the optimal control for one given value of the state, whereas the off-line computation with the Bellman equation computes the optimal control for all possible values of the state.

When $\hat{V}$ is taken to be zero, the most crude approximation imaginable, we get what is usually called "greedy control." This can often result in very poor performance such as instability or very large value of the objective function.

For the case of $f(x, u) = x^T Q x + u^T R u$, but when there still are constraints on the states and/or control signal, one might use $\hat{V}(x) = x^T P x$, where P solves the algebraic Riccati equation in (8.11) for $\gamma = 1$ This solution will be optimal if x_0 is such that there are no constraint violations – otherwise not.

We remark that stability is not guaranteed for any of the approximations of this section without further investigation.

8.7.3 Finite Horizon Approximation

Instead of finding an approximate solution by guessing the value function, one might consider to approximate the optimal control problem in (8.21) with a finite time horizon N and impose a terminal state constraint. We consider

$$
\begin{aligned}
&\text{minimize} \quad \sum_{k=0}^{N-1} f(x_k, u_k), \\
&\text{subject to} \quad x_{k+1} = F(x_k, u_k), \quad k \in \mathbb{Z}_{N-1}, \\
&\qquad\qquad\quad u_k \in U, \quad k \in \mathbb{Z}_{N-1}, \\
&\qquad\qquad\quad x_k \in X, \quad k \in \mathbb{N}_{N-1}, \\
&\qquad\qquad\quad x_N = 0,
\end{aligned}
\tag{8.24}
$$

with variables $(u_0, x_1, \ldots, u_{N-1}, x_N)$, where $x_0 \in X$ is given. In case N is sufficiently large, this should result in a near optimal solution. Notice that u_k can be taken to zero for $k \geq N$, since $x_N = 0$ and $(0, 0)$ is a stationary point. Hence, stability is guaranteed assuming that the approximate optimal control problem is feasible. It does not seem like this should be a simpler problem to solve, since we know that the solution can be obtained from a dynamic programming recursion like in (8.4). However, the solution can alternatively be obtained as the solution of a finite-dimensional optimization problem as in (8.3). We then do not obtain a feedback solution, but only an optimal input sequence. The finite time horizon approximation makes this trade-off possible.

8.7.4 Receding Horizon Approximation

We have so far considered two different approximate solutions to the original optimal control problem in (8.21). The first one was obtained by approximating the value function and could be interpreted as on-line repeatedly solving an optimal control problem with a one time-step horizon. The second approximation was obtained by considering a finite horizon which was not solved repeatedly on-line, i.e. we did not obtain a feedback solution. An obvious remedy for this would be to solve the finite horizon problem repeatedly on-line and use what is called a receding time horizon. To fix the ideas, we solve at each time instant $k = 0, 1, \ldots$ the *receding horizon* optimal control problem

$$
\begin{aligned}
&\text{minimize} \quad J_k = \sum_{i=k}^{k+N-1} f(\tilde{x}_i, \tilde{u}_i), \\
&\text{subject to} \quad \tilde{x}_{i+1} = F(\tilde{x}_i, \tilde{u}_i), \quad i = k, \ldots, k + N - 1, \\
&\qquad\qquad\quad \tilde{u}_i \in U, \quad i = k, \ldots, k + N - 1, \\
&\qquad\qquad\quad \tilde{x}_i \in X, \quad i = k + 1, \ldots, k + N - 1, \\
&\qquad\qquad\quad \tilde{x}_{k+N} = 0,
\end{aligned}
\tag{8.25}
$$

where $\tilde{x}_k = x_k$ is given; how will be defined below. We will call this problem the time k problem. Denote the solution by

$$(\tilde{x}^\star_{k+1}, \ldots, \tilde{x}^\star_{k+N}, \tilde{u}^\star_k, \ldots, \tilde{u}^\star_{k+N-1}),$$

and let $u_k = \tilde{u}^\star_k$. Then the state evolves as before according to

$$x_{k+1} = F(x_k, u_k), \quad k \in \mathbb{Z}_+,$$

with x_0 given. This control strategy is often called model predictive control, since a model that predicts the values of the states are present in the time k problem. The time horizon N is for this reason usually called the *prediction horizon*. Notice that only the optimal control signal corresponding to time instant k in the solution of the time k problem is applied to the system to be controlled in the MPC strategy. All the other computed control signals are disregarded, and the optimization is carried out once again for $k + 1$, and so on. In this way feedback is obtained, even though the control signal is not explicitly given as a feedback policy, i.e. we do not have an explicit function μ relating the optimal u_k to the state x_k. We only have an implicit way to obtain the optimal u_k once the state x_k is given. Hence, we have to resort to the on-line implementation to be carried out in real-time. This can be infeasible for large-scale problems or short sampling times. How to overcome this obstacle will be discussed later. We remark that the receding horizon optimal control problem can be cast as a finite-dimensional optimization problem similarly as in (8.3). There are many different variations of the MPC strategy. One might add a penalty $\phi(\tilde{x}_{k+N})$ to the objective function. The constraint $\tilde{x}_{k+N} = 0$ can be removed or relaxed to a less stringent constraint. This will usually make it possible to use shorter prediction horizons N and still obtain feasibility of the time k problem. Stability of MPC is investigated in Section 8.10. We will now consider an example.

Example 8.9   Consider $F(x, u) = Ax + Bu$ with $A \in \mathbb{R}^{3\times3}$ and $B \in \mathbb{R}^{3\times2}$, where the matrices A and B are chosen randomly with entries drawn from a standard uniform distribution with support on the interval $[0, 1]$. The incremental cost is $f(x, u) = x^T x + u^T u$, and the constraints are $-1 \le x_k \le 1$ and $-0.5 \times 1 \le u_k \le 0.5 \times 1$, where 1 is a vector of ones of appropriate dimension. The initial point is $x_0 = \begin{bmatrix} 0.9 & -0.9 & 0.9 \end{bmatrix}^T$. In Figure 8.3, the cost for the infinite horizon criterion is plotted versus

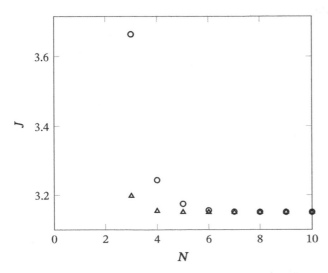

Figure 8.3   The cost for the infinite horizon criterion as a function of the horizon for the finite time horizon approximation (circles), and for MPC (triangle).

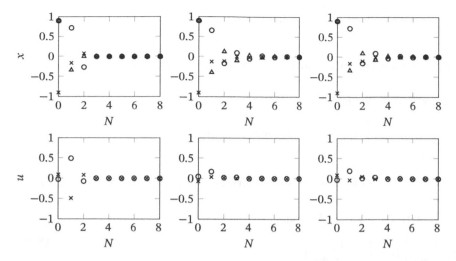

Figure 8.4 The trajectories for the states (top), and the control signals (bottom), for the finite horizon approximation with $N = 3$ (left), and $N = 20$ (middle), and for the MPC with $N = 3$ (right). The different components of the vectors are shown as circles, crosses, and triangles, respectively.

different time horizons N for the finite time horizon approximation. It is seen that the cost decreases as N becomes larger. In fact, it converges to the cost for the infinite time horizon problem. Notice that there is no feasible solution for $N \leq 2$. Also the results for MPC with different time horizons are shown as comparison. We see that MPC is always performing better than the finite-horizon open-loop strategy for the same N. The improvement is larger, the smaller N is.

In Figure 8.4, the trajectories for the states and the control signals for the finite horizon approximation with $N = 3$ and $N = 20$ together with the MPC with $N = 3$ are shown. It is seen that for the finite horizon approximation with $N = 3$ the control signals and the states are zero for $k \geq N$. We also see that the difference in behavior is not so large for the finite horizon approximation with $N = 20$ and the MPC with $N = 3$.

8.8 Explicit MPC

It was clear from the Section 8.7 that we do not obtain an explicit expression for the feedback resulting from MPC, but we are only able to on-line compute the optimal control signal given a measurement of the current state. We will now show how that obstacle can be overcome for a certain important special case using multiparametric programming as in Section 5.6 applied to the time k problem in (8.25), where x_k is the parameter.

We consider the case when the incremental cost is quadratic, i.e. $f(\tilde{x}, \tilde{u}) = \tilde{x}^T Q \tilde{x} + \tilde{u}^T R \tilde{u}$, where Q and R are positive definite matrices. We assume that the dynamics is linear, i.e. $F(\tilde{x}, \tilde{u}) = A\tilde{x} + B\tilde{u}$. For simplicity, we will also assume that the equality constraint on $\tilde{x}_{k+N}$ is not present. Let

$$\tilde{X} = \begin{bmatrix} \tilde{x}_k^T \cdots \tilde{x}_{k+N-1}^T \end{bmatrix}^T, \quad \tilde{U} = \begin{bmatrix} \tilde{u}_k^T \cdots \tilde{u}_{k+N-1}^T \end{bmatrix}^T,$$

and let

$$\Phi = \begin{bmatrix} I \\ A \\ A^2 \\ \vdots \\ A^{N-1} \end{bmatrix}, \quad \Gamma = \begin{bmatrix} 0 & 0 & 0 & \cdots & 0 \\ B & 0 & 0 & \cdots & 0 \\ AB & B & 0 & \cdots & 0 \\ \vdots & \ddots & \ddots & \ddots & \vdots \\ A^{N-2}B & \cdots & AB & B & 0 \end{bmatrix}.$$

Then the equality constraints in (8.25) can be expressed as

$$\tilde{X} = \Phi x_k + \Gamma \tilde{U}.$$

Let Q and R be block-diagonal matrices with Q and R, respectively, on the diagonal such that

$$\sum_{i=k}^{k+N-1} f(x_i, u_i) = \tilde{X}^T Q \tilde{X} + \tilde{U}^T R \tilde{U}.$$

We assume that the sets X and U are defined via the inequalities

$$\mathcal{E}_u \tilde{U} \le \mathcal{F}_u, \quad \mathcal{E}_x \tilde{X} \le \mathcal{F}_x,$$

for some matrices $\mathcal{E}_u$, $\mathcal{F}_u$, $\mathcal{E}_x$, and $\mathcal{F}_x$. It is now straightforward to verify that (8.25) can be reformulated as

$$
\begin{aligned}
&\text{minimize} \quad \tilde{U}^T \mathcal{H} \tilde{U} + 2\tilde{U}^T \mathcal{F} x_k + x_k^T \Phi^T Q \Phi x_k, \\
&\text{subject to} \quad \begin{bmatrix} \mathcal{E}_u \\ \mathcal{E}_x \Gamma \end{bmatrix} \tilde{U} \le \begin{bmatrix} \mathcal{F}_u \\ \mathcal{F}_x - \mathcal{E}_x \Phi x_k \end{bmatrix},
\end{aligned}
\tag{8.26}
$$

with variable $\tilde{U}$, where $\mathcal{H} = \Gamma^T Q \Gamma + R$ and $\mathcal{F} = \Gamma^T Q \Phi$. Note that the last term in the objective function can be omitted when carrying out the optimization. By completing the squares and transforming the variables according to

$$z = \sqrt{2} \left(\tilde{U} + \mathcal{H}^{-1} \mathcal{F} x_k \right),$$

an equivalent optimization problem is obtained, which is a convex multiparametric quadratic program:

$$
\begin{aligned}
&\text{minimize} \quad \tfrac{1}{2} z^T \mathcal{H} z, \\
&\text{subject to} \quad Gz \le w + S\theta,
\end{aligned}
\tag{8.27}
$$

with variable z, where $\theta = x_k$, H is defined as above, and where

$$G = \frac{1}{\sqrt{2}} \begin{bmatrix} \mathcal{E}_u \\ \mathcal{E}_x \Gamma \end{bmatrix}, \quad S = \begin{bmatrix} \mathcal{E}_u \mathcal{H}^{-1} \mathcal{F} \\ \mathcal{E}_x (\Gamma \mathcal{H}^{-1} \mathcal{F} - \Phi) \end{bmatrix}, \quad w = \begin{bmatrix} \mathcal{F}_u \\ \mathcal{F}_x \end{bmatrix}.$$

From this reformulation and the discussion in Section 5.6, we conclude that we can obtain an explicit feedback for MPC in the case of quadratic objective function with affine constraints. For this case, the feedback function is piecewise affine over a polyhedral partitioning of the state-space, and it can be computed off-line. The only major on-line computational effort is related to computing in which polyhedron the current state x_k is in.

8.9 Markov Decision Processes

Here, we will discuss the generalization of the results in Sections 8.1 and 8.3 to a stochastic setting based on the multistage stochastic optimization problem formulation in Section 5.7. We are given a random processes $W = (W_0, W_1, \dots)$ called the *disturbance process*. It is assumed that W_i is independent of W_j for $i \ne j$. This generates another random process $X = (X_0, X_1, \dots)$ via the dynamical recursion

$$X_{k+1} = F_k(X_k, u_k, W_k),\tag{8.28}$$

where $F_k : \mathcal{D}^n \times \mathcal{E}^m \times \mathcal{F}^p \to \mathcal{D}^n, k \in \mathbb{Z}_+$ are functions describing how the *state* X_k evolves depending on the *decision variable* or *control signal* u_k. This is a function $u_k : \mathcal{D}^n \times \cdots \times \mathcal{D}^n \to \mathcal{E}^m$ with

$(X_0, \ldots, X_k) \mapsto u_k(X_0, \ldots, X_k)$, i.e. $u = (u_0, u_1, \ldots)$ is a random process *adapted* to X. The initial state X_0 is assumed to have a probability distribution that is independent of the probability distribution of W_k, $k \in \mathbb{Z}_+$. The sets $\mathcal{D}$ and $\mathcal{E}$ can be finite or infinite as in Section 8.1, but as discussed in Section 5.7 care has to be taken in case these sets are not finite. The control signal is used to control the dynamical system, i.e. the evolution of the states. To this end, we introduce the incremental costs $f_k : \mathcal{D}^n \times \mathcal{E}^m \to \mathbb{R}$ for $k \in \mathbb{Z}_{N-1}$. The incremental costs are functions of the states and the control signal. We also introduce a cost associated with the final state X_N called the final cost or terminal cost $\phi : \mathcal{D}^n \to \mathbb{R}$. The *finite-horizon stochastic optimal control problem* is then the problem of solving

$$\text{minimize} \quad \mathbb{E}\left[\sum_{k=0}^{N-1} f_k(X_k, u_k) + \phi(X_N) \right],$$

$$\text{subject to} \quad X_{k+1} = F_k(X_k, u_k, W_k), \quad k \in \mathbb{Z}_{N-1},$$

(8.29)

with variables $(u_0, X_1, \ldots, u_{N-1}, X_N)$, where u is adapted to X. This is clearly a multistage stochastic optimization problem, and it has a special structure that we will now exploit. We will from now on assume that the constraints have been used to eliminate the variables X_k, and hence, we only have to consider the variables $(u_0, \ldots, u_{N-1})$ in the optimization problem.

8.9.1 Stochastic Dynamic Programming

Similarly, as in Section 5.7, we may rewrite the minimization of the objective function above as

$$\mathbb{E}_{X_0}\left[\min_{u_0} \mathbb{E}_{X_1}\left[\min_{u_1} \mathbb{E}_{X_2}\left[\cdots \min_{u_{N-1}} \mathbb{E}_{X_N}\left[\sum_{k=0}^{N-1} f_k(X_k, u_k) + \phi(X_N) \right] \right] \right] \right],$$

where $\mathbb{E}_{X_k}$ denotes conditional expectation with respect to the probability distribution for X_k given $(X_0, \ldots, X_{k-1})$ for $k \in \mathbb{N}_N$, and where $\mathbb{E}_{X_0}$ is expectation with respect to X_0. Note that we do not have any dependence on u_N in the objective function, and hence, we do not need to minimize over u_N. We will skip the outer expectation since it does not affect the optimization problem, and we then assume that we know the value of the initial state to be $X_0 = x_0$. Hence, we consider

$$\min_{u_0} \mathbb{E}_{X_1}\left[\min_{u_1} \mathbb{E}_{X_2}\left[\cdots \min_{u_{N-1}} \mathbb{E}_{X_N}\left[\sum_{k=0}^{N-1} f_k(X_k, u_k) + \phi(X_N) \right] \right] \right],$$

for all values of x_0. Once the optimal value of the objective function is known for all values of x_0, we can of course compute the mean over the probability distribution for X_0. We make use of the additivity of the objective function and rewrite the minimization of the objective function as

$$\min_{u_0} \mathbb{E}_{X_1}\left[f_0(X_0, u_0) + \min_{u_1} \mathbb{E}_{X_2}\left[f_1(X_1, u_1) \cdots \min_{u_{N-1}} \mathbb{E}_{X_N}\left[f_{N-1}(X_{N-1}, u_{N-1}) + \phi(X_N) \right] \right] \right].$$

We now make use of the properties of the conditional expectations to obtain

$$\min_{u_0} \mathbb{E}_{X_1}\left[f_0(x_0, u_0) + \min_{u_1} \mathbb{E}_{X_2}\left[f_1(x_1, u_1) \cdots \min_{u_{N-1}} \mathbb{E}_{X_N}\left[f_{N-1}(x_{N-1}, u_{N-1}) + \phi(X_N) \right] \right] \right].$$

We consider first the innermost optimization problem which is equivalent to

$$\min_{u_{N-1}} \mathbb{E}_{X_N}\left[f_{N-1}(x_{N-1}, u_{N-1}) + \phi(F_{N-1}(x_{N-1}, u_{N-1}, W_{N-1})) \right],$$

after substitution of the dynamic equation. Since W_k is independent of X_l for $l \leq k$ it follows that we equivalently have

$$\min_{u_{N-1}} \mathbb{E}_{W_{N-1}}\left[f_{N-1}(x_{N-1}, u_{N-1}) + \phi(F_{N-1}(x_{N-1}, u_{N-1}, W_{N-1})) \right],$$

where $\mathbb{E}_{W_k}$ denotes the expectation with respect to the probability distribution for W_k.[7] We now realize a very important fact. When we carry out the optimization above with respect to u_{N-1}, this will only be a function of x_{N-1} and not of any previous values of x_k for $k < N - 1$. This will also be true when we continue to optimize over u_{N-2}, i.e. it will only depend on x_{N-2}, and so on. Hence, the stochastic optimal control problem can be solved with the following *stochastic dynamic programming recursion*. Define the value functions $V_k : \mathcal{D}^n \to \mathbb{R}$ as $V_N(x) = \phi(x)$ and

$$V_k(x) = \min_u \mathbb{E}\left[f_k(x, u) + V_{k+1}(F_k(x, u, W_k))\right], \quad k \in \mathbb{Z}_{N-1},$$

where the optimal control is $u_k^\star = \mu_k(x_k)$ where $\mu_k : \mathcal{D}^n \to \mathcal{E}^m$ for $k \in \mathbb{Z}_{N-1}$ is given by

$$\mu_k(x) = \operatorname*{argmin}_u Q_k(x, u),$$

and where $Q_k : \mathcal{D}^n \times \mathcal{E}^m \to \mathbb{R}$ is the Q-function in the stochastic setting and defined as

$$Q_k(x, u) = \mathbb{E}\left[f_k(x, u) + V_{k+1}(F_k(x, u, W_k))\right], \tag{8.30}$$

for $k \in \mathbb{Z}_{N-1}$. Note that we do not have to write out any subindex for the expectation operator, *cf.* the discussion at the end of Section 3.6. The functions μ_k defines a feedback policy or control policy also for the stochastic optimal control problem. Because of this, X is a Markov process, which we did not define it to be to start off with. Such a Markov process is often called a *Markov decision process*. We remark that we can move the expectation to just affect the function V_k since the incremental cost does not depend on W_k. We then realize that we could have let the incremental cost also depend on W_k, and this is the way Markov decision processes are sometimes presented. However, it should be stressed that for the stochastic setting of the optimal control problem we have to postulate either a feedback policy or that u is adapted to X. This was not the case for the deterministic setting, where we obtained a feedback policy without postulating it or anything else. We summarize the stochastic dynamic programming recursion in Algorithm 8.4. We next consider an example,

---

Algorithm 8.4: Stochastic dynamic programming

Input: Final state penalty ϕ, incremental costs f_k, functions F_k, distributions for W_k

Output: V_k for $k \in \mathbb{Z}_{N-1}$

$V_N(x, u) \leftarrow \phi(x)$
for $k \leftarrow N - 1$ **to** 0 **do**
$\quad | \quad V_k(x) \leftarrow \min_u \mathbb{E}\left[f_k(x, u) + V_{k+1}(F_k(x, u, W_k))\right]$
end

---

which is a stochastic version of the LQ control problem in Example 8.1.

Example 8.10 Consider the problem in (8.29) for the case when the dynamic equation is linear, i.e. $F_k(x, u, w) = A_k x + B_k u + w$, where $A_k \in \mathbb{R}^{n \times n}$ and $B_k \in \mathbb{R}^{n \times m}$ and where $\mathcal{D} = \mathcal{E} = \mathcal{F} = \mathbb{R}$. We assume that W is a zero mean random process with W_i independent of W_j for $i \neq j$ with variance $\mathbb{E}\left[W_k W_k^T\right] = \Sigma_k \in \mathbb{S}_+^n$. We assume that the incremental costs and the final cost are quadratic functions given by $f_k(x, u) = x^T S_k x + u^T R_k u$ and $\phi(x) = x^T S_N x$, respectively, where $R_k \in \mathbb{S}_{++}^m$ and $S_k \in \mathbb{S}_+^n$. Application of the stochastic dynamic programming recursion gives $V_N(x) = x^T S_N x$ and

$$V_k(x) = \min_u Q_k(x, u),$$

---

7 In some applications, it will be convenient to let W_k depend on X_k and then we should take expectation with respect to the conditional probability function for W_k given $X_k = x_k$ instead.

where

$$Q_k(x, u) = x^T S x + u^T R u + \mathbb{E}\left[V_{k+1}(A_k x + B_k u + W_k)\right].$$

We will make the guess that $V_k(x) = x^T P_k x + r_k$ for some $P_k \in \mathbb{S}^n$ and some $r_k \in \mathbb{R}$. This is clearly true for $k = N$ with $P_N = S_N$ and $r_N = 0$. We now assume that it is true for $k + 1$. It then holds that

$$Q_k(x, u) = \begin{bmatrix} x \\ u \end{bmatrix}^T \begin{bmatrix} S_k + A_k^T P_{k+1} A_k & A_k^T P_{k+1} B_k \\ B_k^T P_{k+1} A_k & R_k + B_k^T P_{k+1} B_k \end{bmatrix} \begin{bmatrix} x \\ u \end{bmatrix} + \operatorname{tr} P_{k+1} \Sigma_k + r_{k+1}.$$

As in Example 8.1, it holds that

$$u = -\left(R_k + B_k^T P_{k+1} B_k\right)^{-1} B_k^T P_{k+1} A_k x,$$

minimizes $Q_k(x, u)$. Back substitution of this expression for u results in the following expression for the right-hand side of the dynamic programming recursion:

$$x^T \left\{ S_k + A_k^T P_{k+1} A_k - A_k^T P_{k+1} B_k \left(R_k + B_k^T P_{k+1} B_k\right)^{-1} B_k^T P_{k+1} A_k \right\} x + \operatorname{tr} P_{k+1} \Sigma_k + r_{k+1}.$$

It now follows that our assumption is also true for k if we define P_k as

$$P_k = S_k + A_k^T P_{k+1} A_k - A_k^T P_{k+1} B_k \left(R_k + B_k^T P_{k+1} B_k\right)^{-1} B_k^T P_{k+1} A_k,$$

and $r_k = \operatorname{tr} P_{k+1} \Sigma_k + r_{k+1}$. This defines a recursion for the matrices P_k which are the same discrete-time Riccati recursion that we had in the nonstochastic setting in Example 8.1, which we have already proven is well defined. Hence, the stochastic problem has the same solution as the deterministic problem. The only difference is that the optimal value of the problem is different.

Generally, if the optimal policy is unaffected when a disturbance such as W is replaced with its mean, we say that *certainty equivalence* holds.

Example 8.11 Consider the problem of ordering a quantity $u_k \in \mathbb{R}_+$ of an item at periods k ranging from 0 to $N - 1$ such that a stochastic demand $W_k \in \mathbb{R}$ is met. Denote by $X_k \in \mathbb{R}$ the stock available at the beginning of period k. Assume that W_k are independent random variables. The stock evolves as

$$X_{k+1} = X_k + u_k - W_k.$$

There is a cost $r : \mathbb{R} \to \mathbb{R}$, which is a function of X_k, for keeping stock. Moreover, there is a purchasing cost $c u_k$, where $c > 0$. The objective is to minimize

$$\mathbb{E}\left[r(X_N) + \sum_{k=0}^{N-1} \left(r(X_k) + c u_k\right) \right],$$

with respect to u_k. We assume that r is a convex function that is bounded from below and such that $r(x) \to \infty$ as $|x| \to \infty$. Moreover, we assume that

$$\lim_{x \to -\infty} \frac{dr(x)}{dx} < -c.$$

The stochastic dynamic programming recursion is

$$V_k(x) = \min_{u \geq 0} \mathbb{E}\left[r(x) + cu + V_{k+1}(x + u - W_k) \right].$$

Let us define the functions $\varphi_k : \mathbb{R}^2 \to \mathbb{R}$ as

$$\varphi_k(x, u) = \mathbb{E}\left[V_{k+1}(x + u - W_k) \right].$$

We notice that if V_{k+1} is a convex function in x, then φ_k is a convex function in (x, u). This follows since the argument $z = x + u - W_k$ is an affine transformation and since taking expectation of a function preserves convexity. Under the same assumption, it now follows that

$$r(x) + cu + \varphi_k(x, u)$$

is a convex function of (x, u), since it is the sum of convex functions. Moreover, it is bounded from below if φ_k is bounded from below and

$$\lim_{u \to -\infty} \frac{d\varphi_k(x, u)}{du} < -c.$$

This is implied by V_{k+1} being bounded from below and

$$\lim_{x \to -\infty} \frac{dV_{k+1}(x)}{dx} < -c.$$

Under these assumptions, it follows by Exercise 4.9 that

$$\min_{u \geq 0} \left(cu + \varphi_k(x, u) \right)$$

is a convex function that it is bounded from below. Since V_k is the sum of this function and r, which is convex and bounded from below, it follows that also V_k is convex and bounded from below. The fact that also

$$\lim_{x \to -\infty} \frac{dV_k(x)}{dx} < -c$$

follows from the same property for r and that V_k is the sum of r and a function that is bounded from below. For $k + 1 = N$, we have that $V_N(x) = r(x)$ and hence, V_N is convex, bounded from below and satisfies

$$\lim_{x \to -\infty} \frac{dV_N(x)}{dx} < -c.$$

By induction, it now follows that V_k is convex, bounded from below and satisfies

$$\lim_{x \to -\infty} \frac{dV_k(x)}{dx} < -c,$$

for all k. Since $cu + \varphi_k(x, u)$ is bounded from below, it has an unconstrained minimum that we denote by u_k^0. The optimal $u_k^\star$ is obtained by projecting u_k^0 onto the set $[0, \infty)$. Hence, if the constraint was not present, it would be desirable to order u_k^0. Let $s_k = x_k + u_k^0$, and we can write $u_k^0 = s_k - x_k$. Then we realize that

$$u_k^\star = \mu_k(x_k) = \begin{cases} s_k - x_k, & x_k \leq s_k, \\ 0, & x_k > s_k, \end{cases}$$

where we can interpret s_k as a target value, i.e. as long as the current stock is above the target value we do not need to make any order. Otherwise, we should order the difference.

Example 8.12 We consider the problem of accepting offers $W_k \in \mathbb{R}$ for selling an asset, where W_k are independent random variables for $k \in \mathbb{Z}_N$. If we accept an offer, we can invest the money at a fixed interest rate of $r > 0$ for the remaining period of time. Otherwise, we wait for the next offer, and make a new decision. This is called an *optimal stopping problem*. We can cast this as a stochastic optimal control problem by letting the state space be $\mathcal{D} = \mathbb{R} \cup \{t\}$, where the element t denotes termination of the offers. We let the control space be $\mathcal{E} = \{u^0, u^1\}$, where u^0 denotes keeping the asset, and where u^1 denotes selling the asset. We let the state evolve as

$$X_{k+1} = F(X_k, u_k, W_k) = \begin{cases} t, & \text{if } X_k = t \text{ or } X_k \neq t \text{ and } u_k = u^1, \\ W_k, & \text{otherwise.} \end{cases}$$

The reward at the final time is

$$\phi(x) = \begin{cases} x, & x \neq t, \\ 0, & x = t, \end{cases}$$

which means that we have to accept the last offer if we have rejected all the previous ones. The incremental rewards are

$$f_k(x, u) = \begin{cases} (1 + r)^{N-k}x, & x \neq t \text{ and } u = u^1, \\ 0, & \text{otherwise.} \end{cases}$$

Hence, there will be at most one nonzero incremental reward. The stochastic dynamic programming recursion is

$$V_k(x) = \begin{cases} \max_u \left(f_k(x, u) + \mathbb{E}\left[V_{k+1}(F(x, u, W_k)) \right] \right), & x \neq t, \\ 0, & x = t. \end{cases}$$

Note that the value function must be zero if $x = t$ since it is equal to the optimal cost to go from k, and if we have already sold the asset, we cannot expect any future reward. It now follows that for $x \neq t$, it holds that

$$V_k(x) = \max \left\{ 0 + \mathbb{E}\left[V_{k+1}(W_k) \right], (1 + r)^{N-k}x + 0 \right\},$$

where the first element corresponds to $u = u^0$ and the second element corresponds to $u = u^1$. Hence, we should sell if

$$(1 + r)^{N-k}x \geq \mathbb{E}\left[V_{k+1}(W_k) \right],$$

or equivalently, if the offer x is such that

$$x \geq \frac{\mathbb{E}\left[V_{k+1}(W_k) \right]}{(1 + r)^{N-k}},$$

where the right-hand side is the expected reward discounted to the present time. It can be shown that the right-hand side of the inequality is decreasing with k if W_k are identically distributed, and then if an offer is good at k, it is also good for later times, see Exercise 8.15.

8.9.2 Infinite Time Horizon

It is possible to also extend the results for infinite horizon optimal control in Section 8.3 to a stochastic setting. We consider an *infinite horizon stochastic optimal control problem* defined as

$$\text{minimize } \mathbb{E}\left[\sum_{k=0}^{\infty} \gamma^k f(X_k, u_k) \right], \tag{8.31}$$

$$\text{subject to } X_{k+1} = F(X_k, u_k, W_k), \quad k \in \mathbb{Z}_+,$$

with variables $(u_0, X_1, u_1, X_2, \dots)$, where $0 < \gamma < 1$ is a discount factor. Here, $F : \mathcal{D}^n \times \mathcal{E}^m \times \mathcal{F}^p \rightarrow \mathcal{D}^n$ and $f : \mathcal{D}^n \times \mathcal{E}^m \rightarrow \mathbb{R}$. Note that we do not allow for $\gamma = 1$, which we did in the deterministic setting. We assume that W is a zero mean weakly stationary random process with W_i independent of W_j for $i \neq j$, cf. Section 3.9, and we assume that the initial value $X_0 = x_0$ is known. We now have that the incremental costs and the state-dynamics are independent of the stage index k. It is assumed that $F(0, 0, 0) = 0$ so that zero is a stationary point of the dynamical equation. If this is not the case, we can make a change of coordinates to make this hold. We also assume that the incremental cost f is such that $f(0, 0) = 0$. Without this assumption, it is not possible to obtain *closed loop stability*, in the sense that (X_k, u_k) is bounded in mean square, i.e. the second moments of these random variables are bounded for all $k \geq 0$. To simplify the presentation below, we will also restrict ourselves to the case when the incremental cost is strictly positive definite. We define $J^\star : \mathcal{D}^n \to \mathbb{R}_+$ to be the minimal value for the optimization problem in (8.31) for initial value x, and we define $u_k^\star$ to be the optimizing sequence of decisions or control signals that achieves this minimal value.

8.9.3 Stochastic Bellman Equation

Assume that the above conditions are satisfied and that there exists a solution $V : \mathcal{D}^n \to \mathbb{R}_+$ such that $c_1 \|x\|_2^2 \leq V(x) \leq c_2 \|x\|_2^2 + c_3$ for constants $c_1, c_2, c_3 > 0$ to the *stochastic Bellman equation*

$$V(x) = \min_u \mathbb{E}\left[f(x, u) + \gamma V(F(x, u, W)) \right]. \tag{8.32}$$

Here, $W = W_k$ for an arbitrary k. Then $V(x) = J^\star(x)$ and with the Q-function $Q : \mathcal{D}^n \times \mathcal{E}^m \to \mathbb{R}$ defined as

$$Q(x, u) = \mathbb{E}\left[f(x, u) + \gamma V(F(x, u, W)) \right], \tag{8.33}$$

it holds that $u_k^\star = \mu(X_k)$, where $\mu : \mathcal{D}^n \to \mathcal{E}^m$ is given by

$$\mu(x) = \underset{u}{\operatorname{argmin}}\ Q(x, u) \tag{8.34}$$

is an optimal feedback control. If in addition γ is sufficiently close to one, this feedback results in closed-loop stability in the sense defined above. The proof of this result is given in Section 8.10. We next consider an example which is called *infinite horizon stochastic LQ control*.

Example 8.13 Let us consider the case when $F(x, u, w) = Ax + Bu + w$ for matrices $A \in \mathbb{R}^{n \times n}$ and $B \in \mathbb{R}^{n \times m}$. We also assume that $f(x, u) = x^T S x + u^T R u$, where $S \in \mathbb{S}_{++}^n$ and $R \in \mathbb{S}_{++}^m$, and that W_k are independent and define a weakly stationary random process with zero mean and with covariance $\Sigma \in \mathbb{S}_+^n$. Clearly, we satisfy the assumptions on the functions f and F. We will make the guess that $V(x) = x^T P x + r$ for some $P \in \mathbb{S}_{++}^n$ and some $r \geq 0$. Then

$$Q(x, u) = x^T S x + u^T R u + \gamma (Ax + Bu)^T P (Ax + Bu) + \gamma (\operatorname{tr} P\Sigma + r).$$

As in Example 8.4, Q is minimized for

$$u = -\gamma (R + \gamma B^T P B)^{-1} B^T P A x.$$

Back substitution of this results in

$$x^T P x + r = x^T \left(S + \gamma A^T P A - \gamma^2 A^T P B (R + \gamma B^T P B)^{-1} B^T P A \right) x$$
$$+ \gamma (\operatorname{tr} P\Sigma + r).$$

This equation holds if P is the solution to the discounted algebraic Riccati equation in (8.11), and if $r = \operatorname{tr}(P\Sigma)/(1 - \gamma)$. We see that certainty equivalence holds also for infinite horizon stochastic LQ control.

It is possible to extend the results on VI and PI to the stochastic setting in a straightforward manner. What is needed to prove these extensions are the monotonicity and the contraction properties of the stochastic versions of the Bellman operator and the Bellman policy operator, see Exercise 8.16. Also, the LP formulation can be extended to the stochastic setting.

8.10 Appendix

8.10.1 Stability and Optimality of Infinite Horizon Problem

Consider the infinite time horizon optimal control problem in (8.9). Assume that $F(0, 0) = 0$ and that $f(0, 0) = 0$. Also, assume that f is strictly positive definite. If there exists a strictly positive definite and quadratically bounded solution $V : \mathcal{D}^n \to \mathbb{R}_+$ to

$$V(x) = \min_{u \in U(x)} \{f(x, u) + \gamma V(F(x, u))\}, \tag{8.35}$$

and if γ is sufficiently close to one, then $V(x) = J^\star(x)$, where we recall that $J^\star(x)$ is the optimal value of the problem in (8.9) for initial value x. Moreover, $u_k^\star = \mu(x_k)$, where $\mu : \mathcal{D}^n \to \mathcal{E}^m$ with

$$\mu(x) = \operatorname*{argmin}_{u \in U(x)} \{f(x, u) + \gamma V(F(x, u))\}$$

is an optimal feedback control that results in closed-loop stability.

The proof of this result goes as follows: we have from the Bellman equation that

$$V(x_{k+1}) = \frac{1}{\gamma} \left(V(x_k) - f(x_k, \mu(x_k)) \right), \tag{8.36}$$

where $x_{k+1} = F(x_k, \mu(x_k))$. From this it follows that

$$V(x_{k+1}) \leq \frac{1}{\gamma} \left(V(x_k) - \epsilon \left(\|x_k\|_2^2 + \|\mu(x_k)\|_2^2 \right) \right) \leq \frac{1}{\gamma} \left(1 - \epsilon/c_2 \right) V(x_k),$$

where we in the first inequality have used the fact that there exists $\epsilon > 0$ such that $f(x, u) \geq \epsilon \left(\|x\|_2^2 + \|u\|_2^2 \right)$, and where we in the second inequality have used the fact that there exists c_2 such that $V(x) \leq c_2 \|x\|_2^2$. This implies that

$$V(x_N) \leq \frac{1}{\gamma^N} (1 - \epsilon/c_2)^N V(x_0),$$

which converges to zero as $N \to \infty$ if $\gamma > (1 - \epsilon/c_2)$. This is always true if we take γ sufficiently close to one. Since there exists $c_1 > 0$ such that $c_1 \|x\|_2^2 \leq V(x)$, we obtain that

$$\|x_N\|_2^2 \leq \frac{c_2}{c_1 \gamma^N} (1 - \epsilon/c_2)^N \|x_0\|_2^2.$$

From this, it follows that

$$\|\mu(x_N)\|_2^2 \leq \frac{1}{\epsilon} f(x_N, \mu(x_N)) \leq \frac{1}{\epsilon} V(x_N) \leq \frac{c_2}{\epsilon} \|x_N\|_2^2,$$

where the first inequality follows from f being strictly positive definite, the second inequality follows from (8.36), and the third inequality follows from V being quadratically bounded. This proves closed loop stability. We now consider any stabilizing sequence of u_k, i.e. u_k such that $x_{k+1} = F(x_k, u_k) \to 0$ as $k \to \infty$ for any initial value x_0. From the Bellman equation, it follows that

$$f(x_k, u_k) \geq V(x_k) - \gamma V(x_{k+1}),$$

and that

$$\lim_{N \to \infty} \sum_{k=0}^{N-1} \gamma^k f(x_k, u_k) \geq \lim_{N \to \infty} \sum_{k=0}^{N-1} \gamma^k \left(V(x_k) - \gamma V(x_{k+1}) \right),$$

$$= V(x_0) - \lim_{N \to \infty} \gamma^N V(x_N) = V(x_0),$$

where the last equality follows from the fact that u_k is stabilizing and that $V(0) = 0$. Since the above inequality holds with equality for $u_k = \mu(x_k)$ by the Bellman equation, we have that $u_k = \mu(x_k)$ is optimal for the infinite-horizon optimal control problem.

8.10.2 Stability and Optimality of Stochastic Infinite Time Horizon Problem

We next consider the stochastic case and discuss how the proof has to be adapted to this setting. Consider the infinite time horizon stochastic optimal control problem in (8.31). Assume that $F(0, 0, 0) = 0$ and that $f(0, 0) = 0$. Also assume that f is strictly positive definite. If there exists a solution $V : \mathcal{D}^n \to \mathbb{R}_+$ such that $c_1 \|x\|_2^2 \leq V(x) \leq c_2 \|x\|_2^2 + c_3$ for $c_1, c_2, c_3 > 0$ to (8.32) and if γ is sufficiently close to one, then $V(x) = J^\star(x)$, where we recall that $J^\star(x)$ is the optimal value of the problem in (8.31) for initial value x. Moreover, $u_k^\star$ as defined in (8.34) is an optimal feedback control that results in closed-loop stability in mean square sense.

The proof of this result goes as follows: we have from the stochastic Bellman equation in (8.32) and (8.34) that

$$\mathbb{E}\left[V(X_{k+1})\right] = \frac{1}{\gamma}\left(V(X_k) - f(X_k, \mu(X_k))\right), \tag{8.37}$$

where $X_{k+1} = F(X_k, \mu(X_k), W_k)$, and where $\mathbb{E}$ denotes expectation with respect to W_k. From this it follows that

$$\mathbb{E}\left[V(X_{k+1})\right] \leq \frac{1}{\gamma}\left(V(X_k) - \epsilon\left(\|X_k\|_2^2 + \|\mu(X_k)\|_2^2\right)\right) \leq \frac{1}{\gamma}\left(1 - \epsilon/c_2\right)V(X_k) + \frac{\epsilon c_3}{\gamma c_2},$$

where we in the first inequality have used the fact that there exists $\epsilon > 0$ such that $f(x, u) \geq \epsilon\left(\|x\|_2^2 + \|u\|_2^2\right)$, and where we in the second inequality have used the fact that there exists c_2, c_3 such that $V(x) \leq c_2 \|x\|_2^2 + c_3$. Let us from now on denote the expectation with respect to $(W_0, W_1, \ldots)$ by $\mathbb{E}$. With $V_k = \mathbb{E}\left[V(X_k)\right]$, $a = (1 - \epsilon/c_2)/\gamma$ and $b = \epsilon c_3/(\gamma c_2)$ it then holds that

$$V_{k+1} \leq aV_k + b,$$

where $0 < a < 1$ for γ sufficiently close to one. From this, it follows that

$$V_k \leq a^k V_0 + (a^{k-1} + a^{k-2} + \cdots + 1)b,$$

and hence, by taking the limit of the sum that

$$V_k \leq a^k V_0 + \frac{b}{1-a}, \quad \forall k \geq 0.$$

Since $V_0 = V(x_0)$ is bounded, we have that $V_k \leq c$ for some finite c and all $k \geq 0$. Since there exists $c_1 > 0$ such that $c_1 \|x\|_2^2 \leq V(x)$, we obtain

$$\mathbb{E}\|X_k\|_2^2 \leq \frac{c}{c_1}, \quad \forall k \geq 0.$$

We also have that

$$\|\mu(X_k)\|_2^2 \leq \frac{1}{\epsilon}f(X_k, \mu(X_k)) \leq \frac{1}{\epsilon}V(X_k) \leq \frac{1}{\epsilon}(c_2 \|X_k\|_2^2 + c_3),$$

where the first inequality follows from f being strictly positive definite, the second inequality follows from (8.37) since $\mathbb{E}V(X_{k+1}) \geq 0$, and the third inequality follows from the bound on V. By taking expectation above, it follows that both $\mathbb{E}\|X_k\|_2^2$ and $\mathbb{E}\|\mu(X_k)\|_2^2$ are bounded for all $k \geq 0$, i.e. closed-loop stability holds. We now consider any sequence $u = (u_0, u_1, \ldots)$ such that u_k and $X_{k+1} = F(X_k, u_k, W_k)$ have bounded second moments for all $k \geq 0$. From (8.32), it follows that

$$\mathbb{E}f(X_k, u_k) \geq \mathbb{E}\left[V(X_k) - \gamma V(X_{k+1})\right],$$

and that

$$\lim_{N\to\infty} \mathbb{E}\left[\sum_{k=0}^{N-1} \gamma^k f(X_k, u_k)\right] \geq \lim_{N\to\infty} \mathbb{E}\left[\sum_{k=0}^{N-1} \gamma^k \left(V(X_k) - \gamma V(X_{k+1})\right)\right],$$

$$= V(x_0) - \lim_{N\to\infty} \gamma^N \mathbb{E}V(X_N) = V(x_0),$$

where the last equality follows from the fact that $\mathbb{E}V(X_N) \leq c_2\, \mathbb{E}\|X_N\|_2^2 + c_2 < \infty$ for all $N \geq 0$ and that $\gamma < 1$, and where the penultimate equality follows from the fact that $X_0 = x_0$ is known. Since the above inequality holds with equality for $u_k = \mu(X_k)$ by the stochastic Bellman equation, we have that $u_k = \mu(X_k)$ is optimal for the infinite horizon stochastic optimal control problem.

8.10.3 Stability of MPC

We will now discuss stability of MPC. If $\left(\tilde{u}_k^\star, \dots, \tilde{u}_{k+N-1}^\star\right)$ is optimal for the time k problem in (8.25), then $\left(\tilde{u}_{k+1}^\star, \dots, \tilde{u}_{k+N-1}^\star, 0\right)$ is feasible for the time $k+1$ problem. This follows from the following argument. We notice that the suggested control signal for the time $k+1$ problem together with the optimal choice made at time k will result in exactly the same control signals and states as the ones that are optimal for the time k problem. These are also feasible for the time $k+1$ problem up until and including time $k+N-1$ for the control signal and up until and including time $k+N$ for the state. Then applying $\tilde{u}_{k+N} = 0$, which is feasible, will, since $\tilde{x}_{k+N} = 0$, result in $\tilde{x}_{k+N+1} = 0$, which hence is feasible for the time $k+1$ problem.

We are now able to state the stability result. Assume that (8.25) is feasible for $k = 0$ and that f is strictly positive definite. Then the MPC algorithm results in a convergent closed-loop trajectory, i.e. $(x_k, u_k) \to (0, 0)$ as $k \to \infty$. This is proven in the following way: with the feasible set of controls defined above it holds that

$$J_{k+1} = \sum_{i=k+1}^{k+N} f(\tilde{x}_i, \tilde{u}_i),$$

$$= \sum_{i=k}^{k+N-1} f(\tilde{x}_i, \tilde{u}_i) + f(\tilde{x}_{k+N}, \tilde{u}_{k+N}) - f(\tilde{x}_k, \tilde{u}_k),$$

$$= J_k^\star + f(0, 0) - f(\tilde{x}_k^\star, \tilde{u}_k^\star),$$

where $f(0, 0) = 0$ by assumption. Here, $J_k^\star$ denotes the optimal value of J_k. Hence,

$$0 \leq J_{k+1}^\star \leq J_{k+1} = J_k^\star - f(\tilde{x}_k^\star, \tilde{u}_k^\star).$$

It now follows that $f(\tilde{x}_k^\star, \tilde{u}_k^\star) \to 0$, $k \to \infty$, since otherwise, $J_k^\star \to -\infty$, which is a contradiction. Since f is strictly positive definite we have that

$$f(\tilde{x}_k^\star, \tilde{u}_k^\star) \geq \epsilon \left(\|\tilde{x}_k^\star\|^2 + \|\tilde{u}_k^\star\|^2\right),$$

for some $\epsilon > 0$, and hence, $(\tilde{x}_k^\star, \tilde{u}_k^\star) \to (0, 0)$, $k \to \infty$.

Exercises

8.1 Assume that we have a vessel whose maximum weight capacity is z and whose cargo is to consist of different quantities of N different items. Let v_k denote the value of the kth type of item, and let w_k denote the weight of the kth type of item.

(a) Let x_k be the used weight capacity of the vessel after the first $k-1$ items have been loaded and let the control u_k be the quantity of item k to be loaded on the vessel. Formulate the dynamic equation:

$$x_{k+1} = F_k(x_k, u_k),$$

describing the process.

(b) Determine the constraint set $U(k, x_k)$ for the control signal u_k.

(c) Formulate a DP recursion that solves the problem of finding the most valuable cargo satisfying the maximal weight capacity. Observe that you do *not* need to solve the problem.

8.2 Consider the problem

$$\text{maximize} \ \sum_{k=0}^{N-1} \beta^k \log(u_k),$$

$$\text{subject to} \ x_{k+1} = ax_k^\alpha - u_k \geq 0,$$

$$x_0 \geq 0 \text{ given},$$

where $0 < \alpha, \beta < 1$, and $a > 0$ are some constants. This is commonly referred to as the *consumption problem* in the theory of economics. The variable u_k may be interpreted as the consumption for time period k and x_k the available capital, which is assumed positive, at time period k, respectively. Find the optimal control signal u_k, where $k = 1, 0$, for the problem when the horizon is $N = 2$ using the dynamic programming algorithm.

8.3 A businessman operates out of a van that he sets up in one of the two locations each day. If he operates in location i day k, where $i = 1, 2$, he makes a known and predictable profit denoted r_i^k. However, each time he moves from one location to the other, he pays a setup cost c. The businessman wants to maximize his total profit over N days.

(a) The problem can be formulated as a shortest path problem (SPP), where node (k, i) is representing location i at day k. Let s and e be the start node and the end node, respectively. The costs of all edges are
   - s to i_1 with cost $-r_i^1$
   - i_k to i_{k+1}, i.e. no switch, with cost $-r_{i_{k+1}}^{k+1}, k = 1, \ldots, N-1$
   - i_k to $\bar{i}_{k+1}$, i.e. switch, with cost $c - r_{\bar{i}_{k+1}}^{k+1}, k = 1, \ldots, N-1$
   - i_N to e with cost 0,

   where $\bar{i}$ denotes the location that is not equal to i, i.e. $\bar{1} = 2$ and $\bar{2} = 1$. Draw a figure to illustrate the SPP and the definitions of variables and parameters. Write down the corresponding dynamic programming algorithm. Note that you do not have to solve the problem.

(b) Suppose the businessman is at location i on day $k-1$ and let

$$R_i^k = r_{\bar{i}}^k - r_i^k.$$

Show that if $R_i^k \leq 0$, it is optimal to stay at location i, while if $R_i^k \geq 2c$, it is optimal to switch. You can use the following lemma.

Lemma: For every $k = 1, 2, \ldots, N$, it holds that

$$|V_k(i) - V_k(\bar{i})| \leq c,$$

where $V_k(i)$ is the value of the optimal cost-to-go function at stage k for state i.

8.4 Consider the scalar discrete time system

$$x_{k+1} = x_k + u_k, \quad x_0 = 1,$$

together with the optimization criterion

$$J = |x_N| + \sum_{k=0}^{N-1} |u_k|.$$

(a) Consider the control law

$$u_k = -\alpha x_k, \quad 0 \le \alpha \le 1.$$

Show that the cost is independent of the choice of α. What is the cost?

Hint: The sum of a geometric series is $\sum_{k=0}^{n} a^k = \frac{1-a^{n+1}}{1-a}$.

(b) Use, e.g. dynamic programming to show that the optimal cost-to-go satisfies

$$J_k^\star(x_k) = |x_k|,$$

for all $k \in \mathbb{Z}_N$. What can be concluded about optimality/suboptimality of the control law in (a)?

8.5 Consider the problem of packing a knapsack with items labeled $k \in \mathbb{N}_N$ of different value v_k and weight w_k. There is only one copy of each item, and the task is to pack the knapsack such that the sum of the values of the items in the knapsack is maximized subject to the constraint that the total weight of the items is less than or equal to the weight limit W. The problem can be posed as a multistage decision problem in which at each stage k, it is decided whether item k should be loaded or not. This decision can be coded in terms of the binary variable $u_k \in \{0, 1\}$, where $u_k = 1$ in case item k is loaded and $u_k = 0$, otherwise. If x_k denotes the total weight of the knapsack after the first $k - 1$ items have been loaded, then the following relation holds:

$$x_{k+1} = x_k + w_k u_k, \quad x_1 = 0,$$

for $k \in \mathbb{N}_N$. The constraint that $x_{k+1} \le W$ can be reformulated in terms of u_k as $u_k \le (W - x_k)/w_k$. From this, it follows that it is possible to calculate how to load the knapsack in an optimal way using the dynamic programming recursion

$$V_k(x) = \max_{u \le (W-x)/w_n, u \in \{0,1\}} \{v_n u + V_{k+1}(x + w_n u)\},$$

with final value $V_{N+1}(x) = 0$. We will consider the case when $W = 10$ and when the values of v_k and w_k are defined as in the table below, where $N = 5$.

k	1	2	3	4	5
v_k	2	8	7	1	3
w_k	1	5	4	3	2

For this case, we notice that $x_k \in \{0, 1, \dots, 10\}$.

(a) Compute the values of $V_k(x)$ and the maximizing argument $u = \mu_k(x)$ for each value of $k = 1, \dots, N$ as a table in terms of the values of x. The tables should look like

x	0	1	2	3	4	5	6	7	8	9	10
$V_k(x)$											
$\mu_k(\cdot, x)$											

(b) From the tables derived in (a) compute the optimal loading of the knapsack.

8.6 Consider the optimal control problem

$$\text{minimize } \sum_{k=0}^{\infty} x_k^2 + u_k^2,$$
$$\text{subject to } x_{k+1} = x_k + u_k,$$
$$x_0 \text{ given},$$
$$u_k \in [-1, 1].$$

(a) Compute an optimal feedback policy $u_k = \mu(x_k)$ for this problem when the control signal constraint is neglected.
   Hint: Try the value function $V(x) = px^2$ with $p > 0$.
(b) Now, consider the case with constraints on the control signal. Compute an approximative solution by solving

$$\text{minimize}_{-1 \leq u \leq 1} \left\{ x^2 + u^2 + V(x + u) \right\},$$

   where $V(x)$ is defined as in the hint above.
(c) Prove that the closed-loop system using the feedback of the previous subproblem is stable.

8.7 Show that the Bellman policy operator in (8.15) is a monotone operator and that it is a contraction when $\gamma < 1$.

8.8 Consider the infinite time horizon optimal control problem

$$\text{minimize } \sum_{k=0}^{\infty} f(x_k, u_k),$$
$$\text{subject to } x_{k+1} = F(x_k, u_k),$$

with given initial vale x_0. You may assume that $f(x, u) = \rho x^2 + u^2$ and $F(x, u) = x + u$, where x and u are scalar-valued, and where $\rho > 0$.

(a) Compute the optimal feedback policy using the Bellman equation by guessing that $V(x) = px^2$ for some p.
(b) In general, it is more tricky to solve the Bellman equation, and different iterative procedures are available. Use VI, i.e. let $V_k(x)$ be iteratively computed from

$$V_{k+1}(x) = \min_u \left\{ f(x, u) + V_k(F(x, u)) \right\},$$

   for $k = 1, 2, \ldots$ with $V_0(x) = 0$. Show that $V_k(x) = p_k x^2$ and that the minimizing argument is $l_k x_k$, where $l_k = -p_k/(p_k + 1)$ and where

$$p_{k+1} = \rho + p_k/(p_k + 1), \quad p_0 = 0.$$

(c) Now, instead use PI. In this method, one starts with an initial feedback policy $\mu_0(x)$ and repeats the following two steps iteratively for $k = 0, 1, 2, \ldots$:
   1. Compute $V_k(x)$ such that

$$V_k(x) = f(x, \mu_k(x)) + V_k(f(x, \mu_k(x))).$$

   2. Compute $\mu_{k+1}(x)$ as the minimizing argument in

$$\min_u \left\{ f(x, u) + V_k(F(x, u)) \right\}.$$

Assume that $\mu_0(x) = l_0 x$ and that $V_k(x) = p_k x^2$. Show that $\mu_{k+1}(x) = l_{k+1} x$, where now

$$p_k = \frac{\rho + l_k^2}{1 - (1 + l_k)^2},$$

and

$$l_{k+1} = -p_k/(p_k + 1),$$

with l_0 given.

(d) Compute the sequences p_k and l_k in (b) and (c) for $k = 1, 2, \ldots, 5$ when $\rho = 0.5$. Assume that $l_0 = -0.1$ for the method in (c). The iterates will converge to the solution in (a). Which method converges the fastest?

8.9 In this exercise, we consider the LQ problem in Example 8.4. We will specifically consider the case when $m = 1$ and $n = 2$, and the matrices

$$A = \begin{bmatrix} 0.5 & 1 \\ 0 & 0.5 \end{bmatrix}, \quad B = \begin{bmatrix} 0 \\ 1 \end{bmatrix}, \quad R = 1 \text{ and } S = I.$$

(a) Implement the Hewer iterations in Example 8.6. You may start with $L_0 = 0$. Specifically investigate how many iterations are needed for convergence of L_k.

(b) Implement the approach in Example 8.7. You may start with $L_0 = 0$. How does the choice of initial values x^s and the number r of initial values affect the convergence of L_k. How does the choice of N affect the convergence.

8.10 In this exercise, we investigate how to compute the explicit solution to the MPC problem using the multiparametric toolbox MPT3 for MATLAB; see https://www.mpt3.org/. Consider a second-order system

$$x_{k+1} = \begin{bmatrix} 0.8584 & -0.0928 \\ 0.0464 & 0.9976 \end{bmatrix} x_k + \begin{bmatrix} 0.0928 \\ 0.0024 \end{bmatrix} u_k,$$

$$y_k = \begin{bmatrix} 0 & 1 \end{bmatrix} x_k.$$

(8.38)

This has been obtained from a continuous time system with transfer function

$$\frac{2}{s^2 + 3s + 2},$$

by zero-order hold sampling with a sampling time of 0.05.

(a) Use MPT3 to calculate the explicit MPC for (8.38) using the weighted 2-norm for the incremental costs defined by $Q = I$, $R = 0.01$, when $N = 2$ and for the following constraints

$$-1 \leq u_k \leq 1 \quad \text{and} \quad \begin{bmatrix} -10 \\ -10 \end{bmatrix} \leq x_k \leq \begin{bmatrix} 10 \\ 10 \end{bmatrix}.$$

How many regions are there? Present plots of the partitioning of the state-space for the control signal, and for the time trajectory for the initial condition $x_0 = (1, 1)$.

(b) Find the number of regions in the control law for $N = 1, 2, \ldots, 13$. How does the number of regions depend on N, e.g. is the complexity polynomial or exponential? Estimate the order of the complexity by computing

$$\min_{\alpha, \beta} \|\alpha N^\beta - n_r\|_2^2 \quad \text{and} \quad \min_{\gamma, \delta} \|\gamma(\delta^N - 1) - n_r\|_2^2,$$

where n_r denotes the number of regions. Discuss the result.

> *Hint:* Useful commands: `MPCController`, `ctrl.toExplicit`, `QuadFunction`, `expctrl.evaluate`, `ctrl.evaluate`, `expctrl.feedback.fplot`, `expctrl.cost.fplot`, `expctrl.partition.plot` and `lsqnonlin`.

8.11 [14, Exercise 1.14] A farmer annually producing X_k units of a certain crop stores $(1 - u_k)X_k$ units of his production, where $u_k \in [0, 1]$, and invests the remaining $u_k X_k$ units, which increase the production for next year to a level of X_{k+1} according to

$$X_{k+1} = X_k + W_k u_k X_k,$$

where W_k are i.i.d. random variables that are independent of X_k and u_k. Moreover, it holds that $\mathbb{E}W_k = \bar{w} > 0$. The total expected product stored over N years is given by

$$\mathbb{E}\left[X_N + \sum_{k=1}^{N-1}(1 - u_k)X_k \right].$$

Show that the optimal solution that maximizes the expected product stored is independent of x_k and given by

$$\begin{cases} \mu_0(x_0) = \cdots = \mu_{N-1}(x_{N-1}) = 1, & \bar{w} > 1, \\ \mu_0(x_0) = \cdots = \mu_{N-1}(x_{N-1}) = 0, & 0 < \bar{w} < 1/N, \\ \mu_0(x_0) = \cdots = \mu_{N-\bar{k}-1}(x_{N-\bar{k}-1}) = 1, \ \mu_{N-\bar{k}}(x_{n-\bar{k}}) = \cdots = \mu_{N-1}(x_{N-1}) = 0, & 1/N \leq \bar{w} \leq 1, \end{cases}$$

where $\bar{k}$ is such that $1/(\bar{k} + 1) < \bar{w} < 1/\bar{k}$.

8.12 [14, Exercise 1.15] Consider the following random process:

$$\begin{aligned} X_{k+1} &= (1 - \delta_k)X_k + u_k \gamma_k X_k, \\ Y_{k+1} &= (1 - \delta_k)Y_k + (1 - u_k)\gamma_k X_k, \end{aligned}$$

for $k \in \mathbb{Z}_N$, where δ_k are i.i.d. random variables with values in $[\delta, \delta']$, where $0 < \delta \leq \delta' < 1$, where γ_k are i.i.d. random variables with values in $[a, b]$, where $a > 0$. We assume that $0 < \alpha \leq u_k \leq \beta < 1$. We may interpret X_k as the number of educators in a certain country at time k and Y_k as the number of research scientists. By means of incentives, a science policy maker can determine the proportion u_k of new scientists produced at time k who become educators. The initial values of X_0 and Y_0 are known. We will derive the optimal policy for maximizing $\mathbb{E}Y_N$.
(a) Show that the value functions are given by $V_k(x, y) = c_k x + d_k y$ for some $c_k, d_k \in \mathbb{R}$.
(b) Derive the optimal policy when $\mathbb{E}\gamma_k > \mathbb{E}\delta_k$, and show that the optimal policy $\mu_k(x, y)$ is independent of x and y.

8.13 Consider a scalar linear random process defined by

$$X_{k+1} = a_k X_k + b u_k + W_k, \quad k = 0, 1, \dots, N,$$

where $a_k, b_k \in \mathbb{R}$, and where W_k are i.i.d. random variables with a Gaussian distribution with zero mean and variance $\sigma^2 < 1$. We are interested in minimizing the objective function

$$\mathbb{E}\left[\exp\left(\frac{1}{2}x_N^2 + \frac{1}{2}\sum_{k=0}^{N-1}(x_k^2 + ru_k^2) \right) \right],$$

where $r \in \mathbb{R}_{++}$. Derive the optimal control law assuming that it is adapted to X.

Hint: Start by showing that the following recursion for value functions $V_k : \mathbb{R} \to \mathbb{R}$,

$$V_k(x) = \min_u \mathbb{E}_{W_k} \left[\exp\left(x^2 + ru^2\right) V_{k+1}\left(a_x x + b_k u + W_k\right) \right],$$

with final value $V_N(x) = x^2/2$ provides the optimal solution. Also remember the results in Exercise 3.10.

8.14 [14, Exercise 4.12] We want to run a machine to produce a certain item to meet a known demand $d_k \in \mathbb{R}, k \in \mathbb{Z}_N$. The machine can be in a bad (B) state or a good (G) state. The state of the machine evolves according to the transition probabilities

$$\mathbb{P}[\,G\mid G\,] = \lambda_G, \quad \mathbb{P}[\,B\mid G\,] = 1 - \lambda_G, \quad \mathbb{P}[\,B\mid B\,] = \lambda_B, \quad \mathbb{P}[\,G\mid B\,] = 1 - \lambda_B,$$

where $\lambda_B, \lambda_G \in [0, 1]$. Denote by $X_k \in \mathbb{R}, k \in \mathbb{Z}_N$, the stock in period k. If the machine is in a good state at period k, it can produce $u_k \in [0, \bar{u}]$, where $\bar{u} > 0$ is a known constant, and the stock evolves as

$$X_{k+1} = X_k + u_k - d_k.$$

Otherwise, it evolves as

$$X_{k+1} = X_k - d_k.$$

(a) Let $Y_k \in \{0, 1\}$ be a random process defined as

$$Y_{k+1} = W_k,$$

where $W_k \in \{0, 1\}$ is defined via the conditional probabilities

$$\mathbb{P}\big[\, W_k = 1 \mid Y_k = 1 \,\big] = \lambda_G,$$
$$\mathbb{P}\big[\, W_k = 0 \mid Y_k = 1 \,\big] = 1 - \lambda_G,$$
$$\mathbb{P}\big[\, W_k = 0 \mid Y_k = 0 \,\big] = \lambda_B,$$
$$\mathbb{P}\big[\, W_k = 1 \mid Y_k = 0 \,\big] = 1 - \lambda_B.$$

Show that we may summarize the updates for $Z_k = (X_k, Y_k)$ as

$$Z_{k+1} = F(Z_k, u_k, W_k),$$

where

$$F(Z_k, u_k, W_k) = \begin{bmatrix} X_k + Y_k u_k - d_k \\ W_k \end{bmatrix}.$$

(b) Consider the following objective function:

$$\mathbb{E}\left[\sum_{k=0}^{N} g(X_k) \right],$$

where $g : \mathbb{R} \to \mathbb{R}$ is a convex function bounded from below and such that $g(x) \to \infty$ as $|x| \to \infty$. The objective function should be minimized. Show that the value functions in the stochastic dynamic programming recursion are convex functions in x and that there for each k is a target stock level S_{k+1} such if the machine is in good state, it is optimal to produce $u_k^\star \in [0, \bar{u}]$ that will bring X_{k+1} as close as possible to S_{k+1}.

Hint: Remember the footnote about when W_k depends on X_k in the derivation of the stochastic dynamic programming recursion.

8.15 Let us consider the problem in Example 8.12. Let

$$a_k = \frac{\mathbb{E}_{W_k}\left[V_{k+1}(W_k)\right]}{(1+r)^{N-k}},$$

and show that $a_k \geq a_{k+1}$ for all $k = 0, 1, \ldots, N-1$. We assume that W_k are identically distributed.

8.16 (a) Define the stochastic Bellman operator $\mathcal{T} : \mathbb{R}^{\mathcal{D}^n} \to \mathbb{R}^{\mathcal{D}^n}$ as

$$\mathcal{T}(V) = \min_u \mathbb{E}\left[f(x, u) + \gamma V(F(x, u, W))\right],$$

where the assumptions on f, γ, F, and W are the same as in Section 8.9. Show that if $V_1(x) \leq V_2(x)$ for all $x \in \mathcal{D}^n$, then

$$\mathcal{T}(V_1) - \mathcal{T}(V_2) \leq 0.$$

Also show that $\mathcal{T}$ is a contraction.

(b) Define the stochastic Bellman policy operator $\mathcal{T}_\mu : \mathbb{R}^{\mathcal{D}^n} \to \mathbb{R}^{\mathcal{D}^n}$ as

$$\mathcal{T}_\mu(V) = \mathbb{E}\left[f(x, \mu(x)) + \gamma V(F(x, \mu(x), W))\right],$$

where the assumptions on f, γ, F, and W are the same as in Section 8.9. Show that if $V_1(x) \leq V_2(x)$ for all $x \in \mathcal{D}^n$, then

$$\mathcal{T}_\mu(V_1) - \mathcal{T}_\mu(V_2) \leq 0.$$

Also, show that $\mathcal{T}_\mu$ is a contraction.

Part IV

Learning

9

Unsupervised Learning

We are now going to discuss unsupervised learning. This is about finding lower-dimensional descriptions of a set of data $\{x_1, \ldots, x_N\}$. One simple such lower-dimensional description is the mean of the data. Another one could be to find a probability function from which the data are the outcome. We will see that there are many more lower-dimensional descriptions of data. We will start the chapter by defining entropy, and we will see that many of the probability density functions that are of interest in learning can be derived from the so-called "maximum entropy principle." Specifically, we will derive the categorical distribution, the Ising distribution, and the normal distribution. There is a close relationship between the Lagrange dual function of the maximum entropy problem and maximum likelihood (ML) estimation, which will also be investigated. Other topics that we cover are prediction, graphical models, cross entropy, the expectation maximization algorithm, the Boltzmann machine, principal component analysis, mutual information, and cluster analysis. As a prelude to entropy we will start by discussing the so-called *Chebyshev bounds*.

9.1 Chebyshev Bounds

Consider a probability space $(\Omega, \mathcal{F}, \mathbb{P})$ and a random variable $X : \Omega \to S \subseteq \mathbb{R}^n$. In this section, we will bound $\mathbb{P}[\, X \in C \,]$ for some set C using the knowledge of what some expectations related to X are. We let the pdf of X be $p : S \to \mathbb{R}_+$. We assume that we know that

$$\mathbb{E}[\, f_i(X) \,] = \int_S f_i(x)p(x)dx = a_i, \quad i \in \mathbb{Z}_m,$$

where $f_i : \mathbb{R}^n \to \mathbb{R}$ with $f_0(x) = 1$ and $a_0 = 1$, and where $a_i \in \mathbb{R}$ are given. Now, suppose we are interested in finding a p that maximizes $\mathbb{P}[\, X \in C \,]$, where $C \subseteq S$ is given. This is the same as solving

$$\text{maximize} \quad \int_C p(x)dx,$$
$$\text{subject to} \quad \int_S p(x)f_i(x)dx = a_i, \quad i \in \mathbb{Z}_m,$$
$$p(x) \geq 0, \quad \forall\, x \in S,$$

with variable p. Let $\mathcal{D} = \{p \in \mathbb{R}^{\mathbb{R}^n} \mid \int_S p(x)dx = 1,\ p(x) \geq 0\}$, and let the Lagrangian $L : \mathcal{D} \times \mathbb{R}^{m+1} \to \mathbb{R}$ be defined as

$$L[p, \lambda] = \int_C p(x)dx + \sum_{i=0}^{m} \lambda_i \left(a_i - \int_S p(x)f_i(x)dx \right).$$

Optimization for Learning and Control, First Edition. Anders Hansson and Martin Andersen.
© 2023 John Wiley & Sons, Inc. Published 2023 by John Wiley & Sons, Inc.
Companion Website: www.wiley.com/go/opt4lc

We then have that

$$\sup_p L[p, \lambda] = \sum_{i=0}^m \lambda_i a_i + \sup_p \left(\int_C (1 - f(x, \lambda)) p(x) dx - \int_{S \setminus C} f(x, \lambda) p(x) dx \right),$$

where $f : \mathbb{R}^n \times \mathbb{R}^{m+1} \to \mathbb{R}$ is given by $f(x, \lambda) = \sum_{i=0}^m \lambda_i f_i(x)$. Moreover, we have that

$$\sup_p L[p, \lambda] \le \sum_{i=0}^m \lambda_i a_i,$$

if

$$1 - \inf_{x \in C} f(x, \lambda) \le 0, \quad -\inf_{x \in S \setminus C} f(x, \lambda) \le 0.$$

Since $\mathbb{E}\left[f_i(X) \right] = a_i$, we also have that these conditions imply that $\sup_p \mathbb{P}[X \in C] \le \sum_{i=0}^m \lambda_i a_i$. We can therefore compute the smallest possible such upper bound by solving the dual problem

$$\text{minimize} \quad \sum_{i=0}^m a_i \lambda_i,$$

$$\text{subject to} \quad 1 - \inf_{x \in C} f(x, \lambda) \le 0,$$

$$-\inf_{x \in S \setminus C} f(x, \lambda) \le 0,$$

with variable λ. This is a convex optimization problem in λ, which follows by noting that

$$\inf_{x \in C} f(x, \lambda) = \inf_{x \in C} \sum_{i=0}^m \lambda_i f_i(x)$$

is the infimum over a family of linear functions of λ, and hence, it is a concave function of λ. The same argument applies to the function in the second constraint.

Example 9.1 Let $S = \mathbb{R}_+$, $C = [1, \infty)$, $f_0(x) = 1$ and $f_1(x) = x$. Assume that it is known that $\mathbb{E}f_1(X) = \mathbb{E}X = \mu$, where $0 \le \mu \le 1$. Then the so-called Markov bound

$$\mathbb{P}[X \ge 1] \le \mu$$

holds. The result is derived in Exercise 9.1.

9.2 Entropy

Let us consider a probability space $(\Omega, \mathcal{F}, \mathbb{P})$. Entropy measures the amount of uncertainty in a probability distribution. Assume that we observe the values of a random variable $X : \Omega \to \mathbb{R}$ for outcomes of experiments and estimate the mean of the random variable. What is then the most likely distribution for the random variable? The *maximum entropy principle* says that it is the one that maximizes the entropy among all possible probability distributions that have the same estimated mean. To formalize this, we consider a finite sample space $\Omega = \mathbb{N}_n$ and the set of probability functions defined by

$$\mathcal{D}_n = \left\{ p \in [0, 1]^n \mid p_k \ge 0, \ k \in \mathbb{N}_n; \ \sum_{k=1}^n p_k = 1 \right\},$$

cf. Chapter 3. Then *entropy* $H_n : \mathcal{D}_n \to \mathbb{R}$ can be defined in an axiomatic way. It should satisfy

1. $H_n(p) = H_n(\pi(p))$, where π is any permutation function.
2. $H_n(p) \le H_n(\bar{p})$, where $\bar{p} = (1/n, \dots, 1/n)$.

3. $H_n(p) = H_{n+1}(q)$, where $q = (p, 0)$.
4. If $r \in D_{mn}$ is a joint probability function with marginal probability functions $p \in D_n$ and $q \in D_m$, then $H_{mn}(r) = H_n(p) + \sum_{k:p_k \neq 0} p_k H_m(r_k / p_k)$, where $r_k = (r_{k1}, \ldots, r_{km})$, and where $r = (r_1, \ldots, r_n)$.

The first axiom says that entropy only depends on the distribution and not how we do the labeling when we define the probability function. The second axiom says that the uniform probability function has the highest entropy, i.e. the most uncertainty. The third axiom says that adding elements of probability zero does not change entropy. The fourth axiom says that the entropy for a joint probability function is the sum of the entropy for one of the marginal probability functions plus a weighted average of the entropy of the conditional probability function. This latter term is what is called conditional entropy. It was shown in [62] that any function H_n that satisfies the axioms has to be of the form

$$H_n(p) = -k \sum_{k=1}^{n} p_k \log p_k,$$

where k is a positive constant, and $\log(\cdot)$ is any logarithm. We use the convention $0 \times \log(0) = 0$ to have continuity at the origin. We immediately notice that the function is concave, since it is the sum of concave functions, where the concavity of each term is easily verified by computing its second derivative and verifying that it is negative. Because of this, maximum entropy problems are often tractable. A typical problem reads

$$\text{maximize} \quad - \sum_{k=1}^{n} p_k \ln p_k,$$
$$\text{subject to} \quad Ap = b,$$
$$p \in \Delta^n,$$

(9.1)

with variable $p \in \mathbb{R}^n$, where $A \in \mathbb{R}^{m \times n}$. Every row in the constraint $Ax = b$ is an expectation constraint. To see this, define random variables $X_1, \ldots, X_n$ such that X_i takes the values x_{ij} with probability p_j. Then the expected value of X_i is $\sum_{j=1}^{n} x_{ij} p_j$, which is the ith row of Ap if we let $A_{ij} = x_{ij}$. In applications, the right-hand side b could be obtained from empirical estimates of the expected values. A very simple example of a random variable X_i is one for which $x_{ij} = 1$ if $j = i$ and zero, otherwise. Hence, the expected value is p_j, and if the right-hand side is an empirical estimate of this expected value, we have effectively constrained p_j to be equal to its empirical estimate. If we instead take $A_{ij} = x_{ij}^2$, we see that we define a constraint on the second moment of the random variable. Similarly, we can define constraints on any moment of a random variable. There could also be additional inequality constraints on p, and as long as they are convex constraints, the tractability of the problem is not destroyed. One example is lower bounds on variances of random variables. Note the close connection to Example 5.2.

Example 9.2 In this example, we consider the static ranking of web pages. To formalize this, we let $V = \mathbb{N}_n$ represent a set of n web pages and define a graph $G = (V, E)$, where the edge set $E \subset V \times V$ contains directed edges (i, j) describing that there is a link from web page i to web page j. For an example see Figure 9.1, where $V = \{1, 2, 3, 4\}$, and $E = \{(1, 2), (2, 4), (4, 1), (4, 3), (3, 1)\}$. The most well-known way to model the ranking is called PageRank. It uses a Markov chain model, *cf.* Section 3.10, in which all outgoing links are assigned equal transition probability. The transition probability that a user at page i jumps to page j is given by

$$p_{ij} = \begin{cases} \dfrac{1}{d_i}, & (i, j) \in E, \\ 0, & (i, j) \notin E, \end{cases}$$

where d_i is the out-degree of node i, i.e. the number of edges $(i,j) \in E$, where $j \in V$. The PageRank is then defined as the stationary distribution of the Markov chain, i.e. as the solution of $\pi^T = \pi^T P$, where $P \in \mathbb{R}^{n \times n}$ is the matrix of transition probabilities p_{ij} at position i,j. This is an eigenvalue problem, and π is the normalized eigenvector corresponding to the eigenvalue that is one. For the example in Figure 9.1, we have $d = (1,1,1,2)$ and

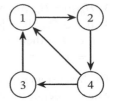

$$P = \begin{bmatrix} 0 & 1 & 0 & 0 \\ 0 & 0 & 0 & 1 \\ 1 & 0 & 0 & 0 \\ 1/2 & 0 & 1/2 & 0 \end{bmatrix},$$

Figure 9.1 Graph showing the links between different web pages.

with eigenvector $\pi = (0.2857, 0.2857, 0.1429, 0.2857)$. We see that web page number three has the lowest ranking.

Another way to model the transition probabilities is based on a network flow approach. Let y_{ij} be the number of users following link $(i,j) \in E$ per unit time. Assume that the web traffic is in a state of equilibrium so that the traffic out of a node is equal to the in-coming traffic per unit time, i.e.

$$\sum_{j:(i,j)\in E} y_{ij} = \sum_{j:(j,i)\in E} y_{ji}, \quad i \in V.$$

The number of "hits" per unit time at node $j \in V$ is then

$$H_j = \sum_{i:(i,j)\in E} y_{ij},$$

and the total number of hits per unit time is

$$Y = \sum_{(i,j)\in E} y_{i,j} = \sum_{j\in V} H_j.$$

We now define the probabilities $p_{ij} = y_{i,j}/Y$, which allow us to write the equilibrium condition as

$$\sum_{j:(i,j)\in E} p_{ij} = \sum_{j:(j,i)\in E} p_{ji}, \quad i \in V. \tag{9.2}$$

Note that these probabilities are not the transition probabilities in the PageRank model. They are obtained by normalization with $\sum_{j:(i,j)\in E} p_{ij}$.

One solution to (9.2) is

$$p_{ij} = \frac{H_i}{Y d_i}, \quad (i,j) \in E,$$

which agrees with the PageRank model after normalization. However, there are many more solutions to the equilibrium condition, which can be interpreted as moment constraints. The maximum entropy solution under the moment constraints is obtained by solving

$$\text{maximize} \quad -\sum_{(i,j)\in E} p_{ij} \ln p_{ij},$$

$$\text{subject to} \quad \sum_{j:(i,j)\in E} p_{ij} = \sum_{j:(j,i)\in E} p_{ji}, \quad i \in V,$$

$$\sum_{(i,j)\in E} p_{ij} = 1,$$

$$p_{ij} \geq 0, \quad (i,j) \in E,$$

with variables p_{ij}, $(i,j) \in E$. We will investigate this optimization problem more in Exercise 9.6.

9.2.1 Categorical Distribution

We will now study a family of distributions known as *categorical* distributions, which can be derived from the problem

$$
\begin{aligned}
\text{maximize} \quad & H_n(p), \\
\text{subject to} \quad & f^T p = b, \\
& \mathbb{1}^T p = 1, \\
& p \geq 0,
\end{aligned}
$$

with variable $p \in \mathbb{R}^n$, where $f \in \mathbb{R}^n$ and $b \in \mathbb{R}$ are given. Define the partial Lagrange function $L : \mathbb{R}^n \times \mathbb{R} \times \mathbb{R} \to \mathbb{R}$ as

$$
L(p, \lambda, \mu) = H_n(p) + \lambda(b - f^T p) + \mu(1 - \mathbb{1}^T p),
$$

where $\lambda \in \mathbb{R}$ and $\mu \in \mathbb{R}$ are Lagrange multipliers for the equality constraints. This is a concave function of p. To find the dual function, we want to maximize this function subject to $p \geq 0$. The partial derivatives of L with respect to p_k vanish if

$$
\frac{\partial L}{\partial p_k} = -\ln p_k - 1 - \lambda f_k - \mu = 0.
$$

Hence,

$$
p_k = \frac{e^{-\lambda f_k}}{e^{1+\mu}},
$$

which satisfies $p_k \geq 0$ for any values of the Lagrange multipliers. Substituting the right-hand for p_k in the Lagrange function results in the Lagrange dual function $g : \mathbb{R} \times \mathbb{R} \to \mathbb{R}$ given by

$$
g(\lambda, \mu) = \frac{\sum_{k=1}^n e^{-\lambda f_k}}{e^{1+\mu}} + \lambda b + \mu.
$$

This is a convex function and minimizing it will provide us with the values of λ and μ that are needed in the expressions for p_k. For reasons that will become clear later, we first minimize with respect to μ. This can be done by setting the partial derivative with respect to μ equal to zero, i.e.

$$
-\frac{\sum_{k=1}^n e^{-\lambda f_k}}{e^{1+\mu}} + 1 = 0.
$$

From this, we obtain $e^{1+\mu} = \sum_{k=1}^n e^{-\lambda f_k} = \Phi(\lambda)$, where $\Phi : \mathbb{R} \to \mathbb{R}_{++}$. Solving for μ and substituting this expression for μ into the Lagrange dual function results in the function $h : \mathbb{R} \to \mathbb{R}$ defined as

$$
h(\lambda) = \lambda b + \ln \Phi(\lambda).
$$

We may now minimize this function by setting the partial derivative equal to zero, i.e.

$$
b + \frac{\partial \Phi}{\partial \lambda} / \Phi(\lambda) = b - \frac{\sum_{k=1}^n f_k e^{-\lambda f_k}}{\Phi(\lambda)} = 0,
$$

which is equivalent to

$$
\sum_{k=1}^n \left(f_k - b \right) e^{-\lambda f_k} = 0.
$$

Solving this equation with respect to λ will give us the pf. We now realize that if we do not want to parameterize the probability function in terms of the expected value b, then we can do it in terms of λ, i.e.

$$
p_k = \frac{e^{-\lambda f_k}}{\sum_{l=1}^n e^{-\lambda f_l}}. \tag{9.3}
$$

The parameter λ is called the *natural parameter*. The probability function we have derived belongs to the family of exponential probability functions. The distribution is known under several different names: the *categorical, Gibbs,* or *Boltzmann* distribution. Note that we may normalize such that

$$
p_k = \begin{cases} \dfrac{e^{z_k}}{1 + \sum_{l=1}^{n-1} e^{z_l}}, & k \in \mathbb{N}_{n-1}, \\[3mm] \dfrac{1}{1 + \sum_{l=1}^{n-1} e^{z_l}}, & k = n, \end{cases}
$$

where $z_k = \lambda(f_n - f_k)$, $k \in \mathbb{N}_{n-1}$, and this is the form of the categorical probability function that is used in logistic regression, which is discussed in Section 10.4.

Example 9.3   A construction company is ordering lumber every month. The lumber comes in three different grades. The construction company cannot decide which quality to order, but it can observe that the prices are different per unit. The different prices are $f_1 = 1$ for the lowest grade, $f_2 = 1.1$ for the middle grade, and $f_3 = 1.2$ for the highest grade, respectively. They have also observed that on average the price is $b = 1.05$. We can then use the maximum entropy principle to estimate what the probabilities are that low-, middle-, or high-grade lumber is delivered. Let p_1 be the probability that low-grade lumber is delivered, p_2 that medium-grade lumber is delivered, and p_3 that high-grade lumber is delivered. We then have the following moment constraint:

$$
1 \times p_1 + 1.1 \times p_2 + 1.2 \times p_3 = 1.05.
$$

From the function

$$
G(\lambda) = (1 - 1.05)e^{-\lambda \times 1} + (1.1 - 1.05)e^{-\lambda \times 1.1} + (1.2 - 1.05)e^{(-\lambda \times 1.2)},
$$

which is shown in Figure 9.2, we see that $G(\lambda)$ is zero for $\lambda = 6.9$, and hence, we can compute the probabilities from (9.3) to be

$$
p_1 = 0.5386, \quad p_2 = 0.2701, \quad p_3 = 0.1913.
$$

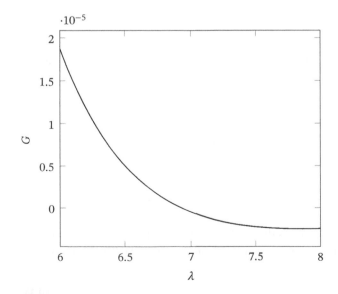

Figure 9.2   Plot of the function $G(\lambda)$.

9.2.2 Ising Distribution

We are now interested in m-dimensional vectors X of binary-valued random variables. To this end, we introduce the bijection $\mathcal{D}_x = \{0, 1\}^m \leftrightarrow \mathbb{N}_n$, where $n = 2^m$. The bijection is defined such that for $x \in \mathcal{D}_x$ and $k \in \mathbb{N}_n$ it holds that $k = \sum_{l=1}^n x_l 2^{l-1}$, where $x = (x_1, \dots, x_m)$ with $x_l \in \{0, 1\}$. With abuse of notation, we let $p(x) = p_k$, where as before $p = (p_1, \dots, p_n) \in \mathcal{D}_n$. From this, we realize that the entropy equivalently can be expressed as $H_n(p) = -\sum_{x \in \mathcal{D}_x} p(x) \ln (p(x))$.

We are now interested in the probability function that maximizes the entropy with constraints on the first moments of each component of X and on the cross-moments between the different components of X. This can be expressed as the problem

$$\text{maximize} \quad H_n(p),$$

$$\text{subject to} \quad \sum_{x \in \mathcal{D}_x} x p(x) = m,$$

$$\sum_{x \in \mathcal{D}_x} x x^T p(x) = M, \tag{9.4}$$

$$\sum_{x \in \mathcal{D}_x} p(x) = 1,$$

$$p(x) \geq 0, \quad x \in \mathcal{D}_x,$$

with variable p, and where, with abuse of notation, m is the vector of first moments, and M is the matrix of second moments, for which we do not specify the diagonal. Ignoring the inequality constraints, we introduce the partial Lagrangian $L : \mathbb{R}^n \times \mathbb{R}^m \times \mathbb{S}^m \times \mathbb{R} \to \mathbb{R}$ via

$$L(p, \lambda, \Lambda, \mu) = H_n(p) + \lambda^T \left(\sum_{x \in \mathcal{D}_x} x p(x) - m \right)$$

$$+ \text{tr} \left(\Lambda \left(\sum_{x \in \mathcal{D}_x} x x^T p(x) - M \right) \right) + \mu \left(\sum_{x \in \mathcal{D}_x} p(x) - 1 \right),$$

where Λ has a zero diagonal because of the unspecified diagonal in the constraints. This is a concave function of p. It holds that

$$\frac{\partial L}{\partial p(x)} = -\ln p(x) - 1 + \lambda^T x + \text{tr} \left(\Lambda x x^T \right) + \mu,$$

from which it follows that

$$p(x) = \exp \left(\lambda^T x + \text{tr}(\Lambda x x^T) - 1 + \mu \right),$$

and hence $p(x) \geq 0$. The Lagrange dual function $g : \mathbb{R}^m \times \mathbb{S}^m \times \mathbb{R} \to \mathbb{R}$ follows by inserting the expression for p into the Lagrangian. Minimizing this function with respect to μ results in choosing μ such that the probabilities sum up to one. With $A = 1 - \mu$ this holds if

$$A = \ln \left(\sum_{x \in \mathcal{D}_x} \exp \left(\lambda^T x + \text{tr} \left(\Lambda x x^T \right) \right) \right).$$

We see that A is a function of λ and Λ, and we, therefore, have $A : \mathbb{R}^m \times \mathbb{S}^m \to \mathbb{R}$. The resulting distribution is called the *Ising distribution*,[1] where λ and Λ are the natural parameters. We introduce

---

1 Strictly speaking, it is the random variable that is defined by the probability function that we have derived which has the Ising distribution.

the so-called *energy function* $E : \{0,1\}^m \rightarrow \mathbb{R}$ given by $E(x) = -\lambda^T x - \mathrm{tr}(\Lambda x x^T)$. We may then write the probability function as $p(x) = \exp\left(-E(x) - A(\lambda, \Lambda)\right)$, where $A(\lambda, \Lambda) = \ln\left(\sum_x \exp(-E(x))\right)$.

In order to relate the natural parameters to the moments m and M, we substitute $\mu = 1 - A(\lambda, \Lambda)$ into the Lagrange dual function and obtain the function $h : \mathbb{R}^m \times \mathbb{S}^m \rightarrow \mathbb{R}$ given by

$$h(\lambda, \Lambda) = A(\lambda, \Lambda) - \lambda^T m - \mathrm{tr}(\Lambda M). \tag{9.5}$$

We proceed to minimize this function with respect to (λ, Λ). The optimality conditions are

$$\frac{\partial h}{\partial \lambda} = \frac{\partial A}{\partial \lambda} - m = 0,$$

$$\frac{\partial h}{\partial \Lambda} = \frac{\partial A}{\partial \Lambda} - M = 0,$$

where

$$\frac{\partial A}{\partial \lambda} = \frac{\sum_{x \in D_x} x \exp\left(-E(x)\right)}{\sum_{x \in D_x} \exp\left(-E(x)\right)} = \frac{\sum_{x \in D_x} x \exp\left(-E(x) - A(\lambda, \Lambda)\right)}{\sum_{x \in D_x} \exp\left(-E(x) - A(\lambda, \Lambda)\right)},$$

$$= \sum_{x \in D_x} x p(x),$$

$$\frac{\partial A}{\partial \Lambda} = \frac{\sum_{x \in D_x} x x^T \exp\left(-E(x)\right)}{\sum_{x \in D_x} \exp\left(-E(x)\right)} = \frac{\sum_{x \in D_x} x x^T \exp\left(-E(x) - A(\lambda, \Lambda)\right)}{\sum_{x \in D_x} \exp\left(-E(x) - A(\lambda, \Lambda)\right)},$$

$$= \sum_{x \in D_x} x x^T p(x).$$

Here, we do not consider the equations related to the diagonal of the second equation. This is because we have constrained the diagonal of Λ to be zero. We see that the equations say that we should match the moments. To solve the above equations with respect to (λ, Λ) is not easy in general, and especially not when the dimension of x is large.

9.2.3 Normal Distribution

The concept of entropy can be generalized to probability distributions for continuous random variables. Given a pdf $p : \mathbb{R} \rightarrow \mathbb{R}_+$, the entropy is defined as

$$H(p) = -\int_{-\infty}^{+\infty} p(x) \log p(x) dx. \tag{9.6}$$

Here, we use p to denote a pdf instead of f as we did in Chapter 3. Most of the results from the discrete case carry over, and the family of distributions that are obtained by maximizing the entropy, irrespective of if they are related to discrete or continuous random variables, is called the *exponential family*. Some examples of distributions that belong to the family are the *Gaussian* or *normal* distribution, the *exponential* distribution and the *Poisson distribution*.

We will now derive the normal distribution as an example. This is obtained by maximizing the entropy subject to constraints on first and second moments of the distribution. Let D be the space of real-valued functions defined on $\mathbb{R}^n$ and consider $p \in D$ subject to

$$p(x) \geq 0, \quad \int_{\mathbb{R}^n} p(x) dx = 1, \quad \int_{\mathbb{R}^n} x p(x) dx = m, \quad \text{and} \quad \int_{\mathbb{R}^n} x x^T p(x) dx = M,$$

where $m \in \mathbb{R}^n$ and $M \in \mathbb{S}_+$ are the first and second moments of the distribution, respectively. The maximum entropy problem is

$$\text{maximize } H(p),$$

$$\text{subject to } \int_{\mathbb{R}^n} x p(x) dx = m,$$

$$\int_{\mathbb{R}^n} x x^T p(x) dx = M, \tag{9.7}$$

$$\int_{\mathbb{R}^n} p(x) dx = 1,$$

$$p(x) \geq 0, \quad x \in \mathbb{R}^n,$$

with variable $p \in \mathcal{D}$, which is a generalization of the problem we discussed in Example 7.3. We define the partial Lagrangian functional $L : \mathcal{D} \times \mathbb{R} \times \mathbb{S}^n \times \mathbb{R} \to \mathbb{R}$ by

$$L[p, \lambda, \Lambda, \mu] = -\int_{\mathbb{R}^n} p(x) \ln p(x) dx + \lambda^T \left(m - \int_{\mathbb{R}^n} x p(x) dx \right)$$

$$+ \frac{1}{2} \text{tr} \left(\Lambda \left(M - \int_{\mathbb{R}^n} x x^T p(x) dx \right) \right)$$

$$+ \mu \left(1 - \int_{\mathbb{R}^n} p(x) dx \right),$$

where we have ignored the constraint $p(x) \geq 0$. The first variation of the Lagrangian is

$$\delta L[\delta p] = -\int_{\mathbb{R}^n} \left(\ln p + 1 + \lambda^T x + \frac{1}{2} x^T \Lambda x + \mu \right) \delta p \, dx,$$

which should be nonpositive for all δp when p is optimal. Hence, the expression in the parenthesis must vanish by the du Bois Raymond lemma, see Section 7.1, and the optimal pdf is

$$p(x) = e^{-1-\mu-\lambda^T x - \frac{1}{2} x^T \Lambda x}, \tag{9.8}$$

which clearly is nonnegative. We will in Section 9.9 verify that p is indeed maximizing the entropy and not merely is a stationary point. The Lagrange dual function $g : \mathbb{R} \times \mathbb{S}^n \times \mathbb{R} \to \mathbb{R}$ is defined by

$$g(\lambda, \Lambda, \mu) = \int_{\mathbb{R}^n} e^{-1-\mu-\lambda^T x - \frac{1}{2} x^T \Lambda x} \, dx + \lambda^T m + \frac{1}{2} \text{tr}(\Lambda M) + \mu,$$

$$= \int_{\mathbb{R}^n} e^{-1-\mu+\frac{1}{2}\lambda^T \Lambda^{-1} \lambda - \frac{1}{2}(x+\Lambda^{-1}\lambda)^T \Lambda (x+\Lambda^{-1}\lambda)} \, dx$$

$$+ \lambda^T m + \frac{1}{2} \text{tr}(\Lambda M) + \mu,$$

where the second equality follows from completing the squares and assuming that Λ is invertible. We will later on see under which assumption we have invertibility. The Lagrange dual function is a convex function, and we determine the μ that minimizes it by setting the partial derivative of g with respect to μ equal to zero. We find that $p(x)$ should integrate to one, i.e.

$$\int_{\mathbb{R}^n} p(x) dx = \int_{\mathbb{R}^n} e^{-1-\mu+\frac{1}{2}\lambda^T \Lambda^{-1} \lambda - \frac{1}{2}(x+\Lambda^{-1}\lambda)^T \Lambda (x+\Lambda^{-1}\lambda)} \, dx,$$

$$= \frac{2^{n/2} e^{-1-\mu+\frac{1}{2}\lambda^T \Lambda^{-1} \lambda}}{\sqrt{\det \Lambda}} \int_{\mathbb{R}} e^{-\bar{x}_1^2} d\bar{x}_1 \times \cdots \times \int_{\mathbb{R}} e^{-\bar{x}_n^2} d\bar{x}_n,$$

$$= \frac{e^{-1-\mu+\frac{1}{2}\lambda^T \Lambda^{-1} \lambda} (2\pi)^{n/2}}{\sqrt{\det \Lambda}} = 1.$$

Hence, $\mu = \frac{1}{2}\lambda^T\Lambda^{-1}\lambda - 1 - \frac{1}{2}\ln\left(\frac{\det\Lambda}{(2\pi)^n}\right)$ from which, we obtain

$$p(x) = \sqrt{\frac{\det\Lambda}{(2\pi)^n}}e^{-\frac{1}{2}(x+\Lambda^{-1}\lambda)^T\Lambda(x+\Lambda^{-1}\lambda)}.$$

If we insert the optimal μ into the Lagrange dual function, we can define a function $h : \mathbb{R}^n \times \mathbb{S}^n \rightarrow \mathbb{R}$ as

$$h(\lambda, \Lambda) = g\left(\lambda, \Lambda, \frac{1}{2}\lambda^T\Lambda^{-1}\lambda - 1 - \frac{1}{2}\ln\left(\frac{\det\Lambda}{(2\pi)^n}\right)\right),$$

$$= \frac{1}{2}\lambda^T\Lambda^{-1}\lambda - \frac{1}{2}\ln\left(\frac{\det\Lambda}{(2\pi)^n}\right) + \lambda^T m + \frac{1}{2}\text{tr}(\Lambda M).$$

Note that this is also a convex function. In order to find λ and Λ, we minimize this function, which can be done by setting the derivatives with respect to λ and Λ equal to zero, i.e.

$$\frac{\partial h}{\partial \lambda} = \Lambda^{-1}\lambda + m = 0,$$

$$\frac{\partial h}{\partial \Lambda} = -\frac{1}{2}\Lambda^{-1}\lambda\lambda^T\Lambda^{-1} - \frac{1}{2}\Lambda^{-1} + \frac{1}{2}M = -\frac{1}{2}mm^T - \frac{1}{2}\Lambda^{-1} + \frac{1}{2}M = 0,$$

where the second equality in the second equation follows from the first equation. Noting that $\Sigma = M - mm^T$ is the covariance matrix, we immediately have that $\Lambda = \Sigma^{-1}$, and hence, the invertibility of Λ is equivalent to the covariance matrix belonging to $\mathbb{S}_{++}^n$. Moreover, we have $\lambda = -\Sigma^{-1}m$ and

$$p(x) = \frac{1}{\sqrt{(2\pi)^n \det\Sigma}}e^{-\frac{1}{2}(x-m)^T\Sigma^{-1}(x-m)}.$$

Note that we may just as well use the natural parameter (λ, Λ) instead of (m, Σ). If h is expressed in terms of (m, Σ) instead of in terms of (λ, Λ) it will not be a convex function. As we will see, this makes it convenient to use the natural parameters.

9.3 Prediction

Let $f_{X,Y} : \mathbb{R}^m \times \mathbb{R}^n \rightarrow \mathbb{R}_+$ be the joint pdf of two random variables X and Y with marginal pdfs $f_X : \mathbb{R}^m \rightarrow \mathbb{R}_+$ and $f_Y : \mathbb{R}^n \rightarrow \mathbb{R}_+$, and suppose we are given an observation x of X and would like to *predict* a value y for Y. The fact that X and Y are not independent will be utilized. Clearly, we could just compute $f_{Y|X} : \mathbb{R}^m \times \mathbb{R}^n \rightarrow \mathbb{R}_+$, the conditional pdf for Y given X, which is defined as

$$f_{Y|X}(y|x) = \frac{f_{X,Y}(x,y)}{f_X(x)},$$

cf. (3.7). This relationship is the foundation for the *Bayesian approach* to statistics, a topic we return to in Section 10.3. In Section 3.11, we derived the conditional pdf for an hidden Markov model (HMM). If one wants a single value for the prediction, one may consider the argument y of $f_{Y|X}(y|x)$ that maximizes it for the observation x, i.e.

$$\hat{y} = \underset{y}{\text{argmax}}\{f_{Y|X}(y|x)\}.$$

This is called the *maximum a posteriori* (MAP) estimate. The reason for the name a posteriori is that the conditional pdf is the pdf for Y resulting after we observe that $X = x$.

9.3.1 Conditional Expectation Predictor

Another possibility is to look for a predictor $g : \mathbb{R}^m \to \mathbb{R}^n$ that minimizes the expected squared error of the prediction, i.e.

$$\text{minimize } \mathbb{E}\left[(Y - g(X))^T (Y - g(X))\right], \tag{9.9}$$

with the function g as variable. This criterion is often called the *mean squared error* (MSE). We will now show that this infinite-dimensional optimization problem has a simple solution in terms of a conditional expectation. To this end, write the above objective function as

$$\int_{\mathbb{R}^{m+n}} f_{X,Y}(x, y)(y - g(x))^T (y - g(x)) \, dx dy,$$

$$= \int_{\mathbb{R}^m} f_X(x) \int_{\mathbb{R}^n} f_{Y|X}(y|x) \left(y^T y - 2 g^T(x) y + g^T(x) g(x)\right) dy dx,$$

$$= \int_{\mathbb{R}^m} f_X(x) \left(\mathbb{E}\left[Y^T Y | X = x\right] - g(x)^T g(x) \right.$$

$$\left. + (g(x) - \mathbb{E}[Y|X = x])^T (g(x) - \mathbb{E}[Y|X = x])\right) dx,$$

where the second equality follows by completing the squares. Thus, the above integral is minimized by $g(x) = \mathbb{E}[Y|X = x]$, i.e. the conditional expectation of Y given $X = x$.

In general, neither the conditional pdf nor the conditional expectation is easy to compute. One important exception is when $f_{X,Y}$ is the normal pdf, i.e.

$$f_{X,Y}(z) = \frac{1}{\sqrt{(2\pi)^{m+n} \det \Sigma}} e^{-\frac{1}{2}(z-\mu)^T \Sigma^{-1}(z-\mu)},$$

where $z = (x, y)$, $\mu = (\mu_x, \mu_y)$ and where

$$\Sigma = \begin{bmatrix} \Sigma_x & \Sigma_{xy} \\ \Sigma_{xy}^T & \Sigma_y \end{bmatrix}.$$

From Example 3.6, we then have that the conditional pdf is given by

$$f_{Y|X}(y|x) = \frac{1}{\sqrt{(2\pi)^n \det \Sigma_{y|x}}} e^{-\frac{1}{2}(y-\mu_{y|x})^T \Sigma_{y|x}^{-1}(y-\mu_{y|x})},$$

where

$$\mu_{y|x} = \mu_y + \Sigma_{xy}^T \Sigma_x^{-1}(x - \mu_x), \quad \Sigma_{y|x} = \Sigma_y - \Sigma_{xy}^T \Sigma_x^{-1} \Sigma_{xy}. \tag{9.10}$$

The conditional expectation is given by $\mu_{y|x}$, which is an affine function of x. Note that the maximizing argument y of $f_{Y|X}(y|x)$ is $y = \mu_{y|x}$, and hence, we have shown that the conditional mean is also the MAP estimate for the normal distribution case. As a consequence, the Kalman filter in Section 3.11 provides the MAP estimate as well as the estimate that minimizes the MSE of the prediction when the noise sequences are Gaussian.

Another special case is a so-called *Gaussian mixture model*, which has a joint pdf of the form

$$f_{X,Y}(z) = \sum_{i=1}^{N} \alpha_i f_{X_i, Y_i}(z),$$

where

$$f_{X_i, Y_i}(z) = \frac{1}{\sqrt{(2\pi)^{m+n} \det \Sigma_i}} e^{-\frac{1}{2}(z-\mu_i)^T \Sigma_i^{-1}(z-\mu_i)},$$

where $z = (x, y)$, $\mu_i = (\mu_{x,i}, \mu_{y,i})$ and where

$$\Sigma_i = \begin{bmatrix} \Sigma_{x,i} & \Sigma_{xy,i} \\ \Sigma_{xy,i}^T & \Sigma_{y,i} \end{bmatrix}.$$

Here, $\alpha_i \geq 0$ and $\sum_{i=1}^{N} \alpha_i = 1$. The marginal pdf for X is then given by

$$f_X(x) = \sum_{i=1}^{N} \alpha_i f_{X_i}(x),$$

where

$$f_{X_i}(x) = \frac{1}{\sqrt{(2\pi)^m \det \Sigma_{x,i}}} e^{-\frac{1}{2}(x-\mu_{x,i})^T \Sigma_{x,i}^{-1}(x-\mu_{x,i})},$$

and hence, the conditional pdf is given by

$$f_{Y|X}(y|x) = \frac{\sum_{i=1}^{N} \alpha_i f_{X_i,Y_i}(x,y)}{\sum_{i=1}^{N} \alpha_i f_{X_i}(x)}.$$

It follows that the optimal predictor is given by

$$\mathbb{E}[Y|X = x] = x = \frac{\sum_{i=1}^{N} \alpha_i \int_{\mathbb{R}^n} y f_{X_i,Y_i}(x,y) dy}{\sum_{i=1}^{N} \alpha_i f_{X_i}(x)}.$$

We again make use of the fact that $f_{X_i,Y_i}(x,y) = f_{X_i}(x)f_{Y_i|X_i}(y|x)$ and obtain

$$\mathbb{E}[Y|X = x] = \frac{\sum_{i=1}^{N} \alpha_i f_{X_i}(x) \int_{\mathbb{R}^n} y f_{Y_i|X_i}(y|x) dy}{\sum_{i=1}^{N} \alpha_i f_{X_i}(x)} = \frac{\sum_{i=1}^{N} \alpha_i f_{X_i}(x) \mu_i(x)}{\sum_{i=1}^{N} \alpha_i f_{X_i}(x)},$$

where

$$\mu_i(x) = \mu_{y,i} + \Sigma_{xy,i}^T \Sigma_{x,i}^{-1}(x - \mu_{x,i})$$

are the linear predictors for Y_i given $X_i = x$. We see that the overall predictor of Y given $X = x$ is a convex combination of these linear predictors, where the weights are functions of x, and hence, it is a nonlinear predictor.

Example 9.4 We now consider the two-dimensional Gaussian mixture defined by

$$f_{X_1,Y_1}(x,y) = \mathcal{N}(0, I),$$
$$f_{X_2,Y_2}(x,y) = \mathcal{N}((1, 2), I),$$
$$f_{X_3,Y_3}(x,y) = \mathcal{N}((-2, 0), I),$$

with $\alpha_i = 1/3$, $i \in \mathbb{N}_3$. Figure 9.3 shows the level curves of the pdf together with the nonlinear predictor given by the conditional expectation.

9.3.2 Affine Predictor

Instead of looking for a general predictor g when minimizing the MSE, we now restrict ourselves to affine predictors $g(x) = Ax + b$ with $A \in \mathbb{R}^{m \times n}$ and $b \in \mathbb{R}^m$. Sometimes the affine predictor is called a *linear predictor*. The problem in (9.9) then becomes a single-stage stochastic optimization problem of the form (5.44). We will do this for a general pdf $f_{X,Y}$ that is not necessarily normal. It is

Figure 9.3 The level curves of the pdf for the Gaussian mixture in Example 9.4 together with the nonlinear predictor given by the conditional expectation.

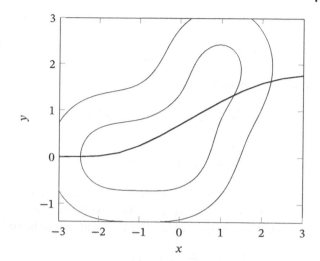

convenient to introduce $m_x = \mathbb{E}X$, $m_y = \mathbb{E}Y$ and new variables $\bar{X} = X - m_x$ and $\bar{Y} = Y - m_y$. We define $J : \mathbb{R}^{m \times n} \times \mathbb{R}^m \to \mathbb{R}_+$ as

$$
\begin{aligned}
J(A, b) &= \frac{1}{2}\mathbb{E}\left[(Y - AX - b)^T (Y - AX - b) \right] = \frac{1}{2}\mathrm{tr}\left(\mathbb{E}\left[(Y - AX - b)(Y - AX - b)^T \right] \right), \\
&= \frac{1}{2}\mathrm{tr}\left(\mathbb{E}\left[(\bar{Y} - A\bar{X} + m_y - Am_x - b)(\bar{Y} - A\bar{X} + m_y - Am_x - b)^T \right] \right), \quad\quad (9.11) \\
&= \frac{1}{2}\mathrm{tr}\left(\mathbb{E}\left[(\bar{Y} - A\bar{X})(\bar{Y} - A\bar{X})^T \right] \right) + \frac{1}{2}(m_y - Am_x - b)^T (m_y - Am_x - b),
\end{aligned}
$$

where we have used the fact that $\mathbb{E}\bar{X} = 0$ and $\mathbb{E}\bar{Y} = 0$. We have with $D_x = \mathbb{E}\left[\bar{X}\bar{X}^T \right]$ and $D_{xy} = \mathbb{E}\left[\bar{X}\bar{Y}^T \right]$ that the optimal A and b must satisfy

$$
\frac{\partial J}{\partial b} = -m_y + Am_x + b = 0,
$$

$$
\frac{\partial J}{\partial A} = -D_{xy}^T + AD_x - m_y m_x^T + b m_x^T + Am_x m_x^T = 0.
$$

Inserting the expression for b from the first equation into the second equation and simplifying results in $A = D_{xy}^T D_x^{-1}$ and $b = m_y - D_{xy}^T D_x^{-1} m_x$. It follows that

$$
Ax + b = m_y + D_{xy}^T D_x^{-1} (x - m_x),
$$

which is in agreement with the normal distribution case in (9.3.1). We have just replaced the moments of the normal distribution with the moments of the general distribution. Thus, if we are satisfied with the best linear predictor, we only need to know the first- and second-order moments of the distribution. The minimal value of J is $\mathrm{tr}\left(D_y - D_{xy}^T D_x^{-1} D_{xy} \right)$. This value is in general larger that the trace of the covariance of Y conditioned on $X = x$. The predictor for Y has the very nice property that

$$
\mathbb{E}\left[AX + b \right] = m_y + D_{xy}^T D_x^{-1}\mathbb{E}\left[X - m_x \right] = m_y,
$$

i.e. its expected value agrees with the expected value of Y. This is what is called an *unbiased* predictor.

Example 9.5   For a general Gaussian mixture, we have that

$$m_x = \sum_{i=1}^{N} \alpha_i \mu_{x,i}, \quad m_y = \sum_{i=1}^{N} \alpha_i \mu_{y,i}.$$

Moreover, we have that

$$D_x = \sum_{i=1}^{N} \alpha_i \left(\Sigma_{x,i} + \mu_{x,i}\mu_{x,i}^T - m_x m_x^T \right),$$

$$D_{xy} = \sum_{i=1}^{N} \alpha_i (\mu_{x,i} - m_x)(\mu_{y,i} - m_y)^T,$$

where the latter formula only holds when X_i and Y_i are independent. A formula for the general case of D_{xy} is more complicated. For the mixture in Example 9.4, it holds that $m_x = -1/3$, $m_y = 2/3$, $D_x = 71/27$, and $D_{xy} = 24/27$. Hence, the affine predictor is given by

$$Ax + b = \frac{2}{3} + \frac{24}{71}\left(x + \frac{1}{3}\right).$$

Figure 9.4 shows the affine predictor together with the nonlinear predictor and the level curves of the pdf.

9.3.3   Linear Regression

If we do not know the pdf for (X, Y), we may replace the expected value with its stochastic averaging approximation (SAA) as in (5.45) using observations (x_i, y_i), $i \in \mathbb{N}_N$, of the random variables (X, Y), i.e. we solve instead

$$\text{minimize } \frac{1}{2}\sum_{i=1}^{N}(y_i - Ax_i - b)^T(y_i - Ax_i - b), \tag{9.12}$$

which is an ordinary least-squares problem. This is actually also a linear regression problem, *cf.* Section 10.1. In Exercise 9.3 we derive the solution to this problem, and we will see that it is closely related to the solution for the affine predictor above. Actually, one just needs to replace the true

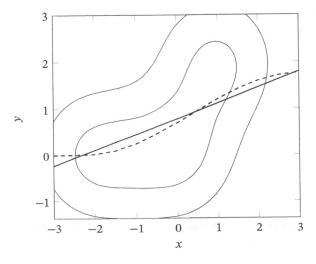

Figure 9.4   Figure showing the level curves of the pdf for the Gaussian mixture in Example 9.4 together with both the nonlinear and affine predictors.

Figure 9.5 Figure showing samples from the Gaussian mixture in Example 9.4 together with both the true affine predictor (solid line), and the affine predictor estimated from samples (dashed line).

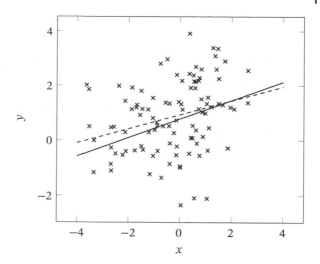

moments of the random variables with their estimates from the observations (x_i, y_i), and then the same formulas for A and b apply. Figure 9.5 shows 100 samples from the Gaussian mixture in Example 9.4 together with both the true affine predictor and the affine predictor estimated from the 100 samples. We see that the two predictors are close to one another.

9.4 The Viterbi Algorithm

We consider an HMM, as defined in Section 3.11. For ease of reference, the definition is repeated. Consider two random processes $X : \Omega \to D^{\mathbb{Z}_+}$ and $Y : \Omega \to \mathcal{E}^{\mathbb{Z}_+}$ that are correlated. The sets D and $\mathcal{E}$ will be defined later. We will assume that X is a Markov process satisfying (3.10), and that Y_j given X_j are independent of Y_k given X_k for $j \neq k$. In Section 3.11, we derived the filtering equations for the conditional pdf $p_{X_k|\bar{Y}_k}$, where $\bar{Y}_k = (Y_0, \ldots, Y_k)$. In this section, we are interested in predicting or estimating $\bar{X}_k = (X_0, \ldots, X_k)$ from the observation $\bar{y}_k$ of $\bar{Y}_k$ using MAP estimation, *cf.* Section 9.3. Note that we are interested in estimating $\bar{X}_k$ rather than just X_k. This is often called a *smoothing problem*. To this end, the joint probability function or pdf $p_{\bar{X}_k, \bar{Y}_k} : D^{k+1} \times \mathcal{E}^{k+1} \to \mathbb{R}_+$ for $(\bar{X}_k, \bar{Y}_k)$ is needed.[2] We also need the conditional probability function or pdf for $\bar{Y}_k$ given $\bar{X}_k$: $p_{\bar{Y}_k|\bar{X}_k} : \mathcal{E}^{k+1} \times D^{k+1} \to \mathbb{R}_+$. Because of the conditional independence, it can be expressed as

$$p_{\bar{Y}_k|\bar{X}_k}(\bar{y}_k|\bar{x}_k) = \prod_{i=0}^{k} p_{Y_i|X_i}(y_i|x_i),$$

where $p_{Y_i|X_i} : \mathcal{E} \times D \to \mathbb{R}_+$ are the conditional probability functions or pdfs for Y_i given X_i. We also define the marginal probability function or pdf for $\bar{X}_k$: $p_{\bar{X}_k} : D^{k+1} \to \mathbb{R}_+$. From the above assumptions, it follows that

$$p_{\bar{X}_k, \bar{Y}_k}(\bar{x}_k, \bar{y}_k) = p_{\bar{Y}_k|\bar{X}_k}(\bar{y}_k|\bar{x}_k) p_{\bar{X}_k}(\bar{x}_k) = \prod_{i=0}^{k} p_{Y_i|X_i}(y_i|x_i) p_{\bar{X}_k}(\bar{x}_k). \tag{9.13}$$

---

2 We notice that maximizing this joint pdf will result in the MAP estimate, since the only difference between the conditional pdf used for MAP estimation and the joint pdf is a normalization with the marginal pdf for the observations.

We now make use of the multiplication theorem, see Section 3.2, and obtain with obvious definitions of the involved functions

$$p_{\bar{X}_k}(\bar{x}_k) = p_{X_0}(x_0)p_{X_1|X_0}(x_1|x_0)p_{X_2|X_0,X_1}(x_2|x_0,x_1)\cdots p_{X_k|X_0,\ldots,X_{k-1}}(x_k|x_0,\ldots,x_{k-1}),$$

$$= p_{X_0}(x_0)p_{X_1|X_0}(x_1|x_0)p_{X_2|X_1}(x_2|x_1)\cdots p_{X_k|X_{k-1}}(x_k|x_{k-1}),$$

where the last equality follows from the Markov property. From (9.13) it then follows that

$$p_{\bar{X}_k,\bar{Y}_k}(\bar{x}_k,\bar{y}_k) = p_{Y_k|X_k}(y_k|x_k)$$

$$\times p_{X_k|X_{k-1}}(x_k|x_{k-1})$$

$$\times p_{Y_{k-1}|X_{k-1}}(y_{k-1}|x_{k-1})$$

$$\times p_{X_{k-1}|X_{k-2}}(x_{k-1}|x_{k-2})$$

$$\times p_{Y_{k-2}|X_{k-2}}(y_{k-2}|x_{k-2})$$

$$\vdots$$

$$\times p_{X_2|X_1}(x_2|x_1)$$

$$\times p_{Y_1|X_1}(y_1|x_1)$$

$$\times p_{X_1|X_0}(x_1|x_0)$$

$$\times p_{Y_0|X_0}(y_0|x_0)p_{X_0}(x_0).$$

Therefore,

$$\max_{\bar{x}_k}\{p_{\bar{X}_k,\bar{Y}_k}(\bar{x}_k,\bar{y}_k)\} = \max_{x_k}\{p_{Y_k|X_k}(y_k|x_k)$$

$$\times \max_{x_{k-1}}\{p_{X_k|X_{k-1}}(x_k|x_{k-1})p_{Y_{k-1}|X_{k-1}}(y_{k-1}|x_{k-1})$$

$$\times \max_{x_{k-2}}\{p_{X_{k-1}|X_{k-2}}(x_{k-1}|x_{k-2})p_{Y_{k-2}|X_{k-2}}(y_{k-2}|x_{k-2})$$

$$\vdots$$

$$\times \max_{x_1}\{p_{X_2|X_1}(x_2|x_1)p_{Y_1|X_1}(y_1|x_1)$$

$$\times \max_{x_0}\{p_{X_1|X_0}(x_1|x_0)p_{Y_0|X_0}(y_0|x_0)p_{X_0}(x_0)\}\}\}\}\}.$$

We now introduce functions $V_k : \mathcal{D} \to \mathbb{R}_+$ for $k \in \mathbb{Z}_N$ defined via $V_0(x) = p_{Y_0|X_0}(y_0|x)p_{X_0}(x)$ and the recursion

$$V_i(x) = p_{Y_i|X_i}(y_i|x)\max_u\{p_{X_i|X_{i-1}}(x|u)V_{i-1}(u)\},$$

for $i = \mathbb{N}_k$. It then follows that

$$\max_{\bar{x}_k}\{p_{\bar{X}_k,\bar{Y}_k}(\bar{x}_k,\bar{y}_k)\} = \max_x\{V_k(x)\},$$

and that the optimal $\bar{x}_k$ is such that x_{i-1} is the maximizing u above for $i \in \mathbb{N}_k$, and x_k is the maximizing x above. The recursions above are summarized as the famous *Viterbi algorithm* in Algorithm 9.1. We remark that it was much easier to derive the Viterbi algorithm than to derive the filtering equations in Section 3.11. The reason for this is that we are only interested in computing the MAP estimate and not in obtaining the conditional probability function or pdf.

There is also a logarithmic version of the Viterbi algorithm which is obtained by defining $J_i : \mathcal{D} \to \mathbb{R}$ for $0 \leq i \leq k$ as $J_i(x) = -\log V_i(x)$. It then follows that the recursion reads

$$J_i(x) = -\log p_{Y_i|X_i}(y_i|x) + \min_u\{-\log p_{X_i|X_{i-1}}(x|u) + J_{i-1}(u)\},$$

Algorithm 9.1: Viterbi algorithm

Input: Time horizon k, initial pdf p_{X_0}, conditional pdfs $p_{Y_i|X_i}$ for $i \in \mathbb{Z}_k$, and $p_{X_i|X_{i-1}}$ for $i \in \mathbb{N}_k$,
measurement data $(y_0, y_1, \ldots, y_k)$

Output: $\bar{x}_k = (x_0, \ldots, x_k)$

$V_0(x) \leftarrow p_{Y_0|X_0}(y_0|x)p_{X_0}(x)$

for $i \leftarrow 1$ **to** k **do**

$\quad x_{i-1} \leftarrow \text{argmax}_u\{p_{X_i|X_{i-1}}(x|u)V_{i-1}(u)\}$

$\quad V_i(x) \leftarrow p_{Y_i|X_i}(y_i|x)\max_u\{p_{X_i|X_{i-1}}(x|u)V_{i-1}(u)\}$

end

$x_k \leftarrow \text{argmax}_x\{V_k(x)\}$

---

with initial value $J_0(x) = -\log p_{Y_0|X_0}(y_0|x) - \log p_{X_0}(x)$. This is known to often have better numerical properties.

A typical example of an HMM for $\mathcal{D} = \mathbb{R}^n$ and $\mathcal{E} = \mathbb{R}^p$ is obtained by considering the random processes defined by the recursion

$$X_{k+1} = F_k(X_k, V_k),$$
$$Y_k = G_k(X_k, E_k), \tag{9.14}$$

where $F_k : \mathbb{R}^n \times \mathbb{R}^n \to \mathbb{R}^n$, $G_k : \mathbb{R}^n \times \mathbb{R}^p \to \mathbb{R}^p$, and where E_k are i.i.d. p-dimensional random vectors and V_k are i.i.d. n-dimensional random vectors, and where $X_0 \in \mathbb{R}^n$ is a random vector with known distributions. We will assume that X_0 is independent of E_k and V_k for all $k \geq 0$. It is straight forward to verify the Markov property and the conditional independence property.

9.5 Kalman Filter on Innovation Form

An important special case of the HMM in (9.14) is obtained when $F_k(x, v) = Ax + v$ and $G_k(x, e) = Cx + e$, where $A \in \mathbb{R}^{n \times n}$, and $C \in \mathbb{R}^{p \times n}$, and where X_0, V_k, and E_k all have Gaussian distributions with expectations $\bar{x}$, 0, and 0, respectively, and covariances R_0, R_1, and R_2, respectively. We derived the Kalman filter in Section 3.11 for this HMM. We will see that we can obtain similar recursions for the MAP estimate, and we will show that they actually provide the same estimate. We have

$$p_{X_0}(x_0) = \mathcal{N}(x_0, \bar{x}_0, R_0),$$
$$p_{X_k|X_{k-1}}(x_k|x_{k-1}) = \mathcal{N}(x_k, Ax_{k-1}, R_1),$$
$$p_{Y_k|X_k}(y_k|x_k) = \mathcal{N}(y_k, Cx_k, R_2),$$

where $\mathcal{N}$ is defined as in Example 3.4, i.e. all the involved pdfs are Gaussian. Applying the logarithmic Viterbi recursion, we find that

$$J_0(x) = \frac{1}{2}(x - \bar{x}_0)^T R_0^{-1}(x - \bar{x}_0) + \frac{1}{2}(y_0 - Cx)^T R_2^{-1}(y_0 - Cx), \tag{9.15}$$

$$J_i(x) = \frac{1}{2}(y_k - Cx)^T R_2^{-1}(y_k - Cx) + \min_u\left\{\frac{1}{2}(x - Au)^T R_1^{-1}(x - Au) + J_{i-1}(u)\right\}, \tag{9.16}$$

for $i \in \mathbb{N}_k$ modulo constant terms. Finally, we obtain the optimal x_k as the solution of

$$\text{minimize} \quad J_k(x).$$

We will now find a more explicit solution to the above recursions by verifying that

$$J_i(x) = \frac{1}{2}x^T P_i x + q_i^T x + r_i,$$

for some $P_i \in \mathbb{S}_+^n$, $q_i \in \mathbb{R}^n$, and $r_i \in \mathbb{R}$. For $i = 0$ this holds with

$$P_0 = R_0^{-1} + C^T R_2^{-1} C, \quad q_0 = -R_0^{-1}\bar{x}_0 - C^T R_2^{-1} y_0. \tag{9.17}$$

The actual value of the constant term r_0 will be of no interest, and this is true for the whole sequence r_i. The argument of the min operator on the right-hand side of (9.16) is a strictly convex function of u, and hence, its minimizer must satisfy the stationary condition

$$-A^T R_1^{-1}(x - Au) + P_{i-1}u + q_{i-1} = 0.$$

The optimal u is therefore

$$u = G_{i-1}^{-1}\left(A^T R_1^{-1}x - q_{i-1}\right),$$

where $G_i = A^T R_1^{-1} A + P_i$, and it satisfies

$$x - Au = \left(R_1 - AG_{i-1}^{-1}A^T\right)R_1^{-1}x + AG_{i-1}^{-1}q_{i-1}.$$

Inserting this on the right-hand side of (9.16) results in

$$\frac{1}{2}(y_i - Cx)^T R_2^{-1}(y_i - Cx)$$
$$+ \frac{1}{2}\left(\left(R_1 - AG_{i-1}^{-1}A^T\right)R_1^{-1}x + AG_{i-1}^{-1}q_{i-1}\right)^T R_1^{-1}\left(\left(R_1 - AG_{i-1}^{-1}A^T\right)R_1^{-1}x + AG_{i-1}^{-1}q_{i-1}\right)$$
$$+ \frac{1}{2}\left(A^T R^{-1}x - q_{i-1}\right)^T G_{i-1}^{-1}P_{i-1}G_{i-1}^{-1}\left(A^T R^{-1}x - q_{i-1}\right)$$
$$+ q_{i-1}^T G_{i-1}^{-1}\left(A^T R^{-1}x - q_{i-1}\right) + r_{i-1}.$$

This is equal to the left-hand side of (9.16) if

$$P_i = C^T R_2^{-1}C + R_1^{-1}\left(R_1 - AG_{i-1}^{-1}A^T\right)R_1^{-1}\left(R_1 - AG_{i-1}^{-1}A^T\right) + R_1^{-1}AG_{i-1}^{-1}P_{i-1}G_{i-1}^{-1}A^T R_1^{-1},$$
$$q_i = -C^T R_2^{-1}y_i + \left(R_1^{-1}\left(R_1 - AG_{i-1}^{-1}A^T\right)R_1^{-1}A - R_1^{-1}AG_{i-1}^{-1}P_{i-1} + R_1^{-1}A\right)G_{i-1}^{-1}q_{i-1},$$

where we again omit the constant terms. By making use of the definition of G_{i-1}, the expressions can be simplified to

$$P_i = C^T R_2^{-1}C + R_1^{-1} - R_1^{-1}AG_{i-1}^{-1}A^T R_1^{-1},$$
$$q_i = -C^T R_2^{-1}y_i + R_1^{-1}AG_{i-1}^{-1}q_{i-1}.$$

Using the Sherman–Morrison–Woodbury (SMW) identity in (2.56), we obtain

$$P_i = C^T R_2^{-1}C + Y_i^f, \tag{9.18}$$

$$q_i = -C^T R_2^{-1}y_i - Y_k^f AP_{i-1}^{-1}q_{i-1}, \tag{9.19}$$

where

$$Y_i^f = \left(R_1 + AP_{i-1}^{-1}A^T\right)^{-1}. \tag{9.20}$$

This also shows that P_i is positive definite. The estimate at the final time $i = k$ is now given by

$$x_k = -P_k^{-1}q_k.$$

The above recursions and this expression can be used to obtain the solution for any value of k, i.e. we can obtain the solution for the problem ending at $k + 1$ from the solution for the problem ending at k with just one more step in the recursion. We summarize the Kalman filter on the innovation form in Algorithm 9.2. It is a good idea to avoid computing the inverse of P_k. It is better to use a Cholesky factorization. Sometimes the algorithm is presented with q_k having the opposite sign. We may also use the fact that $u = x_{i-1}$ to obtain the estimates of x_i for $i \in \mathbb{Z}_{k-1}$ from

$$x_{i-1} = G_{i-1}^{-1} \left(A^T R_1^{-1} x_i - q_{i-1} \right),\tag{9.21}$$

which are the so-called "smoothed estimates." Here, we may use the SMW identity to express G_{i-1}^{-1} as

$$G_{i-1}^{-1} = P_{i-1}^{-1} - P_{i-1}^{-1} A^T \left(R_1 + A P_{i-1}^{-1} A^T \right)^{-1} P_{i-1}^{-1}.$$

The smoothed estimates will be different for different values of k, i.e. we cannot find the smoothed solution for the problem ending at $k+1$ from the solution of the problem ending at k without re-running the backward recursion in (9.21). The more intuitive explanation for this is that the new measurement y_{k+1} affects all the smoothed estimates.

Algorithm 9.2: Innovation form Kalman filter

Input: System matrices A and C, mean $\bar{x}_0$, covariances R_0, R_1, and R_2, measurement data
$\qquad (y_0, y_1, \dots)$
Output: x_k for $k \in \mathbb{Z}_+$

$Y_0^f \leftarrow R_0^{-1}$
$q_0 \leftarrow -R_0^{-1} \bar{x}_0 - C^T R_2^{-1} y_0$
for $k \leftarrow 0$ **to** ∞ **do**
$\quad \left| \; P_k \leftarrow C^T R_2^{-1} C + Y_k^f \right.$
$\quad \left| \; x_k \leftarrow -P_k^{-1} q_k \right.$
$\quad \left| \; Y_{k+1}^f \leftarrow \left(R_1 + A P_k^{-1} A^T \right)^{-1} \right.$
$\quad \left| \; q_{k+1} \leftarrow -Y_{k+1}^f A P_k^{-1} q_k - C^T R_2^{-1} y_{k+1} \right.$
end

The form of the Kalman filter that we have derived is called the information form. It has advantages when the inverses of the covariance matrices are sparse. It is also advantageous when we have little information about the initial value of the state and need to let $R_0^{-1} \to 0$. Then we can initialize with $Y_0^f = 0$. However, the Kalman filter is often presented in another way, *cf.* Section 3.11, which we will now derive from the innovation form summarized in Algorithm 9.2. Let us define $\Sigma_k = P_k^{-1}$ and $\Sigma_k^f = \left(Y_k^f \right)^{-1}$. From the update formula for P_k in (9.18), we then have

$$\Sigma_k^{-1} = C^T R_2^{-1} C + \left(\Sigma_k^f \right)^{-1}.$$

Now, apply the SMW identity to obtain

$$\Sigma_k = \Sigma_k^f - \Sigma_k^f C^T \left(C \Sigma_k^f C^T + R_2 \right)^{-1} C \Sigma_k^f.$$

From the update formula for Y_k^f in (9.18), we obtain

$$\Sigma_k^f = R_1 + A \Sigma_{k-1} A^T,$$

where the initial value is given by $\Sigma_0^f = R_0$. With these definitions, we have agreement with the definition of the initial value P_0 in (9.17).

We will now show that the recursion for q_k in (9.19) is related to recursions for x_k^a and x_k^f, defined as

$$x_k^a = x_k^f + \Sigma_k^f C^T \left(C \Sigma_k^f C^T + R_2 \right)^{-1} \left(y_k - C x_k^f \right),$$

$$x_{k+1}^f = A x_k^a.$$

with $x_0^f = \bar{x}_0$, and where the estimate at time k is given by $x_k = x_k^a$. More precisely, we will show that with the definition $q_k = -P_k x_k^a$, the above recursions are the same as the recursion for q_k that we derived previously in (9.19). It holds that

$$P_k x_k = P_k \left(I - \Sigma_k^f C^T \left(C \Sigma_k^f C^T + R_2 \right)^{-1} C \right) x_k^f$$

$$+ P_k \Sigma_k^f C^T \left(C \Sigma_k^f C^T + R_2 \right)^{-1} y_k,$$

Moreover, it holds that

$$\Sigma_k^f C^T \left(C \Sigma_k^f C^T + R_2 \right)^{-1} = \Sigma_k^f \left(I - C^T R_2^{-1} C \left(Y_k^f + C^T R_2^{-1} C \right)^{-1} \right) C^T R_2^{-1},$$

$$= \left(Y_k^f + C^T R_2^{-1} C \right)^{-1} C^T R_2^{-1},$$

where we in the first equality have used the SMW identity and where we in the second equality have added and subtracted $Y_k^f (Y_k^f + C^T R_2^{-1} C)^{-1}$ inside the parenthesis. Thus, we have that

$$P_k x_k = P_k \left(I - \Sigma_k C^T R_2^{-1} C \right) x_k^f + P_k \Sigma_k C^T R_2^{-1} y_k,$$

$$= Y_k^f A P_{k-1}^{-1} P_{k-1} x_{k-1}^a + C^T R_2^{-1} y_k,$$

which agrees with the recursion for q_k in (9.19). It is also straightforward to show using similar techniques that the initial values agree. The Kalman filter on standard form is summarized in Algorithm 3.1.

9.6 Viterbi Decoder

In this section, we are going to investigate decoding of a coded message. We will specifically consider the so-called *convolutional codes*, for which a signal $u \in \{0, 1\}^{\mathbb{Z}_+}$ is coded into $y \in \{0, 1\}^{\mathbb{Z}_+}$ using a convolution

$$y_k = \sum_{i=1}^{n} c_i u_{k-i},$$

where $c_i \in \{0, 1\}$ for $i \in \mathbb{N}_n$ represents the code. The above summation is carried out as modulo two. We assume that $u_k = 0$ for $k < 0$ to make the convolution well defined. It is then possible to introduce a state $x_k \in \{0, 1\}^n$ such that

$$x_{k+1} = A x_k + B u_k,$$

$$y_k = C x_k,$$

where A is the lower shift matrix of order n, $B = e_1$, $C = \begin{bmatrix} c_1 & \cdots & c_n \end{bmatrix}$, and $x_0 = 0$. We assume a channel model which is such that the received signal is $r_k = y_k + e_k$, where e_k is a realization of a sequence of independent normally distributed random variables with zero mean and unit variance.

It is then straightforward, *cf.* Section 10.1, to see that the ML problem of estimating u_k for $k \in \mathbb{Z}_{N-1}$ is equivalent to solving the optimal control problem

$$\text{minimize} \quad \sum_{k=0}^{N} (r_k - Cx_k)^2,$$

$$\text{subject to} \quad x_{k+1} = Ax_k + Bu_k, \quad k \in \mathbb{Z}_{N-1},$$

with variables $(u_0, x_1, \dots, u_{N-1}, x_N)$, where $x_0 = 0$, see [84]. Omura used dynamic programming as presented in Chapter 8 to solve the problem. This does, however, not result in a very practical algorithm, since a solution for N cannot be used to solve a problem where N is replaced with $N + 1$. This is often of importance in decoding applications. i[109] proposed an algorithm that does not suffer from this limitation. It is based on performing dynamic programming forward in time instead of backward in time. This can be derived from the general approach of partially separable optimization problems as presented in Section 5.5.[3] We introduce for $k \in \mathbb{Z}_{N-1}$ the functions $f_k : \{0,1\}^n \times \{0,1\} \times \{0,1\}^n \to \mathbb{R}$ as

$$f_k(x, u, x^+) = (r_k - Cx)^2 + I_D(x, u, x^+),$$

where $I_D : \{0,1\}^n \times \{0,1\} \times \{0,1\}^n \to \mathbb{R}$ is the indicator function for the set

$$D = \{(x, u, x^+) \in \{0,1\}^n \times \{0,1\} \times \{0,1\}^n \mid x^+ = Ax + Bu\}.$$

We also let $\phi : \{0,1\}^n \to \mathbb{R}$ be defined as $\phi(x) = (y_N - Cx)^2$. With this notation, we may write the optimal control problem above as

$$\text{minimize} \quad \phi(x_N) + \sum_{k=0}^{N-1} f_k(x_k, u_k, x_{k+1}),$$

with variables $(u_0, x_1, \dots, u_{N-1}, x_N)$, where $x_0 = 0$. This is clearly a partially separable optimization problem.

We then introduce the functions $V_k : \{0,1\}^n \to \mathbb{R}$ defined as

$$V_{k+1}(x^+) = \min_{u,x} \left\{ V_k(x) + f_k(x, u, x^+) \right\}, \quad k \in \mathbb{Z}_{N-1},$$

where $V_0(x) = 0$. We also define $\mu_{k+1} : \{0,1\}^n \to \{0,1\} \times \{0,1\}^n$ as the minimizing argument in the above minimization, i.e.

$$\mu_{k+1}(x^+) = \operatorname*{argmin}_{x,u} \left\{ V_k(x) + f_k(x, u, x^+) \right\}.$$

The function $V_N(x) + (r_N - Cx)^2$ is then finally minimized with respect to x to obtain the optimal x_N. After this has been done, all optimal variables can be recovered from the recursion

$$(u_{k-1}, x_{k-1}) = \mu_k(x_k), \quad 1 \leq k \leq N.$$

The minimization in each step in the recursion can be written as

$$\text{minimize} \quad V_k(x) + (r_k - Cx)^2,$$

$$\text{subject to} \quad Ax + Bu = x^+,$$

---

3 We may interpret this as a variation of the Viterbi algorithm if we take y_k in the Viterbi algorithm equal to r_k, consider the state transition probability to be degenerate, and introduce an additional variables u_k to optimize over.

with variables (x, u). Because of the very specific structure of the constraints, we can say much more. To avoid cluttering the notation, we look at $n = 3$. Then we have that the constraints are

$$u = x_1^+,$$
$$x_1 = x_2^+,$$
$$x_2 = x_3^+,$$

and hence, the only optimization variable is x_3. Therefore, the optimization problem is

$$\text{minimize } (r_k - c_1 x_2^+ - c_2 x_3^+ - c_3 x_3)^2 + V_k((x_2^+, x_3^+, x_3)),$$

where the minimizing argument x_3 will be a function of x_2^+ and x_3^+. Also, notice that V_{k+1} will only be a function of x_2^+ and x_3^+. The minimizing argument is therefore

$$\mu_k(x) = \begin{bmatrix} x_1^+ \\ x_2^+ \\ x_3^+ \\ x_3(x_2^+, x_3^+) \end{bmatrix}.$$

Only the last component is nontrivial to compute, and it can be done by enumerating of all possible values of x_2^+ and x_3^+. We now realize that we need tables for V_k and μ_k that have 2^{n-1} entries.

In the practical use of Viterbi decoding, the value of N is not fixed, but it is increasing and represents time. The decoding is done with some fixed delay d measured from N, i.e. it is u_{N-d} that is estimated. Thus, we need to store one table for V_N and d tables for μ_k, where $N - d + 1 \le k \le N$. In case $d2^{n-1}$ is large, this could be costly. Approximations for how to circumvent this were proposed by Viterbi; see, e.g. [110].

9.7 Graphical Models

We will now define distributions on undirected graphs. To this end, we consider a graph $G = (V, E)$, where $V = \mathbb{N}_n$ is the set of vertices and where $E \subset V \times V$ is a set of undirected edges connecting vertices in V. We now let the elements of V index components of an n-dimensional random variable, and we define a probability function by maximizing entropy under moment constraints. However, we will only specify second-order moments for pairs of components that correspond to the edges of the graph. Such a model is called a graphical model.

9.7.1 Ising Distribution

When we consider a graphical model for an n-dimensional Ising distribution similar as in (9.4), we obtain

$$p(x) = \exp\left(\sum_{k \in \mathbb{N}_n} \lambda_k x_k + \sum_{(i,j) \in E} \Lambda_{ij} x_i x_j - A \right),$$

where

$$A(\lambda, \Lambda) = \ln\left(\sum_{x \in D_x} \exp\left(\sum_{k \in \mathbb{N}_n} \lambda_k x_k + \sum_{(i,j) \in E} \Lambda_{ij} x_i x_j \right) \right).$$

We notice that we do not specify all the entries of neither Λ nor the matrix M of second moments. If we let $\Lambda_{ij} = 0$ for $(i,j) \notin E$, then we may express the probability function in terms of $\text{tr}(\Lambda xx^T)$. The

Lagrange dual function will therefore look the same, and so will the optimality conditions, except for the fact that those related to $(i,j) \notin E$ are omitted, i.e. we have

$$\frac{\partial h}{\partial \lambda} = \frac{\partial A}{\partial \lambda} - m = 0,$$

$$\frac{\partial h}{\partial \Lambda_{ij}} = \frac{\partial A}{\partial \Lambda_{ij}} - M_{ij} = 0, \quad (i,j) \in E.$$

Also, for this case, it is in general difficult to solve the optimality conditions.

9.7.2 Normal Distribution

When we derive a graphical model for the normal distribution, *cf.* (9.7), we have that Λ is such that $\Lambda_{ij} = 0$ for $(i,j) \notin E$. Also, we do not specify M_{ij} for $(i,j) \notin E$. The Lagrange dual function will look the same as in (9.2). We first minimize it with respect to λ, which results in $\lambda = -\Lambda m$. We then insert the solution in the Lagrange dual function and obtain the function $f : \mathbb{S}^n \to \mathbb{R}$ given by

$$f(\Lambda) = -\frac{1}{2}\ln\left(\frac{\det \Lambda}{(2\pi)^n}\right) + \frac{1}{2}\mathrm{tr}\,(\Lambda\Sigma), \tag{9.22}$$

where $\Sigma = M - mm^T$. We need to minimize it subject to the constraint that $\Lambda_{ij} = 0$ for $(i,j) \notin E$ in order to express Λ in terms of the moments. We realize that this is a convex optimization problem with a linear constraint on Λ. This optimization problem can be solved with algorithms in Chapter 6. We will see that $\Lambda \in \mathbb{S}^n_+$. However, we would like to analyze the problem a bit more. To this end, we introduce the Lagrangian $\mathcal{L} : \mathbb{S}^n \times \mathbb{S}^n$ given by

$$\mathcal{L}(\Lambda, \Gamma) = -\frac{1}{2}\left(\ln\frac{\det \Lambda}{(2\pi)^n}\right) + \frac{1}{2}\mathrm{tr}\,(\Lambda\Sigma) + \frac{1}{2}\mathrm{tr}\,(\Gamma\Lambda),$$

where $\Gamma_{i,j} = 0$ for $(i,j) \in E$. The optimality conditions are that

$$\frac{\partial \mathcal{L}}{\partial \Lambda} = -\frac{1}{2}\Lambda^{-1} + \frac{1}{2}\Sigma + \frac{1}{2}\Gamma = 0,$$

together with $\Lambda_{ij} = 0$ for $(i,j) \notin E$ and $\Gamma_{ij} = 0$ for $(i,j) \in E$. We realize that we have full freedom in selecting the entries of $\Sigma + \Gamma$ for indexes $(i,j) \notin E$ by choosing Γ. Thus, it does not matter that we did not specify these entries for M. We may take them equal to any value. Because of this, we let $\Sigma_{ij} = 0$ for $(i,j) \notin E$. However, $\Sigma + \Gamma$ is also the covariance matrix of the normal distribution, and hence, has to be positive definite. We can describe the above equation as finding Γ to complete the covariance matrix in such a way that it is positive definite and such that its inverse has zeros for entries not in the set E. This is called the *maximum determinant positive definite completion* of Σ.

9.7.3 Markov Random Field

Let us consider the normal pdf on the form

$$p(x) = \sqrt{\frac{\det \Lambda}{\pi^n}}e^{-(x-m)^T \Lambda(x-m)}.$$

Assume A, B, and C are three mutually disjoint subsets of V and that if we remove the vertices indexed by C, then there are no paths connecting the vertices in A with the vertices in B. Let $D = V \setminus (A \cup B \cup C)$. Also, assume with no loss of generality that we have ordered the components of x in such a way that $x = (x_A, x_B, x_C, x_D)$, where x_A is the subvector of x indexed by A, and so on. We will show that

$$p(x) = p_{A,B|C}(x)p_C(x_C) = p_{A|C}(x_A|x_C)p_{B|C}(x_B|x_C)p_C(x_C), \tag{9.23}$$

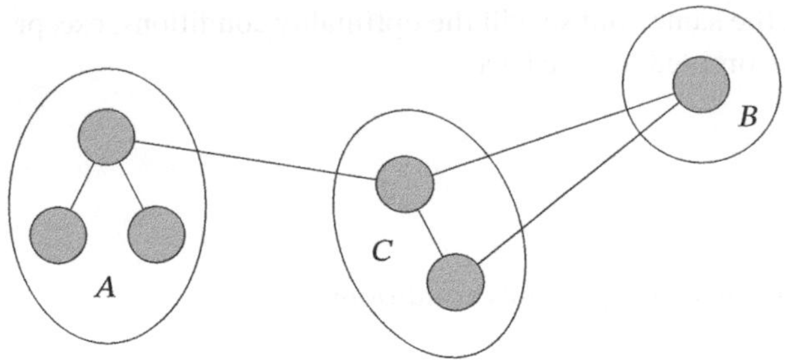

Figure 9.6 A graph where the subset of nodes in C separates the subset of nodes in A from the ones in B.

where $p_{A|C} : \mathbb{R}^{|A|+|C|} \to \mathbb{R}_+$ and $p_{B|C} : \mathbb{R}^{|B|+|C|} \to \mathbb{R}_+$ are the conditional pdfs and where $p_C : \mathbb{R}^{|C|} \to \mathbb{R}_+$ is the marginal pdf for the variables indexed by C. This means that the random variables indexed by A and B conditioned on the random variables indexed by C are independent, which is called *conditional independence*. This is the motivation for the name *Markov random field* for a distribution specified as above.

From the above definitions, it follows that Λ must have the following structure:

$$\Lambda = \begin{bmatrix} \times & 0 & \times & 0 \\ 0 & \times & \times & 0 \\ \times & \times & \times & \times \\ 0 & 0 & \times & \times \end{bmatrix}.$$

We realize that we with no loss of generality may consider the components indexed by D to be part of A and/or B, or we may assume that $V = A \cup B \cup C$, which we from now on do, see Figure 9.6. From this, it follows that Λ must have the following structure:

$$\Lambda = \begin{bmatrix} \times & 0 & \times \\ 0 & \times & \times \\ \times & \times & \times \end{bmatrix},$$

which is called an *arrow structure*. To prove the conditional independence property, we partition the covariance matrix Σ as

$$\Sigma = \begin{bmatrix} \Sigma_A & \Sigma_{AB} & \Sigma_{AC} \\ \Sigma_{AB}^T & \Sigma_B & \Sigma_{BC} \\ \Sigma_{AC}^T & \Sigma_{BC}^T & \Sigma_C \end{bmatrix}.$$

We also let

$$\tilde{\Sigma} = \begin{bmatrix} \tilde{\Sigma}_1 & \tilde{\Sigma}_{12} \\ \tilde{\Sigma}_{12}^T & \tilde{\Sigma}_2 \end{bmatrix} = \begin{bmatrix} \Sigma_A & \Sigma_{AB} \\ \Sigma_{AB}^T & \Sigma_B \end{bmatrix} - \begin{bmatrix} \Sigma_{AC} \\ \Sigma_{BC} \end{bmatrix} \Sigma_C^{-1} \begin{bmatrix} \Sigma_{AC} \\ \Sigma_{BC} \end{bmatrix}^T,$$

which is the covariance matrix for the variables indexed by A and B conditioned on the variables indexed by C. Hence, it only remains to prove that $\tilde{\Sigma}_{12} = 0$. From the formula for the inverse of a blocked matrix in (2.54), it follows that

$$\Sigma^{-1} = \begin{bmatrix} \tilde{\Sigma}^{-1} & \times \\ \times & \times \end{bmatrix}.$$

We once again apply this formula to obtain that

$$\tilde{\Sigma}^{-1} = \begin{bmatrix} \left(\tilde{\Sigma}_1 - \tilde{\Sigma}_{12}\tilde{\Sigma}_2^{-1}\tilde{\Sigma}_{12}^T\right)^{-1} & -\left(\tilde{\Sigma}_1 - \tilde{\Sigma}_{12}\tilde{\Sigma}_2^{-1}\tilde{\Sigma}_{12}^T\right)^{-1}\tilde{\Sigma}_{12}\tilde{\Sigma}_2^{-1} \\ -\tilde{\Sigma}_2^{-1}\tilde{\Sigma}_{12}^T\left(\tilde{\Sigma}_1 - \tilde{\Sigma}_{12}\tilde{\Sigma}_2^{-1}\tilde{\Sigma}_{12}^T\right)^{-1} & \times \end{bmatrix}.$$

From this, we now see that the zero block in Λ corresponds to

$$\left(\tilde{\Sigma}_1 - \tilde{\Sigma}_{12}\tilde{\Sigma}_2^{-1}\tilde{\Sigma}_{12}^T\right)^{-1}\tilde{\Sigma}_{12}\tilde{\Sigma}_2^{-1} = 0,$$

and hence, $\tilde{\Sigma}_{12} = 0$, which is what we wanted to prove. Because of this, (9.23) holds.

There may be many more ways to partition V such that the above property holds. It is possible to show that the pdf in general can be factorized as

$$p(x) = \prod_{C \in \mathcal{C}} f_C(x_C),$$

for some functions $f_C : \mathbb{R}^{|C|} \to \mathbb{R}_+$, where $\mathcal{C}$ is the set of all *cliques* of G, i.e. the set of all complete subgraphs of G, see [112]. Here, $\mathcal{C}$ has a different meaning than before. It is possible to take $\mathcal{C}$ to be the set of maximal cliques of G, where a clique is maximal if it is not contained in any other clique. We will later discuss how this structure may be utilized in more detail. The above Markov property holds for general graphical models defined on undirected graphs, and specifically, also for the Ising model.

9.8 Maximum Likelihood Estimation

We have already seen how we can estimate distributions by maximizing entropy. Now, we will consider the problem of estimating parameters in a distribution by maximizing the so-called "likelihood function." This is called *maximum likelihood* (ML) estimation. For a probability function $p \in \mathcal{D}_n$ that depends on a parameter $\lambda \in \mathbb{R}$, we may define the likelihood function $\mathcal{L} : \mathbb{R}^N \times \mathbb{R} \to [0, 1]$ based on N samples x_k, $i \in \mathbb{N}_N$, of the random variable $X : \mathbb{N}_n \to \mathbb{R}$ as

$$\mathcal{L}(x_1, \ldots, x_N; \lambda) = \prod_{i=1}^{N} p_i,$$

where $p_i = \mathbb{P}[X = x_i]$. Then an estimate of λ is obtained by maximizing $\mathcal{L}$, or equivalently, the logarithm of $\mathcal{L}$.

9.8.1 Categorical Distribution

For the categorical distribution discussed in Section 9.2, we have $x_i \in \{f_1, \ldots, f_n\}$, and we obtain

$$\ln \mathcal{L}(f_{k_1}, \ldots, f_{k_N}; \lambda) = -\sum_{i=1}^{N} \lambda f_{k_i} - N \ln \Phi(\lambda).$$

If we take $b = \frac{1}{N}\sum_{i=1}^{N} f_{k_i}$ in Section 9.2, then minimizing $h(\lambda) = \lambda b + \ln \Phi(\lambda)$ is equivalent to maximizing the likelihood function. Because of this, we realize that entropy maximization generalizes ML estimation in that we also look for the optimal distribution.

9.8.2 Ising Distribution

We now turn to the Ising distribution where we define $\mathcal{L} : \{0, 1\}^{mN} \times \mathbb{R}^m \times \mathbb{S}^m \to [0, 1]$ based on N samples $x_i \in \{0, 1\}^m$, $i \in \mathbb{N}_N$, of the random variable X as

$$\mathcal{L}(x_1, \ldots, x_N; \lambda, \Lambda) = \prod_{i=1}^{N} p(x_i).$$

Then an ML estimate of λ is obtained by maximizing

$$\ln \mathcal{L}(x_1, \dots, x_N; \lambda, \Lambda) = \lambda^T \sum_{i=1}^{N} x_i + \mathrm{tr}\Lambda \sum_{i=1}^{N} x_i x_i^T - NA(\lambda, \Lambda).$$

Note that the subindex i now refers to the ith sample and not the ith component of x. If we take $m = \frac{1}{N} \sum_{i=1}^{N} x_i$ and $M = \frac{1}{N} \sum_{i=1}^{N} x_i x_i^T$ in Section 9.2, then minimizing $h(\lambda, \Lambda) = A(\lambda, \Lambda) - \lambda^T m - \mathrm{tr}(\Lambda M)$ is equivalent to maximizing the likelihood function. This results also hold for the case when we define the Ising model on a graph.

9.8.3 Normal Distribution

We now consider a pdf $p \in \mathbb{R}_+^{\mathbb{R}^n \times \mathbb{R}^M}$ that depends on a parameter $\theta \in \mathbb{R}^M$, and we define the likelihood function $\mathcal{L} : \mathbb{R}^{Nn} \times \mathbb{R}^M \to \mathbb{R}_+$ based on N samples $x_i \in \mathbb{R}^n$ of the random variable X as

$$\mathcal{L}(x_1, \dots, x_N; \theta) = \prod_{i=1}^{N} p(x_i, \theta).$$

As before, the estimate of θ is obtained by maximizing $\mathcal{L}$. Now, consider the normal distribution discussed in Section 9.2, and suppose that $\theta = (\lambda, \Lambda) \in \mathbb{R}^n \times \mathbb{S}^n$. The log-likelihood function is then

$$\begin{aligned}
\ln \mathcal{L}(x_1, \dots, x_N; \theta) &= -\frac{1}{2} \sum_{i=1}^{N} \left(x_i + \Lambda^{-1}\lambda \right)^T \Lambda \left(x_i + \Lambda^{-1}\lambda \right) \\
&\quad + \frac{N}{2} \ln\left(\frac{\det \Lambda}{(2\pi)^n} \right), \\
&= -\frac{1}{2} \mathrm{tr}\left(\Lambda \left(\sum_{i=1}^{N} x_i x_i^T \right) \right) - \lambda^T \sum_{i=1}^{N} x_i - \frac{N}{2} \lambda^T \Lambda^{-1} \lambda \\
&\quad + \frac{N}{2} \ln\left(\frac{\det \Lambda}{(2\pi)^n} \right).
\end{aligned} \tag{9.24}$$

If we take

$$m = \frac{1}{N} \sum_{i=1}^{N} x_i, \quad M = \frac{1}{N} \sum_{i=1}^{N} x_i x_i^T$$

in Section 9.2, then minimizing $h(\lambda, \Lambda) = \frac{1}{2}\lambda^T \Lambda^{-1} \lambda - \frac{1}{2} \ln\left(\frac{\det \Lambda}{(2\pi)^n} \right) + \lambda^T m + \frac{1}{2} \mathrm{tr}(\Lambda M)$ is equivalent to maximizing the likelihood function. Hence, the relationship between maximum entropy and ML estimation is the same also for continuous distributions. Note that the problem of minimizing h is a convex optimization problem. The solution has already been derived in Section 9.2, and with $\Sigma = M - mm^T$, we have $\Lambda = \Sigma^{-1}$ and $\lambda = -\Sigma^{-1}m$. Thus, the solution to the ML problem is simply the sample mean and covariance in case we use the nonnatural parameterization. We may also consider ML estimation for the normal distribution when it is defined on a graph, and we obtain similar results.

9.8.4 Generalizations

There are several ways in which we could generalize the ML problem. For example, if we have prior information such as upper or lower bounds on the matrix Λ, e.g.

$$B_l \preceq \Lambda \preceq B_u,$$

with $B_l, B_u \in \mathbb{S}^n_{++}$, then we have a convex constraint that easily can be incorporated when we minimize $h(\lambda, \Lambda)$. Also, an upper bound $\kappa_{\max}$ on the condition number of Λ can be incorporated by noting that it is equivalent to the existence of $u > 0$ such that $uI \preceq \Lambda \preceq \kappa_{\max} uI$. We may also include prior information by modifying m and M. For example, if we have reason to believe from previous experience that m_0 and M_0 are good values, then we could take

$$m = \alpha m_0 + (1 - \alpha)\frac{1}{N}\sum_{i=1}^{N} x_i,$$

$$M = \beta M_0 + (1 - \beta)\frac{1}{N}\sum_{i=1}^{N} x_i x_i^T,$$

with $\alpha, \beta \in [0, 1]$, where the values of α and β are related to how much we trust our prior information as compared to the information in the data $\{x_1, \ldots, x_N\}$ that we have collected.

9.9 Relative Entropy and Cross Entropy

Sometimes it is of interest to quantify the difference between different distributions, and this is what *relative entropy*, or equivalently the *Kullback–Leibler divergence* does. We will give the definition for two pdfs p and q defined on $\mathbb{R}^n$ with obvious modifications for probability functions. Let

$$\mathcal{D}_n = \left\{ p \in \mathbb{R}^{\mathbb{R}^n} \,\middle|\, \int_{\mathbb{R}^n} p(x)dx = 1, \ p(x) \geq 0, \quad \forall x \in \mathbb{R}^n \right\}.$$

Define the relative entropy $D : \mathcal{D}_n \times \mathcal{D}_n \to \mathbb{R}$ from q to p as

$$D(p, q) = -\int_{\mathbb{R}^n} p(x) \ln \frac{q(x)}{p(x)} dx.$$

9.9.1 Gibbs' Inequality

We are interested in the pdf p that minimizes $D(p, q)$ for a given pdf q. Clearly, the relative entropy is zero for $p = q$. We will prove a well-known result known as *Gibbs' inequality*, which states that

$$D(p, q) \geq 0,$$

with equality if and only if $p(x) = q(x)$ for almost all x. From this, we see that the relative entropy measures the "difference" between two pdfs. However, it is not a metric, since in general $D(p, q) \neq D(q, p)$. The proof of Gibbs' inequality is based on Jensen's inequality that says that for a convex function $\varphi : \mathbb{R} \to \mathbb{R}$, any function $f : \mathbb{R}^n \to \mathbb{R}$, and any pdf p, it holds that

$$\varphi\left(\int_{\mathbb{R}^n} f(x)p(x)dx\right) \leq \int_{\mathbb{R}^n} \varphi\left(f(x)\right) p(x)dx.$$

This is just a slight modification of Jensen's inequality in Exercise 4.8, where dx is replaced with $p(x)dx$. If we let $f(x) = q(x)/p(x)$, and $\varphi(f) = -\ln(f)$ in Jensen's inequality, we find that

$$D(p, q) \geq -\ln \int_{\mathbb{R}^n} \frac{q(x)}{p(x)} p(x)dx = -\ln \int_{\mathbb{R}^n} q(x)dx = 0.$$

We will now use Gibbs' inequality to show that the pdf

$$p(x) = \exp\left(-1 - \mu - \lambda^T x - \frac{1}{2}x^T \Lambda x\right)$$

in (9.8) indeed maximizes the entropy. Suppose that q is another pdf that satisfies the same moment constraints, i.e.

$$\int_{\mathbb{R}^n} xp(x)dx = \int_{\mathbb{R}^n} xq(x)dx = m,$$

$$\int_{\mathbb{R}^n} xx^T p(x)dx = \int_{\mathbb{R}^n} xx^T q(x)dx = M.$$

It then follows that

$$H(q) = -\int_{\mathbb{R}^n} q(x)\ln(q(x))dx,$$

$$= -\int_{\mathbb{R}^n} q(x)\ln\left(\frac{q(x)}{p(x)}\right)dx - \int_{\mathbb{R}^n} q(x)\ln(p(x))dx,$$

$$= -D(q,p) - \int_{\mathbb{R}^n} q(x)\left(-1 - \mu - \lambda^T x - \frac{1}{2}x^T \Lambda x\right)dx,$$

$$= -D(q,p) - \int_{\mathbb{R}^n} p(x)\left(-1 - \mu - \lambda^T x - \frac{1}{2}x^T \Lambda x\right)dx,$$

$$= -D(q,p) + H(p),$$

and since $D(q,p) \geq 0$ with equality when $q = p$, it follows that $H(q) \leq H(p)$ with equality when $q = p$. Thus, p maximizes the entropy under the given moment constraints.

9.9.2 Cross Entropy

Another concept of relevance is *cross entropy*, which is defined as $C : \mathcal{D}_n \times \mathcal{D}_n \to \mathbb{R}$, where

$$C(p,q) = -\int_{\mathbb{R}^n} p(x)\ln(q(x))dx.$$

We immediately realize that

$$C(p,q) = -\int_{\mathbb{R}^n} p(x)\ln\left(\frac{q(x)}{p(x)}\right)dx - \int_{\mathbb{R}^n} p(x)\ln(p(x))dx = D(p,q) + H(p),$$

where $H(p)$ is the entropy of p. Hence, it follows from Gibbs' inequality that

$$C(p,q) \geq H(p),$$

with equality if and only if $p(x) = q(x)$ for almost all x. We also realize that cross entropy is the expected value of $-\ln q$ with respect to the pdf p, i.e.

$$C(p,q) = -\mathbb{E}\big[\ln q\big],$$

where $\mathbb{E}$ denotes expectation with respect to p. If we consider q to be parameterized with a parameter $\theta \in \mathbb{R}^m$, i.e. $q : \mathbb{R}^n \times \mathbb{R}^m \to \mathbb{R}_+$, then the ML problem is equivalent to

$$\text{minimize} \quad -\sum_{k=1}^{N} \ln q(x_k; \theta),$$

with variable θ, where x_k, $k \in \mathbb{N}_N$, are the observed data. If we assume that x_k are observations of a random variable with pdf p, then the objective function is proportional to an SAA of the cross entropy.

9.10 The Expectation Maximization Algorithm

Let us consider an ML problem with a likelihood function $\mathcal{L} : \mathbb{R}^N \times \mathbb{R}^p \to \mathbb{R}_+$ based on observations $y \in \mathbb{R}^N$ parameterized with $\theta \in \mathbb{R}^p$, i.e. we are given $\mathcal{L}(y; \theta)$ for known y and we want to solve the problem

$$\text{maximize } \ln \mathcal{L}(y; \theta),$$

with variable θ. Unfortunately, it is sometimes complicated to write down $\mathcal{L}$, and the expression for its gradient with respect to θ. Hence, the evaluations of function and gradients might be time-consuming and/or error-prone to implement. However, in case we had some more observations $z \in \mathbb{R}^M$, then sometimes the ML problem with $x = (y, z)$ as observation happens to not suffer from the above difficulties. This is often the case when some data are missing because of errors in the collection of the data, so-called *missing data*, or in case there are data that are difficult to measure, so-called *latent variables*.

We will now show how to circumvent the above problems. To this end, let $f_X : \mathbb{R}^N \times \mathbb{R}^M \times \mathbb{R}^p \to \mathbb{R}_+$ be the joint pdf for the random variable $X = (Y, Z)$ with parameter θ, of which we have the partial observation $Y = y$.[4] We also define the conditional pdf $f_{Z|Y} : \mathbb{R}^N \times \mathbb{R}^M \times \mathbb{R}^p \to \mathbb{R}_+$ via

$$f_{Z|Y}(z|y; \theta) = \frac{f_X(y, z; \theta)}{f_Y(y; \theta)} = \frac{f_X(y, z; \theta)}{\mathcal{L}(y; \theta)}.$$

This follows since $\mathcal{L}$ is the marginal pdf for the observation y. Moreover, we consider an arbitrary pdf $q \in \mathcal{D}_M$. Note that this is not the marginal pdf for the latent variable z. We now consider the infinite-dimensional optimization problem

$$\text{maximize } \ln \mathcal{L}(y; \theta) - D(q, f_{Z|Y}),$$

with variables θ and q, where D is the relative entropy. The second term in the objective function depends on θ, since $f_{Z|Y}$ is a function of θ. We have that only the second term in the objective function depends on q, and therefore, we may first carry out the maximization with respect to this term over q with the trivial maximum $q = f_{Z|Y}$ with $D(f_{Z|Y}, f_{Z|Y}) = 0$. Then we are left with the ML problem originally defined, and hence, the problems are equivalent, i.e. we can trivially obtain the solution for one of them from the other. Furthermore, it holds that

$$\ln \mathcal{L}(y; \theta) - D(q, f_{X|Y}) = \ln \mathcal{L}(y; \theta) + H(q) - C(q, f_{Z|Y}) = H(q) - C(q, f_X),$$

where H is the entropy and C is the cross-entropy. The above results follow from the fact that cross entropy is the sum of entropy and relative entropy, and from the definition of the conditional pdf for Z given Y. We now apply block-coordinate accent to the optimization problem above, i.e. we repeat the following steps:

1. Fix θ and optimize with respect to q.
2. Fix q and optimize with respect to θ.

This is not a procedure that in general is guaranteed to converge to the optimal solution. However, our derivation shows that we can never obtain a worse value of the objective function by iterating

---

4 In this section, we do not have several observations of a scalar-valued random variable. Instead, we have one observation of a vector-valued random variable. The former case can be treated as a special case of the vector-valued case by taking each component of the vector-valued random variable to be a random variable with the same distribution.

as above. A detailed discussion of the convergence properties is given in, e.g. [117]. The first step above is trivial, since by the original formulation of the objective function, we immediately have that $q = f_{Z|Y}$ is optimal. In the second step q is fixed, and therefore, it is by the reformulation of the objective function equivalent to maximizing $-C(f_{Z|Y}, f_X)$ with respect to θ, since the term $H(q)$ only depends on the previous value of θ which we will denote $\bar{\theta}$. Hence, we first need to evaluate the function $Q : \mathbb{R}^p \times \mathbb{R}^p \to \mathbb{R}$ given by

$$Q(\theta, \bar{\theta}) = -C(f_{Z|Y}, f_X) = \mathbb{E}\left[\ln(f_X(Y, Z; \theta)) | Y = y; \bar{\theta}\right],$$

i.e. we need to compute the conditional expected value of the log of the joint pdf f_X. This is often easy, and it is called the *expectation step* in the expectation maximization (EM) algorithm. After this, we need to solve

$$\text{maximize} \quad Q(\theta, \bar{\theta}),$$

with variable θ, which is often easier to solve than the original ML problem. This is called the *maximization step*. We will later exemplify the claims regarding what is difficult and easy to compute. The E-step is sometimes approximated using an SAA also called *empirical cross-entropy*, i.e.

$$\mathbb{E}\left[\ln f_X(Y, Z; \theta) | Y = y; \bar{\theta}\right] \approx \frac{1}{M} \sum_{i=1}^{M} \ln f_i,$$

where $f_i = f_X(y, z_i; \theta)$ is obtained by drawing a sample z_i from the conditional pdf $f_{Z|Y}$. This is called *Monte Carlo EM*.

9.11 Mixture Models

A mixture model is a probabilistic model for representing a random variable whose distribution function is a convex combination of a number of other distribution functions, each of which represents a so-called "mixture component." Specifically, given a collection of k random variables $C_1, \ldots, C_k$ with distribution functions $F_{C_i} : \mathbb{R}^d \to [0, 1]$, $i \in \mathbb{N}_k$, we define the mixture distribution function $F_Y : \mathbb{R}^d \to [0, 1]$ as

$$F_Y(y) = \sum_{j=1}^{k} \alpha_j F_{C_j}(y),$$

where $\alpha \in \Delta^k$. In other words, the random variable Y may be viewed as an overall population that is derived from a mixture of subpopulations.

An important special case is when the k components are Gaussian random variables. The corresponding mixture model is called a *Gaussian mixture model* (GMM), and the mixture pdf can be expressed as

$$f_Y(y; \theta) = \sum_{j=1}^{k} \alpha_j \mathcal{N}(y, \mu_j, \Sigma_j), \tag{9.25}$$

where θ represents the model parameters $(\alpha_j, \mu_j, \Sigma_j)$, $j \in \mathbb{N}_k$. The problem of computing a ML estimate of the model parameters based on a given set of m independent observations, $y_1, \ldots, y_m \in \mathbb{R}^d$, can be expressed as the nonlinear optimization problem

$$\text{maximize} \quad \sum_{i=1}^{m} \ln\left(\sum_{j=1}^{k} \alpha_j \mathcal{N}(y_i, \mu_j, \Sigma_j)\right),$$

subject to $\alpha \in \Delta^k$,

$$\Sigma_j \geq 0, \ j \in \mathbb{N}_k,$$

with variables $(\alpha_j, \mu_j, \Sigma_j), j \in \mathbb{N}_k$. This problem is generally nonconvex and intractable, but local optimization methods may be used in the pursuit of a local maximum. We note that the ML estimation problem becomes much easier if, in addition to the m observations $y_1, \ldots, y_m$, we are given labels $z_1, \ldots, z_m \in \mathbb{N}_k$ such that z_i identifies which of the k components the ith observation originates from. However, such labels are typically not available.

We will now show how the EM algorithm can be used to derive a relatively simple iterative procedure that converges to a local maximum. To this end, we introduce a discrete random variable Z which takes the value j with probability $\alpha_j, j \in \mathbb{N}_k$, i.e. Z is a latent variable that identifies one of the k components. Moreover, we define the pdf of Y given $Z = z$ as

$$f_{Y|Z}(y|z; \theta) = \sum_{j=1}^{k} \delta_j(z) \mathcal{N}(y, \mu_j, \Sigma_j),$$

where $\delta_j(z) = 1$ if $z = j$ and 0 otherwise. The joint pdf[5] of Y and Z is then

$$f_{Y,Z}(y, z; \theta) = \sum_{j=1}^{k} \delta_j(z) \alpha_j \mathcal{N}(y, \mu_j, \Sigma_j) = \prod_{j=1}^{k} \left[\alpha_j \mathcal{N}(y, \mu_j, \Sigma_j)\right]^{\delta_j(z)}. \tag{9.26}$$

It is easy to check that (9.25) follows from (9.26) by marginalizing over Z. Moreover, the probability function of Z conditioned on Y may be expressed as

$$f_{Z|Y}(z|y; \theta) = \frac{f_{Y,Z}(y, z; \theta)}{f_Y(y; \theta)} = \frac{\sum_{j=1}^{k} \delta_j(z) \alpha_j \mathcal{N}(y, \mu_j, \Sigma_j)}{\sum_{l=1}^{k} \alpha_l \mathcal{N}(y, \mu_l, \Sigma_l)}. \tag{9.27}$$

Now, given m observations $y_1, \ldots, y_m$ and model parameters $\bar{\theta}$, e.g. an initial guess or the parameters from the previous iteration of the EM algorithm, the E-step of the EM algorithm may be expressed as

$$Q(\theta, \bar{\theta}) = \mathbb{E}\left[\sum_{i=1}^{m} \ln(f_{Y,Z}(Y_i, Z_i; \theta)) \mid Y_1 = y_1, \ldots, Y_m = y_m; \bar{\theta}\right],$$

$$= \sum_{i=1}^{m} \mathbb{E}\left[\ln(f_{Y,Z}(Y_i, Z_i; \theta)) \mid Y_i = y_i; \bar{\theta}\right],$$

$$= \sum_{i=1}^{m} \sum_{j=1}^{k} f_{Z|Y}(j|y_i; \bar{\theta}) \ln(f_{Y,Z}(y_i, j; \theta)),$$

where $(Y_i, Z_i), i \in \mathbb{N}_m$, are independent pairs of random variables with the same joint pdf as that of (Y, Z). Using (9.27), we can write $Q(\theta, \bar{\theta})$ as

$$Q(\theta, \bar{\theta}) = \sum_{i=1}^{m} \sum_{j=1}^{k} \bar{w}_{ij} \ln\left(\alpha_j \mathcal{N}(y_i, \mu_j, \Sigma_j)\right), \tag{9.28}$$

---

5 Note that Y is a continuous random variable and Z is a discrete random variable, for which we define a joint pdf.

where $\bar{w}_{ij} = f_{Z|Y}(j|y_i, \bar{\theta})$ is the probability that y_i originates from the jth mixture component under the mixture model defined by the model parameters $\bar{\theta}$.

The M-step of the EM algorithm is the problem of maximizing $Q(\theta, \bar{\theta})$ with respect to the model parameters θ. This is a block separable optimization problem, which follows by writing $Q(\theta, \bar{\theta})$ as

$$Q(\theta, \bar{\theta}) = \sum_{j=1}^{k} \bar{c}_j \ln(\alpha_j) + \sum_{j=1}^{k} \left(\sum_{i=1}^{m} \bar{w}_{ij} \ln\left(\mathcal{N}(y_i, \mu_j, \Sigma_j)\right) \right),$$

where we define $\bar{c}_j = \sum_{i=1}^{m} \bar{w}_{ij}$. Thus, the update of α may be expressed as

$$\alpha^+ = \underset{\alpha}{\operatorname{argmax}} \left\{ \sum_{j=1}^{k} \bar{c}_j \ln(\alpha_j) \,|\, \alpha \in \Delta^k \right\},$$

$$= (\bar{c}_1/m, \dots, \bar{c}_k/m), \tag{9.29}$$

where we have used the fact that $\sum_{j=1}^{k} \bar{c}_j = m$. Similarly, the update of μ_j and $\Sigma_j, j \in \mathbb{N}_k$, amounts to solving the problem

$$\text{maximize } \sum_{i=1}^{m} \bar{w}_{ij} \left(-(y_i - \mu_j)^T \Sigma_j^{-1}(y_i - \mu_j) + \ln \det\left(\Sigma_j^{-1}\right) \right), \tag{9.30}$$

which is concave in μ_j for a fixed Σ_j and concave in Σ_j^{-1} for a fixed μ_j. The first-order optimality conditions are

$$\sum_{i=1}^{m} \bar{w}_{ij} \Sigma_j^{-1}(y_i - \mu_j) = 0, \qquad \sum_{i=1}^{m} \bar{w}_{ij} \left(\Sigma_j - (y_i - \mu_j)(y_i - \mu_j)^T \right) = 0,$$

from which it follows that the updated mean is

$$\mu_j^+ = \frac{1}{\bar{c}_j} \sum_{i=1}^{m} \bar{w}_{ij} y_i, \quad j \in \mathbb{N}_k, \tag{9.31}$$

and the update covariance matrix is

$$\Sigma_j^+ = \frac{1}{\bar{c}_j} \sum_{i=1}^{m} \bar{w}_{ij}(y_i - \mu_j^+)(y_i - \mu_j^+)^T, \quad j \in \mathbb{N}_k. \tag{9.32}$$

We note that the weights $\bar{w}_{ij}$ are positive, and hence, Σ_j^+ is nonsingular if and only if $\operatorname{span}(\{y_1 - \mu_j^+, \dots, y_m - \mu_j^+\}) = \mathbb{R}^d$. Equivalently, Σ_j^+ is singular if and only if

$$\dim \operatorname{aff}(\{y_1, \dots, y_m\}) < d.$$

For example, this is the case if the number of observations m is less than or equal to d or if all observations lie on a hyperplane.

Example 9.6 We now illustrate the use of GMMs to approximate the pdf of a random variable based on m independent observations. Figure 9.7 shows examples in one and two dimensions. The one-dimensional example in Figure 9.7a shows the GMM pdf for a model with k components and with parameter estimates based on $m = 1000$ observations and computed using the EM algorithm. The observations are shown as a normalized histogram. The model seemingly fits the histogram well when we use a mixture model with three components. The two-dimensional example in Figure 9.7b shows $m = 500$ observations as dots and an ellipse for each component of the GMM obtained using the EM algorithm. Each ellipse defines the superlevel set containing 95% of the probability mass for the corresponding component. We see that the model with four components visually appears to be a reasonable approximation.

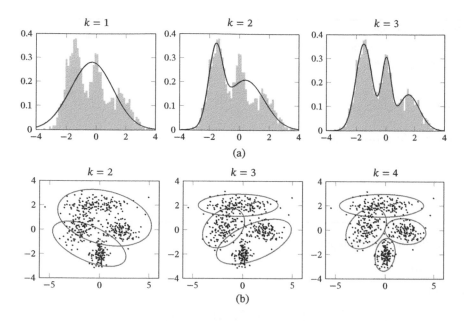

Figure 9.7 Observations and estimated GMM with k components based on maximum likelihood estimation: (a) one-dimensional GMMs and (b) two-dimensional GMMs.

9.12 Gibbs Sampling

Given a random variable $X : \mathbb{R}^n \to \mathbb{R}^n$ with pdf $f_X : \mathbb{R}^n \to \mathbb{R}_+$, it can be very expensive to evaluate quantities of interest that involve an integral when n is large, e.g. a marginal distribution or the expected value of some function of the random variable. One approach to addressing this issue is to use *Monte Carlo* techniques. The basic principle is to compute an SAA of the quantity of interest by drawing a set of random samples from the pdf.

The class of *Markov chain Monte Carlo* (MCMC) methods construct a Markov chain that has the desired distribution as its equilibrium distribution. However, the sequence generated by the Markov chain is autocorrelated, and this must be taking into account when estimating the error introduced by the SAA. A very long sequence is typically needed to compute a reasonably accurate approximation when the chain is highly autocorrelated. Moreover, the initialization of the Markov chain can have a strong influence on the first part of the sequence, which is typically discarded and is referred to as the *burn-in* phase.

Gibbs sampling is an example of an MCMC method that is sometimes useful when there is a simple way to sample from the one-dimensional conditional distributions $f_{X_i|X_{\bar{i}^c}} : \mathbb{R} \to \mathbb{R}_+, i \in \mathbb{N}_n$, where $X_{\bar{i}^c} = (X_1, \dots, X_{i-1}, X_{i+1}, \dots, X_n)$ is the random variable obtained from X by discarding X_i. Specifically, the Gibbs sampler starts with some initialization $x^{(0)} \in \mathbb{R}^n$ and generates a sequence $x^{(1)}, x^{(2)}, \dots$ by updating the n variables in a cyclic or randomized manner. Algorithm 9.3 shows the Gibbs sampler with a cyclic order, which is also known as the *deterministic-scan* Gibbs sampler. We note that $x_{\bar{i}^c}$ is defined analogously to $X_{\bar{i}^c}$. It can be shown that under mild conditions on f_X, the Gibbs sampler generates a realization of a Markov chain that converges to the target distribution as $k \to \infty$. Further details and a thorough introduction to sampling methods can be found in, e.g. [18, Chapter 11].

To illustrate the principle behind Gibbs sampling and its inherent limitations, we now consider the case where f_X is a two-dimension Gaussian pdf.

Algorithm 9.3: Gibbs sampler

Input: Starting point $x^{(0)} \in \mathbb{R}^n$ and number of samples K.
Output: Sequence $x^{(1)}, \ldots, x^{(K)}$.

$x \leftarrow x^{(0)}$
for $k \leftarrow 1$ **to** K **do**
    for $i \leftarrow 1$ **to** n **do**
        Draw a sample $\tilde{x}_i$ from $f_{X_i|X_{i^c}}(\cdot \,|x_{i^c})$.
        $x_i \leftarrow \tilde{x}_i$
    end
    $x^{(k)} \leftarrow x$
end

Example 9.7 Consider the Gaussian pdf

$$f_X(x) = \mathcal{N}(x, \mu, \Sigma),$$

with $X = (X_1, X_2)$ and

$$\mu = \begin{bmatrix} 2 \\ 2 \end{bmatrix}, \quad \Sigma = \begin{bmatrix} c & s \\ -s & c \end{bmatrix}^T \begin{bmatrix} 1 & 0 \\ 0 & \sigma^2 \end{bmatrix} \begin{bmatrix} c & s \\ -s & c \end{bmatrix} = \begin{bmatrix} c^2 + \sigma^2 s^2 & (1 - \sigma^2)cs \\ (1 - \sigma^2)cs & s^2 + \sigma^2 c^2 \end{bmatrix},$$

where $c = \cos(\theta)$ and $s = -\sin(\theta)$ are parameterized by $\theta \in [0, \pi/2]$. We note that the condition number of Σ is $\kappa = 1/\sigma^2$, and the correlation between X_1 and X_2 is

$$\rho = \frac{\Sigma_{12}}{\sqrt{\Sigma_{11}\Sigma_{22}}} = \frac{(1 - \sigma^2)cs}{\sqrt{(c^2 + \sigma^2 s^2)(s^2 + \sigma^2 c^2)}} = \frac{(\kappa - 1)cs}{\sqrt{\kappa + (\kappa - 1)^2 s^2 c^2}}.$$

Moreover, the conditional distribution of X_i given X_j, $i \neq j$, is given by

$$f_{X_i|X_j}(x_i|x_j) = \mathcal{N}(x_i, \mu_{i|j}, \Sigma_{i|j}),$$

where

$$\mu_{i|j} = \mu_i + \Sigma_{ij}\Sigma_{jj}^{-1}(x_j - \mu_j), \quad \Sigma_{i|j} = \Sigma_{ii} - \Sigma_{ij}\Sigma_{jj}^{-1}\Sigma_{ji}.$$

Figure 9.8 compares the Gibbs sampler in Algorithm 9.3 to direct sampling for different values of σ and θ. The plots clearly show that unlike the direct sampler, which produces independent samples, the Gibbs sampler generates an autocorrelated sequence. This is especially noticeable when X_1 and X_2 are highly correlated.

9.13 Boltzmann Machine

As an application of the EM algorithm, we will discuss the so-called *Boltzmann machine* (BM). To this end, we revisit the Ising distribution from Section 9.2 with pdf $p(x) = \exp\left(-E(x) - A(\lambda, \Lambda)\right)$, where $A(\lambda, \Lambda) = \ln\left(\sum_x \exp(-E(x))\right)$. The Boltzmann machine is obtained by splitting the variable x into *visible* and *hidden* or latent variables, i.e. $x = (v, h)$. We now consider N observations $v_1, \ldots, v_N$, of the visible variable. The corresponding hidden variables are $h_1, \ldots, h_N$. We let $x_i = (v_i, h_i)$, and with abuse of notation we let $v = (v_1, \ldots, v_N)$, $h = (h_1, \ldots, h_N)$ and $x = (v, h)$. We also define $r : \mathcal{D}_x^N \times \mathbb{R}^m \times \mathbb{S}^m \to \mathbb{R}$ via

$$r(x, \lambda, \Lambda) = \prod_{i=1}^{N} p(x_i) = e^{-\bar{E}(x,\lambda,\Lambda) - NA(\lambda,\Lambda)},$$

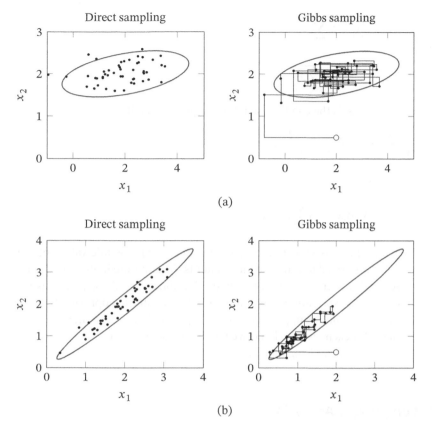

Figure 9.8 Realizations of $K = 50$ samples obtained via direct sampling and Gibbs sampling; the ellipse marks the superlevel set of f_X that contains 95% of the probability mass, and o marks the starting point $x^{(0)} = (2, 1/2)$ used in the Gibbs sampler. (a) Low correlation: $\sigma = 0.2$, $\theta = \pi/30$, and $\rho \approx 0.45$. (b) High correlation: $\sigma = 0.1$, $\theta = \pi/4$, and $\rho = 12/13$.

where $\bar{E} : \mathcal{D}_x^N \times \mathbb{R}^m \times \mathbb{S}^m \to \mathbb{R}$ is given by $\bar{E}(x, \lambda, \Lambda) = \sum_{i=1}^N E(x_i)$. The conditional pdf for h given v is $s : \mathcal{D}_v^N \times \mathcal{D}_h^N \times \mathbb{R}^m \times \mathbb{S}^m \to \mathbb{R}$ given by

$$s(v, h, \lambda, \Lambda) = \frac{r(v, h, \lambda, \Lambda)}{\sum_h r(v, h, \lambda, \Lambda)} = \frac{e^{-\bar{E}(v,h,\lambda,\Lambda)}}{\sum_h e^{-\bar{E}(v,h,\lambda,\Lambda)}},$$

where $\mathcal{D}_v$ and $\mathcal{D}_h$ are defined such that $\mathcal{D}_v \times \mathcal{D}_v = \mathcal{D}_x$. The summation over h above is carried out over all $h \in \mathcal{D}_h$. From now on, we tacitly assume that all summations are carried over all the elements in the corresponding sets unless otherwise stated. We define $Q : \mathbb{R}^m \times \mathbb{S}^m \to \mathbb{R}$ via

$$Q(\lambda, \Lambda) = E_s \ln r = \sum_h s(v, h, \lambda^-, \Lambda^-) \left(-\bar{E}(v, h, \lambda, \Lambda) - NA(\lambda, \Lambda) \right),$$

$$= -\sum_h s(v, h, \lambda^-, \Lambda^-) \bar{E}(v, h, \lambda, \Lambda) - NA(\lambda, \Lambda),$$

where (λ^-, Λ^-) are the values of the parameters from the previous iteration in the EM algorithm. In order to maximize Q, the gradient with respect to (λ, Λ) is typically needed. We have

$$\frac{\partial Q}{\partial \lambda} = -\sum_h s(h, v, \lambda^-, \Lambda^-) \frac{\partial \bar{E}}{\partial \lambda} - N \frac{\partial A}{\partial \lambda},$$

$$= \sum_h s(h, v, \lambda^-, \Lambda^-) \sum_{i=1}^N x_i - N \sum_\xi \xi p(\xi),$$

$$= \sum_{i=1}^N \sum_{h_i} s_i(h_i, v_i, \lambda^-, \Lambda^-) x_i - N \sum_\xi \xi p(\xi),$$

where $s_i : D_v \times D_h \times \mathbb{R}^m \times \mathbb{S}^m \to \mathbb{R}$ is the conditional pdf for h_i given v_i defined as

$$s_i(v_i, h_i, \lambda, \Lambda) = \frac{p(x_i)}{\sum_{h_i} p(x_i)} = \frac{e^{-E(v_i, h_i, \lambda, \Lambda)}}{\sum_{h_i} e^{-E(v_i, h_i, \lambda, \Lambda)}}.$$

Similarly, we obtain

$$\frac{\partial Q}{\partial \Lambda} = \sum_{i=1}^N \sum_{h_i} s_i(h_i, v_i, \lambda^-, \Lambda^-) x_i x_i^T - N \sum_\xi \xi \xi^T p(\xi).$$

Note that above $x = (v, h)$, and that v is known and that we sum over all possible values of the hidden variables h. Hence, it is not trivial to compute the gradients if h and ξ are high dimensional because of the fact that we need to sum over many variables. We notice that the expressions for the gradients of Q are differences of expectations, and therefore, it is possible to approximate them using Monte Carlo techniques like Gibbs sampling; see, e.g. [38]. Note that the gradients can be computed as sums of gradients for each observation i. We will later discuss what further structure can be utilized when we have a graphical Ising model.

9.14 Principal Component Analysis

We are given a collection of data $x_i \in \mathbb{R}^n$, $i \in \mathbb{N}_N$. We are interested in approximating this data with $W^T c_i$, where $W \in \mathbb{R}^{m \times n}$, $m < n$, and where $c_i \in \mathbb{R}^m$, $i \in \mathbb{N}_N$. We require the rows of W to be orthonormal, i.e. $WW^T = I$. We will use a least-squares criterion to formalize what we mean with approximation. To this end, we let $J : \mathbb{R}^{m \times n} \times \mathbb{R}^n \times \cdots \times \mathbb{R}^n \to \mathbb{R}_+$ be defined by

$$J(W, c_1, \ldots, c_N) = \frac{1}{2} \sum_{i=1}^N \|x_i - W^T c_i\|_2^2.$$

The optimization problem is

$$\text{minimize } J(W, c_1, \ldots, c_N),$$
$$\text{subject to } WW^T = I,$$

with variables $W, c_1, \ldots, c_N$. The solution to this problem will give us a lower-dimensional description of the original data. The rows of W are called the principal components, and the analysis we carry out is called *principal component analysis* (PCA).

9.14.1 Solution

This is a convex problem with respect to c_i for fixed value of W. Hence, we first carry out this minimization. The optimal values are obtained by setting the gradient equal to zero, i.e.

$$\frac{\partial J}{\partial c_i} = WW^T c_i - W x_i = 0, \quad i \in \mathbb{N}_N.$$

Since $WW^T = I$, we get that $c_i = Wx_i$. We realize that the principal components are used to compress the original data x_i to the lower-dimensional data c_i. Back substitution into J gives

$$J(W, Wx_1, \ldots, Wx_N) = \frac{1}{2}\sum_{i=1}^{N} x_i^T\left(I - W^TW\right)^2 x_i.$$

Since $I - W^TW$ is a projection matrix, we may remove the square. From this, we see that it is equivalent to maximize

$$\frac{1}{2}\sum_{i=1}^{N} x_i^T W^T W x_i = \frac{1}{2}\left(\text{tr} X^T X W^T W\right) = \frac{1}{2}\text{tr}\left(WX^TXW^T\right),$$

where

$$X = \begin{bmatrix} x_1^T \\ \vdots \\ x_N^T \end{bmatrix}.$$

Let $X = UDV^T$ be a singular value decomposition such that the diagonal matrix D has the elements sorted in decreasing order. Let $Y = V^TW^T$. Then we may define the criterion above as $\tilde{J} : \mathbb{R}^{n\times m} \to \mathbb{R}_+$, where

$$\tilde{J}(Y) = \frac{1}{2}\text{tr}\left(Y^TD^2Y\right),$$

and define the optimization problem equivalently as

$$\text{maximize } \tilde{J}(Y),$$
$$\text{subject to } Y^TY = I,$$

where we have made use of the fact that $WW^T = I$ if and only if $Y^TY = I$. This is a nonconvex optimization problem. It can be shown that the gradient of the constraint function $h : \mathbb{R}^{n\times n} \to \mathbb{S}^n$ defined by $h(Y) = I - Y^TY$ is full rank for all orthogonal Y, and hence, the linear independence condition is satisfied for the necessary optimality conditions in Section 4.7. To see this notice that it can be shown that

$$\frac{\partial \text{ svec } h(Y)}{\partial(\text{vec } Y)^T}\left(\frac{\partial \text{ svec } h(Y)}{\partial(\text{vec } Y)^T}\right)^T$$

is a diagonal matrix with positive diagonal. It is important to only consider the symmetric vectorization of h since otherwise, the condition does not hold. Because of this the Lagrange multiplier has to be a symmetric matrix. Define the Lagrangian $L : \mathbb{R}^{n\times m} \times \mathbb{S}^m \to \mathbb{R}$ as

$$L(Y, \Lambda) = \tilde{J}(L) + \text{tr}\left(\Lambda\left(I - Y^TY\right)\right).$$

Then a necessary condition for Y to be optimal is that there exist Λ such that the gradient of the Lagrangian vanishes, i.e.

$$\frac{\partial L}{\partial Y} = D^2Y - Y\Lambda = 0.$$

Let Z be such that $\begin{bmatrix} Y & Z \end{bmatrix}$ is orthogonal, i.e. $\begin{bmatrix} Y & Z \end{bmatrix}^T \begin{bmatrix} Y & Z \end{bmatrix} = I$, and multiply the above equation with $\begin{bmatrix} Y & Z \end{bmatrix}^T$ from the left to obtain the equivalent equations

$$Y^TD^2Y - \Lambda = 0,$$
$$Z^TD^2Y = 0.$$

Note that the first equation always has a solution Λ for any Y, and hence, the necessary conditions of optimality are equivalent to existence of Z such that

$$\begin{bmatrix} Y & Z \end{bmatrix}^T \begin{bmatrix} Y & Z \end{bmatrix} = I,$$
$$Z^T D^2 Y = 0.$$

Clearly, one solution to these equations is $\begin{bmatrix} Y & Z \end{bmatrix} = I$. It is actually optimal. Moreover, $\begin{bmatrix} Y & Z \end{bmatrix} = \text{blkdiag}(X_1, X_2)$, for any orthogonal $X_1 \in \mathbb{R}^{m \times m}$ and $X_2 \in \mathbb{R}^{(n-m) \times m}$ are also optimal. We will show this by noting that the objective function can be written as

$$\frac{1}{2} \sum_{i=1}^{N} \|Dy_i\|_2^2,$$

where $Y = \begin{bmatrix} y_1 & \cdots & y_m \end{bmatrix}$. Now we start by optimizing with respect to y_1. We have

$$Dy_1 = \begin{bmatrix} d_1 y_{11} \\ d_2 y_{21} \\ \vdots \\ d_n y_{n1} \end{bmatrix},$$

where d_i are the diagonal elements of D, which we remember are ordered such that $d_i \geq d_j$ when $i < j$. The constraint $Y^T T = I$ implies that $y_1^T y_1 = 1$, and therefore, it is optimal to take $y_1 = e_1$, which is the first basis vector. All the other y_i, for $2 \leq i \leq m$ has to have its first component equal to zero in order to be orthogonal to y_1. Because of this, we by repeating the arguments above, find that $y_2 = e_2$, i.e. we pick out the second largest diagonal element of D. The remaining y_i now have to have the first two components equal to zero in order to be orthogonal to y_1 and y_2. We thus conclude that $Y = \begin{bmatrix} I \\ 0 \end{bmatrix}$ is optimal. We then notice that $Y = \begin{bmatrix} X_1 \\ 0 \end{bmatrix}$ with X_1 orthogonal will result in the same objective function value. This follows from the fact that

$$\tilde{J}\left(\begin{bmatrix} X_1 \\ 0 \end{bmatrix}\right) = \frac{1}{2} \text{tr}\left(\begin{bmatrix} X_1 \\ 0 \end{bmatrix}^T D^2 \begin{bmatrix} X_1 \\ 0 \end{bmatrix}\right) = \frac{1}{2} \text{tr}\left(X_1^T D_1^2 X_1\right) = \frac{1}{2} \text{tr}\left(X_1 X_1^T D_1^2\right),$$

where $D = \text{blkdiag}(D_1, D_2)$. Hence, the PCA picks out the components of x_i corresponding to the m largest singular values of X^T.

Example 9.8 In this example, we will perform PCA analysis on the Fisher Iris flower data set [39]. It contains measurements of four different characteristics of three different iris species. There are 150 rows in the data set. Each subset of 50 rows corresponds to three different iris species. Each of the four columns corresponds to the a different characteristic. We preprocess the data by subtracting the mean value of each column from all the values in that column. We also divide all values in a column with the standard deviation of its column values. This is then the X-matrix that we use for PCA. Hence, each row of it is a scaled observation x_i^T. We compute the two principal components corresponding to the largest singular values, and then compute the compressed data $c_i \in \mathbb{R}^2$. In Figure 9.9, we have plotted the first component of each c_i versus its second component. The different species are marked differently in the plot. The PCA analysis makes it possible to visualize high-dimensional data in low-dimensional plots and, hence, makes the data more understandable.

We will now relate the solution $(W, c_1, \dots, c_N)$ to X. We approximated x_i with $W^T c_i$, where we found that $c_i = Wx_i$. Because of this X is approximated with $XW^T W$. Since W has orthonormal rows, $W^T W$ is a projection matrix, and hence, the rank of $XW^T W$ is equal to m. This shows

Figure 9.9 Plot of compressed data c resulting from PCA. The difference species are displayed with different markers.

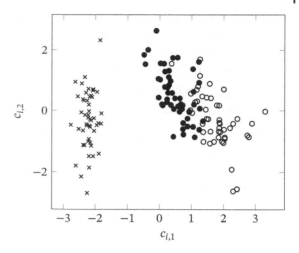

that PCA computes a low-rank approximation of X which minimizes the Frobenius norm of $X(I - W^T W) = X - X W^T W$. Furthermore, we have $X W^T W = U D V^T V_1 V_1^T = U D \begin{bmatrix} I \\ 0 \end{bmatrix} V_1^T = U_1 D_1 V_1^T$, where $U = \begin{bmatrix} U_1 & U_2 \end{bmatrix}$ and $V = \begin{bmatrix} V_1 & V_2 \end{bmatrix}$.

9.14.2 Relation to Rank-Constrained Optimization

We will now relate the above solution to the rank-constrained optimization problem in (5.23). There the solution is $Z = U_1 D_1 V_1^T$, and for PCA it is $X W^T W = U_1 D_1 V_1^T$. Hence, the solutions agree. The objective function that is minimized in PCA is the Frobenius norm of $X(I - W^T W) = X - X W^T W$ with respect to W. In the rank-constrained problem, we instead minimize the induced norm of $X - Z$. In the rank-constrained optimization, we have an explicit constraint on the rank of Z, whereas we in PCA constrain the rank implicitly with the number of rows in W, which are equal to the constraint on the rank.

9.15 Mutual Information

A quantity that is closely related to entropy is *mutual information*. For a joint pdf $r : \mathbb{R}^m \times \mathbb{R}^n \to \mathbb{R}_+$ of two random variables with marginal pdfs $p : \mathbb{R}^m \to \mathbb{R}_+$ and $q : \mathbb{R}^n \to \mathbb{R}_+$ we define the mutual information $I : \mathcal{D}_{m+n} \to \mathbb{R}_+$ as

$$I(r) = \int_{\mathbb{R}^m \times \mathbb{R}^n} r(x, y) \ln \frac{r(x, y)}{p(x) q(y)} dx dy,$$

where $\mathcal{D}_{m+n}$ is defined as in Section 9.9.

9.15.1 Channel Model

We will show how mutual information can be used to generalize principal component analysis. To this end, we consider two zero mean independent random variables $X \in \mathbb{R}^n$ and $E \in \mathbb{R}^m$ that both have a normal density with covariances $\Sigma \in \mathbb{S}_+^n$ and I, respectively. We also define the random variables $Z = WX$ and $Y = Z + E$, where $W \in \mathbb{R}^{m \times n}$ with $m < n$.[6] We can interpret Z as information that is transmitted over a channel W with additive noise E. We would like to choose W to maximize the mutual information between Y and Z, i.e. between what is transmitted and what is received. We realize that (Y, Z) has a zero mean normal pdf r with covariance

$$\begin{bmatrix} W\Sigma W^T + I & W\Sigma W^T \\ W\Sigma W^T & W\Sigma W^T \end{bmatrix}.$$

We let p and q be the marginal pdfs for Y and Z and define $J : \mathbb{R}^{m \times n} \to \mathbb{R}_+$ as

$$J(W) = I(r).$$

Tedious calculations show that

$$J(W) = \frac{1}{2} \ln \, \det \left(I + W\Sigma W^T \right).$$

9.15.2 Orthogonal Case

The function J is not bounded from above, and therefore, we introduce a constraint $WW^T = I$ that the function should satisfy, i.e. its rows should be orthonormal. We now let V be such that $V^T V = I$ and $\Sigma = VD^2V^T$ with D a diagonal matrix. From this, it follows that

$$J(W) = \frac{1}{2} \ln \, \det \left(I + \bar{Y}^T D^2 \bar{Y} \right),$$

where $\bar{Y} = V^T W^T$. We assume that the diagonal elements d_i, $i \in \mathbb{N}_n$ of D are positive and ordered such that $d_i \geq d_j$ if $i \geq j$.

We realize that it is equivalent to maximize over $\bar{Y}$ since V is invertible. We therefore define $\tilde{J} : \mathbb{R}^{n \times m} \to \mathbb{R}_+$ via $\tilde{J}(\bar{Y}) = J(W)$. Notice that $WW^T = I$ is equivalent to $\bar{Y}^T \bar{Y} = I$. We define the Lagrangian $L : \mathbb{R}^{n \times m} \times \mathbb{S}^m \to \mathbb{R}$ as

$$L(\bar{Y}, \Lambda) = \frac{1}{2} \ln \, \det \left(I + \bar{Y}^T D^2 \bar{Y} \right) + \frac{1}{2} \mathrm{tr} \left(\Lambda \left(I - \bar{Y}^T \bar{Y} \right) \right).$$

We have that the existence of Λ such that

$$\frac{\partial L}{\partial \bar{Y}} = D^2 \bar{Y} \left(I + \bar{Y}^T D^2 \bar{Y} \right)^{-1} - \bar{Y}\lambda = 0$$

is a necessary condition for optimality of $\bar{Y}$. Let $\bar{Z}$ be such that $\begin{bmatrix} \bar{Y} & \bar{Z} \end{bmatrix}$ is square and orthogonal. Multiply the above equation with the transpose of this matrix from the left to obtain that equivalent conditions for optimality are existence of $\bar{Z}$ and Λ such that

$$\bar{Y}^T D^2 \bar{Y} \left(I + \bar{Y}^T D^2 \bar{Y} \right)^{-1} - \Lambda = 0,$$

$$\bar{Z}^T D^2 \bar{Y} = 0,$$

where $\bar{Z}$ is such that $\begin{bmatrix} \bar{Y} & \bar{Z} \end{bmatrix}$ is orthogonal. The second equation follows since $I + \bar{Y}^T D^2 \bar{Y}$ is invertible for all $\bar{Y}$. From the formula $A(I + A)^{-1} = I - (I + A)^{-1}$, we may rewrite the above equations as

$$I - \left(I + \bar{Y}^T D^2 \bar{Y} \right)^{-1} - \Lambda = 0,$$

$$\bar{Z}^T D^2 \bar{Y} = 0.$$

---

6 Note that the dimensions of X and Y are not the same as the dimensions of x and y in the definition of mutual information in Section 9.14.

This now shows that the first equation has a solution in terms of Λ for any $\bar{Y}$, since $I - \left(I + \bar{Y}^T D^2 \bar{Y}\right)^{-1}$ is symmetric for any $\bar{Y}$. Hence, the optimality conditions simplify to the existence of $\bar{Z} \in \mathbb{R}^{n \times (n-m)}$ such that

$$\bar{Z}^T D^2 \bar{Y} = 0,$$

$$\begin{bmatrix} \bar{Y} & \bar{Z} \end{bmatrix}^T \begin{bmatrix} \bar{Y} & \bar{Z} \end{bmatrix} = I.$$

These are similar to the optimality conditions for PCA if we identify Σ with $X^T X$.[7] However, the objective functions are not the same, and therefore, we have to proceed slightly differently.

As in Section 9.14, $\begin{bmatrix} \bar{Y} & \bar{Z} \end{bmatrix} = I$ is a solution to the optimality conditions and so is $\begin{bmatrix} \bar{Y} & \bar{Z} \end{bmatrix} = \text{blkdiag}(X_1, X_2)$ with $X_1 \in \mathbb{R}^{m \times m}$ and $X_2 \in \mathbb{R}^{(n-m) \times (n-m)}$ orthonormal. It is straightforward to verify that the objective function evaluates to

$$\tilde{J}(\bar{Y}) = \frac{1}{2} \sum_{k=1}^{m} \ln\left(1 + d_k^2\right).$$

Can we consider more general $\begin{bmatrix} \bar{Y} & \bar{Z} \end{bmatrix}$? Any orthogonal matrix with determinant equal to one can be written as a product of Givens rotations. Notice that there is no restriction in assuming the determinant to be equal to one, since we multiply both from left and right. A Givens rotation is defined as $G : \mathbb{N}_n \times \mathbb{N}_n \times [0, 2\pi] \to \mathbb{R}^{n \times n}$, where

$$G(i, j, \theta) = \begin{bmatrix} 1 & \cdots & 0 & \cdots & 0 & \cdots & 0 \\ \vdots & \ddots & \vdots & & \vdots & & \vdots \\ 0 & \cdots & c & \cdots & -s & \cdots & 0 \\ \vdots & & \vdots & \ddots & \vdots & & \vdots \\ 0 & \cdots & s & \cdots & c & \cdots & 0 \\ \vdots & & \vdots & & \vdots & \ddots & \vdots \\ 0 & \cdots & 0 & \cdots & 0 & \cdots & 1 \end{bmatrix},$$

where $c = \cos\theta$, $s = \sin\theta$, and where c and s are positioned on the ith and jth rows and columns, $cf.$ (2.17). We assume that $i < j$. From this, it follows that

$$G(i, j, \theta)^T D^2 G(i, j, \theta) = \begin{bmatrix} d_1^2 & \cdots & 0 & \cdots & 0 & \cdots & 0 \\ \vdots & \ddots & \vdots & & \vdots & & \vdots \\ 0 & \cdots & c^2 d_i^2 + s^2 d_j^2 & \cdots & -scd_i^2 + scd_j^2 & \cdots & 0 \\ \vdots & & \vdots & \ddots & \vdots & & \vdots \\ 0 & \cdots & -scd_i^2 + scd_j^2 & \cdots & s^2 d_i^2 + c^2 d_j^2 & \cdots & 0 \\ \vdots & & \vdots & & \vdots & \ddots & \vdots \\ 0 & \cdots & 0 & \cdots & 0 & \cdots & d_n^2 \end{bmatrix}.$$

We understand that it is only if $i \leq m$ and $j > m$ that we do not have the case $G(i, j, \theta) = \text{blkdiag}(X_1, X_2)$ discussed above. For these values of i and j, the top left $m \times m$-dimensional block of the above matrix is given by

$$\begin{bmatrix} d_1^2 & \cdots & 0 & \cdots & 0 \\ \vdots & \ddots & \vdots & & \vdots \\ 0 & \cdots & c^2 d_i^2 + s^2 d_j^2 & \cdots & 0 \\ \vdots & & \vdots & \ddots & \vdots \\ 0 & \cdots & 0 & \cdots & d_m^2 \end{bmatrix}.$$

---

7 We will discuss this in more detail later on. There should actually be a normalization with N.

Hence, the objective function value is given by

$$\tilde{J}(\bar{Y}) = \frac{1}{2} \sum_{k \in \mathbb{N}_m \setminus \{i\}} \ln\left(1 + d_k^2\right) + \ln\left(1 + c^2 d_i^2 + s^2 d_j^2\right).$$

We realize that the Givens rotation replaces the ith diagonal element d_i^2 with $c^2 d_i^2 + s^2 d_j$, which is a convex combination of the values d_i^2 and d_j^2. Since $d_j < d_i$ it follows that $c^2 d_i^2 + s^2 d_j < d_i^2$, and hence, the Givens rotation has resulted in a smaller value of the objective function value. Any further rotations for which $i \leq m$ and $j > m$ will only further decrease the objective function value. Any other rotation will not affect the objective function value. From this, we realize that all $\bar{Y} = \begin{bmatrix} X_1 \\ 0 \end{bmatrix}$ with $X_1^T X_1 = I$ are optimal, and hence, maximizes the mutual information under the constraint that $\bar{Y}^T \bar{Y} = I$. There are actually several local stationary points. The reason for this is that for $\theta = k\pi/2$, $k \in \mathbb{N}$ it holds that $-scd_i^2 + scd_j^2 = 0$, and hence, $\bar{Z}^T D \bar{Y} = 0$ irrespective of what values i and j have.

9.15.3 Nonorthogonal Case

We will now investigate what happens if we drop the orthogonality constraint and only require that $\|y_i\|_2^2 = 1, i \in \mathbb{N}_m$, where $\bar{Y} = \begin{bmatrix} y_1 & \cdots & y_m \end{bmatrix}$. We define the Lagrangian $M : \mathbb{R}^{n \times m} \times \mathbb{R}^m \to \mathbb{R}$ as

$$M(\bar{Y}, \lambda) = \frac{1}{2} \ln \, \det\left(I + \bar{Y}^T D^2 \bar{Y}\right) + \frac{1}{2} \sum_{i=1}^m \lambda_i \left(1 - \|y_i\|_2^2\right).$$

We see that

$$\frac{\partial M}{\partial \bar{Y}} = D^2 \bar{Y} \left(I + \bar{Y}^T D^2 \bar{Y}\right)^{-1} - \bar{Y} \mathrm{diag}(\lambda),$$

and hence, necessary conditions for optimality are existence of Z and $\lambda \in \mathbb{R}^m$ such that

$$\bar{Y}^T D^2 \bar{Y} \left(I + \bar{Y}^T D^2 \bar{Y}\right)^{-1} - \bar{Y}^T \bar{Y} \mathrm{diag}(\lambda) = 0,$$

$$\bar{Z}^T D^2 \bar{Y} = 0,$$

where $\bar{Z}$ is such that $\begin{bmatrix} \bar{Y} & \bar{Z} \end{bmatrix}$ has full column rank. Notice that $\begin{bmatrix} \bar{Y} & \bar{Z} \end{bmatrix}$ is not necessarily a square matrix. Similarly as above, we may rewrite the above equations as

$$I - \left(I + \bar{Y}^T D^2 \bar{Y}\right)^{-1} - \bar{Y}^T \bar{Y} \mathrm{diag}(\lambda) = 0,$$

$$\bar{Z}^T D^2 \bar{Y} = 0.$$

It is straightforward to verify that $\begin{bmatrix} \bar{Y} & \bar{Z} \end{bmatrix} = I$ satisfy also these optimality conditions. However, any orthogonal $\begin{bmatrix} \bar{Y} & \bar{Z} \end{bmatrix}$ will not satisfy them, since for orthogonal $\bar{Y}$ it must hold that $I + \bar{Y}^T D \bar{Y}$ is diagonal for there to exist λ that satisfies the equation. There are however nonorthogonal $\bar{Y}$ that satisfy them. Consider $\bar{Y} = \begin{bmatrix} \mathbb{1} & 0 \end{bmatrix}^T$, where $\mathbb{1} \in \mathbb{R}^m$ is a vector of all ones. This means that all $y_i = e_1$, where e_1 is the first basis vector in $\mathbb{R}^n$. It is straightforward to verify that $\lambda = d_1^2/(1 + m d_1^2)\mathbb{1}$ satisfies the necessary optimality conditions for this $\bar{Y}$. It actually holds that we may take y_i equal to any basis vector for $\mathbb{R}^n$, and that they may be linearly dependent.

Now, the question arises if such a $\bar{Y}$ where some of the columns are not linearly independent may be optimal and not only constitute a stationary point. To investigate this, we consider the case when $m = 2$, $d_1 > d_2$ and let $\bar{Y} = \begin{bmatrix} \mathbb{1} & 0 \end{bmatrix}^T$. Then $\ln \, \det(I + \bar{Y}^T D^2 \bar{Y}) = \ln(1 + 2d_1^2) \approx 2d_1^2$ for small values of d_1^2. For $\bar{Y} = \begin{bmatrix} I \\ 0 \end{bmatrix}$ we have $\ln \, \det(I + Y^T D^2 Y) = \ln(1 + d_1^2 + d_2^2 + d_1^2 d_2^2) \approx d_1^2 + d_2^2 + d_1^2 d_2^2$ for small values of d_1^2 and d_2^2. It is easy to find values of $d_1 > d_2$ for which the second approximation is smaller

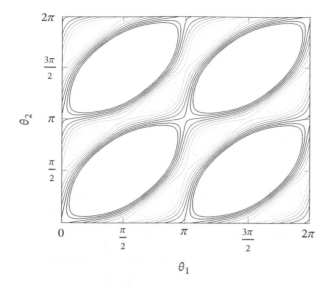

Figure 9.10 Level curves for the objective function between 2.15 (dark gray) and 2.4 (light gray).

than the first. Hence, we realize that when the signal-to- noise ratio is small, i.e. when Σ is small as compared to I, then it is better to only consider the signal with the largest variance. In general there are cases in-between, where one should pick out more than one of the largest, but not all m. Also there are cases when it optimal for the vectors y_i to have an angle in-between them such that they are neither aligned nor orthogonal [28]. A small example with $m = n = 2, d_1 = 2$, and $d_2 = 1$ is visualized in Figure 9.10, where we let $y_1 = \begin{bmatrix} \cos \theta_1 \\ \sin \theta_1 \end{bmatrix}$ and $y_2 = \begin{bmatrix} \cos \theta_2 \\ \sin \theta_2 \end{bmatrix}$ for $\theta_1, \theta_2 \in [0, 2\pi]$ and plot the level curves of the objective function values. It is seen that

$$\bar{Y} = \begin{bmatrix} 1 & 1 \\ 0 & 0 \end{bmatrix} ; \quad \begin{bmatrix} 1 & 0 \\ 0 & 1 \end{bmatrix} ; \quad \begin{bmatrix} 0 & 1 \\ 1 & 0 \end{bmatrix} ; \quad \begin{bmatrix} 0 & 0 \\ 1 & 1 \end{bmatrix}$$

correspond to saddle points. There are several optima, and one is given by

$$\bar{Y} = \begin{bmatrix} 0.83 & 0.83 \\ -0.56 & 0.56 \end{bmatrix},$$

which corresponds to $\theta_1 = 5.7$ and $\theta_2 = 0.6$. Note that the other optima have the property that the angle between y_1 and y_2 are all the same. If the signal-to-noise ratio is large, then $\bar{Y} = \begin{bmatrix} X_1 \\ 0 \end{bmatrix}$ with X_1 having orthonormal columns satisfies the necessary optimality conditions also without orthogonality constraints. This follows from the fact that the first equation of the optimality conditions in the limit when D is much larger than I is given by

$$I - \bar{Y}^T \bar{Y} \text{diag}(\lambda) = 0.$$

The problem discussed is a special case of optimization on a matrix manifold, which is discussed in more detail in [1].

9.15.4 Relationship to PCA

The relation to PCA become apparent if we consider the case where we have observations $x_i, i \in \mathbb{N}_N$ of X and we want to approximate them with $W^T z_i$, where $z_i = W x_i$. We estimate the covariance matrix of X with $\Sigma = \frac{1}{N}\bar{X}^T\bar{X}$, where

$$\bar{X} = \begin{bmatrix} x_1^T \\ x_2^T \\ \vdots \\ x_N^T \end{bmatrix}.$$

We then let $\bar{X} = UDV^T$ be a singular value decomposition of X. From this it follows that

$$J(W) = \frac{1}{2}\ln\,\det\left(I + \bar{Y}^T\frac{D^2}{N}\bar{Y}\right),$$

where $\bar{Y} = V^TW^T$. The approximation of $\bar{X}$ will be $\bar{X}W^TW$, where W^TW is not necessarily a projection matrix. However, the rank of $\bar{X}W^TW$ will still be m. Notice that we cannot relax the condition on orthogonality in the principal component analysis without obtaining $\bar{Y} = \begin{bmatrix} \mathbb{1}^T \\ 0 \end{bmatrix}$ as the optimal solution.[8] The signal-to-noise ratio will not help in making the correct choice from an information point of view. This is why optimizing mutual information seems to be more appropriate.

9.16   Cluster Analysis

Cluster analysis is about partitioning observations into groups called *clusters* so that the observations in each cluster are close to one another. For observations $x_i \in \mathbb{R}^n$, $i \in \mathbb{N}_N$, we may measure the closeness with some norm and define the distance $d : \mathbb{R}^n \times \mathbb{R}^n \to \mathbb{R}_+$ as, e.g. the squared norm of the difference between two observations, i.e. $d(x_i, x_j) = \|x_i - x_j\|^2$. We will assume that there are $K < N$ clusters, and we are looking for an *encoder* $C : \mathbb{N}_N \to \mathbb{N}_K$ that assigns each observation to a cluster. This should be done in such a way that the sum of all distances within each cluster is minimized, i.e. we have the following combinatorial optimization problem

$$\text{minimize } f(C),$$

with variable C, where $f : \mathbb{N}_N \times \mathbb{N}_K \to \mathbb{R}_+$ is defined by

$$f(C) = \frac{1}{2}\sum_{k=1}^{K}\sum_{i\in C_k}\sum_{j\in C_k}d(x_i, x_j),$$

where $C_k = \{i \in \mathbb{N}_N : C(i) = k\}$. Define the mean vectors for each cluster as

$$m_k = \frac{1}{N_k}\sum_{i\in C_k}x_i, \quad k \in \mathbb{N}_K,$$

where $N_k = |C_k|$. It then holds that

$$f(C) = \frac{1}{2}\sum_{k=1}^{K}\sum_{i\in C_k}\sum_{j\in C_k}d(x_i, x_j) = \sum_{k=1}^{K}N_k\sum_{i\in C_k}d(x_i, m_k),$$

if d is the squared Euclidian norm. Notice that for any $y_k \in \mathbb{R}^n$, it holds that $\sum_{i\in C_k}d(x_i, m_k) \leq \sum_{i\in C_k}d(x_i, y_k)$ with equality for $y_k = m_k$ and hence, an equivalent optimization problem is

$$\text{minimize } F(C, y_1, \ldots, y_K),$$

---

8 Notice that what is called $\bar{X}$ and $\bar{Y}$ in this section is called X and Y in the section on PCA. The reason is that we used X and Y for random variables in this section.

Figure 9.11 Plot of resulting clustering. The original data x_i is marked with + and the y_k are marked with o.

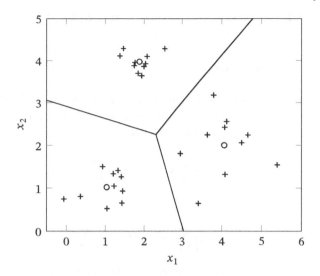

with variables $C, y_1, \ldots, y_K$, where $F : \mathbb{N}_K \times \mathbb{R}^n \times \cdots \times \mathbb{R}^n \to \mathbb{R}_+$ is defined as

$$F(C, y_1, \ldots, y_K) = \sum_{k=1}^{K} N_k \sum_{i \in C_k} d(x_i, y_k).$$

The sets C_k are functions of the encoder C. The so-called K-means algorithm tries to solve the above formulation using block coordinate descent, i.e. it iteratively solves the following two optimization problems:

1. For fixed C, minimize $F(C, y_1, \ldots, y_K)$ with respect to $y_1, \ldots, y_K$.
2. For fixed $(y_1, \ldots, y_K)$ minimize $F(C, y_1, \ldots, y_K)$ with respect to C.

The first problem is a least-squares problem with solution equal to the average of the x_i over $i \in C_k$, i.e.

$$y_k = \frac{1}{N_k} \sum_{i \in C_k} x_i.$$

This is the reason for the name K-means. The second problem also has an explicit solution given by assigning observation x_i to cluster k if $d(x_i, y_k) \le d(x_i, y_j)$ for all $j \ne k$. The algorithm can unfortunately be trapped in local minima.

Example 9.9   We now consider a problem in two dimensions for which $N = 30$. We want to perform the clustering using $K = 3$. We initialize y_k as the first three values of x_i that we are given. They actually come from the same cluster, showing that initialization is not extremely critical. The result is shown in Figure 9.11, where we see that we get excellent clustering.

Exercises

9.1   Consider the special case of finding a Chebyshev bound as detailed in Section 9.1 when $S = \mathbb{R}_+, C = [1, \infty), f_0(x) = 1$ and $f_1(x) = x$. Assume that it is known that $\mathbb{E} f_1(X) = \mathbb{E} X = \mu$, where $0 \le \mu \le 1$. Show that this implies the so-called Markov bound

$$\mathbb{P}[\, X \ge 1 \,] \le \mu.$$

9.2 Show how the function h in (9.5) for the Ising distribution can be derived by back-substitution of $\mu = 1 - A(\lambda, \Lambda)$ into the Lagrange dual function.

Hint: You need to first obtain the Lagrange dual function by back substitution of p into the Lagrangian.

9.3 Consider minimizing (9.12) with respect to (A, b). Show that the solution is the same as the solution obtained by minimizing the criterion in (9.11) if m_x, m_y, D_x, and D_{xy} are replaced with their sample averages

$$\hat{m}_x = \frac{1}{N}\sum_{i=1}^{N} x_i, \quad \hat{m}_y = \frac{1}{N}\sum_{i=1}^{N} y_i,$$

and

$$\hat{D}_x = \frac{1}{N}\sum_{i=1}^{N} \bar{x}_i\bar{x}_i^T, \quad \hat{D}_{xy} = \frac{1}{N}\sum_{i=1}^{N} \bar{x}_i\bar{y}_i^T,$$

respectively, where $\bar{x}_i = x_i - \hat{m}_x$ and $\bar{y}_i = y_i - \hat{m}_y$.

9.4 [22, Exercise 4.57] Consider two discrete random variables X and Y, where $(X, Y) : \mathbb{N}_n \times \mathbb{N}_m \to \mathbb{R}$. The mutual information $I : \mathcal{D}_{mn} \to \mathbb{R}$ is defined as

$$I(p_{X,Y}) = \sum_{i=1}^{n}\sum_{j=1}^{m} p_{X,Y}(i,j) \log \frac{p_{X,Y}(i,j)}{p_X(i)p_Y(j)},$$

where $p_{X,Y} : \mathbb{N}_n \times \mathbb{N}_m \to [0,1]$ is the joint probability function, and where $p_X : \mathbb{N}_n \to [0,1]$ and $p_Y : \mathbb{N}_m \to [0,1]$ are the marginal probability functions for X and Y, respectively.

(a) Show that the mutual information can be expressed as

$$I(p_{X,Y}) = \sum_{i=1}^{n}\sum_{j=1}^{m} p_X(i)p_{Y|X}(j|i) \log \frac{p_{Y|X}(j|i)}{\sum_{k=1}^{n} p_X(k)p_{Y|X}(j|k)},$$

where $p_{Y|X} : \mathbb{N}_m \times \mathbb{N}_n \to [0,1]$ is the conditional probability function for Y given $X = i$.

(b) We define the matrix $P \in [0,1]^{m\times n}$ to have elements (j, i) given by $p_{Y|X}(j|i)$. Then let $x \in [0,1]^n$ have elements i given by $p_X(i)$ and let $y \in [0,1]^m$ have elements j given by $p_Y(j)$. Show that $y = Px$ and that

$$I(p_{X,Y}) = c^T x - \sum_{j=1}^{m} y_j \log y_j,$$

where $c_i = \sum_{j=1}^{m} P_{ji} \log P_{ji}$.

(c) We now interpret X as the input and Y as the output of a communication channel, respectively. The capacity C of the channel is according to Shannon equal to the largest value of the mutual information over all probability functions for X, and hence, defined as the optimal value of

$$\text{maximize} \quad c^T x - \sum_{j=1}^{m} y_j \log y_j,$$

$$\text{subject to} \quad y = Px,$$

$$\mathbb{1}^T x = 1,$$

$$x \geq 0,$$

with variables (x, y). Here, is it assumed that we use $\log_2$. Let $m = 2, n = 2$ and consider the channel defined by

$$P = \begin{bmatrix} 1 - p & p \\ p & 1 - p \end{bmatrix},$$

and show that

$$C = 1 + p\log_2 p + (1 - p)\log_2(1 - p).$$

9.5 Let f be a pdf defined on $\mathbb{R}_+$ and let F be the corresponding distribution function $F(x) = \int_0^x f(t)dt$. The expected value is $\mu = \int_0^\infty xf(x)dx$. The *Lorenz curve* $L_F : [0, 1] \to \mathbb{R}$ is defined as

$$L_F(u) = \frac{1}{\mu} \int_0^u F^{-1}(x)dx.$$

If F is the income distribution in a country, then $L(u)$ represents the fraction of the total income which is in the hands of the uth fraction of the population with the lowest income. The *Gini index* $G : [0, 1]^{\mathbb{R}} \to [0, 1]$, defined as

$$G(F) = 2 \int_0^1 \left(u - L_F(u) \right) du = 1 - 2 \int_0^1 L_F(u)du$$

is twice the area between the Lorenz curve of the line of perfect equality $L_{F_0}(u) = u$. The index is zero for perfect equality and close to one if most of the income is with a very small portion of the population. We are interested in finding the probability density function f that maximizes the entropy for given values of the mean μ and of the Gini index γ, i.e. we want to solve

$$\text{minimize} \int_0^\infty f(x) \ln f(x)dx,$$

$$\text{subject to} \int_0^\infty f(x)dx = 1,$$

$$\int_0^\infty xf(x)dx = \mu,$$

$$G(F) = \gamma.$$

(a) Show that the Gini index can be expressed as

$$G(F) = 1 - \frac{1}{\mu} \int_0^\infty \bar{F}(x)^2 \, dx,$$

where $\bar{F}(x) = 1 - F(x)$.

(b) Now, consider the equivalent maximum entropy problem

$$\text{minimize} \int_0^\infty f(x) \ln f(x)dx,$$

$$\text{subject to} \int_0^\infty f(x)dx = 1,$$

$$\int_0^\infty xf(x)dx = \mu,$$

$$\int_0^\infty \bar{F}(x)^2 \, dx = \eta,$$

with variable f, where $\eta = \mu(1 - \gamma)$ and show that an optimal $\bar{F}$ satisfies

$$\bar{F}' + c_1\bar{F} + c_2\bar{F}^2 = c_3,$$

where c_i are real constants.

Hint: You may use the following necessary conditions of optimality. Let $L_0 :$ $\mathbb{R} \times \mathbb{R} \times \mathbb{R} \to \mathbb{R}$, $L_1 : \mathbb{R} \times \mathbb{R} \times \mathbb{R} \to \mathbb{R}$, and $M_i : \mathbb{R} \times \mathbb{R} \times \mathbb{R} \to \mathbb{R}$, $i \in \mathbb{N}_n$. Define $H : \mathbb{R} \times \mathbb{R} \times \mathbb{R} \times \mathbb{R} \times \mathbb{R}^n \to \mathbb{R}$ as

$$H(x, y, z, \lambda, v) = L_0(x, y, z) + \lambda L_1(x, y, z) + \sum_{i=1}^{n} v_i M_i(x, y, z).$$

Then the existence of $\lambda : \mathbb{R} \to \mathbb{R}$ such that

$$\frac{d\lambda(x)}{dx} = -\frac{\partial H(x, y(x), z(x), \lambda(x), v)}{\partial y},$$

$$0 = \frac{\partial H(x, y(x), z(x), \Lambda(x), v)}{\partial z}$$

is a necessary condition of optimality for

$$\text{minimize} \int_0^\infty L_0(x, y(x), z(x))dx,$$

$$\text{subject to} \quad \frac{dy(x)}{dx} = L_1(x, y(x), z(x)),$$

$$\int_0^\infty M_i(x, y(x), z(x)) = c_i, \quad i \in \mathbb{N}_n.$$

(c) Verify that

$$\bar{F}(x) = \frac{1}{\sigma e^{\rho x} + 1 - \sigma}$$

satisfies the above differential equation and that the corresponding f is a probability density function for any $\sigma > 0$ and $\rho > 0$.

(d) Show that the parameters of $\bar{F}$ and the parameters of the problem are related as

$$\gamma = 1 + \frac{1}{\sigma - 1} - \frac{1}{\ln \sigma}, \quad \rho = \frac{\ln \sigma}{(\sigma - 1)\mu}.$$

(e) Consider data from the following countries, which are approximate.

Country	Gini index	Average annual income (USD)
United states	0.414	63 093
United Kingdom	0.348	44 505
Sweden	0.288	41 748

Plot the probability density functions obtained from the maximum entropy solution for the three countries.

9.6 In this exercise we consider the static ranking of web pages in Example 9.2 and the corresponding maximum entropy problem.

(a) Show that the solution of the maximum entropy problem is $p_{ij} = \exp(-\lambda_i + \lambda_j)/\Phi(\lambda)$, $(i, j) \in E$, where $\Phi(\lambda) = \sum_{(i,j) \in E} \exp(-\lambda_i + \lambda_j)$, and where $\lambda \in \mathbb{R}^n$ are the natural parameters. Notice that you get different answers depending on how you rewrite the equilibrium conditions as equations with a zero right-hand side.

(b) Let $a_i = \exp(-\lambda_i)$, $i \in V$ and let $A = \text{diag}(a_i)$. Moreover, define the sparse matrix $M \in \mathbb{R}^{n \times n}$ as

$$M_{ij} = \begin{cases} \Phi(\lambda)^{-1}, & (i,j) \in E, \\ 0, & (i,j) \notin E. \end{cases}$$

Show that $P = AMA^{-1}$.

(c) The so-called iterative scaling or matrix balancing approach for finding the optimal solution to the maximum entropy problem is based on the expression above for P and the fact that $\Phi(\lambda) = \sum_{(i,j) \in E} a_i/a_j$. It is an iterative algorithm and can be described as follows, where superscripts are used to denote iteration index. Given a tolerance $\epsilon > 0$, initialize as $a_i^{(0)} = 1.0$, $Z^{(0)} = \sum_{(i,j) \in E} 1$, and then repeat starting with $k = 0$:

1. $p_{ij}^{(k)} = a_i^{(k)} / \left(a_j^{(k)} Z^{(k)} \right)$, $(i,j) \in E$

2. $\rho_i^{(k)} = \sum_{j:(i,j) \in E} p_{ij}^{(k)}$, $i \in V$

3. $\sigma_i^{(k)} = \sum_{j:(j,i) \in E} p_{ji}^{(k)}$, $i \in V$

4. $\eta_i^{(k)} = \left(\sigma_i^{(k)} / \rho_i^{(k)} \right)^{1/2}$, $i \in V$

5. If $1 - \epsilon \le \eta_i^{(k)} \le 1 + \epsilon$ for all $i \in V$ go to Step 7.

6. For those $i \in V$ for which $1 - \epsilon \le \eta_i^{(k)} \le 1 + \epsilon$ does not hold, let $a_i^{(k+1)} = a_i^{(k)} \eta_i^{(k)}$, increase k, and then go to Step 1.

7. Break if $1 - \epsilon \le \sum_{(i,j) \in E} p_{ij} \le 1 + \epsilon$. Otherwise, let $Z^{(k+1)} = \sum_{(i,j) \in E} a_i^{(k)} / a_j^{(k)}$, increase k, and then go to Step 1

Implement the above algorithm and test it on random graphs generated by the MATLAB function `webgraph.m`. This function generates the edges for a graph with `dim` edges, for which the first node is connected to all other nodes in both directions. In this way the constructed graph is strongly connected, modeling the fact that every page potentially can be visited from every page despite that there is not a link from the current page. The other nodes are connected to at most 14 other nodes. Notice that you are not asked to compute the page rank. It is enough to compute the unnormalized p_{ij} for the edges. Make a plot of the computational time as a function of the variable `dim`. Average over 10 random graphs for each dimension. You should at least consider graphs for which `dim=10000`. You should use $\epsilon = 10^{-12}$.

(d) Show that the entropy problem in (9.1) can be equivalently expressed as a conic optimization problem involving the exponential cone, i.e. as

$$\begin{aligned} \text{minimize} \quad & -\mathbb{1}^T t, \\ \text{subject to} \quad & Ap = b, \\ & \mathbb{1}^T p = 1, \\ & (t_k, p_k, 1) \in K_{\exp}, \quad k \in \mathbb{N}_n, \end{aligned}$$

with variables p, t, where $t = (t_1, \dots, t_n)$.

(e) Use the conic solver MOSEK to solve the conic formulation of the static ranking of web pages and compare the solution times with your own solver. How big problems can you solve with the two different solvers? Make sure to compare exactly the same problems, i.e. do not use different random realizations for the two different solvers but the same realizations of the edges for the graph. To make the comparison fair try to choose tolerances in MOSEK so that $1 - \epsilon \le \sum_{(i,j) \in E} p_{ij} \le 1 + \epsilon$, or do it the other way around, i.e. choose ϵ for the iterative scaling approach to match the accuracy you get from MOSEK with its default settings.

(f) It is also possible to solve the maximum entropy problem using an interior-point solver. An especially good solver for this problem can be downloaded at http://web.stanford .edu/group/SOL/software/pdco/. Compare the performance of this solver with the two previous ones for the static ranking of web pages. Also, here try to make the comparison fair with respect to tolerances. Check the link to the presentation at ISMP2003 on the above web page. It seems that they are able to solve maximum entropy problems for web traffic with roughly 50 000 nodes using MATLAB. Are you able to?

9.7 Consider a primitive clinic in a village. People in the village have the property that they are either healthy or have a fever. They can only tell if they have a fever by asking the doctor in the clinic. The doctor makes a diagnosis of fever by asking patients how they feel. Villagers only answer that they feel normal, dizzy, or cold. This defines a HMM (X, Y) with $X_k \in D = \{\alpha_1, \alpha_2\}$ and $Y_k \in \mathcal{E} = \{\beta_1, \beta_2, \beta_3\}$, where α_1 =healthy, α_2 =fever, β_1 =normal, β_2 =cold, and β_3 =dizzy. Introduce the notation:

$$a_{ij} = \mathbb{P}[X_{k+1} = \alpha_j | X_k = \alpha_i],$$
$$b_{ij} = \mathbb{P}[Y_k = \beta_j | X_k = \alpha_i].$$

Also, let the matrices A and B be defined such that element (i, j) of the matrix is equal to a_{ij} and b_{ij}, respectively. Consider the case when

$$A = \begin{bmatrix} 0.7 & 0.3 \\ 0.4 & 0.6 \end{bmatrix}, \quad B = \begin{bmatrix} 0.5 & 0.4 & 0.1 \\ 0.1 & 0.3 & 0.6 \end{bmatrix},$$

and assume that $\mathbb{P}[X_0 = \alpha_1] = 0.6$ and $\mathbb{P}[X_0 = \alpha_2] = 0.4$. The doctor has for a patient observed the first day normal, the second day cold, and the third day dizzy. What is the most likely value of the condition for the patient for the different days?

9.8 Show that the smoothing problem in Section 9.4 can be formulated as an optimal control problem

$$\text{minimize} \quad \phi(x_0) + \sum_{k=1}^{N} f_k(x_k, u_k),$$
$$\text{subject to} \quad x_{k-1} = F_k(x_k, u_k), \quad k \in \mathbb{N}_N,$$

with variables $(x_0, u_1, \ldots, u_N, x_N)$ for some f_k and F_k. Notice that the time index is running in reverse order as compared to a standard control problem.

Hint: Take the logarithm of the joint pdf $p_{\bar{X}_N, \bar{Y}_N}$.

9.9 Recall the Gaussian mixture model from Section 9.11, which involves a pdf of the form

$$f_Y(y; \theta) = \sum_{j=1}^{k} \alpha_j \mathcal{N}(y, \mu_j, \Sigma_j),$$

where θ represents the model parameters $\alpha_j, \mu_j, \Sigma_j, j \in \mathbb{N}_k$. Now, suppose that we fix $\Sigma_j = \sigma^2 I$ for some $\sigma^2 > 0$ and let $\tilde{\theta}$ represent the reduced set of model parameters $\alpha_j, \mu_j, j \in \mathbb{N}_k$. Show that the EM algorithm for estimating $\tilde{\theta}$ reduces to the K-means algorithm in the limit as $\sigma^2 \to 0$.

9.10 Implement the EM algorithm for estimating the model parameters of a Gaussian mixture model; *cf.* Section 9.11. Test your implementation on a set of samples from a one-dimensional Gaussian mixture model with three components. The following MATLAB code illustrates how to generate $m = 1000$ samples $y_1, \ldots, y_m$ from a mixture of three univariate Gaussian distributions:

```
alpha = [0.5 0.2 0.3];    % component weights
mu = [-1.5 0 1.5];         % component means
Sigma = [0.3 0.1 0.6];    % component variance
k = length(alpha);         % number of components
m = 1000;                  % number of samples

% Generate subpopulation labeling
[~,z] = histc(rand(m,1),cumsum([0,alpha]));

% Generate samples from GMM
N = sum(z==1:k);
y = zeros(m,1);
for j=1:k
    y(z==j) = mu(j) + sqrt(Sigma(j))*randn(N(j),1);
end
```

9.11 Determine γ such that the matrix

$$\Sigma = \begin{bmatrix} 1 & \gamma & 1 \\ \gamma & 2 & 1 \\ 1 & 1 & 3 \end{bmatrix}$$

is positive semidefinite and such that its inverse has a zero in the position of the variable γ.

9.12 In this exercise, we will perform PCA analysis on the Fisher Iris flower data that we investigated in Example 9.8. The data set can be download in MATLAB with the command `load fisheriris.mat`. It contains measurements of four different characteristics of three different iris species in each column. There are 150 rows in the data set. Each subset of 50 rows corresponds to three different iris species. You should preprocess the data by subtracting the mean value of each column from all the values in that column. Then you should divide all values in a column with the standard deviation of its column values. This is then the X-matrix that you should use for PCA. Hence, each row of it is an observation x_i^T. You should compute the two principal components corresponding to the largest singular values, and then compute the compressed data $c_i \in \mathbb{R}^2$. Finally, plot for each c_i its second component versus its first component. Make sure to mark the different species differently in your plot. You should obtain the same plot as in Figure 9.9. Would it have been possible to separate the different species from one another using this plot in case you had not known from which species the data originated?

9.13 In this exercise you are asked to implement the K-means algorithm in Section 9.16 for cluster analysis. You should try out the algorithm on the Fisher Iris flower data set that you used in Exercise 9.12. Preprocess the data set in the same way as you did in that exercise, i.e. make sure that all columns have zero mean and unit standard deviation.

(a) Try the algorithm using two and three clusters, respectively, i.e. for the cases $K = 2, 3$. How does the algorithm cope? You do not need to make any plots. It is enough to investigate the resulting encoders, and how good they perform.

(b) Instead of using the 4-dimensional data that you used in (a) instead use the 2-dimensional data c_i that resulted from the PCA analysis in Exercise 9.12. Try the cases $K = 2, 3$.

(c) Instead of using the 2-dimensional data that you used in (b) instead use a 3-dimensional data c_i that you obtain from PCA analysis similarly as in Exercise 9.12 by instead using the three principal components corresponding to the three largest singular values. Try the cases $K = 2, 3$ again.

(d) Relate your results in (a)–(c) to what you visually observed in Exercise 9.12.

10

Supervised Learning

In this chapter, we will discuss supervised learning. What distinguishes supervised learning problems from unsupervised learning problems is that the data come in pairs, i.e. we may say $(x_k, y_k) \in \mathbb{R} \times \mathbb{R}$ for $k \in \mathbb{N}_N$ and we would like to find a relationship between the pairs of data. We will start with linear regression. This does not mean that the data pairs are related to one another in a linear way. Instead, it is the class of functions that we consider that is parameterized in a linear way. First, we will do this in a finite-dimensional space, and there we will also discuss statistical interpretations and generalizations such as maximum likelihood estimation, maximum a posteriori estimation, and regularization. We will then also do regression in an infinite-dimensional space, i.e. in a Hilbert space. We will see that this is equivalent to maximum a posteriori estimation for so-called Gaussian processes. Then we will discuss classification both using linear regression, logistic regression, support vector machines, and the restricted Boltzmann machine. The chapter is finished off with artificial neural networks and the so-called back-propagation algorithm. We also discuss a form of implicit regularization known as dropout.

10.1 Linear Regression

We start by considering the problem of curve fitting. Assume that we want to fit the polynomial function $f : \mathbb{R} \to \mathbb{R}$ defined by

$$f(x) = a_1 + a_2 x + \cdots a_m x^{m-1},$$

to pairs $(x_k, y_k) \in \mathbb{R} \times \mathbb{R}$ for $k \in \mathbb{N}_N$. Define $a = (a_1, \ldots, a_m) \in \mathbb{R}^m$ and $\beta : \mathbb{R} \to \mathbb{R}^m$ by $\beta(x) = (1, x, \ldots, x^{m-1})$. It then follows that

$$f(x) = a^T \beta(x). \tag{10.1}$$

This is called a *linear regression* model, since the right-hand side is linear in the coefficient vector a.

10.1.1 Least-Squares Estimation

We now would like to find a common value of a which is such that y_k is close to $f(x_k) = a^T \beta(x_k)$ for all pairs of data (x_k, y_k). This so-called *fit* is obtained by minimizing the sum of the squared distances between $f(x_k)$ and y_k, i.e. we solve

$$\text{minimize} \quad \frac{1}{2} \sum_{k=1}^{N} \left(a^T \beta(x_k) - y_k \right)^2 \tag{10.2}$$

Optimization for Learning and Control, First Edition. Anders Hansson and Martin Andersen.
© 2023 John Wiley & Sons, Inc. Published 2023 by John Wiley & Sons, Inc.
Companion Website: www.wiley.com/go/opt4lc

with variable $a \in \mathbb{R}^m$. This is a linear LS problem. We may consider more general functions than monomials, i.e. we let $\varphi_j : \mathbb{R} \to \mathbb{R}, j \in \mathbb{N}_m$, be any functions and define

$$f(x) = \sum_{j=1}^{m} a_j \varphi_j(x) = a^T \beta(x),$$

with $\beta(x) = \big(\varphi_1(x), \ldots, \varphi_m(x)\big)$. Often, one lets $\varphi_1(x) = 1$. It is also possible to generalize the regression model to $x \in \mathbb{R}^n$. We just need to define $\varphi_j : \mathbb{R}^n \to \mathbb{R}, j \in \mathbb{N}_m$, and $\beta : \mathbb{R}^n \to \mathbb{R}^m$. An important special case is when $n = m$ and $\beta(x) = x$.

Another generalization is when $f(x)$ is vector-valued. We consider the case when $f : \mathbb{R}^n \to \mathbb{R}^p$ is given by

$$f(x) = A\beta(x),$$

where $A \in \mathbb{R}^{p \times m}$, $\beta : \mathbb{R}^n \to \mathbb{R}^m$ with $\beta(x) = \big(\varphi_1(x), \ldots, \varphi_m(x)\big)$ and $\varphi_j : \mathbb{R}^n \to \mathbb{R}, j \in \mathbb{N}_m$. The LS criterion is then the sum of the LS criteria for each row in the regression model, and the LS problem can be written as

$$\text{minimize} \quad \frac{1}{2} \sum_{k=1}^{N} \|y_k - A\beta(x_k)\|_2^2,$$

with variable A. The solution to this problem is closely related to the SAA approximation of the single-stage stochastic optimization problem for the affine predictor in Section 9.3. This is one of the motivations for calling $f(x)$ a *predictor* – it can be used to predict values of y_k when only x_k is known.

Example 10.1 We are given pairs of data (x_k, y_k) that happen to satisfy $y_k = \sin x_k$. We are not aware of this relationship, and instead, we want to find a linear regression model as in (10.1) that solves (10.2) for $\beta(x) = (1, x, x^2, x^3, x^4)$. We solve the resulting normal equations as in (5.3), and then plot the resulting polynomial and compare it with the 11 points we used for fitting, see Figure 10.1. The fit is pretty good inside the interval of the available data. We do not expect it to be very good outside this interval. The polynomial is only of fourth degree.

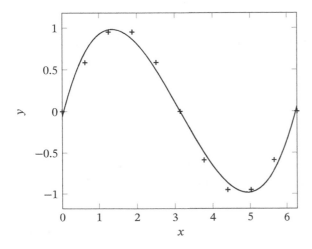

Figure 10.1 Plot showing the fourth degree polynomial (solid line), and the 11 data points (+), used to fit the polynomial to a sinusoidal.

10.1.2 Maximum Likelihood Estimation

The linear LS problem may be viewed as a maximum likelihood problem. Consider e_k to be the outcomes of independent, normally distributed random variables $E_1, \ldots, E_N$ with zero mean and variance σ^2, i.e. they have the pdf $f_{E_k} : \mathbb{R} \to \mathbb{R}_+$ defined by

$$f_{E_k}(e_k) = \frac{1}{\sqrt{2\pi}\sigma} e^{-\frac{1}{2\sigma^2} e_k^2}.$$

Now, suppose that

$$y_k = a^T \beta(x_k) + e_k, \quad k \in \mathbb{N}_N,$$

with $a \in \mathbb{R}^m$ and $\beta : \mathbb{R}^n \to \mathbb{R}^m$ model N noisy observations y_k. The pdf for the kth observation $f_{Y_k} : \mathbb{R} \to \mathbb{R}_+$ is then $f_{Y_k}(y_k) = f_{E_k}(y_k - a^T \beta(x_k))$, and the likelihood function $\mathcal{L} : \mathbb{R}^N \times \mathbb{R}^m \to \mathbb{R}_+$ is

$$\mathcal{L}(y_1, \ldots y_N; a) = \prod_{k=1}^{N} f_{Y_k}(y_k) = \prod_{k=1}^{N} f_{E_k}\left(y_k - a^T \beta(x_k)\right).$$

The negative log-likelihood function is given by

$$\frac{1}{2\sigma^2} \sum_{k=1}^{N} \left(y_k - a^T \beta(x_k)\right)^2 + N \ln\left(\sqrt{2\pi}\sigma\right).$$

For any fixed value of $\sigma > 0$, the minimum of this function is obtained for the same a that solves the LS problem in (10.2),[1] and hence, maximum likelihood estimation is equivalent to LS estimation for this problem formulation. We may also consider σ to be an unknown parameter and maximize the likelihood function jointly for (a, σ), see Exercise 10.1.

We now consider correlated residuals. Let $e = (e_1, \ldots, e_N)$ be the outcome of a normally distributed random variable E with zero mean and covariance $\Sigma \in \mathbb{S}_{++}^N$. Then the joint pdf $f_E : \mathbb{R}^N \to \mathbb{R}_+$ for E is given by

$$f_E(e) = \frac{1}{\sqrt{(2\pi)^N \det \Sigma}} e^{-\frac{1}{2} e^T \Sigma^{-1} e}.$$

Define

$$X = \begin{bmatrix} \beta(x_1)^T \\ \vdots \\ \beta(x_N)^T \end{bmatrix},$$

and $y = (y_1, \ldots, y_N)$. Then it is straightforward to see that the ML problem is equivalent to the weighted LS problem

$$\text{minimize } \frac{1}{2}(y - Xa)^T \Sigma^{-1} (y - Xa),$$

with variable a, if we assume that Σ is known. Notice that we do not have enough data to also estimate Σ. The reason for this is that there would then be more parameters to estimate than the number of data points N, since the symmetric matrix Σ has $N(N+1)/2$ entries in its upper triangular part.

10.1.3 Maximum a Posteriori Estimation

Recall from Section 9.3 that it is possible to estimate the parameters in a regression model using a MAP estimate. To this end, we assume that a is also an outcome of a random variable A with some

---

1 The minimum is not necessarily unique without further assumptions.

pdf $f_A : \mathbb{R}^m \to \mathbb{R}_+$. It is natural to assume that A and E_k are independent, in which case the joint pdf $f : \mathbb{R}^N \times \mathbb{R}^m \to \mathbb{R}_+$ is given by

$$f_{Y,A}(y, a) = f_A(a) \prod_{k=1}^{N} f_{E_k}\left(y_k - a^T \beta(x_k)\right),$$

where $y = (y_1, \dots, y_N)$. In MAP estimation, the density of A conditioned on the observations $Y = y$ is maximized. Since this density is obtained from $f_{Y,A}(y; a)$ by dividing by $\int_{\mathbb{R}^m} f_{Y,A}(y; a)\, da$, which is a constant for known y, we may equivalently maximize $f_{Y,A}$, or minimize the negative logarithm of it, with respect to a. This results in the optimization problem

$$\text{minimize} \quad \frac{1}{2\sigma^2} \sum_{k=1}^{N} \left(y_k - a^T \beta(x_k)\right)^2 + N \ln\left(\sqrt{2\pi}\sigma\right) - \ln f_A(a) \tag{10.3}$$

with variable a. Compared to the ML problem, the only difference is that we have added the term $-\ln f_A(a)$ to the objective function. This can be interpreted as a *regularization* of the LS problem. Different pdfs f_A result in different regularizations, and hence, reflect different prior knowledge.

10.2 Regression in Hilbert Spaces

We now consider the generalization of regression to infinite dimension, i.e. we let $\beta(x) = \left(\varphi_1(x), \varphi_2(x), \dots\right)$, where $\varphi_i : \mathbb{R}^n \to \mathbb{R}, i \in \mathbb{N}$. For $a = \left(a_1, a_2, \dots\right)$, where $a_i \in \mathbb{R}, i \in \mathbb{N}$, we say that $a \in \ell_2$ if $\sum_{i=1}^{\infty} a_i^2 < \infty$. We define the inner product $\langle \cdot, \cdot \rangle_{\ell_2} : \ell_2 \times \ell_2 \to \mathbb{R}$ as $\langle a, b \rangle_{\ell_2} = \sum_{i=1}^{\infty} a_i b_i$. The corresponding norm $\| \cdot \|_{\ell_2} : \ell_2 \to \mathbb{R}_+$ is defined as $\|a\|_{\ell_2}^2 = \langle a, a \rangle_{\ell_2}$. For functions $f : \mathbb{R}^n \to \mathbb{R}$, we define the space of square integrable functions L_2, i.e. the set of functions such that

$$\int_{\mathbb{R}^n} f(x)^2 dx < \infty.$$

We define the inner product $\langle \cdot, \cdot \rangle_{L_2} : L_2 \times L_2 \to \mathbb{R}$ by

$$\langle f, g \rangle_{L_2} = \int_{\mathbb{R}^n} f(x) g(x) dx.$$

The corresponding norm $\| \cdot \|_{L_2} : L_2 \to \mathbb{R}_+$ is defined by $\|f\|_{L_2}^2 = \langle f, f \rangle_{L_2}$. We remark that both ℓ_2 and L_2 are *Hilbert spaces*. Now, suppose that $\varphi_i \in L_2, i \in \mathbb{N}$ is a family of orthonormal functions, i.e.

$$\langle \varphi_i, \varphi_j \rangle_{L_2} = 0, \qquad i \neq j$$
$$\langle \varphi_i, \varphi_j \rangle_{L_2} = 1, \qquad i = j.$$

It is well known that a sum of the form $f(x) = \sum_{i=1}^{\infty} a_i \varphi_i(x)$ is convergent and belongs to L_2 if and only if $a \in \ell_2$, e.g. [Theorem 1, p. 59][73]. In that case, it also holds that $a_i = \langle f, \varphi_i \rangle_{L_2}$, and hence, we realize that $\|f\|_{L_2} = \|a\|_{\ell_2}$.

10.2.1 Infinite-Dimensional LS Problem

We will now consider $f(x) = \sum_{i=1}^{\infty} a_i \varphi_i(x)$ as a regressor, and we want to solve the infinite-dimensional regularized LS problem

$$\text{minimize} \quad \frac{1}{2} \sum_{k=1}^{N} \left(y_k - f(x_k)\right)^2 + \frac{\nu}{2} \|f\|_{L_2}^2,$$

with variable a. We reformulate this problem as a constrained problem

$$\text{minimize} \quad \frac{1}{2}\sum_{k=1}^{N}e_k^2 + \frac{v}{2}\|a\|_{\ell_2}^2$$

$$\text{subject to} \quad e_k = y_k - \sum_{i=1}^{\infty}a_i\varphi_i(x), \quad k \in \mathbb{N}_K$$

with variables (a, e), where $e = (e_1, \dots, e_N)$. To ease the notation, we will write this as

$$\text{minimize} \quad \frac{1}{2}e^Te + \frac{v}{2}a^Ta$$

$$\text{subject to} \quad e = y - Xa,$$

where X is the infinite-dimensional matrix given by

$$X = \begin{bmatrix} \beta^T(x_1) \\ \vdots \\ \beta^T(x_N) \end{bmatrix}.$$

We then introduce the Lagrangian $L : \mathbb{R}^N \times \ell_2 \times \mathbb{R}^N \to \mathbb{R}$ defined by

$$L(e, a, \lambda) = \frac{1}{2}e^Te + \frac{v}{2}a^Ta + \lambda^T(e - y + Xa).$$

Completing the squares in the Lagrangian results in

$$L(e, a, \lambda) = \frac{1}{2}(e + \lambda)^T(e + \lambda) + \frac{v}{2}\left(a + \frac{1}{v}X^T\lambda\right)^T\left(a + \frac{1}{v}X^T\lambda\right)$$
$$- \frac{1}{2}\lambda^T\lambda - \frac{1}{2v}\lambda^TXX^T\lambda - \lambda^Ty.$$

Thus, for a given λ, the Lagrangian is minimized by taking

$$e = -\lambda, \qquad a = -\frac{1}{v}X^T\lambda,$$

which yields the Lagrange dual function $g : \mathbb{R}^N \to \mathbb{R}$ defined by

$$g(\lambda) = -\frac{1}{2}\lambda^T\lambda - \frac{1}{2v}\lambda^TXX^T\lambda - \lambda^Ty.$$

10.2.2 The Kernel Trick

Let us introduce the function $K : \mathbb{R}^n \times \mathbb{R}^n \to \mathbb{R}$ defined by

$$K(x, \bar{x}) = \beta(x)^T\beta(\bar{x}) = \sum_{i=1}^{\infty}\varphi_i(x)\varphi_i(\bar{x}).$$

We call this function the *kernel function*. We then define the matrix $\mathcal{K} \in \mathbb{S}^N$ with elements $\mathcal{K}_{ij} = K(x_i, x_j)$. We may then write the dual optimization problem as

$$\text{maximize} -\frac{1}{2}\lambda^T\left(I + \frac{1}{v}\mathcal{K}\right)\lambda + \lambda^Ty,$$

with variable λ, which has solution

$$\lambda = -\left(I + \frac{1}{v}\mathcal{K}\right)^{-1}y.$$

We may also write $f(x)$ in terms of the Kernel function as

$$f(x) = a^T\beta(x) = -\frac{1}{v}\lambda^TX\beta(x) = -\frac{1}{v}\sum_{k=1}^{N}\lambda_kK(x_k, x).$$

The fact that the solution to this infinite-dimensional regression problem can be obtained from a finite-dimensional optimization problem by introducing the kernel function is sometimes called the *kernel trick*.

From the above expression for the regressor $f(x)$, we realize that one could just as well start with a series expansion in terms of the kernel function. A natural question that arises is for what functions K there are orthonormal φ_i. The answer is that K should be a *positive semidefinite kernel*, i.e. for any $N \in \mathbb{N}$, and any $c_i \in \mathbb{R}$ and $x_i \in \mathbb{R}^n$, $i \in \mathbb{N}_N$, it should hold that

$$\sum_{i=1}^{N} \sum_{j=1}^{N} c_i c_j K(x_i, x_j) \geq 0.$$

This result is known as *Mercer's theorem*, [94, p. 96]. Popular choices are dth degree polynomials given by $K(x, \bar{x}) = \left(1 + x^T \bar{x}\right)^d$ and *radial basis functions* that are functions that only depend on $\|x - \bar{x}\|_2$. The dth degree polynomials correspond to finitely many orthonormal φ_i. The *Gaussian radial basis kernel*

$$K(x, \bar{x}) = \exp\left(-\frac{1}{2\sigma^2}\|x - \bar{x}\|_2^2\right),$$

where $\sigma \in \mathbb{R}_{++}$, corresponds to an infinite series of orthonormal φ_i [94].

10.3 Gaussian Processes

The MAP estimate in Section 10.1 is related to the Bayesian approach to statistics in which the so-called *posterior distribution* is computed, *cf*. Section 9.3. To formalize this, we consider

$$Y_k = A^T x_k + E_k, \qquad k \in \mathbb{N}_N, \tag{10.4}$$

where E_k and Y_k are random variables, and where A is a random vector of the same dimension as the vector x_k. We assume that E_k and A are independent. The posterior distribution is the distribution for A conditioned on the observation that $Y_k = y_k$. For Gaussian or normal distributions for A and E_k the posterior distribution is also Gaussian, and hence, it is sufficient to compute the posterior mean and covariance. The MAP estimate is given by the conditional mean *cf*. Section 9.3. From the above model, we obtain the following linear regression model for the outcomes (y_k, e_k, a) of the experiment

$$y_k = a^T x_k + e_k, \qquad k \in \mathbb{N}_N,$$

where $y_k \in \mathbb{R}$ and $x_k \in \mathbb{R}^n$ are the observations we have made and where $a \in \mathbb{R}^n$ is the parameter we would like to estimate.

10.3.1 Gaussian MAP Estimate

Suppose that A is Gaussian random variable with zero mean and covariance $\Sigma_a \in \mathbb{S}_{++}^n$, and that $E = (E_1, \ldots, E_N)$ is also Gaussian with zero mean and covariance $\Sigma_e \in \mathbb{S}_{++}^N$. If we let $Y = (Y_1, \ldots, Y_N)$, and

$$X = \begin{bmatrix} x_1^T \\ \vdots \\ x_N^T \end{bmatrix},$$

it follows that

$$Y = XA + E.$$

It is straightforward to show that (Y, A) is Gaussian with zero mean and covariance

$$\begin{bmatrix} R & X\Sigma_a \\ \Sigma_a X^T & \Sigma_a \end{bmatrix},$$

where $R = \Sigma_e + X\Sigma_a X^T$. Let us define $f_{Y,A} : \mathbb{R}^N \times \mathbb{R}^n \to \mathbb{R}_+$ as the joint pdf for (Y, A), $f_{A|Y} : \mathbb{R}^N \times \mathbb{R}^n \to \mathbb{R}_+$ as the conditional pdf for A given Y, and $f_Y : \mathbb{R}^N \to \mathbb{R}_+$ as the marginal pdf for Y. From the results in Section 9.3, it now follows with $\Sigma_{a|y} = \Sigma_a - \Sigma_a X^T R^{-1} X \Sigma_a$ that the joint pdf factorizes as $f_{Y,A}(y, a) = f_Y(y) f_{A|Y}(a|y)$, where

$$f_Y(y) = \frac{1}{\sqrt{(2\pi)^N \det R}} e^{-\frac{1}{2} y^T R^{-1} y}$$

$$f_{A|Y}(a|y) = \frac{1}{\sqrt{(2\pi)^n \det \Sigma_{a|y}}} e^{-\frac{1}{2}(a - \mu_{a|y})^T \Sigma_{a|y}^{-1} (a - \mu_{a|y})},$$

where $\mu_{a|y} = \Sigma_a X^T R^{-1} y$ is the conditional mean for A given $Y = y$. Clearly, the conditional pdf is maximized by $a = \mu_{a|y}$, which proves that the MAP estimate is the conditional mean for Gaussian distributions, as we saw in Section 9.3.

10.3.2 The Kernel Trick

A MAP prediction with a new value x is then given by

$$\mu_{a|y}^T x = x^T \mu_{a|y} = x^T \Sigma_a X^T \left(\Sigma_e + X \Sigma_a X^T \right)^{-1} y.$$

We realize that Σ_a only appears in terms of expressions of the form $z^T \Sigma_a \bar{z}$ for $z \in \mathbb{R}^n$ and $\bar{z} \in \mathbb{R}^n$. Because of this we could instead write the predictor in terms of the covariance function $K : \mathbb{R}^n \times \mathbb{R}^n \to \mathbb{R}_+$ defined by $K(z, \bar{z}) = z^T \Sigma_a \bar{z}$. We then let $\mathcal{K} \in \mathbb{S}_+^N$ be defined by $\mathcal{K}_{ij} = K(x_i, x_j)$, and we let $\eta_k, k \in \mathbb{N}_N$, be defined by $\eta_k = K(x_k, x)$. Then the predictor can be written as

$$f(x) = \eta^T \left(\Sigma_e + \mathcal{K} \right)^{-1} y.$$

This is the same predictor as we obtain when we do regression in Hilbert spaces in Section 10.2 if we let $\Sigma_e = \nu I$ and if we define the orthonormal functions φ_i such that the resulting kernel function is $K(z, \bar{z}) = z^T \Sigma_a \bar{z}$. We notice that this is the covariance between $A^T z$ and $A^T \bar{z}$. Thus, we can generalize the regression model in (10.4) by instead considering

$$Y_k = \mathcal{A}(x_k) + E_k,$$

with $\mathcal{A} : \mathbb{R}^n \to \mathbb{R}$, where we specify the covariance between $\mathcal{A}(z)$ and $\mathcal{A}(\bar{z})$ by specifying the covariance function for any $z \in \mathbb{R}^n$ and $\bar{z} \in \mathbb{R}^n$. We still assume that the joint distribution is Gaussian. This defines a zero mean real-valued *Gaussian random process* $\mathcal{A}$ on $\mathbb{R}^n$. Such a process is sometimes called a *random field*, cf. Section 3.9. Notice that any of the kernel functions discussed in Section 10.2 may be used. The Gaussian radial basis function is called the squared exponential covariance function. Another common covariance function is

$$K(z, \bar{z}) = \exp\left(-\frac{1}{\sigma} \|z - \bar{z}\|_2 \right),$$

which defines the so-called *Ornstein–Uhlenbeck* process, where $\sigma \in \mathbb{R}_{++}$. When the covariance function only depends on $z - \bar{z}$, the process is *stationary*, and when it only depends on $\|z - \bar{z}\|_2$ is also *isotropic*. Stationary and isotropic together is sometimes called *homogeneous*. In practice, these properties reflect the differences, or rather the lack of them, in the behavior of the process given the location of the observer. Actually, there are many more possibilities, see [94]. We specifically

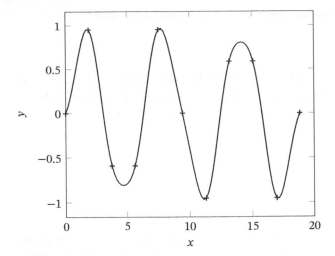

Figure 10.2 Plot showing the Gaussian regression model (solid line), and the 11 data points (+), used to compute the model.

realize that infinite-dimensional regression in Hilbert spaces can be equivalently expressed as MAP estimation for Gaussian processes.

Example 10.2 Here we again consider data from a sinusoidal, but this time collected from three periods of the sinusoidal. We take $\Sigma_e = 0$, since we have no measurement errors of the sinusoidal. We use the squared exponential covariance function with parameter $\sigma = 1$. In Figure 10.2, the predicted values together with the 11 data points used to compute the predictor are shown.

10.4 Classification

In *classification problems*, we are given pairs of data (x_k, y_k) for $k \in \mathbb{N}_N$ with $x_k \in \mathbb{R}^n$, and where y_k is *qualitative* in the sense that is it belongs to a discrete finite set with cardinality K. Without loss of generality, we may take this set to be $\mathbb{N}_K$. We say that the data (x_k, y_k) belongs to class l if $y_k = l$. These types of data are sometimes also called *categorical* or *discrete* as well as *factors*. We are interested in finding functions $f_l : \mathbb{R}^n \to \mathbb{R}$, $l \in \mathbb{N}_K$, that are such that $f_l(x_k) > 0$ if $y_k = l$ and $f_l(x_k) < 0$ if $y_k \neq l$. The set $\{x \mid f_l(x) = 0\}$ then separates class l from the other classes. To the left in Figure 10.3, we are shown two classes of data that can easily be separated with, e.g. a straight line. In the same figure to the right, we see two classes that cannot be separated with any connected line.

10.4.1 Linear Regression

A simple approach to classification is to model each class $l \in \mathbb{N}_K$ as a linear regression $z_l : \mathbb{R}^n \to \mathbb{R}$, where

$$z_l(x) = a_l^T x + b_l,$$

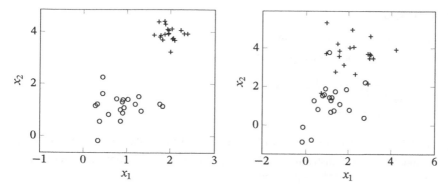

Figure 10.3 Plots showing data points from two classes marked with + for the first class and o for the second class. In the left plot, the data points for the two classes are well separated. In the right plot, the data points for the two classes are mixed.

where $a_l \in \mathbb{R}^n$ and $b_l \in \mathbb{R}$. The objective is to choose the regression parameters (a_l, b_l) such that z_l is close to one if x belongs to class l and otherwise close to zero. We obtain this by solving the LS problem

$$\text{minimize} \quad \frac{1}{2}\sum_{l=1}^{K} \left(\sum_{k:y_k=l}\left(a_l^T x_k + b_l - 1\right)^2 + \sum_{k:y_k \neq l}\left(a_l^T x_k + b_l\right)^2 \right),$$

with variables $a = (a_1, \dots, a_K)$ and $b = (b_1, \dots, b_K)$. Then we use the functions $\delta_l : \mathbb{R}^n \to \mathbb{R}$ defined by $\delta_l(x) = z_l(x)$ as *discriminant functions*, i.e. we classify x to belong to class l if $\delta_l(x) > \delta_k(x)$ for all $k \neq l$. Hence, we get $f_l(x) = \delta_l(x) - \max_{k \neq l} \delta_k(x)$. Unfortunately, this method is prone to give bad results for $K \geq 3$. However, there is a simple remedy which is to consider polynomial regression models, and it might be necessary to have polynomial terms up to degree $K - 1$, see [54]. More general basis functions may also be considered in this framework just as for the curve fitting problem in Section 10.1. The LS problems will still be linear.

Example 10.3 We consider an example where $K = 2$ and where $n = 2$. There are in total 20 data points from each class, and hence, $N = 40$. In Figure 10.4, the data points are shown together with the line defined by $\delta_1(x) = \delta_2(x)$, which separates the two classes.

Figure 10.4 Plot showing the data points from the two classes marked with + for the first class and o for the second class. The straight line separates the two classes from one another.

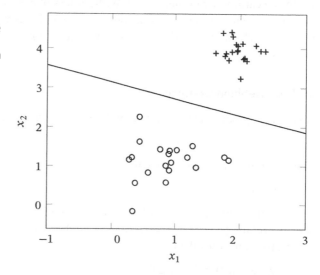

10.4.2 Logistic Regression

Another popular approach to classification is called *logistic regression*. It is based on the *categorical distribution*, *cf*. Example 3.2. One of its forms is given by

$$p_l(z_1, z_2, \ldots, z_{K-1}) = \begin{cases} \frac{e^{z_l}}{1+\sum_{k=1}^{K-1} e^{z_k}}, & l \in \mathbb{N}_{K-1} \\ \frac{1}{1+\sum_{k=1}^{K-1} e^{z_k}}, & l = K, \end{cases}$$

where we consider $p_l : \mathbb{R}^{K-1} \to [0, 1]$ for $l \in \mathbb{N}_K$ to be functions of the scalars $z_l \in \mathbb{R}$. Notice that any value of z_l will make this a valid distribution. We let $z_l(x) = a_l^T x + b_l$ as above and define the likelihood function $\mathcal{L} : \mathbb{R}^{Nn} \times \mathbb{R}^{(K-1)(n+1)} \to [0, 1]$ by

$$\mathcal{L}(x_1, \ldots, x_N; a_1, \ldots, a_{K-1}, b_1, \ldots, b_{K-1}) = \prod_{l=1}^{K} \prod_{k:y_k=l} p_l(z_1(x), \ldots, z_{K-1}(x)).$$

Then the log-likelihood function is given by

$$\sum_{l=1}^{K-1} \sum_{k:y_k=l} a_l^T x_k + b_l - \sum_{k=1}^{N} \ln\left(1 + \sum_{l=1}^{K-1} e^{a_l^T x_k + b_l}\right),$$

which is a concave function. This follows from Exercises 4.4 and the fact that concavity is preserved under an affine transformation. We use the functions $\delta_l : \mathbb{R}^n \to \mathbb{R}$ defined by $\delta_l(x) = p_l(a_1^T x + b_1, \ldots, a_{K-1}^T x + b_{K-1})$ as discriminant functions similarly as above, i.e. we let $f_l(x) = \delta_l(x) - \max_{k \neq l} \delta_k(x)$.

The classifiers discussed so far work with discriminant functions, and then the function f_l that separates the classes is obtained indirectly as $f_l(x) = \delta_l(x) - \max_{k \neq l} \delta_k(x)$. In Section 10.5, we will discuss how to work directly with the functions f_l.

10.5 Support Vector Machines

We are as previously given pairs of data (x_k, y_k) for $k \in \mathbb{N}_N$ with $x_k \in \mathbb{R}^n$, and where y_k is qualitative in the sense that is it belongs to a discrete finite set. However, we now restrict ourselves to the case when the cardinality of this set is 2. Without loss of generality, we may take the set to be $\{-1, 1\}$. We are interested in finding a function $f : \mathbb{R}^n \to \mathbb{R}$ that is such that $f(x_k) > 0$ if $y_k = 1$ and $f(x_k) < 0$ if $y_k = -1$, which may also be expressed as $y_k f(x_k; a, b) > 0$. The set $\{x \mid f(x) = 0\}$ then separates the two classes.

10.5.1 Hebbian Learning

A simple approach is to take $f : \mathbb{R}^n \times \mathbb{R}^n \times \mathbb{R} \to \mathbb{R}$ as a linear regression

$$f(x; a, b) = a^T x + b.$$

The objective is to choose the regression parameters (a, b) such that

$$y_k f(x_k; a, b) > 0, \quad k \in \mathbb{N}_N.$$

This is a feasibility problem. It has a solution if and only if the two classes can be separated with the hyperplane $\{x \mid a^T x + b = 0\}$. Notice that the above inequalities are homogeneous in (a, b), and hence, they are equivalent to

$$y_k f(x_k; a, b) \geq 1, \quad k \in \mathbb{N}_N.$$

It is always possible to formulate feasibility problems as optimization problems, and there are many ways of doing this. One is to define the function $g : \mathbb{R} \to \mathbb{R}_+$ by

$$g(y) = \max(0, y),$$

and define the optimization problem

$$\text{minimize } \sum_{k=1}^{N} g(1 - y_k f(x_k; a, b)) \tag{10.5}$$

with variables (a, b). Clearly, the optimal value is zero if and only if the two classes can be separated by a hyperplane. This formulation is the basis for *Hebbian learning*, *cf*. Exercise 10.6, where we carry out the minimization using a subgradient algorithm. The function g is called the *rectifier function* or the *rectified linear unit* (ReLU) function.

10.5.2 Quadratic Programming Formulation

We notice that the objective function in (10.5) is not differentiable, and this limits the possible class of optimization methods that may be used. Because of this it might be better to just consider the feasibility problem above. It is actually an LP feasibility problem, which makes it tractable. However, it suffers from the problem that the solution is generally not unique, i.e. there are often many hyperplanes that separate the two classes. This problem it shares with the optimization formulation since they are equivalent formulations of the same problem. A remedy to the nonuniqueness problem is to maximize the minimum distance from x_k to the hyperplane. Since the distance is given by $y_k f(x_k; a, b)/\|a\|_2$, and since f is homogeneous in (a, b) this can be accomplished by minimizing $\|a\|_2^2$, see Exercise 10.7 for details. This results in the QP

$$\text{minimize } \frac{1}{2}\|a\|_2^2$$
$$\text{subject to } y_k f(x_k; a, b) \geq 1, \quad k \in \mathbb{N}_N,$$

with variables (a, b).

10.5.3 Soft Margin Classification

The QP above can only deliver a solution in case it is feasible, i.e. if there exists a separating hyperplane for the two classes. The nondifferentiable formulation above does not suffer from this problem. In case, the optimal objective value is not zero, it has instead found a hyperplane that minimizes the sum of a measure of the misfit of the x_k that are not separated correctly by the hyperplane. An epigraph reformulation of (10.5) can be obtained by introducing variables $\xi = (\xi_1, \ldots, \xi_N) \in \mathbb{R}^N$, resulting in the equivalent optimization problem

$$\text{minimize } \sum_{k=1}^{N} \xi_k$$
$$\text{subject to } y_k f(x_k; a, b) \geq 1 - \xi_k, \quad k \in \mathbb{N}_N$$
$$\xi_k \geq 0, \quad k \in \mathbb{N}_N$$

with variables (a, b, ξ). This, however, suffers from the problem of possible nonuniqueness. A remedy to this is to consider

$$\text{minimize } \sum_{k=1}^{N} \xi_k + \frac{v}{2}\|a\|_2^2$$
$$\text{subject to } y_k f(x_k; a, b) \geq 1 - \xi_k, \quad k \in \mathbb{N}_N \tag{10.6}$$
$$\xi_k \geq 0, \quad k \in \mathbb{N}_N$$

with variables (a, b, ξ), where $v \geq 0$ can be used to make a trade-off between the amount of misfit and the distance to the hyperplane. This is the formulation that is the basis of *support vector machines* (SVMs). The optimization problem is a convex optimization problem and because of this it is also tractable. We remark that in the linear regression we may use nonlinear functions of x_k instead of x_k itself without changing the type of optimization problems. However, the two classes will then not be separated by a hyperplane but by a general surface.

10.5.4 The Dual Problem

We will now investigate the above optimization problem further using duality. We consider the regressor $f : \mathbb{R}^m \times \mathbb{R}^n \times \mathbb{R} \to \mathbb{R}$ given by

$$f(x; a, b) = a^T \beta(x) + b,$$

where $\beta : \mathbb{R}^m \to \mathbb{R}^n$. The Lagrangian $L : \mathbb{R}^n \times \mathbb{R} \times \mathbb{R}^N \times \mathbb{R}^N \times \mathbb{R}^N \to \mathbb{R}$ for (10.6) is given by

$$L(a, b, \xi, \lambda, \mu) = \sum_{k=1}^{N} \xi_k + \frac{v}{2} \|a\|_2^2 + \sum_{k=1}^{N} \lambda_k \left(1 - \xi_k - y_k f(x_k; a, b)\right) - \sum_{k=1}^{N} \mu_k \xi_k$$

$$= \frac{v}{2} \|a\|_2^2 - \sum_{k=1}^{N} \lambda_k y_k a^T \beta(x_k) - \sum_{k=1}^{N} \lambda_k y_k b + \sum_{k=1}^{N} \left(1 - \lambda_k - \mu_k\right) \xi_k + \sum_{k=1}^{N} \lambda_k.$$

The minimum of the Lagrangian is unbounded from below unless $1 - \lambda_k - \mu_k = 0$, $k \in \mathbb{N}_N$, and $\sum_{k=1}^{N} \lambda_k y_k = 0$. When bounded, the minimizing a is given by the solution of

$$\frac{\partial L}{\partial a} = va - \sum_{k=1}^{N} \lambda_k y_k \beta(x_k) = 0,$$

i.e.

$$a = \frac{1}{v} \sum_{k=1}^{N} \lambda_k y_k \beta(x_k).$$

This follows from the fact that the Lagrangian is convex in a. We know from complementary slackness that $\lambda_k = 0$, if the first constraint in (10.6) is satisfied strictly at optimality. This means that the optimal solution $a^\star$ only depends on x_k if $y_k f(x_k; a^\star, b^\star) = 1 - \xi_k^\star$, and these vectors x_k are called *support vectors*. Back substitution into the Lagrangian gives the Lagrange dual function $g : \mathbb{R}^N \times \mathbb{R}^N \to \mathbb{R}$ given by

$$g(\lambda, \mu) = \sum_{k=1}^{N} \lambda_k - \frac{1}{2v} \sum_{k=1}^{N} \sum_{l=1}^{N} \lambda_k \lambda_l y_k y_l \beta^T(x_k), \beta(x_l)$$

with dom $g = \{(\lambda, \mu) | 1 - \lambda_k - \mu_k = 0, \ k \in \mathbb{N}_N, \ \sum_{k=1}^{N} \lambda_k y_k = 0\}$. Since Slater's condition is fulfilled for (10.6), we may obtain the optimal (λ, μ) by solving the dual optimization problem

$$\text{maximize} \ \ g(\lambda, \mu)$$

$$\text{subject to} \ \ \lambda_k + \mu_k = 1, \quad k \in \mathbb{N}_N$$

$$\sum_{k=1}^{N} \lambda_k y_k = 0$$

$$\lambda_k \geq 0, \quad \mu_k \geq 0, \quad k \in \mathbb{N}_N.$$

with variables (λ, μ). This is a convex optimization problem with quadratic objective function and simple inequality constraints, which can be solved efficiently. There are actually special purpose algorithms that are very efficient, e.g. the so-called *sequential minimal optimization* algorithm [91].

10.5.5 Recovering the Primal Solution

The solution to the dual problem also provides the solution to the primal problem. We have already seen the expression for a. However, it still remains to compute the value of b from the solution of the dual problem. For any $\lambda_k > 0$, which we obtain from the dual problem, we know from complementary slackness that $1 - \xi_k - y_k f(x_k; a, b) = 0$, i.e.

$$b = \frac{\xi_k + 1}{y_k} - a^T \beta(x_k), \quad k \in \mathcal{I},$$

where $\mathcal{I} = \{k \in \mathbb{N}_N | \lambda_k > 0\}$. We also know that we will have $\xi_k = 0$ for some values of $k \in \mathcal{I}$, i.e. there is a nonempty set $\mathcal{J} = \{k \in \mathcal{I} | \xi_k = 0\}$ for which it holds that

$$b = \frac{1}{y_k} - a^T \beta(x_k), \quad k \in \mathcal{J}.$$

Any such k can be used to compute b, and they should all give the same result. However, for numerical stability, one may average over them. We can find these values of k from the complementary slackness condition $\mu_k \xi_k = 0$. All k such that $\mu_k > 0$ will do. To summarize, the value of b can be computed by identifying the indexes k for which both the optimal λ_k and μ_k are strictly positive. A generalization of the SVM to more than two classes is given in [29].

Example 10.4 We consider an example where we have 20 points $x_1, \ldots, x_{20} \in \mathbb{R}^2$ of one class and 20 points $x_{21}, \ldots, x_{40} \in \mathbb{R}^2$ of another class. We use the parameter value $v = 20$ and compute a and b for the function $f(x) = a^T x + b$ that should separate the two classes. In Figure 10.5, we see the points together with the line $f(x) = 0$, which almost separates the two classes of points. It is easy to see that they cannot be separated with a straight line, but the SVM solution only misclassifies a few points.

10.5.6 The Kernel Trick

We realize that the dual problem does not depend explicitly on the function β but only on the inner products $\beta(x_k)^T \beta(x_l)$. This is also the case for the resulting regressor, since it can be written as

$$f(x; a, b) = \frac{1}{v} \sum_{k=1}^N \lambda_k y_k \beta(x_k)^T \beta(x) + b,$$

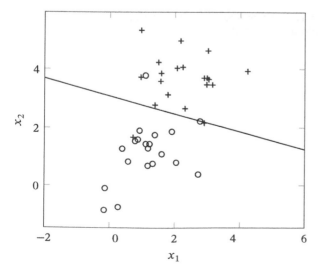

Figure 10.5 Plot showing the data points from the two classes marked with + for the first class and o for the second class. Also shown is the line $f(x) = 0$ that almost separates the two classes.

and hence, it is sufficient to know $K : \mathbb{R}^m \times \mathbb{R}^m \to \mathbb{R}_+$ defined by $K(x, \bar{x}) = \beta(x)^T \beta(\bar{x})$. In other words, the kernel trick also applies for SVM, so we may also consider any of the kernels discussed in relation to regression in Hilbert spaces and regression using Gaussian processes, *cf.* sections 10.2 and 10.3.

10.6 Restricted Boltzmann Machine

The BM discussed in Section 9.13 can also be used for classification. It is applicable when the pairs of data (x_k, y_k), $k \in \mathbb{N}_N$, are such that x_k is a binary vector, i.e. $x_k \in \{0, 1\}^M$. We still have that y_k is categorical, but we do not have $y_k \in \mathbb{N}_K$. Instead, we say that (x_k, y_k) belongs to class $l \in \mathbb{N}_K$ if $y_k \in \{0, 1\}^K$ is a vector of all zeros except for a one in position l, i.e. $y_k = e_l$, where e_l is the lth unit vector. We then let $v_k = (x_k, y_k) \in \{0, 1\}^{M+N}$ be the visible variables in the Boltzmann machine.

Once the available data (x_k, y_k), $k \in \mathbb{N}_N$, has been used to find optimal values of the parameters (λ, Λ), e.g. using ML estimation, we may use the resulting Ising probability function $p(y, x, h)$, where h are the hidden variables for discrimination. We let $\delta_l(x) = \sum_h p(e_l, x, h)$ be the discriminant function and say that x belongs to class l if $\delta_l(x) > \delta_k(x)$ for all $k \neq l$. This means that we assign the class with the largest probability to each observation.

10.6.1 Graphical Ising Distribution

As mentioned in Section 9.13, it is not easy to compute the ML estimate. Because of this, one often uses the so-called *Restricted Botzmann Machine* (RBM). This is based on a special case of the graphical Ising distribution, which is obtained by considering a graph on the variables (v, h) such that there are no edges between the variables within v or h, respectively, see Figure 10.6. This results in the following structure for the parameter Λ:

$$\Lambda = \begin{bmatrix} 0 & \Lambda_{12} \\ \Lambda_{12}^T & 0 \end{bmatrix},$$

which means that the probability function for the Ising distribution can be written as

$$p(v, h) = \frac{1}{Z} e^{\lambda_1^T v + \lambda_2^T h + 2v^T \Lambda_{12} h},$$

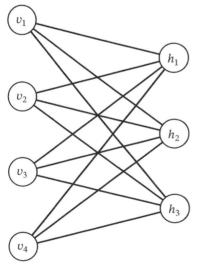

Figure 10.6 Graph showing how the visible and hidden layers in an RBM are connected.

where

$$Z = \sum_{h,v} e^{\lambda_1^T v + \lambda_2^T h + 2v^T \Lambda_{12} h}.$$

The convention is that if there are no limits for the summations, the sum should be taken over all possible values of the variables. It follows that the conditional probability function for h given v may be factorized. To see this, notice that the marginal probability function for v can be written as

$$\sum_h p(v, h) = \frac{e^{\lambda_1^T v}}{Z} \sum_h e^{(\lambda_2 + 2\Lambda_{12}^T v)^T h} = \frac{e^{\lambda_1^T v}}{Z} \sum_h \prod_i e^{(\lambda_2 + 2\Lambda_{12}^T v)_i^T h_i}$$

$$= \frac{e^{\lambda_1^T v}}{Z} \prod_i \sum_{h_i} e^{(\lambda_2 + 2\Lambda_{12}^T v)_i^T h_i} = \frac{e^{\lambda_1^T v}}{Z} \prod_i \left(1 + e^{(\lambda_2 + 2\Lambda_{12}^T v)_i^T} \right).$$

Because of this, the conditional probability function is given by

$$\frac{p(v, h)}{\sum_h p(v, h)} = \frac{\frac{1}{Z} e^{\lambda_1^T v + \lambda_2^T h + 2v^T \Lambda_{12} h}}{\frac{e^{\lambda_1^T v}}{Z} \prod_i \left(1 + e^{(\lambda_2 + 2\Lambda_{12}^T v)_i^T} \right)} = \frac{\prod_i e^{(\lambda_2 + 2\Lambda_{12}^T v)_i h_i}}{\prod_i \left(1 + e^{(\lambda_2 + 2\Lambda_{12}^T v)_i^T} \right)}.$$

The conditional probability function for h_j given v is then

$$s_j(h_j, v) = \frac{e^{(\lambda_2 + 2\Lambda_{12}^T v)_j h_j}}{\prod_i \left(1 + e^{(\lambda_2 + 2\Lambda_{12}^T v)_i^T} \right)},$$

and the overall conditional probability function is obtained as the product of these factors. The expression of the factors can be simplified considerably. Notice that they are for $h_j = 0$ and $h_j = 1$ given by

$$s_j(0, v) = \frac{1}{\prod_i (1 + e^{x_i})}, \qquad s_j(1, v) = \frac{e^{x_j}}{\prod_i (1 + e^{x_i})},$$

where $x_i = (\lambda_2 + 2\Lambda_{12}^T v)_i$. Since $s_j(0, v) + s_j(1, v) = 1$, we have that $\prod_i (1 + e^{x_i}) = e^{x_j} + 1$, and hence

$$s_j(1, v) = \frac{e^{x_j}}{1 + e^{x_j}} = \frac{1}{1 + e^{-x_j}} = \sigma(x_j),$$

where $\sigma : \mathbb{R} \to \mathbb{R}$ is the *logistic function*, which is an example of a so-called *sigmoid function*.

10.6.2 Gradient Expressions for EM Algorithm

We will now show that the expressions for the gradients of the Q-function in the EM-algorithm are simpler for the RBM than for the BM, *cf.* Section 9.13. For the sake of simplicity, we assume that there is only one observation v. Notice that the case of several observations is obtained by summing up the gradients for the different observations just as for the general Boltzmann machine. Subindices from now will on refer to component of vectors. The expressions for the gradients for the RBM are

$$\frac{\partial Q}{\partial \lambda} = \sum_h s(h, v, \lambda^-, \Lambda^-) \begin{bmatrix} v \\ h \end{bmatrix} - \sum_\xi p(\xi_v, \xi_h) \begin{bmatrix} \xi_v \\ \xi_h \end{bmatrix}$$

$$\frac{\partial Q}{\partial \Lambda_{12}} = 2 \sum_h s(h, v, \lambda^-, \Lambda^-) v h^T - \sum_\xi p(\xi_v, \xi_h) \xi_v \xi_h^T,$$

where $s(h, v, \lambda^-, \Lambda^-) = \prod_k s_k(h_k, v)$, and where $\xi = (\xi_v, \xi_h)$. We let $p_v(\xi_v) = \sum_{\xi_h} p(\xi_v, \xi_h)$ be the marginal probability function for the observations. We may then write the gradients as

$$\frac{\partial Q}{\partial \lambda} = \sum_h s(h, v, \lambda^-, \Lambda^-) \begin{bmatrix} v \\ h \end{bmatrix} - \sum_{\xi_v} p_v(\xi_v) \sum_{\xi_h} s(\xi_h, \xi_v, \lambda^-, \Lambda^-) \begin{bmatrix} \xi_v \\ \xi_h \end{bmatrix}$$

$$\frac{\partial Q}{\partial \Lambda_{12}} = 2 \sum_h s(h, v, \lambda^-, \Lambda^-) v h^T - \sum_{\xi_v} p_v(\xi_v) \sum_{\xi_h} s(\xi_h, \xi_v, \lambda^-, \Lambda^-) \xi_v \xi_h^T.$$

We realize that

$$\sum_h s(h, v, \lambda^-, \Lambda^-) h_l = \sum_h \prod_k s_k(h_k, v) h_l = \sum_h \prod_{k \neq l} s_k(h_k, v) s_l(h_l, v) h_l$$

$$= \prod_{k \neq l} \sum_{h_k} s_k(h_k, v) \sum_{h_l} s_l(h_l, v) h_l = s_l(1, v),$$

where the last equality follows from the fact that $s_k(0, v) + s_k(1, v) = 1$ and $s_l(0, v) \times 0 = 0$. From this, it follows that

$$\frac{\partial Q}{\partial \lambda_1} = v - \sum_{\xi_v} p_v(\xi_v) \xi_v$$

$$\left(\frac{\partial Q}{\partial \lambda_2} \right)_i = s_i(1, v) - \sum_{\xi_v} p_v(\xi_v) s_i(1, \xi_v)$$

$$\left(\frac{\partial Q}{\partial \Lambda_{12}} \right)_{i,j} = 2 s_j(1, v) v_i - \sum_{\xi_v} p_v(\xi_v) s_j(1, \xi_v)(\xi_v)_i.$$

The first term in each of the expressions is cheap to evaluate. The second terms can be approximated using Monte Carlo methods. In [38], the so-called contrastive divergence method based on Gibbs sampling is described. We want to draw a sample ξ_v from p_v. We initialize the Gibbs sampler with $\xi_v^{(0)} = v$, cf. Section 9.12. Because of the graphical structure of the RBM, we then draw a sample $h^{(1)}$ from the conditional distribution of h given v, i.e. from $s_j(h_j, \xi_v^{(0)})$ above. We then finally draw a sample $\xi_v^{(1)}$ from the conditional distribution of v given h, where we use $h = h^{(1)}$. This conditional distribution can be obtained similarly to $s_j(h_j, v)$ above. Here we have run the Gibbs sampler for only one step, but it is, of course, possible to run more steps. However, empirical evidence suggests that one step is enough. After the samples have been obtained, we approximate the above sums as

$$\sum_{\xi_v} \xi_v^{(1)}, \quad \sum_{\xi_v} s_i(1, \xi_v^{(1)}), \quad \sum_{\xi_v} s_j(1, \xi_v^{(1)})(\xi_v^{(1)})_i,$$

respectively. This is a very simplistic approximation.

10.7 Artificial Neural Networks

The regression models we have discussed so far have been linear, i.e. they have been linear in the parameters to be estimated. *Artificial neural networks* (ANN)s can be used to specify nonlinear regression models. It can be seen as a generalization of the RBM.

10.7.1 The Network

An ANN with L layers and n_i neurons for layer i is defined as follows: let $x_i \in \mathbb{R}^{n_i}$ for $i \in \mathbb{Z}_{L-1}$ be the *input activation* of layer i, and let

$$z_i = \Phi_i(x_{i-1}), \quad i \in \mathbb{N}_L,$$

where $z_i \in \mathbb{R}^{n_i}$ is the *output* of layer i. The function $\Phi_i : \mathbb{R}^{n_{i-1}} \to \mathbb{R}^{n_i}$ is called the *propagation function*. Typically, it can be a linear or an affine function, and then we may write

$$\Phi_i(x) = W_i x + v_i, \quad i \in \mathbb{N}_L, \tag{10.7}$$

where $W_i \in \mathbb{R}^{n_i \times n_{i-1}}$ and $v_i \in \mathbb{R}^{n_i}$. The input to the next layer is obtained by

$$x_i = h_i(z_i), \quad i \in \mathbb{N}_{L-1}, \tag{10.8}$$

where $h_i : \mathbb{R}^{n_i} \to \mathbb{R}^{n_i}$ is called the *activation function*. It is often the case that this function can be written as

$$h_i(z) = \left(h_{i1}(z_1), \dots, h_{in_i}(z_{n_i}) \right), \tag{10.9}$$

and each component h_{ij} is typically a saturation function or a sigmoid function, i.e. a function $\sigma : \mathbb{R} \to [0, 1]$ such that $\lim_{t \to \infty} \sigma(t) = 1$ and $\lim_{t \to -\infty} \sigma(t) = 0$. Another popular choice is the ReLU, which was defined in Section 10.5. Figure 10.7 shows an example of a neural network.

We now define the functions $f_i : \mathbb{R}^{n_{i-1}} \times \mathbb{R}^{p_i} \to \mathbb{R}^{n_i}, i \in \mathbb{N}_L$, as

$$f_i(x_{i-1}, \theta_i) = h_i(\Phi_i(x)) = h_i(W_i x_{i-1} + v_i), \tag{10.10}$$

where

$$\theta_i = \mathrm{vec}\left(\begin{bmatrix} W_i & v_i \end{bmatrix} \right) \in \mathbb{R}^{p_i}. \tag{10.11}$$

It follows that

$$x_i = f_i(x_{i-1}; \theta_i), \quad i \in \mathbb{N}_L. \tag{10.12}$$

Finally, we define $f : \mathbb{R}^{n_0} \times \mathbb{R}^{p_1} \times \cdots \times \mathbb{R}^{p_L} \to \mathbb{R}^{n_L}$ as $f(x_0; \theta_1, \dots, \theta_L) = x_L$. With $x = x_0 \in \mathbb{R}^{n_0}$ and $\theta = (\theta_1, \dots, \theta_L) \in \mathbb{R}^p$, where $p = \sum_{i=1}^{L} p_i$, we have hence defined a nonlinear regression model or predictor $f(x; \theta)$. The recursive structure of the predictor is illustrated in Figure 10.8.

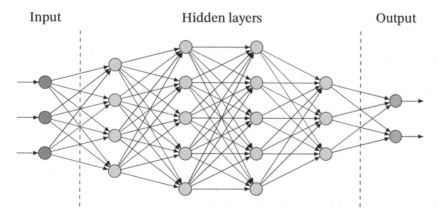

Input   Hidden layers   Output

Figure 10.7 A neural network with four hidden layers. The nodes represent the activation functions, and the edges illustrate how the output from one layer is propagated to the next.

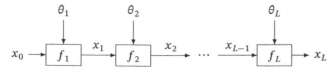

Figure 10.8 Figure illustrating the recursive definition of the ANN predictor.

10.7.2 Approximation Potential

It is interesting to note that any continuous function $f : [0,1]^n \to \mathbb{R}$ can be represented as

$$f(x) = \sum_{q=0}^{2n} g\left(\sum_{p=1}^{n} \phi_{p,q}(x_p)\right),$$

where $\phi_{p,q} : [0,1] \to [0,1]$ are continuous increasing functions, and where $g : \mathbb{R} \to \mathbb{R}$ is a continuous function [64, 72]. As a result, a two-layer ANN is sufficient to represent any continuous function. However, very little is known about how to choose g. A result by Cybenko [30] says that any continuous function $f : [0,1]^n \to \mathbb{R}$ can be approximated arbitrarily well as

$$f(x) \approx \sum_{j=1}^{N} \alpha_j \sigma(a_j^T x + b_j) \tag{10.13}$$

if N is sufficiently large, and where σ is any continuous sigmoid function, and $\alpha_j \in \mathbb{R}, a_j \in \mathbb{R}^n$ and $b_j \in \mathbb{R}, j \in \mathbb{N}_N$.

10.7.3 Regression Problem

Regression using ANNs is about selecting a $\theta \in \mathbb{R}^p$ such that for pairs of data $(x_k, y_k) \in \mathbb{R}^{n_0} \times \mathbb{R}^{n_L}$, $k \in \mathbb{N}_N$, it holds that the output $f(x_k; \theta)$ of the ANN is close to y_k. The goodness of the closeness is measured with a function $V : \mathbb{R}^p \to \mathbb{R}$ which could typically be the sum of squared norms of $y_k - f(x_k; \theta)$, e.g.

$$V(\theta) = \frac{1}{2}\sum_{k=1}^{N} \|y_k - f(x_k; \theta)\|_2^2, \tag{10.14}$$

the minimization of which is a nonlinear LS problem.

10.7.4 Special Cases

It is possible to interpret logistic regression as a one-layer ANN. This follows immediately by taking

$$W_1 = \begin{bmatrix} a_1^T \\ \vdots \\ a_{K-1}^T \end{bmatrix}, \qquad v_1 = \begin{bmatrix} b_1 \\ \vdots \\ b_{K-1} \end{bmatrix},$$

and $h(z) = (z_1/s, \ldots, z_{K-1}/s, 1/s)$, where $s = 1 + \sum_{k=1}^{K-1} e^{z_k}$. Hence the output x_1 of the ANN are the probabilities $p_l(z_1, z_2, \ldots, z_{K-1}), l \in \mathbb{N}_K$ of the categorical distribution, i.e. $x_{1l} = p_l(z_1, z_2, \ldots, z_{K-1})$. The training data are $(x_k, y_k) \in \mathbb{R}^n \times \{0,1\}$ for $k \in \mathbb{N}_N$, and the function to maximize is the likelihood function for the data.

It is also possible to interpret an RBM as a one-layer ANN with binary input activation. To see this, let $x_0 = v, v_1 = \lambda_2, W_1 = 2\Lambda_{12}^T$, and $h_i(z_i) = \sigma(z_i)$. Then $x_{1i} = s_i(1, v)$, and hence, the outputs of the ANN are the conditional probabilities for $h_i = 1$ given v, or equivalently the expected values of the hidden variables given the visible variables. The training data are the data $v = (v_1, \ldots, v_N)$ for the visible layer of the RBM, and the function to maximize is the likelihood function for the data. This optimization can be performed using the EM algorithm which makes use of the output of the ANN to compute gradients.

Finally, we interpret the optimization problem for Hebbian learning in (10.5) as a one-layer ANN. Let $v_1 = b$, $W_1 = a^T$ and $h(z_1) = g(z_1)$. The training data is $(x_k, y_k) \in \mathbb{R}^n \times \{-1, 1\}$, $k \in \mathbb{N}_N$ and the function to maximize is the sum of the outputs of the ANN over all training data.

Because of these interpretations, it easy to see how one can generalize logistic regression, the RBM, and Hebbian learning using multilayer ANNs.

10.7.5 Back Propagation

In order to minimize the function V in, e.g. (10.7) with respect to the parameter θ its gradient needs to be computed efficiently. This will, of course, depend on the specific choice of the function V, but the main computational burden will be related to computing the gradient of the predictor $f(x; \theta)$ with respect to θ.

From (10.12), it follows that for $i, j \in \mathbb{N}_L$,

$$
\frac{\partial x_i}{\partial \theta_j^T} = \begin{cases} 0, & j > i, \\ \frac{\partial f_i}{\partial \theta_i^T}, & i = j, \\ \frac{\partial f_i}{\partial x_{i-1}^T} \frac{\partial x_{i-1}}{\partial \theta_j^T}, & j < i, \end{cases}
$$

from which we may conclude that

$$
\frac{\partial f}{\partial \theta_i^T} = \frac{\partial f_L}{\partial x_{L-1}^T} \frac{\partial f_{L-1}}{\partial x_{L-2}^T} \cdots \frac{\partial f_i}{\partial \theta_i^T}, \quad i \in \mathbb{N}_L.
$$

We realize from the above formula that the gradients of f are computed by starting with the last layer and then progressing backward through the layers of the ANN. This has given rise to the name *back propagation* for the algorithm for computing the gradients for ANNs. We also realize that not all partial derivatives have to be stored, but only cumulative products in the above expression.

We will now provide some more details on how the partial derivatives are computed for the case of an affine propagation function. We have from (10.11) that

$$
W_i x_{i-1} + v_i = \left(\begin{bmatrix} x_{i-1}^T & 1 \end{bmatrix} \otimes I \right) \theta_i.
$$

Hence, it follows from (10.10) that

$$
\frac{\partial f_i}{\partial \theta_i^T} = \frac{\partial h_i}{\partial z_i^T} \left(\begin{bmatrix} x_{i-1}^T & 1 \end{bmatrix} \otimes I \right),
$$

and

$$
\frac{\partial f_i}{\partial x_{i-1}^T} = \frac{\partial h_i}{\partial z_i^T} W_i.
$$

In case, h_i has the structure in (10.9), it follows that $\frac{\partial h_i}{\partial z_i^T}$ is a diagonal matrix. The back propagation algorithm for the case of affine propagation function is summarized in Algorithm 10.1. We have included also the forward propagation of x in the algorithm. In case $\frac{\partial h_i}{\partial z_i^T}$ is diagonal, special care should be taken to make use of this in the implementation of the algorithm.

It should be mentioned that back propagation is related to what is called *automatic differentiation*, and specifically what is called *reverse mode accumulation*. For more information about back propagation, we refer to [45, Section 6.5] and [9].

Algorithm 10.1: Back propagation algorithm

Input: input x to ANN, number of Layers L, activation functions h_i, weights W_i, bias terms v_i,
for $i \in \mathbb{N}_L$
Output: $\frac{\partial f}{\partial \theta_i^T}$ for $i \in \mathbb{N}_L$

$x_0 \leftarrow x$
for $i \leftarrow 1$ **to** L **do**
$\quad \Big| \quad z_i \leftarrow W_i x_{i-1} + v_i$
$\quad \Big| \quad x_i \leftarrow h_i(z_i)$
end
$H_L \leftarrow \frac{\partial h_L(z_L)}{\partial z_L^T}$
$\frac{\partial f}{\partial \theta_L} \leftarrow H_L \left(\begin{bmatrix} x_{L-1}^T & 1 \end{bmatrix} \oplus I \right)$
$P \leftarrow H_L W_L$
for $i \leftarrow L-1$ **to** 1 **do**
$\quad \Big| \quad H_i \leftarrow \frac{\partial h_i(z_i)}{\partial z_i^T}$
$\quad \Big| \quad P \leftarrow P H_i$
$\quad \Big| \quad \frac{\partial f}{\partial \theta_i^T} \leftarrow P \left(\begin{bmatrix} x_{i-1}^T & 1 \end{bmatrix} \oplus I \right)$
$\quad \Big| \quad P \leftarrow P W_i$
end

10.8 Implicit Regularization

Regularization is also important when training ANNs. This is especially true, since often the number of parameters in ANN exceeds the number of data points. Sometimes it is not easy to find suitable regularization functions. Fortunately, there are implicit ways of obtaining regularization, which is the topic of this section.

10.8.1 Least-Norm Solution

To provide insight into how a stochastic gradient method can provide regularization for free, we consider a linear regression problem where the pairs of data (x_k, y_k) with $x_k \in \mathbb{R}^n$ and $y_k \in \mathbb{R}$, $k \in \mathbb{N}_N$, should satisfy the linear regression model

$$y_k = a^T x_k. \tag{10.15}$$

This means that we are able to *interpolate* the data. This is possible if $N \leq n$. Before, when we have discussed regression, we have tacitly assumed that $N > n$. However, as mentioned above, common practice when using ANNs is to be in this interpolation regime. We can collect all pairs of data in

$$X = \begin{bmatrix} x_1^T \\ \vdots \\ x_N^T \end{bmatrix} \in \mathbb{R}^{N \times n}, \qquad y = \begin{bmatrix} y_1 \\ \vdots \\ y_N \end{bmatrix} \in \mathbb{R}^N.$$

Then a should satisfy

$$Xa = y,$$

which has infinitely many solutions if the system of equations is consistent. It would then be natural to look for a solution that minimizes the norm $\|a\|_2$. In Exercise 10.11, we show that if X has full row rank, then this solution is given by

$$a = X^T \left(XX^T\right)^{-1} y.$$

We will now see how the incremental stochastic optimization method in (6.72) can be used to obtain this least-norm solution by solving the optimization problem

$$\text{minimize } \frac{1}{N} \sum_{k=1}^{N} V_k(a^T x_k, y_k),$$

with variable a, [120]. Here $V_k : \mathbb{R} \times \mathbb{R} \to \mathbb{R}$ are functions such that $V_k(0, 0) = 0$ and such that the incremental method converges to a vector a that satisfies $Xa = y$. This holds true for many convex functions. One example is

$$V_k(z_k, y_k) = \frac{1}{2}(y_k - z_k)^2, \tag{10.16}$$

which corresponds to the LS criterion. We realize that $f_k(a) = V_k(a^T x_k, y_k)$ to get agreement with the optimization problem in (5.46). The incremental method reads

$$a_{k+1} = a_k - t_k \nabla f_{i_k}(a_k).$$

We have that

$$\nabla f_{i_k}(a) = \frac{\partial V_{i_k}(a^T x_{i_k}, y_k)}{\partial z_{i_k}} x_{i_k},$$

and hence, if we start with $a_0 = 0$, we find that the incremental method converges to $a = X^T \alpha$ for some $\alpha \in \mathbb{R}^N$. Together with $Xa = y$, we obtain the equation

$$XX^T \alpha = y,$$

which has a unique solution if X has full row rank. It then follows that

$$a = X^T \left(XX^T\right)^{-1} y,$$

which is the least-norm solution. This could also have been obtained by solving the optimization problem

$$\text{minimize } \sum_{k=1}^{N} (y_k - a^T x_k)^2 + \gamma \|a\|_2^2,$$

with variable a, which is a regularized LS problem, and then letting $\gamma \to 0$, see Exercise 10.11.

10.8.2 Dropout

Another mechanism that results in implicit regularization is the concept of *dropout*. We will study it for the linear regressor in (10.15). We consider $V_k(z, y)$ given by (10.16), and the resulting LS problem is

$$\text{minimize } \frac{1}{2} \sum_{k=1}^{N} (y_k - a^T x_k)^2,$$

with variable $a \in \mathbb{R}^n$. Dropout is achieved by replacing a_i, i.e. the ith component of a with $\delta_i a_i$, where δ_i is an outcome of a random variable Δ_i with a Bernoulli distribution, i.e. $P(\Delta_i = 1) = p_i$ and $P(\Delta_i = 0) = 1 - p_i$. We assume that Δ_i is independent of Δ_j for $i \neq j$. Let $\Delta = \text{diag}(\Delta_1, \ldots, \Delta_n)$ and $\delta = \text{diag}(\delta_1, \ldots, \delta_n)$. We may then write the kth term in the objective function as the outcome of $\frac{1}{2}(y_k - x_k^T \Delta a)^2$. We actually also assume that we have different outcomes δ of Δ for each k that are independent of one another, but since we only need to analyze on fixed value of k to understand how the incremental method performs with dropout, we will neglect all dependence on k and consider $V^d : \mathbb{R}^n \to \mathbb{R}$ defined as

$$V^d(a) = \frac{1}{2}(y - x^T \Delta a)_2^2,$$

for $y \in \mathbb{R}$ and $x \in \mathbb{R}^n$. Once we have determined a, we would like to use this parameter to predict y given a new value of x. We therefore define the predictor $\hat{Y} : \mathbb{R}^n \times \{0, 1\}^n \to \mathbb{R}$ as $\hat{Y}(x; \Delta_1, \dots, \Delta_n) = x^T \Delta a$. This is, however, not so useful, since it involves the random variable Δ_i. We are more interested in the expected value of this predictor, which is often called the *ensemble average predictor* $\hat{y} : \mathbb{R}^n \to \mathbb{R}$ defined as

$$\hat{y}(x) = \mathbb{E}\left[\hat{Y}(x; \Delta_1, \dots, \Delta_n) \right] = x^T P a,$$

where $P = \text{diag}(p_1, \dots, p_n)$. The interesting question is if running the incremental method on the problem

$$\text{minimize } \mathbb{E}\left[V^d(a) \right],$$

with variable a will result in a good ensemble average predictor. We therefore introduce the function $V^e : \mathbb{R}^n \to \mathbb{R}$ as

$$V^e(a) = \frac{1}{2}(y - x^T P a)^2,$$

which we like to be small for the ensemble average predictor to be good. We will now compare the gradient of this function with the expected value of the gradient of V^d. Remember that this expected value is what determines the behavior of the incremental method, *cf.* Section 6.8. We have

$$\frac{\partial V^e(a)}{\partial a} = -(y - x^T P a)P x$$

$$\frac{\partial V^d(a)}{\partial a} = -(y - x^T \Delta a)\Delta x.$$

From this, we obtain

$$\mathbb{E}\left[\frac{\partial V^d(a)}{\partial a} \right] = -yPx + \mathbb{E}\left[\Delta x x^T \Delta \right] a,$$

where element (i, j) of $\mathbb{E}\left[\Delta x x^T \Delta \right]$ is given by

$$\mathbb{E}\left[\Delta_i \Delta_j x_i x_j \right] = \begin{cases} p_i p_j x_i x_j, & i \neq j \\ p_i x_i^2, & i = j. \end{cases}$$

This implies that

$$\mathbb{E}\left[\Delta x x^T \Delta \right] = P x x^T P + \text{diag}(\sigma_1^2 x_1^2, \dots, \sigma_n^2 x_n^2),$$

where $\sigma_i^2 = p_i(1 - p_i)$ is the variance of Δ_i. This means that we have

$$\mathbb{E}\left[\frac{\partial V^d(a)}{\partial a} \right] = \frac{\partial V^e(a)}{\partial a} + \text{diag}(\sigma_1^2 x_1^2, \dots, \sigma_n^2 x_n^2)a,$$

which is the gradient of

$$V^e(a) + \frac{1}{2}a^T \text{diag}(\sigma_1^2 x_1^2, \dots, \sigma_n^2 x_n^2)a. \tag{10.17}$$

We see that the second term is a *ridge* regularization, and this explains how dropout implicitly provides regularization. The largest possible regularization is obtained when $p_i = 1/2$, since this value maximizes σ_i^2. For more information about dropout see [8].

Exercises

10.1 Consider the linear regression model

$$y_k = a^T \beta(x_k) + e_k,$$

for pairs of data (x_k, y_k), $k \in \mathbb{N}_N$, where $a \in \mathbb{R}^m$ and $\beta : \mathbb{R}^n \to \mathbb{R}^m$. We assume that the pdf related to e_k is

$$f_{E_k}(e_k) = \frac{1}{\sqrt{2\pi}\sigma} e^{-\frac{1}{2\sigma^2} e_k^2}.$$

Define the likelihood function $\mathcal{L} : \mathbb{R}^N \times \mathbb{R}^m \times \mathbb{R} \to \mathbb{R}_+$

$$\mathcal{L}\left(y_1, \ldots y_N; a, \sigma\right) = \prod_{k=1}^{N} f_{E_k}\left(y_k - a^T \beta(x_k)\right),$$

where we not only consider a to be a parameter but also the standard deviation σ. Show that the ML estimate of (a, σ) is given by

$$a = \left(X^T X\right)^{-1} X^T y$$

$$\sigma = \sqrt{\frac{1}{N}(y - Xa)^T(y - Xa)},$$

where $y = (y_1, \ldots, y_N)$ and

$$X = \begin{bmatrix} \beta^T(x_1) \\ \vdots \\ \beta^T(x_N) \end{bmatrix}.$$

10.2 Consider the LS problem

$$\text{minimize } \frac{1}{2}\|Y_N - X_N a\|_2^2,$$

with variable a, where

$$Y_N = \begin{bmatrix} y_1 \\ \vdots \\ y_N \end{bmatrix}, \quad X_N = \begin{bmatrix} x_1^T \\ \vdots \\ x_N^T \end{bmatrix}.$$

You may assume that measurements y_k are scalar. The regressors x_k and the parameter a have dimension $n > 1$. We assume that we have solved this problem, i.e. we have the solution to the above problem, and we denote it with a_N. We then obtain a new measurement y_{N+1} and a new regressor x_{N+1}, and we would like to solve the above LS problem with N replaced with $N + 1$, i.e. we would like to compute a_{N+1} such that it solves

$$\text{minimize } \frac{1}{2}\|Y_{N+1} - X_{N+1} a\|_2^2.$$

(a) Show that this can be done with the following updated formula:

$$a_{N+1} = a_N + P_{N+1} x_{N+1}\left(y_{N+1} - x_{N+1}^T a_N\right),$$

where $P_N^{-1} = X_N^T X_N$ and where $P_{N+1}^{-1} = P_N^{-1} + x_{N+1} x_{N+1}^T$.

(b) Show that the recursion above for P_N can be equivalently written as

$$P_{N+1} = P_N - \frac{1}{1 + x_{N+1}^T P_N x_{N+1}} P_N x_{N+1} x_{N+1}^T P_N.$$

You may use the matrix inversion lemma which says that for matrices A, C, U, and V such that the dimensions are compatible and such that the inverses below exist it holds that

$$(A + UCV)^{-1} = A^{-1} - A^{-1}U(C^{-1} + VA^{-1}U)^{-1}VA^{-1}.$$

10.3 Consider the MAP optimization problem in (10.3) restated for ease of reference below

$$\text{minimize } \sum_{k=1}^{N} \frac{1}{2\sigma^2}(y_k - a^T\beta(x_k))^2 + N \ln\left(\sqrt{2\pi}\sigma\right) - \ln f_A(a),$$

with variable a, where f_A is the pdf for the prior.
  (a) Consider the case of a double-sided exponential distribution given by

$$f_A(a) = \prod_{i=1}^{m} \frac{1}{2\lambda} e^{-|a_i|/\lambda},$$

for some $\lambda > 0$, and show that an equivalent optimization problem is

$$\text{minimize } \sum_{k=1}^{N} \frac{1}{2}(y_k - a^T\beta(x_k))^2 + c\sum_{i=1}^{m} |a_i|,$$

with variable a for some constant $c > 0$. This is known as *lasso regularization*.
  (b) Consider the case when each component A_i of the random variable A has a uniform distribution on the interval $[-\lambda, \lambda]$, where $\lambda > 0$. Show that an equivalent MAP optimization problem is

$$\text{minimize } \sum_{k=1}^{N} \frac{1}{2}(y_k - a^T\beta(x_k))^2$$

$$\text{subject to } \|a\|_\infty \le \lambda,$$

with variable a.

10.4 Show that the logistic regression problem in Section 10.4 can be equivalently written as a conic optimization problem involving the exponential cone.

10.5 Consider Example 10.2 and write a MATLAB code the reproduces the result in the example.

10.6 Consider the optimization problem in (10.5).
  (a) Show that an incremental subgradient method with step length one for updating the parameters a and b results in the Hebbian learning algorithm

$$a_{k+1} = a_k + y_{i_k}x_{i_k}$$
$$b_{k+1} = b_k + y_{i_k},$$

if $1 - y_i f(x_{i_k}; a_k, b_k) > 0$ and that there is no update otherwise. Above the index i_k could be picked randomly with equal probability from $\mathbb{N}_N$ at each iteration k or cyclic as $i_k = (k \bmod N) + 1$.
  (b) Make a MATLAB implementation of the cyclic Hebbian learning algorithm. Generate two-dimensional data in $\mathbb{R}^2$ that can be classified with a linear classifier and try out the algorithm. How fast does it converge?

10.7 Consider the following optimization problem:

$$\text{maximize} \quad y_k \frac{a^T x_k + b}{\|a\|_2}$$

$$\text{subject to} \quad y_k(a^T x_k + b) \geq 1, \quad k \in \mathbb{N}_N$$

with variables (a, b), where $b, y_k \in \mathbb{R}$ and $a, x_k \in \mathbb{R}^n$. Show that it is equivalent to

$$\text{maximize} \quad \|a\|_2$$

$$\text{subject to} \quad y_k(a^T x_k + b) \geq 1, \quad k \in \mathbb{N}_N.$$

10.8 Consider the Fisher Iris data set that you investigated in Exercise 9.13 using PCA. You are going to use the compressed data c_i that you obtained using the two principal components corresponding to the two largest singular values. You are asked to classify these using an SVM as in (10.6) with f being an affine function, i.e. $f(c) = a^T c + b$, where $a \in \mathbb{R}^2$ and $b \in \mathbb{R}$ are the parameters that define the separating straight line. The data set has three classes, and the SVM is only able to separate between two classes. However, you will instead carry out three classifications, where you classify species 1 against species 2 and 3, species 2 against species 1 and 3, and finally species 3 against species 1 and 2. In this way, you will obtain three separating straight lines. Plot the lines on top of the figure you plotted in Exercise 9.13. How well do the lines separate the different species? Are there species that cannot be classified? Are there species that are not uniquely classified? How does the results depend on the choice of the regularization parameter v? Notice that you may use different values of v in the three different classifications.
Hint: You can directly solve the primal problems using, e.g. YALMIP, since it is of low dimension.

10.9 Consider the empirical risk minimization problem

$$\text{minimize} \quad \frac{1}{n}\sum_{i=1}^{n}\phi(a_i^T w) + \gamma\, g(w) \tag{10.18}$$

with variable $w \in \mathbb{R}^d$ and where $\phi(t) = \max(0, 1 - t)$ and $g(w) = (1/2)\|w\|_2^2$. The problem data are $a_i \in \mathbb{R}^d$ for $i = 1, \ldots, n$, and $\gamma > 0$ is a parameter.
(a) We start by deriving a dual coordinate ascent method for the problem (10.18).
   (1) Show that the dual problem is equivalent to the problem

$$\text{maximize} \quad -\sum_{i=1}^{n}\phi^*(-x_i) - \frac{1}{2n\gamma}\|A^T x\|_2^2 \tag{10.19}$$

   with variable $x \in \mathbb{R}^n$ and where $a_1 \ldots, a_n$ are the columns of A^T.
   Hint: Introduce a new variable z and the constraint $z = Aw$.
   (2) Show that $\phi^*(s)$, the conjugate of $\phi(t) = \max(0, 1 - t)$, is given by

$$\phi^*(s) = \begin{cases} s & -1 \leq s \leq 0 \\ \infty & \text{otherwise.} \end{cases}$$

   (3) Derive the update for a step of (dual) coordinate ascent with exact line search, i.e.

$$x_i \leftarrow \text{argmin}_s\left\{\phi^*(-s) + \frac{1}{2n\gamma}\|A^T(x - x_i e_i) + s a_i\|_2^2\right\}.$$

   (4) Show that the optimal primal variable $w^\star$ can be recovered from a dual optimal $x^\star$.

(b) Show that the proximal operator associated with $g_i(w) = \phi(a_i^T w) + \gamma g(w)$ can be expressed as

$$\text{prox}_{tg_i}(w) = \underset{u}{\text{argmin}} \left\{ \phi(a_i^T u) + \frac{\gamma}{2}\|u\|_2^2 + \frac{1}{2t}\|u - w\|_2^2 \right\}$$

$$= \begin{cases} \frac{1}{1+t\gamma}(w + ta_i) & t(\|a_i\|_2^2 - \gamma) < 1 - a_i^T w \\ \frac{1}{1+t\gamma}\left(w + \frac{1 - a_i^T w + t\gamma}{\|a_i\|_2^2}a_i\right) & -t\gamma \le 1 - a_i^T w \le t(\|a_i\|_2^2 - \gamma) \\ \frac{1}{1+t\gamma}w & 1 - a_i^T w < -t\gamma. \end{cases}$$

Hint: Derive the optimality condition associated with the prox-problem and show that $u^\star$ has the form $u^\star = (1 + t\gamma)^{-1}(w - \beta ta_i)$, where $\beta \in \partial\phi(a_i^T u^\star)$.

(c) The so-called linear softmargin support vector machine training problem is a special case of the problem (10.18). Specifically, if we partition w as $w = (\tilde{w}, b) \in \mathbb{R}^{d-1} \times \mathbb{R}$ and define $a_i = (y_i x_i, y_i)$, where $x_i \in \mathbb{R}^{d-1}$ corresponds to a so-called feature vector with corresponding label $y_i \in \{-1, 1\}$, then $a_i^T w = y_i(x_i^T \tilde{w} + b)$, and hence, $\phi(a_i^T w) = \max(0, 1 - y_i(x_i^T \tilde{w} + b))$. Given a vector of labels $y = (y_1, \dots, y_n)$ and a matrix $X \in \mathbb{R}^{n \times (d-1)}$ with rows $x_1^T \dots, x_n^T$, we can express A as

$$A = \text{diag}(y)\begin{bmatrix} X & 1 \end{bmatrix}.$$

The file `classification.mat` is an example of such a data set.

The solution to the SVM training problem defines a hyperplane that can be used to define a classifier: the function $f(x) = \text{sgn}(\tilde{w}^T x + b)$ provides a label prediction for an unlabeled feature vector x.

(1) Implement an incremental proximal method for solving (10.18), and test your implementation on the provided dataset.

(2) Implement a dual coordinate ascent method for solving (10.19), and test your implementation on the provided dataset.

(3) Compare the two methods: plot the objective value as a function of the number of iterations, e.g. integer multiples of n iterations.

10.10 The soft margin support vector machine-training problem is a convex quadratic problem:

$$\begin{aligned} \text{minimize} \quad & 1^T v + \frac{\gamma}{2}\|w\|_2^2 \\ \text{subject to} \quad & \text{diag}(y)(Aw + 1b) \ge 1 - v \\ & v \ge 0 \end{aligned} \tag{10.20}$$

with variables $w \in \mathbb{R}^n$, $b \in \mathbb{R}$, and $v \in \mathbb{R}^n$, and with problem data $A \in \mathbb{R}^{m \times n}$ and $y \in \{-1, 1\}^m$. The optimal v can be expressed in terms of w and b as

$$v = \max(0, 1 - \text{diag}(y)(Aw + 1b)).$$

The Lagrangian is given by

$$L(v, w, b, z, \lambda) = 1^T v + \frac{\gamma}{2}\|w\|_2^2 + z^T(1 - v - \text{diag}(y)(Aw + 1b)) - \lambda^T v,$$

and the dual problem can be expressed as

$$\begin{aligned} \text{maximize} \quad & -\frac{1}{2\gamma}z^T \text{diag}(y)AA^T \text{diag}(y)z + 1^T z \\ \text{subject to} \quad & y^T z = 0 \\ & 0 \le z \le 1. \end{aligned} \tag{10.21}$$

The Lagrange multiplier associated with $v \geq 0$ is $\lambda = 1 - z$, so the conditions $\lambda \geq 0$ and $z \geq 0$ are equivalent to $0 \leq z \leq 1$. It follows from the optimality conditions that

$$\gamma w = A^T \text{diag}(y)z,$$

and complementary slackness implies that

$$\left. \begin{array}{rcl} z_i^{\star} = 0 & \Leftrightarrow & y_i u_i \geq 1 \\ 0 < z_i^{\star} < 1 & \Leftrightarrow & y_i u_i = 1 \\ z_i^{\star} = 1 & \Leftrightarrow & y_i u_i \leq 1 \end{array} \right\} \qquad \text{for } i = 1, \ldots, m \tag{10.22}$$

where $u = Aw + 1b = \gamma^{-1}AA^T\text{diag}(y)z + 1b$.

The dual problem (10.21) can be solved using the so-called *sequential minimal optimization* (SMO) method. Suppose z_k is the value of the dual variable at the beginning of iteration k. The SMO iteration then consists of the following three steps:

- *Working set selection*: select two dual variables $z_{k,i}$ and $z_{k,j}$ ($i \neq j$) where at least one of the two variables violates the optimality conditions (10.22).
- *Two-variable subproblem*: compute $z_{k+1,i}$ and $z_{k+1,j}$ by solving the dual problem (10.21) with

$$z = z_k + (z^i - z_{k,i})e_i + (z^j - z_{k,j})e_j. \tag{10.23}$$

- *Update intercept*: compute b_{k+1} such that the updated dual variables $z_{k+1,i}$ and $z_{k+1,j}$ satisfy the optimality conditions (10.22).

The main advantage of this seemingly simple method is that the two-variable subproblem is cheap to solve. Each iteration increases the dual objective, and the method stops when all dual variables satisfy the optimality conditions within some tolerance. The method can be shown to converge, but the working set selection can have a significant impact on performance: there are $m(m-1)/2$ potential working sets at each iteration, so a good working set selection heuristic is crucial. A popular heuristic is the so-called *maximal violating pair* working set selection rule which chooses a pair (i, j) as

$$i \in \underset{l \in I_1}{\text{argmin}}\{y_l - u_l\}, \qquad j \in \underset{l \in I_2}{\text{argmax}}\{y_l - u_l\}, \tag{10.24}$$

where

$$I_1 = \{l \mid z_l > 0, y_l = 1\} \cup \{l \mid z_l < 1, y_l = -1\}$$
$$I_2 = \{l \mid z_l > 0, y_l = -1\} \cup \{l \mid z_l < 1, y_l = 1\}.$$

This selection heuristic can be motivated from the optimality conditions (10.22): $y_l u_l$ must satisfy $y_l u_l \leq 1$ if $z_l > 0$, and similarly, $y_l u_l \geq 1$ if $z_l < 1$. By multiplying both sides of both $y_l u_l \leq 1$ and $y_l u_l \geq 1$ by y_l, we can express the optimality conditions as

$$0 \leq y_l - u_l, \qquad l \in I_1$$
$$0 \geq y_l - u_l, \qquad l \in I_2.$$

It follows that z satisfies the optimality conditions when (i, j) is chosen according to maximal violating pair heuristic (10.24) and $y_i - u_i \geq 0 \geq y_j - u_j$.

(a) Derive a simple method for solving the two variable subproblem with working set (i, j). You may assume that z_k is feasible.

Hint: The constraint $y^T z = 0$ implies that z^i and z^j can be parameterized as $z^i(t) = z_{k,i} + ty_i$ and $z^j(t) = z_{k,j} - ty_j$, and hence, $z(t) = z_k + t(e_i y_i - e_j y_j)$.

(b) Derive an expression for b_{k+1} so that $(z_{k+1,i}, u_{k+1,i})$ and $(z_{k+1,j}, u_{k+1,j})$ satisfy the optimality conditions (10.22).

(c) Derive a recursive update of u, i.e. express u_{k+1} in terms of u_k, b_k, and b_{k+1}.

(d) Implement the SMO algorithm with the working set selection heuristic (10.24) and test it on the same data as in the previous exercise.

10.11 We consider the under-determined linear system of equations

$$Ax = b,$$

where $A \in \mathbb{R}^{m \times n}$ has full row rank with $m < n$, and where $b \in \mathbb{R}^m$ and $x \in \mathbb{R}^n$.

(a) Compute the solution that satisfies the linear system of equations and that has the smallest value of the norm $\|x\|_2$. You should determine an explicit formula for the solution.

(b) Compute the solution to the LS problem

$$\text{minimize} \, \|Ax - b\|_2^2,$$

with variable x using a gradient method, and compare it with the solution above. Generate random problem data for the matrix and vector with entries from a standardized normal distribution.

(c) Consider the optimization problem

$$\text{minimize} \, \|Ax - b\|_2^2 + \gamma \|x\|_2^2,$$

where $\gamma > 0$. Compute the optimal solution and show that the optimal solution approaches the least-norm solution in (a) when $\gamma \to 0$.

Hint: Use the matrix inversion lemma in (2.56).

10.12 Implement a stochastic gradient algorithm that uses dropout. Generate pairs of data $(x_k, y_k) \in \mathbb{R}^n \times \mathbb{R}$ for $k \in \mathbb{N}_N$ randomly in the following way. First, generate $a_0 \in \mathbb{R}^n$ randomly from a standardized Gaussian distribution. Then generate x_k randomly from a standardized Gaussian distribution and also generate e_k randomly from a standardized Gaussian distribution. Then let $y_k = x_k^T a_0 + 0.1 e_k$. Use dropout probabilities $p_i = 0.5$ in the following stochastic gradient method

$$a^{k+1} = a^k + t_k (y_{i_k} - x_{i_k}^T \delta a) \delta x_{i_k},$$

to estimate an ensemble average predictor $\hat{y}(x) = x^T P a$, where $P = \text{diag}(p_i)$. Above δ is a diagonal matrix for which you should choose the diagonal elements equal to 0 or 1 with probability p_i, independently for each value of k. You may take the values i_k as $i_k = 1 + (k-1) \bmod N$. You may take $t_k = 1/k$. Compare the resulting value of a that you obtain with the solution you obtain by minimizing the regularized LS criterion

$$\frac{1}{2} \sum_{k=1}^N \|y_k - x_k^T P a\|_2^2 + \frac{\sigma^2}{2} a^T \left(\sum_{k=1}^N \text{diag}(x_k)^2 \right) a,$$

where $\sigma^2 = p_i(1 - p_i)$.

10.13 When dropout was used with an incremental method for solving the LS problem in Section 10.8, we realized that the predictor was given by the expected value of the stochastic predictor. This could be computed exactly. For general ANNs, it is only possible to approximately compute the expectation as will be discussed in this exercise. Let Δ_i, $i \in \mathbb{N}_n$, be random variables with a Bernoulli distribution, i.e. $P(\Delta_i = 1) = p_i$ and $P(\Delta_i = 0) = 1 - p_i$.

We assume that Δ_i is independent of Δ_j for $i \neq j$. Let $\Delta = \mathrm{diag}(\Delta_i, \ldots, \Delta_n)$. Consider a stochastic predictor $\hat{Y} : \mathbb{R}^n \times \{0,1\}^n \to \mathbb{R}$ given by

$$\hat{Y}(x; \Delta_1, \ldots, \Delta_n) = x^T \Delta a,$$

where $x, a \in \mathbb{R}^n$. Let $\sigma : \mathbb{R} \to \mathbb{R}$ be the logistic function defined as

$$\sigma(y) = \frac{1}{1 + e^{-y}},$$

and let $\hat{Z} : \mathbb{R}^n \times \{0,1\}^n \to \mathbb{R}$ be defined as

$$\hat{Z}(x; \Delta_1, \ldots, \Delta_n) = \sigma(\hat{Y}(x; \Delta_1, \ldots, \Delta_n)).$$

The joint probability function for $p : \{0,1\}^n \to [0,1]$ for $(\Delta_1, \ldots, \Delta_n)$, can be expressed in terms of p_i. As an example for $n = 3$, we have

$$p(1,0,0) = P(\Delta_1 = 1, \Delta_2 = 0, \Delta_3 = 0) = p_1(1 - p_2)(1 - p_3).$$

We now introduce a bijection between $\{0,1\}^n$ and $\mathbb{N}_m$, where $m = 2^n$, such that we can collect the m probabilities defining the probability function above in the vector $P \in [0,1]^m$, for which it hold that $P_j \geq 0$ and $\sum_{j=1}^m P_j = 1$. We will from now on also consider $\hat{Y}$ and $\hat{Z}$ to be defined on $\mathbb{N}_m$, and we will use subscripts to refer to the values of the random variables. We will also drop the dependence on x. This means that the expected value of $\hat{Y}$ is given by $\mathbb{E}\hat{Y} = \sum_{j=1}^m \hat{Y}_j P_j$. Luckily, we do not have to use this formula to compute the expectation, since we may instead use the fact that $\mathbb{E}\hat{Y} = x^T \mathrm{diag}(p_i)a$. However, for the expectation of $\hat{Z}$, there is no such shortcut. Hence, we need to use $\mathbb{E}\hat{Z} = \sum_{j=1}^m \sigma(\hat{Y}_j)P_j$. Even for moderate values of n it holds that $m = 2^n$ is huge. Therefore, it is necessary to find approximate values for this expectation.

(a) Let us define the weighted geometric mean $G : \mathbb{R}^m \to \mathbb{R}$ as

$$G(x) = \prod_{j=1}^m x_j^{P_j},$$

and the weighted geometric mean of the complement $G' : \mathbb{R}^m \to \mathbb{R}$ as

$$G'(x) = \prod_{j=1}^m (1 - x_j)^{P_j}.$$

Show that

$$\frac{G(\hat{Z})}{G(\hat{Z}) + G'(\hat{Z})} = \sigma\left(\mathbb{E}\hat{Y}\right),$$

where the left-hand side is called the normalized weighted geometric mean.

(b) Show that

$$G(\hat{Z}) \leq \frac{G(\hat{Z})}{G(\hat{Z}) + G'(\hat{Z})} \leq \mathbb{E}\hat{Z},$$

if $0 < \hat{Z}_i \leq 1/2$. Hence, $\sigma(\mathbb{E}\hat{Y})$ is a good approximation for $\mathbb{E}\hat{Z} = \mathbb{E}\left[\sigma(\hat{Y})\right]$ for the case when $0 < \hat{Z}_i \leq 1/2$.

Hint: Use the Ky Fan inequality from Exercise 4.10.

10.14 We will use the Deep Learning Toolbox in MATLAB to fit an ANN to pairs of data (x_i, y_i). We will use a two-layer network implementing the function in (10.13). The following commands read in the data, define a net with two layers, three hidden neurons, train the net, display the net, compute the predicted outputs of the net, and plot a comparison of the predicted values $\hat{y}_i$ and the original values of y_i versus the values x_i:

```
[x,y] = simplefit_dataset;
net = fitnet(3);
net = train(net,x,y);
view(net),
haty = net(x);
plot(x,haty,'-',x,y,'--');
```

Play around with the number of hidden neurons and try to figure out how many are needed to get an almost perfect fit.

11

Reinforcement Learning

Reinforcement learning is an area of machine learning concerned with how agents should take actions in an environment in order to maximize the notion of a so-called cumulative reward. The idea goes back to the work by the Russian Nobel laureate Ivan Pavlov on classical conditioning where he showed that you can train dogs by rewarding or punishing them. To formalize this, we consider the dog to be in different *states* x_k at different *stages* k in time. The value of the state x_{k+1} depends on the current state x_k and the *action* u_k we take during our training. Each and every state x_k is also related to a *reward* r_k. It is then desirable to maximize the sum, possibly discounted, of rewards over some set of stages, say $\{0, 1, \ldots, N-1\}$, where N could be infinity. In reinforcement learning, the action is determined by an *agent*, and the next state and the reward are determined by the *environment*. One can view reinforcement learning as a way of solving the optimal control problems in Chapter 8 without explicitly knowing the details of the mathematical model that describes how the state evolves. The control signal in optimal control corresponds to the action in reinforcement learning. The incremental costs in optimal control are the negative of the rewards in reinforcement learning.

11.1 Finite Horizon Value Iteration

Let $f_k : \mathcal{D}^n \times \mathcal{E}^m \to \mathbb{R}$ and $F_k : \mathcal{D}^n \times \mathcal{E}^m \to \mathcal{D}^n$ be given. Consider the finite horizon optimal control problem

$$\text{minimize} \quad \phi(x_N) + \sum_{k=0}^{N-1} f_k(x_k, u_k) \tag{11.1}$$

$$\text{subject to} \quad x_{k+1} = F_k(x_k, u_k), \quad k \in \mathbb{Z}_{N-1},$$

for a given initial value x_0 with variables $(u_0, x_1, \ldots, u_{N-1}, x_N)$. Notice that here we do not consider constraints on x_k or u_k, but the results can easily be generalized to the constrained case. Reinforcement learning for this optimal control problem is closely related to the parametric approximations used for the dynamic programming recursion in Section 8.2. However, we cannot directly use this recursion, since when defining the approximate Q-function, an analytical expression for the value of the state at the next iterate was needed in terms of the function F. In reinforcement learning, only the *value of the state* at the next iterate is available and not the function, i.e. we have an oracle or simulator that for given values of (x_k, u_k) computes $x_{k+1} = F_k(x_k, u_k)$ without us knowing what F_k is. We will show how this can be accommodated by instead of using the dynamic programming recursion in (8.4), use a recursion for the Q-function in (8.5). This will make it possible for us

Optimization for Learning and Control, First Edition. Anders Hansson and Martin Andersen.
© 2023 John Wiley & Sons, Inc. Published 2023 by John Wiley & Sons, Inc.
Companion Website: www.wiley.com/go/opt4lc

to directly learn the Q-function instead of computing it using F_k from a learned value function V_k. We will show that the following recursion holds

$$Q_k(x, \bar{u}) = f_k(x, \bar{u}) + \min_u Q_{k+1}\left(F_k(x, \bar{u}), u\right), \tag{11.2}$$

starting with $k = N - 2$ and finishing with $k = 0$, and where $Q_{N-1}(x, u) = f_{N-1}(x, u) + \phi(F_{N-1}(x, u))$. The above recursion follows from the fact that (8.4) and (8.5) implies that

$$V_{k+1}(x) = \min_u Q_{k+1}(x, u),$$

and hence,

$$V_{k+1}(F_k(x, \bar{u})) = \min_u Q_{k+1}(F_k(x, \bar{u}), u),$$

By adding $f_k(x, \bar{u})$ to both sides the recursion is obtained, which is summarized in Algorithm 11.1.

Algorithm 11.1: Finite horizon value iteration for Q-function

Input: Final state penalty ϕ, incremental costs f_k, oracle for F_k
Output: Q_k for $k \in \mathbb{Z}_{N-1}$

$Q_{N-1}(x, \bar{u}) \leftarrow f_{N-1}(x, \bar{u}) + \phi(F_{N-1}(x, \bar{u}))$
for $k \leftarrow N - 2$ **to** 0 **do**
$\quad\big|\quad Q_k(x, \bar{u}) \leftarrow f_k(x, \bar{u}) + \min_u Q_{k+1}(F_k(x, \bar{u}), u)$
end

11.1.1 Fitted Value Iteration

We define feature vectors $\varphi : \mathcal{D}^n \times \mathcal{E}^m \to \mathbb{R}^p$ and directly approximate Q_k with $\tilde{Q}_k : \mathcal{D}^n \times \mathcal{E}^m \times \mathbb{R}^p \to \mathbb{R}$ as

$$\tilde{Q}_k(x, u, a_k) = a_k^T \varphi(x, u),$$

for $k \in \mathbb{Z}_{N-1}$. This is a linear regression model, but we could also have considered a more general model as we did in Section 8.2. We leave the details of generalizing to this case for the reader to carry out. We then consider samples $(x_k^s, u_k^s) \in \mathcal{D}^n \times \mathcal{E}^m$, where $s \in \mathbb{N}_r$, and define

$$\beta_{N-1}^s = \phi(F_{N-1}(x_{N-1}^s, u_{N-1}^s)),$$

and

$$\beta_k^s = \min_u \tilde{Q}_{k+1}(F_k(x_k^s, u_k^s), u, a_{k+1}), \tag{11.3}$$

where a_{k+1} is a known value from a previous iterate defined below. We see that we do not need an analytical expression for F_k in order to define β_k^s for $k \in \mathbb{Z}_{N-1}$. It is enough to know the value of $F_k(x_k^s, u_k^s)$ which is the next value of the state in a simulation. Moreover, depending on how the feature vectors are chosen, the minimization above could become very tractable. This can be a great advantage for the approximation method based directly on the Q-function. After this, we define the following LS problem:

$$\text{minimize} \quad \frac{1}{2} \sum_{s=1}^{r} \left(\tilde{Q}_k(x_k^s, u_k^s, a) - f_k(x_k^s, u_k^s) - \beta_k^s\right)^2, \tag{11.4}$$

with variable a for $k \in \mathbb{Z}_{N-1}$. Denote the optimal solution by a_k. The iterations start with $k = N - 1$ and go down to $k = 0$, where we alternate between solving (11.4) and (11.3). Once all parameters

a_k have been computed, the approximate optimal control is $u_k^\star = \mu_k(x_k)$, where $\mu_k : \mathcal{D}^n \to \mathcal{E}^m$ is given by

$$\mu_k(x) = \underset{u}{\text{argmin}} \ \tilde{Q}_k\left(x, u, a_k\right). \tag{11.5}$$

We notice that using the Q-function instead of using the value function as in Section 8.2 comes at the price of also having to sample the control signal space $\mathcal{E}^m$.

Example 11.1 We will in this example perform fitted value iteration for the optimal control problem in Example 8.1 based on the iteration of the Q-functions. We will as in Example 8.3 specifically consider the case when $m = 1$ and $n = 2$, and we let A_k, B_k, R_k, S_k be independent of k and we write them as A, B, R, S. Since we know that the Q-functions are quadratic, we will use a feature vector that is $\varphi(x, u) = (x_1^2, x_2^2, u^2, 2x_1x_2, 2x_1u, 2x_2u)$, where $x = (x_1, x_2) \in \mathbb{R}^2$ and $u \in \mathbb{R}$. Notice that the indices refer to components of the vector and not to time. We let

$$\tilde{Q}_k(x, u, a) = a^T \varphi(x, u),$$

where $a \in \mathbb{R}^6$. With $\tilde{P} \in \mathbb{S}^2, \tilde{r} \in \mathbb{R}^2$ and $\tilde{q} \in \mathbb{R}$ defined as

$$\begin{bmatrix} \tilde{P} & \tilde{r} \\ \tilde{r}^T & \tilde{q} \end{bmatrix} = \begin{bmatrix} a_1 & a_4 & a_5 \\ a_4 & a_2 & a_6 \\ a_5 & a_6 & a_3 \end{bmatrix},$$

we may write

$$\tilde{Q}_k(x, u, a) = \begin{bmatrix} x \\ u \end{bmatrix}^T \begin{bmatrix} \tilde{P} & \tilde{r} \\ \tilde{r}^T & \tilde{q} \end{bmatrix} \begin{bmatrix} x \\ u \end{bmatrix}. \tag{11.6}$$

For $k = N - 1$, we define

$$\beta_{N-1}^s = \left(x_+^s\right)^T S x_+^s,$$

where $x_+^s = A x_{N-1}^s + B u_{N-1}^s$. For $k = N - 2$ down to $k = 0$, we solve the linear LS problem in (11.3) to obtain β_k^s. From Example 5.7, it follows that the solution is

$$\beta_k^s = \left(x_+^s\right)^T \left(\tilde{P}_{k+1} - \tilde{r}_{k+1} \tilde{q}_{k+1}^{-1} \tilde{r}_{k+1}^T\right) x_+^s,$$

where $x_+^s = A x_k^s + B u_k^s$. We then obtain a_k for $k \in \mathbb{Z}_{N-1}$ as the solution to the linear LS problem in (11.4):

$$\text{minimize} \ \frac{1}{2} \sum_{s=1}^r \left(\varphi^T(x_k^s, u_k^s) a - \left(x_k^s\right)^T S x_k^s - \left(u_k^s\right)^T R u_k^s - \beta_k^s\right)^2.$$

Notice that the subindex k for a now refers to stage index. The solution a_k satisfies the normal equations, *cf.* (5.3),

$$\Phi_k^T \Phi_k a_k = \Phi_k^T \gamma_k,$$

where

$$\Phi_k = \begin{bmatrix} \varphi^T(x_k^1, u_k^1) \\ \vdots \\ \varphi^T(x_k^r, u_k^r) \end{bmatrix}, \qquad \gamma_k = \begin{bmatrix} \left(x_k^1\right)^T S x_k^1 + \left(u_k^1\right)^T R u_k^1 + \beta_k^1 \\ \vdots \\ \left(x_k^r\right)^T S x_k^r + \left(u_k^r\right)^T R u_k^r + \beta_k^r \end{bmatrix}.$$

It is here crucial to choose (x_k^s, u_k^s) such that $\Phi_k^T \Phi_k$ is invertible. We realize that we need $r \geq 6$ for this to hold. Moreover, we need to choose (x_k^s, u_k^s) sufficiently different to have $\Phi_k^T \Phi_k$ well conditioned.

For a general, n we need $r \geq (m + n)(m + n + 1)/2$. Compared to Example 8.3 we require a larger value of r. The optimal feedback function is by (11.5) and (11.6) given by

$$\mu_k(x) = -\tilde{q}_k^{-1}\tilde{r}_k^T x,$$

where we again have used the results in Example 5.7.

11.2 Infinite Horizon Value Iteration

We will now address the infinite horizon optimal control problem in (8.9) for the case of no state constraints, restated below for ease of reference:

$$\text{minimize} \quad \sum_{k=0}^{\infty} \gamma^k f(x_k, u_k)$$

$$\text{subject to} \quad x_{k+1} = F(x_k, u_k), \quad k \in \mathbb{Z}_+,$$

with variables $(u_0, x_1, \ldots)$, where x_0 is given. Algorithms for reinforcement learning for the infinite horizon problem are based on approximate version of value iteration (VI) and policy iteration (PI) for the Q-function. When VI is used, it is called *Q-learning*.

The Q-function $Q : \mathcal{D}^n \times \mathcal{E}^m \to \mathbb{R}$ is defined as $Q(x, u) = f(x, u) + \gamma V(F(x, u))$, and it satisfies

$$Q(x, \bar{u}) = f(x, \bar{u}) + \min_u \gamma Q(F(x, \bar{u}), u). \tag{11.7}$$

To see this we have from the Bellman equation in (8.10) that

$$V(x) = \min_u Q(x, u),$$

and

$$\gamma V(F(x, \bar{u})) = \min_u \gamma Q(F(x, \bar{u}), u).$$

By adding $f(x, \bar{u})$ to both sides of the above equation, the desired equation for Q is obtained.

We introduce the *Bellman Q-operator* $T_Q : \mathbb{R}^{\mathcal{D}^n \times \mathcal{E}^m} \to \mathbb{R}^{\mathcal{D}^n \times \mathcal{E}^m}$ defined as

$$T_Q(Q)(x, \bar{u}) = f(x, \bar{u}) + \min_u \gamma Q(F(x, \bar{u}), u). \tag{11.8}$$

Here $\mathbb{R}^{\mathcal{D}^n \times \mathcal{E}^m}$ is the set of real-valued functions defined on $\mathcal{D}^n \times \mathcal{E}^m$, cf. notation section in the appendix. It can then be shown that the following VI

$$Q_{k+1} = T_Q(Q_k), \tag{11.9}$$

for $Q_k : \mathcal{D}^n \times \mathcal{E}^m \to \mathbb{R}$ with boundary condition $Q_0(x, u) = f(x, u)$ converges to $Q(x, u)$ satisfying (11.7) as $k \to \infty$, see Exercise 11.6. It is summarized in Algorithm 11.2. This algorithm is not easy to implement in practice unless (x, u) takes only finitely many values.

Algorithm 11.2: Infinite horizon value iteration for Q-function

Input: Incremental cost f, oracle for F, $\gamma \in (0, 1]$, tolerance $\varepsilon > 0$
Output: Q solving $Q = T_Q(Q)$

$Q \leftarrow f$
while $\|Q - T_Q(Q)\|_\infty \geq \varepsilon$ **do**
$\quad\mid\ Q \leftarrow T_Q(Q)$
end

11.2.1 Q-Learning

We will now see how VI can be generalized by developing an alternative iterative method for solving (11.7). To this end, we introduce the error function $e : \mathbb{R}^{D^n \times \mathcal{E}^m} \to \mathbb{R}^{D^n \times \mathcal{E}^m}$ defined as

$$e(Q) = Q - T_Q(Q),$$

where T_Q is defined in (11.8). With this definition, we may write the equation in (11.7) as $e(Q) = 0$. We now apply the following standard algorithm

$$Q_{k+1} = Q_k - t_k e(Q_k), \quad k \in \mathbb{Z}_+, \tag{11.10}$$

to find a root of the equation $e(Q) = 0$. You can initialize with $Q_0 = f$, but there are better ways. The step lengths t_k should satisfy $t_k \in (0, 1]$. We will prove convergence of the above iterations to a solution $Q^\star$ of $e(Q) = 0$, i.e. $Q^\star = T_Q(Q^\star)$. Let $\Delta_k = Q_k - Q^\star$. Then it holds that

$$\Delta_{k+1} = (1 - t_k)\Delta_k + t_k \left(T_Q(Q_k) - T_Q(Q^\star) \right).$$

From this it follows for $t_k \in (0, 1]$ that

$$\|\Delta_{k+1}\|_\infty \le (1 - t_k)\|\Delta_k\|_\infty + t_k\|T_Q(Q_k) - T_Q(Q^\star)\|_\infty \le (1 - t_k + t_k \gamma)\|\Delta_k\|_\infty,$$

where the last inequality follows from the contraction property of T_Q for $\gamma \in [0, 1)$ shown in Exercise 11.6. We see that we have a contraction for Δ_k for $t_k \in (0, 1]$, if $\gamma \in (0, 1)$, and therefore, Δ_k converges to zero proving that Q_k converges to $Q^\star$. We remark that we recover VI for $t_k = 1$. The algorithm is summarized in Algorithm 11.3.

Algorithm 11.3: Generalized value iteration for Q-function

Input: Incremental cost f, oracle for F, $\gamma \in (0, 1]$, tolerance $\varepsilon > 0$, step-lengths $t_k \in (0, 1]$ for $k \in \mathbb{N}$

Output: Q solving $Q = T_Q(Q)$

$Q \leftarrow f$
$k \leftarrow 1$
while $\|Q - T_Q(Q)\|_\infty \ge \varepsilon$ **do**
  $\quad Q \leftarrow Q - t_k(Q - T_Q(Q))$
  $\quad k \leftarrow k + 1$
end

It is possible to instead of in each iteration k consider all values of $(x, u) \in D^n \times \mathcal{E}^m$ only consider one sample (x_k, u_k) at a time. These samples could be generated in a cyclic order or in a randomized cyclic order such that each sample is visited infinitely many times. We assume that $D^n \times \mathcal{E}^m$ is a finite set. Then it holds that

$$Q_{k+1}(x_k, u_k) = Q_k(x_k, u_k) - t_k \left[Q(x_k, u_k) - f(x_k, u_k) - \min_u \gamma Q(F(x_k, u_k), u) \right], \tag{11.11}$$

converges to a solution of $e(Q) = 0$ as k goes to infinity when $t_k \in (0, 1]$ and $\gamma \in (0, 1)$. The results follow trivially from the convergence proof above. The algorithm has similarities with the coordinate descent method in Section 6.9, since Q is only updated for one discrete value in each iteration. This algorithm is often referred to as *Q-learning*, and it is summarized in Algorithm 11.4.

It is possible to develop VI methods for the stochastic infinite horizon optimal control problem in (8.33) using the Q-function in (8.35) which satisfies the Bellman equation for the Q-function in (11.32). The above algorithm can be used by replacing $F(x_k, u_k)$ with $F(x_k, u_k, w_k)$, where w_k is a realization of the random process W. If the step sizes satisfies $\sum_{k=0}^\infty t_k = \infty$ and $\sum_{k=0}^\infty t_k^2 < \infty$, then the algorithm can be shown to converge in a stochastic sense [114].

Algorithm 11.4: *Q*-learning

Input: Incremental cost f, oracle for F, parameter $\gamma \in (0, 1]$, tolerance $\varepsilon > 0$, step-lengths
$\quad\quad t_k \in (0, 1]$ for $k \in \mathbb{N}$
Output: Q solving $Q = T_Q(Q)$

$Q(x, u) \leftarrow f(x, u), \; \forall (x, u) \in \mathcal{D}^n \times \mathcal{E}^m$
$k \leftarrow 1$
while $\|Q - T_Q(Q)\|_\infty \geq \varepsilon$ **do**
$\quad\quad (x_k, u_k)$ sampled in randomized cyclic order from $\mathcal{D}^n \times \mathcal{E}^m$
$\quad\quad Q(x_k, u_k) \leftarrow Q(x_k, u_k) - t_k \left[Q(x_k, u_k) - f(x_k, u_k) - \min_u \gamma Q(F(x_k, u_k), u) \right]$
$\quad\quad k \leftarrow k + 1$
end

11.3 Policy Iteration

Reinforcement learning based on PI is called *self-learning*. The policy evaluation step is referred to as a *critic* and the policy improvement is referred to as an *actor*. These types of methods are called *actor-critic* methods. In case parametric approximations using artificial neural networks ANNs are involved, the actor and critic are called *actor networks* and *critic networks*, respectively. In order to do PI for the Q-function formulation in (11.7), we define $Q_k : \mathcal{D}^n \times \mathcal{E}^m \to \mathbb{R}$ as $Q_k(x, u) = f(x, u) + \gamma V_k(F(x, u))$. It then follows that

$$V_k(x) = Q_k\left(x, \mu_k(x)\right),$$

from (8.16), and therefore

$$V_k(F(x, u)) = Q_k\left(F(x, u), \mu_k(F(x, u))\right).$$

Multiply with γ and add $f(x, u)$ to obtain that Q_k is the solution of

$$Q_k(x, u) = f(x, u) + \gamma Q_k\left(F(x, u), \mu_k(F(x, u))\right). \tag{11.12}$$

This is the policy evaluation step in terms of the Q-function. We then obtain a new feedback policy by solving

$$\mu_{k+1}(x) = \operatorname*{argmin}_u Q_k(x, u), \tag{11.13}$$

which is the policy improvement step in terms of the Q-function. These iterations are exactly the same as the iterations in (8.16)–(8.17) except that we need to solve for a function Q_k that depends also on the control signal in the policy iteration step. The PI algorithm for the Q-function is summarized in Algorithm 11.5.

Algorithm 11.5: PI for Q-function

Input: Incremental cost f, oracle for F, parameter $\gamma \in (0, 1]$, tolerance $\varepsilon > 0$, initial policy μ
Output: Q solving $Q = T_Q(Q)$

while $\|Q - T_Q(Q)\|_\infty \geq \varepsilon$ **do**
$\quad\quad$ Solve $Q(x, u) = f(x, u) + \gamma Q(F(x, u), \mu(F(x, u)))$ w.r.t. Q
$\quad\quad \mu \leftarrow \operatorname{argmin}_u Q(x, u)$
end

Example 11.2 We consider the infinite horizon LQ control problem in Example 8.4. We guess that

$$Q_k(x, u) = \begin{bmatrix} x \\ u \end{bmatrix}^T \begin{bmatrix} U_k & W_k \\ W_k^T & V_k \end{bmatrix} \begin{bmatrix} x \\ u \end{bmatrix},$$

for some

$$\begin{bmatrix} U_k & W_k \\ W_k^T & V_k \end{bmatrix} \in \mathbb{S}_+^{m+n},$$

where $V_k \in \mathbb{S}_{++}^m$. It then follows from (11.13) that

$$\mu_k(x) = -L_{k+1}x,$$

where $L_{k+1} = V_k^{-1}W_k^T$. The recursion for Q_k in (11.12) is seen to be satisfied if

$$\begin{bmatrix} U_k & W_k \\ W_k^T & V_k \end{bmatrix} = \begin{bmatrix} S & 0 \\ 0 & R \end{bmatrix} + \gamma [A \ B]^T \begin{bmatrix} I \\ -L_k \end{bmatrix}^T \begin{bmatrix} U_k & W_k \\ W_k^T & V_k \end{bmatrix} \begin{bmatrix} I \\ -L_k \end{bmatrix} [A \ B],$$

for a given L_k. This is an algebraic Lyapunov equation which has a positive semidefinite solution since

$$\begin{bmatrix} S & 0 \\ 0 & R \end{bmatrix},$$

is positive semidefinite. This assumes that

$$\sqrt{\gamma} \begin{bmatrix} I \\ -L_k \end{bmatrix} [A \ B],$$

has all its eigenvalues strictly inside the unit disc. This is true if $\sqrt{\gamma}(A - BL_k)$ has all its eigenvalues strictly inside the unit disc by Exercise 11.1. As in Example 8.6, it can be shown that if we start with a stabilizing L_0, then all subsequent L_k, will also be stabilizing. Moreover, it can be shown that all V_k are positive definite so that the inverse in the formula for L_k exists.

11.3.1 Critic Network

We now approximate the policy evaluation step in (11.12) for PI. This is based on defining $\tilde{Q} : D^n \times \mathcal{E}^m \times \mathbb{R}^p \to \mathbb{R}$ using an ANN or as a linear regression with p parameters similarly as we have done in Section 8.2. This function will be used to approximate Q_k in (11.12). Before we do that we notice that (11.12) implies

$$\begin{aligned} Q_k(x_0, u_0) &= f(x_0, u_0) + \gamma Q_k(F(x_0, u_0), \mu_k(F(x_0, u_0))) \\ &= f(x_0, u_0) + \gamma Q_k(x_1, \mu_k(x_1)) \\ &= f(x_0, u_0) + \gamma f(x_1, \mu_k(x_1)) + \gamma^2 Q_k(x_2, \mu_k(x_2)) \\ &\vdots \\ &= f(x_0, u_0) + \sum_{i=1}^{N-1} \gamma^i f(x_i, \mu_k(x_i)) + \gamma^N Q_k(x_N, \mu_k(x_N)), \end{aligned}$$

where $x_{i+1} = F(x_i, \mu_k(x_i))$ for $i \in \mathbb{N}_{N-1}$, and $x_1 = F(x_0, u_0)$. In case N is large and μ_k is stabilizing, we have that x_N is close to zero and that also $Q_k(x_N)$ is close to zero. We realize that the only difference for the approximate evaluation of the Q-function as compared to the approximate evaluation of the value function in (8.18) is that the first incremental cost is evaluated using u_0 and not $\mu_k(x_0)$.

We denote these approximations for different initial values (x^s, u^s) for $s \in \mathbb{N}_r$ as

$$\beta_k^s = f(x^s, u^s) + \sum_{i=1}^{N-1} \gamma^i f(x_i, \mu_k(x_i)),$$

where $x_{i+1} = F(x_i, \mu_k(x_i))$ for $i \in \mathbb{N}_{N-1}$, and $x_1 = F(x^s, u^s)$. We then find the approximation of Q_k by solving

$$\text{minimize } \frac{1}{2} \sum_{s=1}^{r} \left(\tilde{Q}(x^s, u^s, a_k) - \beta_k^s \right)^2,$$

with variable a_k. After this, we use the following exact policy improvement step

$$\mu_{k+1}(x) = \underset{u}{\text{argmin}} \, \tilde{Q}(x, u, a_k). \tag{11.14}$$

A drawback of this method compared to when using value functions is that the reuse of trajectories is more problematic, since sufficiently many different values of the control signal might then not be explored. For a more detailed discussion on this see [17, Section 5.3].

Example 11.3 We will in this example consider the optimal control problem in Example 8.4. We will specifically consider the case when $m = 1$ and $n = 2$. Since we know that the value function is quadratic, we will use a feature vector that is $\varphi(x, u) = (x_1^2, x_2^2, u^2, 2x_1 x_2, 2x_1 u, 2x_2 u)$, where $x = (x_1, x_2) \in \mathbb{R}^2$ and $u \in \mathbb{R}$. Notice that the indices refer to components of the vector and not to time. We let

$$\tilde{Q}(x, u, a) = a^T \varphi(x, u),$$

where $a \in \mathbb{R}^6$. With $\tilde{P} \in \mathbb{S}^2$, $\tilde{r} \in \mathbb{R}^2$ and $\tilde{q} \in \mathbb{R}$ defined as

$$\begin{bmatrix} \tilde{P} & \tilde{r} \\ \tilde{r}^T & \tilde{q} \end{bmatrix} = \begin{bmatrix} a_1 & a_4 & a_5 \\ a_4 & a_2 & a_6 \\ a_5 & a_6 & a_3 \end{bmatrix},$$

we may write

$$\tilde{Q}_k(x, u, a) = \begin{bmatrix} x \\ u \end{bmatrix}^T \begin{bmatrix} \tilde{P} & \tilde{r} \\ \tilde{r}^T & \tilde{q} \end{bmatrix} \begin{bmatrix} x \\ u \end{bmatrix}. \tag{11.15}$$

Then $a_k \in \mathbb{R}^6$ with an abuse of notation is obtained as the solution to the linear LS problem

$$\text{minimize } \frac{1}{2} \sum_{s=1}^{r} \left(\varphi^T(x^s, u^s) a - \beta_k^s \right)^2,$$

with variable a. The solution a_k satisfies the normal equations, *cf.* (5.3),

$$\Phi_k^T \Phi_k a_k = \Phi_k^T \beta_k,$$

where

$$\Phi_k = \begin{bmatrix} \varphi^T(x^1, u^1) \\ \vdots \\ \varphi^T(x^r, u^r) \end{bmatrix}, \qquad \beta_k = \begin{bmatrix} \beta_k^1 \\ \vdots \\ \beta_k^r \end{bmatrix},$$

with

$$\beta_k^s = (x^s)^T S x^s + (u^s)^T R u^s + \sum_{i=1}^{N-1} \gamma^i \left(x_i^T S x_i + \mu_k(x_i)^T R \mu_k(x_i) \right),$$

where $x_1 = Ax^s + Bu^s$ and $x_{i+1} = Ax_i + B\mu_k(x_i)$ for $i \in \mathbb{N}_{N-2}$ with initial values x^s, $s \in \mathbb{N}_r$. It is here crucial to choose (x^s, u^s) such that $\Phi_k^T\Phi_k$ is invertible. We realize that we need $r \geq 6$ for this hold. Moreover, we need to choose (x^s, u^s) sufficiently different for $\Phi_k^T\Phi_k$ to be well conditioned. For a general n, we will need $r \geq (m+n)(m+n+1)/2$.

From Example 5.7, we realize that the solution to (11.14) is given by

$$\mu_{k+1}(x) = \operatorname*{argmin}_u \tilde{Q}_k(x, u, a_k) = -\tilde{q}_k^{-1}\tilde{r}_k^T x,$$

assuming that $\tilde{q}$ is positive. Here $\tilde{q}_k$ and $\tilde{r}_k$ are defined from a_k. We may hence write

$$\mu_{k+1}(x) = -L_{k+1}x,$$

where $L_{k+1} = \tilde{q}_k^{-1}\tilde{r}_k^T$. It is a good idea to start with some L_0 such that μ_0 is stabilizing.

11.3.2 SARSA-Algorithm

We will now present an alternative approach. We define $\varepsilon : \mathcal{D}^n \times \mathcal{E}^m \times \mathcal{E}^m \times \mathbb{R}^p \to \mathbb{R}$ as

$$\varepsilon(x, u, v, a) = \tilde{Q}(x, u, a) - f(x, u) - \gamma\tilde{Q}(F(x, u), v, a).$$

Then the error we obtain when we approximate Q_k in (11.12) with $\tilde{Q}$ can be written as $\varepsilon(x, u, \mu(F(x, u)), a)$, where $\mu = \mu_k$. We then define the LS problem

$$\text{minimize } \frac{1}{2}\sum_{i=0}^{N-1}\varepsilon(x_i, u_i, \mu_k(x_{i+1})), a)^2,$$

with variable a and solution a_k, where

$$x_{i+1} = F\left(x_i, u_i\right)$$
$$u_i = \mu_k(x_i)$$

with x_0 given. Then μ is updated as in (11.14). For each value of k, we should choose different initial values x_0. This should be done such that many different states and controls are explored. It is common to add a realization w of a zero mean white noise random process W to the control signal in order to obtain more exploration, i.e. we use

$$x_{i+1} = F\left(x_i, u_i + w_i\right)$$
$$u_i = \mu_k(x_i).$$

It is important to gather as much information that the above LS problems has a unique solution. Also when updating a, the information from the previous batches should not be disregarded. Hence, batch-recursive LS should be used.

Sometimes $k = i$ is used, i.e. the policy improvement step is carried out for each i. Then a pure recursive LS technique can be used to solve the LS problem. These types of algorithms are often referred to as *state-action-reward-state-action* (SARSA) algorithms. They are possible to run in real time by letting x_{i+1} be given by measurements from a real system when the control signal u_i has been applied. This will result in what is known as *adaptive control*, see [7].

Exact recursive LS techniques are only available for linear LS problems, but there are approximate techniques available for nonlinear LS problems for the case when $\tilde{Q}$ as a nonlinear function. It should be stressed that the theoretical convergence properties of the above schemes are not well understood.

Example 11.4 We will in this example consider the optimal control problem in Example 8.4. We will specifically consider the case when $m = 1$ and $n = 2$. Since we know that the value function is quadratic, we will use a feature vector that is $\varphi(x, u) = (x_1^2, x_2^2, u^2, 2x_1x_2, 2x_1u, 2x_2u, 1)$, where $x = (x_1, x_2) \in \mathbb{R}^2$ and $u \in \mathbb{R}$. Notice that the indices refer to components of the vector and not to time. We let

$$\tilde{Q}(x, u, a) = a^T \varphi(x, u),$$

where $a \in \mathbb{R}^7$. With $\tilde{P} \in \mathbb{S}^2$, $\tilde{r} \in \mathbb{R}^2$ and $\tilde{q} \in \mathbb{R}$ defined as

$$\begin{bmatrix} \tilde{P} & \tilde{r} \\ \tilde{r}^T & \tilde{q} \end{bmatrix} = \begin{bmatrix} a_1 & a_4 & a_5 \\ a_4 & a_2 & a_6 \\ a_5 & a_6 & a_3 \end{bmatrix},$$

we may write

$$\tilde{Q}_k(x, u, a) = \begin{bmatrix} x \\ u \end{bmatrix}^T \begin{bmatrix} \tilde{P} & \tilde{r} \\ \tilde{r}^T & \tilde{q} \end{bmatrix} \begin{bmatrix} x \\ u \end{bmatrix} + a_7. \tag{11.16}$$

Here we have added a constant term compared to what we did in Example 11.3, since we have a random process W involved. From Example 5.7, we realize that the solution to (11.14) is given by

$$\mu_{k+1}(x) = \underset{u}{\arg\min} \, \tilde{Q}_k(x, u, a_k) = -\tilde{q}_k^{-1} \tilde{r}_k^T x,$$

assuming that $\tilde{q}_k$ is positive. Here $\tilde{q}_k$ and $\tilde{r}_k$ are defined from a_k. We may hence write

$$\mu_{k+1}(x) = -L_{k+1}x,$$

where $L_{k+1} = \tilde{q}_k^{-1} \tilde{r}_k^T$. With

$$\bar{x}_{i+1} = \varphi(x_i, u_i) - \gamma \varphi(x_{i+1}, -L_i x_{i+1}),$$

and

$$\bar{y}_{i+1} = x_i^T S x_i + u_i^T R u_i,$$

we may write the residual as

$$\varepsilon(x_i, u_i, -L_i x_{i+1}, a) = \bar{x}_{i+1}^T a - \bar{y}_{i+1},$$

and hence, the LS problem reads

$$\text{minimize } \frac{1}{2} \sum_{i=1}^{N} (\bar{x}_i^T a - \bar{y}_i)^2,$$

where we have shifted the summation index. We now use recursive LS as in Exercise 10.2 to update the parameter a for an LS problem with N terms to an LS problem with $N + 1$ terms in order to obtain adaptive control. The recursion for this reads

$$a_{N+1} = a_N + P_{N+1}\bar{x}_{N+1}\left(\bar{y}_{N+1} - \bar{x}_{N+1}^T a_N\right),$$

where

$$P_{N+1} = P_N - \frac{1}{1 + \bar{x}_{N+1}^T P_N \bar{x}_{N+1}} P_N \bar{x}_{N+1}\bar{x}_{N+1}^T P_N.$$

For the definition of P_N see the solution of Exercise 10.2. We then have that $L_{N+1} = \tilde{q}_{N+1}^{-1} \tilde{r}_{N+1}^T$, where $\tilde{q}_{N+1}$ and $\tilde{r}_{N+1}$ are defined by a_{N+1}. We may start with $L_0 = 0$ and $a_0 = 0$ if we have no better initial guess. We should take P_0 to have a large value, e.g. a large multiple of the identity matrix.

11.3.3 Actor Network

It is possible to also approximate the policy with a linear regression or an ANN. Instead of optimizing with respect to u in the policy improvement step, we then optimize with respect to the parameters of the linear regression or the ANN, see Section 11.5. One advantage of this is that no optimization has to be carried out in real time when the controller is running, since the feedback function will be known explicitly in terms of the linear regression or ANN. Also, it will in many cases improve the speed of the learning.

11.3.4 Variational Form

We remark that it may be highly beneficial to consider the variational form of the Bellman equation for the Q-function by replacing the incremental cost with the temporal difference in (8.12). Since the variational form of the Bellman equation for the Q-function looks the same as the original Bellman equation for the Q-function, with the only difference that the incremental cost is replaced with the temporal difference, all formulas in this section also remain the same when the temporal difference is used as incremental cost. The reason that the variational form may be beneficial is that $W(x)$ may be taken as an initial guess of the value function, and then we only need to model the difference between the true solution and the initial guess, which might require less parameters.

11.4 Linear Programming Formulation

In Section 8.6, we showed how to formulate the Bellman equation as an LP. Our intention is now to show that this is also possible to do for the Bellman equation for the Q-function. To this end, we start the VI in (11.9) with a Q_0 such that $Q_1 = T_Q(Q_0) \geq Q_0$ for all $(x, u) \in \mathcal{D}^n \times \mathcal{E}^m$.[1] One possible choice is $Q_0 = 0$. We obtain $T_Q(Q_1) \geq T_Q(Q_0)$ from the monotonicity property, see Exercise 11.6, and hence $Q_2 = T_Q(Q_1) \geq Q_1 \geq Q_0$. If we repeat this, we obtain $Q_k = T_Q(Q_{k-1}) \geq Q_{k-1} \geq Q_0$. Since Q_k converges to the solution Q of (11.7), we have shown that Q is the maximum element, see [22, p. 45] of the set of functions Q that satisfy the linear inequalities

$$Q(x, u) \leq f(x, u) + \gamma Q(F(x, u), v), \quad \forall (x, u, v) \in \mathcal{D}^n \times \mathcal{E}^m \times \mathcal{E}^m.$$

The maximum element can be obtained by solving the LP

$$\text{maximize} \quad \sum_{(x,u) \in \mathcal{D}^n \times \mathcal{E}^m} c(x, u) Q(x, u) \tag{11.17}$$

$$\text{subject to} \quad Q(x, u) \leq f(x, u) + \gamma Q(F(x, u), v), \ \forall (x, u, v) \in \mathcal{D}^n \times \mathcal{E}^m \times \mathcal{E}^m \tag{11.18}$$

where $c(x, u) > 0$ is arbitrary. This follows from the dual characterization of maximum elements, see [22, p. 54]. The optimization variable is $Q(x, u)$ for all values of x and u.

The LP formulation is often not tractable in general, since there might be many variables and constraints. As in Section 8.6, it is possible to approximate $Q(x, u)$ with, e.g. a feature-based linear regression. Sampling of constraints may also be used. Moreover, we may use the LP to approximately evaluate a fixed policy μ, which may then be used together with PI.

We remark that the stochastic infinite horizon optimal control problem in (8.33) can also be solved using an LP. This is because we just need to replace $Q(F(x, v), v)$ in the right-hand side in the inequalities with $\mathbb{E}[\, Q(F(x, v, W), v)\,]$, which if W only takes finitely many values is a finite sum, which is linear in Q.

---

1 We assume that $\mathcal{D}$ and $\mathcal{E}$ are finite sets.

11.5 Approximation in Policy Space

So far we have discussed how to approximate the optimal cost-to-go function for the finite horizon case or the optimal cost function for the infinite horizon case. This has been done indirectly by approximating the corresponding optimal Q-functions. From these functions, the optimal policy has been computed by minimizing Q with respect to u parametrically for all values of x. This can be challenging, and as has been hinted above, it is possible to also obtain a parametric solution to this optimization problem. Hence, we assume that we are given a function $Q : \mathcal{D}^n \times \mathcal{E}^m$, and that we would like to solve the problem

$$\mu(x) = \operatorname*{argmin}_u Q(x, u), \tag{11.19}$$

approximately. To this end, we define $\tilde{\mu} : \mathcal{D}^n \times \mathbb{R}^p \to \mathcal{E}^m$ using an ANN or a linear regression with p parameters similarly as we previously have approximated value functions and Q-functions. We then solve the following LS problem:

$$\text{minimize } \frac{1}{2} \sum_{k=1}^{N} \|u_k - \tilde{\mu}(x_k, a)\|_2^2,$$

with respect to a, where $u_k, k \in \mathbb{N}_N$ are solutions to (11.19) for the samples $x = x_k$. One benefit of the approach is that less optimization problems need to be solved when the policy is implemented in real time, and moreover evaluating $\tilde{\mu}$ might be done much faster than solving the optimization problem. Hence, smaller sampling times in real-time applications are possible, which may be important for some applications. Notice that this approach can be used for any policy for which we know its values for a discrete number of samples, i.e. it does not necessarily have to be related to (11.19).

11.5.1 Iterative Learning Control

We will now discuss approximation in policy space which is not based on the knowledge of Q-functions. We start by discussing an alternative way of solving (11.1) without a model. We define $J : \mathbb{R}^{Nm} \to \mathbb{R}$ as the function defined by the objective function in (11.1) when the constraints are used to substitute away x_k, i.e. J is a function of $u = (u_0, \ldots, u_{N-1})$. Similarly, we consider x_k to be functions of u. Obviously, x_k does not depend on u_l for $l \geq k$, but we do not bother about that for now, since it would only clutter the notation. Then in order to minimize J with respect to u with some gradient method we need to compute the gradient of J with respect to u. This can be done using the chain rule. We obtain

$$\frac{dJ(u)}{du^T} = \frac{d\phi(x_N)}{dx_N^T} \frac{dx_N}{du^T} + \sum_{k=0}^{N-1} \frac{\partial f_k(x_k, u_k)}{\partial x_k^T} \frac{dx_k}{du^T} + \frac{\partial f_k(x_k, u_k)}{\partial u_k^T} \frac{du_k}{du^T},$$

where

$$\frac{dx_{k+1}}{du_l^T} = \frac{\partial F_k(x_k, u_k)}{\partial x_k^T} \frac{dx_k}{du_l^T} + \frac{\partial F_k(x_k, u_k)}{\partial u_k^T} \delta(k - l),$$

and where

$$\delta(k) = \begin{cases} 1, & k = 0 \\ 0, & k \neq 0. \end{cases}$$

The initial value is zero. Notice that $\frac{du_k}{du^T}$ is a trivial matrix consisting of ones and zeros. For the linear case when $F_k(x, u) = A_k x + B_k u$, where $A_k \in \mathbb{R}^{n \times n}$ and $B_k \in \mathbb{R}^{n \times m}$, we have

$$\frac{dx_{k+1}}{du_l^T} = A_k \frac{dx_k}{du_l^T} + B_k \delta(k - l).$$

From this we realize that not only the states but also their derivatives can be obtained from simulation using the same dynamical equations or an experiment involving the dynamical system. For the gradient, one simulation or experiment has to be carried out for each value of l and each component of the control signal. In case A_k and B_k do not depend on k, the time-invariant case, we realize that

$$\frac{dx_k}{du_l^T} = \frac{dx_{k-l}}{du_0^T},$$

since the initial value is zero. For the linear time-invariant case, this means that one just needs to obtain the so-called "impulse response" of the dynamical system. It is of course debatable if it is a good idea to carry out impulse response experiments to obtain the impulse response. The reason is that the impulse response is given by $A^{k-1}B, k \geq 1$, and hence it can be computed from knowledge of the system matrices (A, B). They can as will be discussed in Chapter 12 be obtained using system identification. To use an impulse as the input signal for system identification is normally not a good experiment design, see Section 12.9. However, using system identification in conjunction with the methods of Chapter 8 is a more involved approach than just using an impulse response which can be obtained from a very simple experiment.

For the case of nonlinear F_k, the nonlinear dynamical equations may be used instead. The result will then not be exact, but only hold approximately, i.e.

$$\frac{dx_{k+1}}{du_{i,j}} \approx F_k\left(\frac{dx_k}{du_{i,j}}, \frac{du_k}{du_{i,j}}\right).$$

Here index i refers to stage index and index j to component index. This all assumes that the coordinate system has been chosen such that $F_k(0,0) = 0$. For the nonlinear time-invariant case one simulation for each component of u_k is enough. What has been presented above is strongly related to what is called *iterative learning control* (ILC), see, e.g. [49, 77]. This approach is typically used in applications where the same repeated task is carried out over and over again. Any reference value is incorporated in the definition of the incremental costs f_k. In case the whole state cannot be measured, it is still possible to use impulse responses for obtaining derivatives assuming that only what is measured is penalized in the objective function, see Exercise 11.12

11.5.2 ILC Using Root-Finding

What has been presented above is just one form of ILC. There is another more popular method which does not rely on obtaining the impulse response. To set the stage, we consider the dynamical system

$$x_{k+1} = F_k(x_k, u_k)$$
$$y_k = G_k(x_k, u_k)$$

for $k \in \mathbb{Z}_+$, where $x_k \in \mathbb{R}^n$, $u_k \in \mathbb{R}^m$, and $y_k \in \mathbb{R}^m$. Here $F_k : \mathbb{R}^n \times \mathbb{R}^m \to \mathbb{R}^n$ and $G_k : \mathbb{R}^n \times \mathbb{R}^m \to \mathbb{R}^m$. We assume that the initial value x_0 is zero. We define $y = (y_0, y_1, \ldots, y_{N-1})$ and $u = (u_0, u_1, \ldots, u_{N-1})$. We then define $H : \mathbb{R}^{Nm} \to \mathbb{R}^{Nm}$ such that

$$y = H(u).$$

In order to do this, we use the equation $x_{k+1} = F_k(x_k, u_k)$ to recursively eliminate the variables x_k from the dynamical system equations. We are now interested in finding u such that the error signal $\varepsilon : \mathbb{R}^{Nm} \to \mathbb{R}^{Nm}$ defined as

$$\varepsilon(u) = y - r = H(u) - r,$$

is zero for a given so-called *reference value* $r \in \mathbb{R}^{Nm}$ for y. The following root-finding algorithm, see Appendix 11.6:

$$u_{k+1} = u_k - t\varepsilon(u_k),$$

will be used. Here the subindex k refers to iteration index and not to stage index or time. Notice that the computations involved in the root-finding algorithm only require us to be able to evaluate the error e for a known input u. This can be done with simulations or with experiments on a real dynamical system, i.e. no explicit knowledge of the functions F_k and G_k is needed.

The assumptions for convergence are that ε is Lipschitz continuous and strongly monotone. These conditions are not easy to investigate for a general H. However, we may phrase them in systems theory terms. The Lipschitz constant β is the *incremental gain* of the dynamical system, i.e. the smallest β such that

$$\|H(u) - H(v)\|_2 \le \beta \|u - v\|_2, \quad \forall u, v \in \mathbb{R}^{Nm}.$$

The strong monotonicity condition is the same as saying that the dynamical system is *incrementally strictly passive* with dissipation α, i.e.

$$(H(u) - H(v))^T (u - v) \ge \alpha \|u - v\|_2^2, \quad \forall u, v \in \mathbb{R}^{Nm}.$$

Example 11.5 We now assume that $F_k(x, u) = Ax + Bu$ and that $G_k = Cx + Du$ for matrices (A, B, C, D) of compatible dimensions. Then H is a linear function and we may instead write $y = Hu$. Here H now is the matrix defined as

$$H = \begin{bmatrix} h_0 & 0 & \cdots & 0 \\ h_1 & \ddots & \ddots & \vdots \\ \vdots & \ddots & \ddots & 0 \\ h_{N-1} & \cdots & h_1 & h_0 \end{bmatrix},$$

where $h_0 = D$ and $h_i = CA^i B \in \mathbb{R}^m$ for $i \in \mathbb{N}$ are the Markov parameters, or equivalently the impulse response coefficients for the linear dynamical system. It is straightforward to see that the Lipschitz constant is $\beta = \|H\|_2$, i.e. the largest singular value of H. Moreover, the criterion for strong monotonicity is that

$$(u - v)^T H^T (u - v) \ge \alpha \|u - v\|_2^2, \quad \forall u, v \in \mathbb{R}^{Nm}.$$

We realize that this is equivalent to

$$\frac{1}{2} x^T \left(H + H^T \right) x \ge \alpha \|x\|_2^2, \quad \forall x \in \mathbb{R}^{Nm},$$

which is satisfied if and only if the smallest eigenvalue $\lambda_{\min}$ of $H + H^T$ is greater than or equal to 2α. If we then denote the largest eigenvalue of $H^T H$ with $\lambda_{\max}$ it follows that the above algorithm converges for $t \in (0, 4/(\lambda_{\min} + 2\lambda_{\max})]$, assuming that $\lambda_{\min} > 0$, see Appendix 11.6.

It is possible to instead of working directly on ε as defined above consider an invertible function $T : \mathbb{R}^{Nm} \to \mathbb{R}^{Nm}$ defined as $T(\varepsilon(u))$, and apply the root finding algorithm to T. This assumes that T is chosen such that $T(\varepsilon) = 0$ if and only if $\varepsilon = 0$. One possible choice is to take T as a linear function defined by an invertible matrix $T \in \mathbb{R}^{Nm \times Nm}$, i.e. we consider $T\varepsilon(u) = 0$. We realize that for the linear case, the matrix H above is then replaced by the matrix TH, and it is for this matrix, we need to compute α and β. In case we know the impulse response, we may take $T = H^{-1}$ and we obtain convergence in one step with $t_k = 1$. In case we have some approximate knowledge of the impulse response, we can still make use of this to make TH close to the identity matrix. Instead of this, it is also possible to use feedback in order to make H itself close to the identity matrix, see *e.g.* [23].

We realize that there is a strong advantage of using the root-finding algorithm, since no gradients are needed, which is the case for formulations involving minimization of, e.g. $\varepsilon^T \varepsilon$. However, this comes at the price of assumptions on Lipschitz continuity and strong monotonicity.

11.5.3 Iterative Feedback Tuning

We will now consider the case when the control signal is given by a policy $\mu : \mathbb{R}^n \times \mathbb{R}^p \to \mathbb{R}^m$, i.e. $u_k = \mu(x_k, a)$, where a is a parameter that we would like to learn. Typically, we would like it to be the solution of a problem like

$$\text{minimize} \quad \phi(x_N) + \sum_{k=0}^{N-1} f_k(x_k, u_k) \tag{11.20}$$

$$\text{subject to} \quad x_{k+1} = F_k(x_k, u_k), \quad k \in \mathbb{Z}_{N-1} \tag{11.21}$$

$$u_k = \mu(x_k, a), \quad k \in \mathbb{Z}_{N-1} \tag{11.22}$$

with variable a for a given initial value x_0, where ϕ, f_k and F_k are defined as before in this chapter. We define $J : \mathbb{R}^p \to \mathbb{R}$ as the function defined by (11.20) when (11.21–11.22) are used to substitute away x_k and u_k. Similarly, we consider x_k and u_k to be functions of a. Then in order to minimize J with respect to a with some gradient method, we need to compute the gradient of J with respect to a. This can again be done using the chain rule. We obtain

$$\frac{dJ(a)}{da^T} = \frac{d\phi(x_N)}{dx_N^T}\frac{dx_N}{da^T} + \sum_{k=0}^{N-1} \frac{\partial f_k(x_k, u_k)}{\partial x_k^T}\frac{dx_k}{da^T} + \frac{\partial f_k(x_k, u_k)}{\partial u_k^T}\frac{du_k}{da^T},$$

where

$$\frac{dx_{k+1}}{da^T} = \frac{\partial F(x_k, u_k)}{\partial x_k^T}\frac{dx_k}{da^T} + \frac{\partial F(x_k, u_k)}{\partial u_k^T}\frac{du_k}{da^T}$$

$$\frac{du_k}{da^T} = \frac{\partial \mu(x_k, a)}{\partial x_k^T}\frac{dx_k}{da^T} + \frac{\partial \mu(x_k, a)}{\partial a^T}.$$

The initial value is zero.

Example 11.6 For the case when $F_k(x, u) = A_k x + B_k u$, where $A_k \in \mathbb{R}^{n \times n}$ and $B_k \in \mathbb{R}^{n \times m}$, and when $\mu(x, a) = Lx$, where $L \in \mathbb{R}^{m \times n}$ with $a^T = \begin{bmatrix} L_1 & \cdots & L_m \end{bmatrix}$, where L_i are the rows of L, we have

$$\frac{dx_{k+1}}{da^T} = A_k \frac{dx_k}{da^T} + B_k \frac{du_k}{da^T}$$

$$\frac{du_k}{da^T} = L\frac{dx_k}{da^T} + \text{blkdiag}(x_k^T, \ldots, x_k^T).$$

We realize that the derivatives are obtained by simulation of the closed-loop system or from experiments involving the closed-loop system with the current x_k as an additional input.

The results of the example also hold true approximately for the nonlinear case when $F_k(0, 0)$ is zero. Clearly, many simulations have to be carried out since one simulation is needed for every component of a, i.e. there are $p = mn$ simulations. This idea is the basis for what is called *iterative feedback tuning* (IFT). There often output feedback is considered, and reference values for the output signal are defined explicitly. Then it is possible with transfer function manipulations to show that only two additional simulations are needed for the case when $m = 1$ and when there is only one output signal, see e.g. [56].

Before we end this section, it should be mentioned that just using different initial values to obtain good experimental conditions for the simulations might not be a good idea. It might be the case that the simulations are experiments where one does not control the initial values. Also, big initial values might be needed to get good information and then the approximations for nonlinear state dynamics might not be good. A remedy to this is to inject a perturbation by changing the dynamical equations to

$$x_{k+1} = F_k(x_k, u_k + w_k),$$

where w_k is a realization of white noise. This should only be done in the first experiment in each iteration for IFT. Sometimes the experimental conditions are such that one has to live with perturbations also in the other experiments. One should then resort to stochastic optimization methods as discussed in Section 6.8. We will discuss this more in an exercise.

11.6   Appendix – Root-Finding Algorithms

In this appendix, we will discuss root-finding algorithms for solving nonlinear equations. We will consider both the deterministic case and the stochastic case.

Let us consider the function $g : \mathbb{R}^n \to \mathbb{R}^n$ and the equation

$$g(x) = 0. \tag{11.23}$$

We assume that g is *strongly monotone* with parameter $\alpha \in \mathbb{R}_{++}$, i.e. it satisfies

$$(g(y) - g(x))^T(y - x) \geq \alpha \|y - x\|_2^2, \quad \forall x, y \in \mathbb{R}^n. \tag{11.24}$$

We also assume that it is Lipschitz continuous with Lipschitz constant $\beta \in \mathbb{R}_{++}$, i.e.

$$\|g(y) - g(x)\| \leq \beta \|y - x\|_2, \quad \forall x, y \in \mathbb{R}^n. \tag{11.25}$$

It then follows from the proof of convergence preceding (6.19) for the gradient method for α-strongly convex and β-smooth functions that the algorithm

$$x_{k+1} = x_k - tg(x_k), \quad k \in \mathbb{Z}_+, \tag{11.26}$$

converges for $t \in (0, 2/(\alpha + \beta)]$ by replacing ∇f with g. To see this we notice that β-smoothness for convex f is equivalent to Lipschitz continuity of $g = \nabla f$, and that strong monotonicity of $g = \nabla f$ is implied by strong convexity of f.

We are now interested in the stochastic version of the above problem. To this end, we consider a probability space $(\Omega, \mathcal{F}, \mathbb{P})$, and we define a random variable ξ from the sample space Ω to some set D. The random variable could be discrete or continuous. We are given the function $G : \mathbb{R}^n \times D \to \mathbb{R}^n$ and define the function $g : \mathbb{R}^n \to \mathbb{R}^n$ as $g(x) = \mathbb{E}[G(x, \xi)]$ and the equation

$$g(x) = 0. \tag{11.27}$$

We will show that the above equation can be solved with an algorithm of the form

$$x_{k+1} = x_k - t_k g_k, \quad k \in \mathbb{Z}_+, \tag{11.28}$$

where x_0 is an initial guess, $t_k > 0$ is the step size at iteration k, and g_k is a realization of an estimator of G_k of $g(x_k)$. It should be stressed that G_k is a random variable for each k. Hence, the iteration (11.28) is a realization of a stochastic process

$$X_{k+1} = X_k - t_k G_k, \quad k \in \mathbb{Z}_+, \tag{11.29}$$

where the random variable X_0 is the initial state of the process. We consider an unbiased estimator G_k of $g(x_k)$ with bounded variance, i.e.

$$\mathbb{E}[G_k|X_k = x_k] = g(x_k), \quad \mathbb{E}[\|G_k - g(x_k)\|_2^2|X_k = x_k] \le c^2, \tag{11.30}$$

for all k and for some scalar $c \ge 0$. Note that in the special case where $c = 0$, the iteration (11.28) is essentially the root-finding method (11.26). We note that $G(x, \xi)$ is a random variable, and it is an unbiased estimator of $g(x)$ and hence, it is natural to choose $G_k = G(X_k, \xi_k)$, where ξ_k has the same distribution as ξ for all $k \ge 0$, and where ξ_k and ξ_j are independent for $j \ne k$. We assume that g satisfies (11.24) and (11.25). For a constant step size $t_k = t$, it then follows from the proof preceding (6.71) that (11.28) converges in mean square to a ball centered at $x^\star$, where $g(x^\star) = 0$, and with a radius that is proportional to the constant step size t. A more elaborate analysis can be done to show that we have convergence in means square to $x^\star$ for a diminishing step-size.

Exercises

11.1 Let $A \in \mathbb{R}^{m \times n}$ and $B \in \mathbb{R}^{n \times m}$, where $m \le n$. Show that the eigenvalues of BA are the eigenvalues of AB together with $n - m$ zero eigenvalues.

11.2 We will in this exercise compare the fitted value iterations in Example 8.3 with the fitted Q-function iterations in Example 11.1 for a finite horizon LQ control problem. Implement the two algorithms in MATLAB and investigate their performance for the case when

$$A_k = \begin{bmatrix} 0.5 & 1 \\ 0 & 0.5 \end{bmatrix}; \quad B_k = \begin{bmatrix} 0 \\ 1 \end{bmatrix},$$

and $R_k = 1$ and $S_k = I$. Consider a time horizon of $N = 10$. Also compare the resulting feedback gains L_k you obtain with the ones obtained from the Riccati recursion in Example 8.1.

11.3 Show that the Q-functions defined in (8.32) satisfy the recursion

$$Q_k(x, \bar{u}) = \mathbb{E}\left[f_k(x, \bar{u}) + \min_u Q_{k+1}\left(F_k(x, \bar{u}, W_k), u\right) \right]. \tag{11.31}$$

11.4 We now investigate how to do fitted Q-function iterations for the stochastic LQ problem in Example 8.10. We then need to extend $\tilde{Q}_k$ in Example 11.1 with a term a_7, i.e.

$$\tilde{Q}_k(x, u, a) = \begin{bmatrix} x \\ u \end{bmatrix}^T \begin{bmatrix} \bar{P} & \bar{r} \\ \bar{r}^T & \bar{q} \end{bmatrix} \begin{bmatrix} x \\ u \end{bmatrix} + a_7.$$

Make the necessary additional modification to accommodate for this in the algorithm. Specifically you should replace x_+^s with $x_+^s = Ax_k^s + Bu_k^s + w_k$, where w_k is a realization from a white noise random process W with unit variance. Use the same problem data as in Exercise 11.2 and run the algorithm in MATLAB. We know that the same feedback gains L_k are optimal for both the stochastic and deterministic case by certainty equivalence. Do you get the same solution? Does the quality of the solution depend on the number of samples you consider?

11.5 Show that the Q-function in (8.35) satisfies

$$Q(x, \bar{u}) = \mathbb{E}\left[f(x, \bar{u}) + \gamma Q(F(x, \bar{u}, W), u) \right]. \tag{11.32}$$

11.6 (a) Show that the Bellman Q-operator in (11.8) is such that

$$T_Q(Q_1)(x, \bar{u}) \le T_Q(Q_2)(x, \bar{u}),$$

for all $(x, \bar{u}) \in \mathcal{D}^n \times \mathcal{E}^m$ if $Q_1(x, \bar{u}) \le Q_2(x, \bar{u})$ for all $(x, \bar{u}) \in \mathcal{D}^n \times \mathcal{E}^m$. Also, show that it is a contraction.

(b) We let the *Stochastic Bellman Q-operator* $T_Q : \mathbb{R}^{\mathcal{D}^n} \to \mathbb{R}^{\mathcal{D}^n}$ be defined as

$$T_Q(Q)(x, \bar{u}) = \mathbb{E}\left[f(x, \bar{u}) + \min_u \gamma Q_k (F(x, \bar{u}, W), u) \right], \tag{11.33}$$

where $F : \mathcal{D}^n \times \mathcal{E}^m \times \mathcal{F}^p \to \mathcal{D}^n$. Expectation is with respect to W. Show the same results as in the previous subexercise hold also for the stochastic case.

11.7 Consider the optimal control problem in Example 8.8.

(a) Solve it numerically using the LP formulation for the Q-function in Section 11.4. Solve it for different values of γ, e.g. $\gamma = 0.25$ and $\gamma = 0.9$.

(b) Solve the same problems using (11.9).

(c) Solve the same problems using (11.10).

(d) Solve the same problems using (11.11).

(e) Discuss pros and cons with the different methods used above to solve the problem.

11.8 We now consider a stochastic version of the problem in the previous exercise. This is obtained by defining $F(x, u, w)$ as

x \ u	−1	0	1
−1	−1	−1	0
0	−1	0	1
1	0	1	1

for $w = 0$, as

x \ u	−1	0	1
−1	−1	0	1
0	0	1	1
1	1	1	1

for $w = 1$, and as

x \ u	−1	0	1
−1	−1	−1	−1
0	−1	−1	0
1	−1	0	1

for $w = -1$. We assume that the random variable W_k takes the values $(-1, 0, 1)$ with probabilities $(p, 1 - 2p, p)$, where $p \in [0, 0.5]$. Notice that the deterministic case is recovered for $p = 0$.

(a) Solve the problem using (11.11) modified to the stochastic setting as discussed at the end of Section 11.2. For what values of γ and p do you obtain a nontrivial solution?

(b) Solve the problem using the LP formulation in Section 11.4 for the Q-function. See the end of Section 11.4 for how to modify the LP to the stochastic case.

11.9 (a) Define the Bellman Q policy operator $T_{Q\mu} : \mathbb{R}^{D^n \times \mathcal{E}^m} \to \mathbb{R}^{D^n \times \mathcal{E}^m}$ as

$$T_{Q\mu}(Q) = f(x, u) + \gamma Q(F(x, u), \mu(F(x, u))),$$

where $F : D^n \times \mathcal{E}^m \to D^n$. Show that if $Q_1(x, u) \leq Q_2(x, u)$ for all $(x, u) \in D^n \times \mathcal{E}^m$, then

$$T_{Q\mu}(Q_1) - T_{Q\mu}(Q_2) \leq 0.$$

Also, show that $T_{Q\mu}$ is a contraction.

(b) Define the stochastic Bellman Q policy operator $T_{Q\mu} : \mathbb{R}^{D^n \times \mathcal{E}^m} \to \mathbb{R}^{D^n \times \mathcal{E}^m}$ as

$$T_{Q\mu}(Q) = \mathbb{E}\left[f(x, u) + \gamma Q(F(x, u, W), \mu(F(x, u, W))) \right],$$

where $F : D^n \times \mathcal{E}^m \times F^p \to D^n$. Expectation is with respect to W. Show that if $Q_1(x) \leq Q_2(x)$ for all $(x, u) \in D^n \times \mathcal{E}^m$, then

$$T_{Q\mu}(Q_1) - T_{Q\mu}(Q_2) \leq 0.$$

Also show that $T_{Q\mu}$ is a contraction.

11.10 In this exercise, we consider the same LQ problem as in Exercise 8.9.

(a) Implement the approach in Example 11.3. You may start with $L_0 = 0$. How does the choice of initial values (x^s, u^s) and the number r of initial values affect the convergence of L_k? How does the choice of N affect the convergence? Compare your results with what was obtained in Exercise 8.9.

(b) Consider the stochastic policy evaluation step for the Q function given by

$$Q_k = T_{Q\mu_k}(Q_k),$$

where $T_{Q\mu}$ is defined in Exercise 11.9. Show that it implies that

$$Q_k(x_0, u_0) = f(x_0, u_0) + \sum_{i=1}^{N-1} \gamma^i \mathbb{E}\left[f\left(X_i, \mu_k(X_i)\right) \right] + \gamma^N \mathbb{E}\left[Q_k\left(X_N, \mu_k(X_N)\right) \right]$$

assuming that $X_{k+1} = F(X_k, \mu_k(X_k), W_k)$ and that $X_0 = x_0$ is given. It is tempting to try to obtain a SAA of $Q_k(x_0, u_0)$ by using β_k^s as defined in Section 11.3, where x_i now are realizations of X_i. Hence, the idea is that the approach in that section can be used also for the stochastic case. Therefore, repeat what was done in the previous subexercise, but now let $x_{k+1} = Ax_k + Bu_k + w_k$, where w_k is a realization of W_k, which is assumed to have a Gaussian distribution with zero mean and covariance equal to the identity matrix. We also assume that W_j is independent of W_k for $j \neq k$. You may start with $L_0 = 0$. How does the choice of initial values (x^s, u^s) and the number r of initial values affect the convergence of L_k? How does the choice of N affect the convergence? Compare your results with what was obtained in the previous subexercise. Are the assumptions for an SAA valid?

Hint: You need to modify the approach in Example 11.3 to let

$$\varphi(x, u) = (x_1^2, x_2^2, u^2, 2x_1x_2, 2x_1u, 2x_2u, 1),$$

and hence, $a \in \mathbb{R}^7$. This means that the last component of a will model the constant term in the value function which we know is present for the stochastic case, c.f. Example 8.13.

11.11 In this exercise, we consider the same stochastic LQ problem as in the previous subexercise. Implement the approach in Example 11.4. Are you able to obtain convergence of L_k to the correct value? Does it depend on the initial value P_0?

11.12 We are given matrices $A \in \mathbb{R}^{n \times n}, B \in \mathbb{R}^{n \times 1}, C \in \mathbb{R}^{1 \times n}$ and $D \in \mathbb{R}$. We consider the following optimal control problem:

$$\text{minimize} \quad \frac{1}{2} \sum_{k=0}^{N} (y_k - r_k)^2$$

$$\text{subject to} \quad x_{k+1} = A x_k + B u_k, \quad k \in \mathbb{Z}_{N-1}$$

$$y_k = C x_k + D u_k, \quad k \in \mathbb{Z}_N,$$

for a given initial value x_0 with variables $(u_0, x_1, y_1 \ldots, u_{N-1}, x_N, y_N)$. Denote by J the objective function. Let $u = (u_0, \ldots, u_N)$ and $y = (y_0, \ldots, y_N)$.

(a) Show that

$$\frac{dy}{du^T} = H,$$

where

$$H = \begin{bmatrix} D & 0 & \cdots & 0 \\ CB & \ddots & \ddots & \vdots \\ \vdots & \ddots & \ddots & 0 \\ CA^{N-1}B & \cdots & CB & D \end{bmatrix}.$$

(b) Show that the optimality conditions for the optimal control problem are

$$\frac{dJ}{du} = \left(\frac{dy}{du^T} \right)^T (y - r) = 0,$$

where

$$y = \mathcal{O} x_0 + H u,$$

with

$$\mathcal{O} = \begin{bmatrix} C \\ CA \\ \vdots \\ CA^N \end{bmatrix}.$$

(c) Show that the optimal u is the solution to

$$H^T H u = H^T r,$$

if $x_0 = 0$.

11.13 We are given matrices $A \in \mathbb{R}^{n \times n}, B \in \mathbb{R}^{n \times 1}, C \in \mathbb{R}^{1 \times n}$ and $D \in \mathbb{R}$. We consider the following stochastic optimal control problem:

$$\text{minimize} \quad \frac{1}{2} \mathbb{E} \left[\sum_{k=0}^{N} (Y_k - r_k)^2 \right]$$

$$\text{subject to} \quad X_{k+1} = A X_k + B u_k + W_k, \quad k \in \mathbb{Z}_{N-1}$$

$$Y_k = C X_k + D u_k + E_k, \quad k \in \mathbb{Z}_N$$

for a given initial value $X_0 = x_0$ with variables $(u_0, X_1, Y_1 \ldots, u_{N-1}, X_N, Y_N)$. The sequence r_k is a desired reference value. Denote by J the objective function. Let $u = (u_0, \ldots, u_N)$, $Y = (Y_0, \ldots, Y_N)$, $W = (W_0, \ldots, W_{N-1})$, and $E = (E_0, \ldots, E_N)$. We assume that W_k and E_k are zero mean random variables. This is a stochastic version of the problem in the previous exercise. We now want to use a stochastic gradient method to find a solution to the problem.

(a) Show that for arbitrary u we have that

$$Y = \mathcal{O}x_0 + Hu + \mathcal{T}W + E,$$

where

$$
\mathcal{T} = \begin{bmatrix}
0 & 0 & \cdots & 0 \\
C & \ddots & \ddots & \vdots \\
\vdots & & \ddots & 0 \\
CA^{N-1} & \cdots & C & 0
\end{bmatrix},
$$

Here $\mathcal{O}$ and H are defined as in the previous exercise.

(b) We now define

$$\frac{\widehat{dY}}{du^T} = \begin{bmatrix} Y_{\text{imp}} & SY_{\text{imp}} & \cdots & S^{N-1}Y_{\text{imp}} \end{bmatrix},$$

where S is a shift matrix that is all zeros except for ones on the first sub-diagonal, where

$$Y_{\text{imp}} = \mathcal{O}x_0 + H_1 + \mathcal{T}W + E$$

is the impulse response, and where H_1 is the first column of H. Notice that this is obviously not the true value of the Jacobian $\frac{dY}{du^T} = H$. However, it can be obtained from an impulse response experiment. Let

$$G = \left(\frac{\widehat{dY}}{du^T} \right)^T (Y - r),$$

and show that this is an unbiased estimator of the gradient of the objective function for the stochastic optimal control problem when $x_0 = 0$. You may assume that we have evaluated Y and Y_{imp} for different uncorrelated noise sequences.

(c) Consider a simple example where

$$A = \begin{bmatrix} 0.5 & 1 \\ 0 & 0.5 \end{bmatrix}, \quad B = \begin{bmatrix} 1 \\ 0 \end{bmatrix}, \quad C = \begin{bmatrix} 0 & 1 \end{bmatrix}, \quad D = 2.$$

Let $N = 50$, generate a reference value given by

$$r_k = \begin{cases} 10, & 0 \le k \le 24 \\ -10. & 25 \le k \le 50, \end{cases}$$

assume that the initial value x_0 is zero, and let the noise sequences both have zero mean, unit variance and have Gaussian distribution. Solve the stochastic optimal control problem with a stochastic gradient method using the gradient suggested above.

11.14 We will now investigate a stochastic version of Example 11.5. To this end, we consider the Markov process in the previous exercise for which we have that

$$Y = \mathcal{O}x_0 + Hu + \mathcal{T}W + E,$$

where W and W are zero mean random vectors. We define the error function

$$\varepsilon(u, W, E) = Y - r = \mathcal{O}x_0 + Hu + \mathcal{T}W + E - r,$$

similarly as for the deterministic case, and then we apply the root-finding algorithm

$$u_{k+1} = u_k - t g_k,$$

where $g_k = y_k - r$, and y_k is a realization of Y for the kth iteration. Notice that the subindex k does not refer to the stage index for the Markov process.

(a) Show that $\varepsilon(u, W, E)$ is an unbiased estimator of $\mathbb{E}[\varepsilon(u, W, E)]$.

(b) Consider the same numerical example as in the last subexercise of the previous exercise. Compute the largest singular value of H and the smallest eigenvalue of $H + H^T$. What is the largest possible step-length that can be used in order to guarantee convergence in m.s. to a ball centered around the true solution?

(c) Use the stochastic root-finding algorithm applied to the same Markov process as in the last subexercise in the previous exercise. Use the same reference value. Compare the convergence for the root-finding algorithm with that of the stochastic gradient algorithm of the previous exercise. Experiment both with fixed step-size and with diminishing step-size for the root-finding algorithm. You may assume that $x_0 = 0$.

11.15 We are given matrices $A \in \mathbb{R}^{n\times n}, B \in \mathbb{R}^{n\times 1}, C \in \mathbb{R}^{1\times n}$ and $D \in \mathbb{R}$. We consider the following stochastic optimal control problem:

$$\text{minimize } \frac{1}{2}\mathbb{E}\left[\sum_{k=0}^{N}(Y_k - r_k)^2 \right]$$

$$\text{subject to } X_{k+1} = AX_k + Bu_k + W_k, \quad k \in \mathbb{Z}_{N-1}$$

$$Y_k = CX_k + Du_k + E_k, \quad k \in \mathbb{Z}_N,$$

for a given initial value $X_0 = x_0$ with variables $(u_0, X_1, Y_1 \ldots, u_{N-1}, X_N, Y_N)$. Let $u = (u_0, \ldots, u_N)$, $Y = (Y_0, \ldots, Y_N)$, $W = (W_0, \ldots W_{N-1})$, and $E = (E_0, \ldots, E_N)$. We assume that W_k and E_k are zero mean random variables. The sequence r_k is a desired reference value. We consider a policy $\mu(x, a) = Lx$, where $a^T = L \in \mathbb{R}1 \times n$. Let $J(a)$ be the value of the objective function when the above constraints and $u_k = \mu(X_k, a)$ are used to eliminate the variables X_k, u_k and Y_k.

(a) Show that

$$\frac{dJ}{da^T} = \sum_{k=0}^{N}(Y_k - r_k)\left(C\frac{dX_k}{da^T} + D\frac{du_k}{da^T}\right),$$

where

$$\frac{dX_{k+1}}{da_i} = A\frac{dX_k}{da_i} + B\frac{du_k}{da_i} + W_k$$

$$\frac{du_k}{da_i} = L\frac{dX_k}{da_i} + X_{i,k},$$

where $X_{i,k}$ is the ith component of X_k for $i \in \mathbb{Z}_n$. The initial value is $\frac{dX_0}{da^T} = 0$.

(b) Solve again the same stochastic control problem numerically that we have considered in the previous two exercises. This time use a stochastic gradient method to compute the optimal policy discussed above. You may assume that $x_0 = 0$. You should in each iteration use three independent noise sequences for computing X_k and $\frac{dX_k}{da_i}$, and the noise sequences should of course also be independent for each iteration. Are you able to obtain as small value of the objective function as you were before? If not, why? How does the optimal policy perform for a different reference value as the one it was trained for? How do the approaches in the previous two exercises perform for another reference value in case no new training is performed for this new reference value?

11.16 Use the reinforcement learning toolbox in MATLAB to solve the shortest path problem in Example 8.2 using Q-learning. Use the MDP environment. You need to code the states and actions differently as compared to what is done in the example. You should label the states from 1 to 14. You also need to label the actions with positive integers, say 1, 2, and 3. Moreover, you need to code what happens, if you, for example choose action 1, when you are at stage $k = 1$ in node 1. A good idea is to say that you stay in the same node and that the distance is very long, say 10. In this way, this action cannot be optimal. Also remember that the rewards in reinforcement learning are the negative of the costs in optimal control, and hence, all the rewards you code have to be the negative of the distances.

12

System Identification

System identification is about learning models for dynamical systems. Examples include the flow of water in a river, the planetary motions, and the number of cars on a segment of a freeway. We will in this chapter limit ourselves to discrete-time dynamical systems. However, many of the results can be generalized to continuous-time dynamical systems using sampling. The origin of system identification goes back to the work by Karl Åström and Torsten Bolin in 1965.

We start by defining what we mean by a dynamical system in state-space form. We define a regression problem for learning/estimating the dynamical system, and specifically, we define it as an ML problem. From this, we then derive input–output models and the corresponding ML problem. We discuss in detail how the parameters can be estimated by solving a nonlinear LS problem. Then, we discuss how to estimate the model when some of the data are missing. Nuclear norm system identification is also discussed in this context. Prior information can be incorporated into system identification easily using Gaussian processes and empirical Bayes. We show how this can be implemented using the sequential convex optimization technique based on the majorization minimization principle. Recurrent neural networks and temporal convolutional neural networks are shown to be generalizations of the linear dynamical models to nonlinear dynamical models. The chapter is finished off with a discussion on experiment design for system identification.

12.1 Dynamical System Models

A discrete-time dynamical system is described by the so-called *state equation*

$$x_{k+1} = F_k(x_k, u_k), \tag{12.1}$$

where k is a nonnegative integer describing *time*, $x_k \in \mathbb{R}^n$ is the *state*, $u_k \in \mathbb{R}^m$ is the *input signal* and $F_k : \mathbb{R}^n \times \mathbb{R}^m \to \mathbb{R}^n$ is describing how the state evolves, *cf.* (8.1). We assume that the *initial value* x_0 is given. Then given the values of u_k for $k \in \mathbb{Z}_+$ all values of x_k can be computed recursively using the state equation. The system above is called *time-varying* if F_k depends explicitly on k. Otherwise, it is called *time-invariant*. Also, it is called *deterministic* in case the input u is fully known. Often, it is not possible to observe or measure the full state x but only some function of it, and this can be modeled as

$$y_k = G_k(x_k, u_k), \tag{12.2}$$

where $y_k \in \mathbb{R}^p$ is called the *output signal* and $G_k : \mathbb{R}^n \times \mathbb{R}^m \to \mathbb{R}^p$. We have assumed that the input signal, the output signal, and the state are real-valued. However, it is possible to let them, or some components of them, take values in finite sets.

Optimization for Learning and Control, First Edition. Anders Hansson and Martin Andersen.
© 2023 John Wiley & Sons, Inc. Published 2023 by John Wiley & Sons, Inc.
Companion Website: www.wiley.com/go/opt4lc

A *linear time-invariant system* is obtained by taking F_k and G_k as linear functions independent of k, i.e.

$$x_{k+1} = Ax_k + Bu_k, \tag{12.3}$$

$$y_k = Cx_k + Du_k, \tag{12.4}$$

where $A \in \mathbb{R}^{n \times n}, B \in \mathbb{R}^{n \times m}, C \in \mathbb{R}^{p \times n}$, and $D \in \mathbb{R}^{p \times m}$ are called the *system matrices*.

12.2 Regression Problem

For the linear system in (12.3)–(12.4), the system identification problem is to find the system matrices and the initial value for given sequences of inputs and outputs, i.e. we are given $(y_k, u_k), k \in \mathbb{Z}_N$ and would like to compute $\theta = (A, B, C, D, x_0)$. Normally, the measured sequences do not satisfy (12.3)–(12.4) exactly for any θ due to measurement errors and/or other limitations in the model. Hence, it is reasonable to define what is called a *predictor* $\hat{y}_k \in \mathbb{R}^p$ of the output as

$$x_{k+1} = Ax_k + Bu_k$$

$$\hat{y}_k = Cx_k + Du_k,$$

and define the *prediction error* $e_k \in \mathbb{R}^p$ as

$$e_k = y_k - \hat{y}_k.$$

Here y_k is our measured output, and $\hat{y}_k$ is recursively defined by the measured inputs. Naturally, we would like the prediction error to be small. We realize that an equivalent way of writing the equations is

$$x_{k+1} = Ax_k + Bu_k$$

$$y_k = Cx_k + Du_k + e_k.$$

This way of defining the prediction error assumes that all errors in the model are related to measurement errors. More generally, we may consider

$$x_{k+1} = Ax_k + Bu_k + v_k$$

$$y_k = Cx_k + Du_k + e_k,$$

where $v_k \in \mathbb{R}^n$ models errors in the state equation. If we assume that (v_k, e_k) are realizations of a sequence of Gaussian zero mean random variables (V_k, E_k) with covariance

$$R = \begin{bmatrix} R_1 & R_{12} \\ R_{12}^T & R_2 \end{bmatrix} \in \mathbb{S}_{++}^{n+p},$$

such that (V_j, E_j) and (V_k, E_k) are independent for $j \neq k$, then the ML problem of estimating $\theta = (A, B, C, D, R, x_0)$ is equivalent to

$$\text{minimize} \quad \frac{1}{2} \sum_{k=0}^{N-1} \begin{bmatrix} v_k \\ e_k \end{bmatrix}^T R^{-1} \begin{bmatrix} v_k \\ e_k \end{bmatrix} + \frac{1}{2} e_N^T R_2^{-1} e_N + \frac{N}{2} \ln \det R + \frac{1}{2} \ln \det R_2$$

$$\text{subject to} \quad x_{k+1} = Ax_k + Bu_k + v_k, \quad 0 \leq k \leq N-1$$

$$y_k = Cx_k + Du_k + e_k, \quad 0 \leq k \leq N$$

with variables (θ, x, v, e), where $x = (x_1, \ldots, x_N)$, $e = (e_0, \ldots, e_N)$, and $v = (v_0, \ldots, v_{N-1})$. There are many parameters to estimate. We can reduce them by considering the innovation form as discussed in Exercise 3.17, i.e.

$$x_{k+1} = Ax_k + Bu_k + Ke_k \tag{12.5}$$

$$y_k = Cx_k + Du_k + e_k, \tag{12.6}$$

where $K \in \mathbb{R}^{n \times p}$. The variables x and e in the innovation model are not the same x and e as used in the previous model. For details about how they are related, see Exercise 3.17. There is no loss in generality to consider the innovations form. Then we define $\theta = (A, B, C, D, K, R_2, x_0)$, and the ML problem for identification is now to solve

$$\text{minimize} \quad \frac{1}{2} \sum_{k=0}^{N} e_k^T R_2^{-1} e_k + \frac{N+1}{2} \ln \det R_2$$

$$\text{subject to} \quad x_{k+1} = A x_k + B u_k + K e_k, \quad k \in \mathbb{Z}_{N-1}$$

$$y_k = C x_k + D u_k + e_k, \quad k \in \mathbb{Z}_N$$

(12.7)

with variables (θ, x, e). For the case when $p = 1$, it follows that we can solve a constrained LS problem with no weighting with R_2, and then estimate R_2 as $e^T e/(N+1)$, where e is the optimal solution, *cf.* Section 9.8. The optimization problem is however not convex due to bilinearity of the variables in the constraints. Even worse is the fact that there are uncountably many solutions to the problem because of the fact that the input–output relations are unaffected by state transformations, *cf.* Exercise 2.12. How to circumvent the latter problem will be discussed next. We will from now on restrict ourselves to the case when $m = p = 1$. Most of the results presented below carry over to the general case with some slight modifications.

12.3 Input–Output Models

One remedy to the nonuniqueness of solutions is to use input–output models. Introduce the Z-transform of a signal $x = (x_0, x_1, \ldots)$, where $x_k \in \mathbb{R}^n$, as the function $X : \mathbb{C} \to \mathbb{C}^n$ defined by

$$X(z) = \sum_{k=0}^{\infty} x_k z^{-k}.$$

Applied to (12.5–12.6) this results in

$$z X(z) - z x_0 = A X(z) + B U(z) + K E(z)$$

$$Y(z) = C X(z) + D U(z) + E(z),$$

where Y, U, and E are the Z-transforms of y, u, and e, respectively. If we eliminate X using the first equation and substitute into the second equation, we obtain

$$Y(z) = C(zI - A)^{-1} z x_0 + \mathcal{G}(z) U(z) + \mathcal{H}(z) E(z),$$

where $\mathcal{G}(z) = C(zI - A)^{-1} B + D$ and $\mathcal{H}(z) = C(zI - A)^{-1} K + I$ are rational and proper functions of z with degree n in the denominator. The rational functions may be written as

$$\mathcal{G}(z) = \frac{\mathcal{B}(z)}{\mathcal{A}(z)}; \qquad \mathcal{H}(z) = \frac{\mathcal{C}(z)}{\mathcal{A}(z)},$$

where we have defined the polynomials

$$\mathcal{A}(z) = z^n + a_1 z^{n-1} + \cdots + a_{n-1} z + a_n$$

$$\mathcal{B}(z) = b_0 z^n + b_1 z^{n-1} + \cdots + b_{n-1} z + b_n$$

$$\mathcal{C}(z) = z^n + c_1 z^{n-1} + \cdots + c_{n-1} z + c_n$$

in the variable z. Notice that $\det (zI - A) = \mathcal{A}(z)$ and that $\mathcal{A}(z)(zI - A)^{-1} = \text{adj}(zI - A)$. Because of this, we may write

$$\mathcal{A}(z) Y(z) = C \text{adj}(zI - A) z x_0 + \mathcal{B}(z) U(z) + \mathcal{C}(z) E(z).$$

Multiply this equation with $z^{-k}/(2\pi i)$ for $k = -n, -n+1, \ldots N - n$ and integrate along the unit circle C in the complex plane.[1] Then, since

$$\oint_C z^k dz = \begin{cases} 2\pi i, & k = -1 \\ 0, & k \neq -1 \end{cases},$$

it follows that

$$T_a y = \xi + T_b u + T_c e, \tag{12.8}$$

where $y = (y_0, \ldots, y_N)$, $u = (u_0, \ldots, u_N)$, and $e = (e_0, \ldots, e_N)$, and where

$$T_a = I + a_1 S + \cdots + a_{n-1} S^{n-1} + a_n S^n$$
$$T_b = b_0 I + b_1 S + \cdots + b_{n-1} S^{n-1} + b_n S^n$$
$$T_c = I + c_1 S + \cdots + c_{n-1} S^{n-1} + c_n S^n$$

are Toeplitz matrices with the shift matrix $S \in \mathbb{R}^{(N+1) \times (N+1)}$ being a matrix of all zeros except for ones on the first subdiagonal, *cf.* (2.24). The vector $\xi \in \mathbb{R}^{N+1}$ is a vector of all zeros except for the first n elements, which are functions of the initial value x_0. We may hence write $\xi = (\xi_0, 0)$, where $\xi_0 \in \mathbb{R}^n$. Each and every row in the above equation is related to a specific k above. Except for the first n rows in the above equation, one may equivalently write

$$A(q)y_k = B(q)u_k + C(q)e_k,$$

where we have introduced the shift operator q which shifts the time index of a signal, i.e. $q y_k = y_{k+1}$. This is called an *autoregressive-moving-average model with exogenous terms* (ARMAX). The case when $C(q) = 1$ is called an *autoregressive model with exogenous terms* (ARX), the case when $A(q) = C(q)$ is called an *output error* (OE) model, and the case when $A(q) = C(q) = 1$ is called a *finite impulse response* (FIR) model.

12.3.1 Optimization Problem

We now consider the coefficients defining the Toeplitz matrices together with ξ_0 as the model parameters. Then the ML problem in (12.7) can equivalently be stated as

$$\begin{aligned} \text{minimize} \quad & \tfrac{1}{2}\|e\|_2^2 \\ \text{subject to} \quad & T_a y = \xi + T_b u + T_c e \end{aligned} \tag{12.9}$$

with variables (θ, e), where $\theta = (a, b, c, \xi_0)$, $a = (a_1, \ldots, a_n)$, $b = (b_0, \ldots, a_n)$, and $c = (c_1, \ldots, c_n)$. In this way, we have removed nonuniqueness related to the fact that there are infinitely many realizations of an input–output model. However, the optimization problem may still not have a unique solution, and this is related to the fact that the data (y_k, u_k) may not contain enough information to uniquely determine θ. Criteria that ensure uniqueness are in system identification called *persistence of excitation*, see e.g. [68].

It might be a good idea to try to remove the constraint in the optimization problem by eliminating e. For the case of an ARX model, this is trivial, and then the optimization problem is a linear LS problem. For the ARMAX model, we see that T_c is still invertible, *cf.* (2.26), and thus

$$e = T_c^{-1}\left(T_a y - \xi - T_b u\right).$$

Hence, the unconstrained regression problem can be written as

$$\text{minimize} \quad \frac{1}{2}\left\|T_c^{-1}\left(T_a y - \xi - T_b u\right)\right\|_2^2,$$

with variable θ. This is a nonlinear LS problem. Actually, it is a separable LS problem as in (6.59), i.e. if we fix c, then the problem is a linear LS in the remaining variables.

---

1 This is nothing but the inverse Z-transform.

12.3.2 Implementation Details

It holds that T_c^{-1} is also a Toeplitz matrix, see Exercise 2.6. This is useful when computing the Jacobian of e with respect to θ, which is needed in the optimization methods in Section 6.7. It holds that

$$\frac{\partial e}{\partial a_k} = T_c^{-1} S^k y, \quad k \in \mathbb{N}_n$$

$$\frac{\partial e}{\partial b_k} = -T_c^{-1} S^k u, \quad k \in \mathbb{Z}_n$$

$$\frac{\partial e}{\partial c_k} = -T_c^{-1} S^k e, \quad k \in \mathbb{N}_n$$

$$\frac{\partial e}{\partial \xi_0^T} = -T_c^{-1} \begin{bmatrix} I \\ 0 \end{bmatrix}$$

with the convention that $S^0 = I$. Moreover, any Toeplitz matrix commutes with any shift matrix, *cf*. Exercise 2.7. Hence, the gradients for different values of k can be obtained from the gradient for $k = 0$ using trivial shifts. We now realize that all the gradients are obtained by multiplying some vector with T_c^{-1} from the left or equivalently solving the linear system of equations:

$$T_c z = r,$$

for some right-hand side r that is either the residuals e or the given data (y, u). Since T_c is lower triangular, this system of equations can be solved recursively, and it can also be interpreted as a filtration, since it is equivalent to

$$z_k = \frac{1}{C(q)} r_k.$$

When there are zeros of the polynomial $C(z)$ such that $|z| > 1$, where $z \in \mathbb{C}$, then this filtration is not numerically well behaved, and it is also the case that T_c is an ill-conditioned matrix. It is a common practice to only look for solutions to the optimization problem such that the zeros of $C(z)$ are inside the unit circle. The reason for this is that the spectrum of $C(q)E_k$, when E_k is white noise, is the same as the spectrum of $\bar{C}(q)E_k$ if $C(z) = 0$ implies either $\bar{C}(z) = 0$ or $\bar{C}(1/z) = 0$, [6, pp 141–143]. Therefore, one should in a numerical implementation of an optimization algorithm for identification of an ARMAX model always mirror any iterate of c such that the corresponding zeros of C are inside the unit circle. It should also be mentioned that solving linear systems of equations involving Toeplitz matrices can be done efficiently using the discrete Fourier transform [44]. However, this only pays off in case n is much greater than $\log n$. It is in general tricky to initialize the optimization solver in such a way that it is not trapped in a local minimum. Typically, higher-order models and other system identification techniques are used initially to find an approximate model that can be used for initialization; see [68, Chapter 10.4] for details.

12.3.3 State Space Realization

When the model has been obtained for the input–output model, it is possible to go back and obtain a state description in terms of (A, B, C, K, x_0). This description is, of course, not unique. One possibility is to use the observable canonical form, which yields

$$x_{k+1} = \begin{bmatrix} 0 & 0 & \cdots & 0 & -a_n \\ 1 & 0 & \cdots & 0 & -a_{n-1} \\ \vdots & \vdots & & \vdots & \vdots \\ 0 & 0 & \cdots & 1 & -a_1 \end{bmatrix} x_k + \begin{bmatrix} b_n - a_n b_0 \\ b_{n-1} - a_{n-1} b_0 \\ \vdots \\ b_1 - a_1 b_0 \end{bmatrix} u_k + \begin{bmatrix} c_n - a_n \\ c_{n-1} - a_{n-1} \\ \vdots \\ c_1 - a_1 \end{bmatrix} e_k$$

$$y_k = \begin{bmatrix} 0 & 0 & \cdots & 0 & 1 \end{bmatrix} x_k + b_0 u_k + e_k.$$

It only remains to compute the initial value x_0, which must satisfy

$$Cx_0 = y_0 - Du_0 - e_0,$$

where the right-hand side is known from the above definitions. This equation is not enough, but we also have

$$CAx_0 = Cx_1 - CBu_1 - CKe_1 = y_1 - Du_1 - e_1 - CBu_1 - CKe_1,$$

with known right-hand side. Continuing like this, we can obtain the equation:

$$\mathcal{O}x_0 = r,$$

where

$$\mathcal{O} = \begin{bmatrix} C \\ CA \\ CA^2 \\ \vdots \\ CA^{n-1}, \end{bmatrix}$$

is the so-called *observability matrix*, and where r is some known quantity that can be computed recursively. Since we have used the observer canonical form, it is known that the observability matrix is invertible, see, e.g. [97], and hence, we can solve for x_0.

12.4 Missing Data

Sometimes not all the data (y_k, u_k), $k \in \mathbb{Z}_N$ are available for estimating the model. The natural idea then is to consider the missing data as additional variables to optimize over in (12.7). This will result in more bilinear terms in the constraints. Clearly, it will be a more challenging optimization problem. We will in this section only consider the input–output model formulation. State-space models will be considered in Section 12.5.

12.4.1 Block Coordinate Descent

We will restrict ourselves to the case of missing output data. This is the most common case, but it is easy to generalize the results to missing input data. The most natural formulation is as mentioned above to consider the optimization problem

$$\begin{aligned} &\text{minimize} \quad \tfrac{1}{2}\|e\|_2^2 \\ &\text{subject to} \quad T_a y = \xi + T_b u + T_c e \end{aligned} \tag{12.10}$$

with variables (θ, e, y_m), which is the same as (12.9) except that we also have the missing outputs y_m as optimization variables. Here $y_m = T_m y$, where $T = \begin{bmatrix} T_m & T_o \end{bmatrix} \in \mathbb{R}^{(N+1)\times(N+1)}$ is a permutation matrix, *cf.* (2.5). This can be written as an unconstrained problem:

$$\text{minimize} \quad \frac{1}{2}\left\| T_c^{-1}\left(T_a y - \xi - T_b u\right)\right\|_2^2 \tag{12.11}$$

with variables (θ, y_m). This is also a separable nonlinear LS problem, since the residual e is linear in the remaining variables when c is fixed. Unfortunately, the solution obtained will not be unbiased[2] unless $T_c^{-1} T_a = I$, which is true for OE models, for which $T_a = T_c$, and for FIR models, for which $T_a = T_c = I$. A common way of solving (12.11) is to consider a block-coordinate method,

---

2 An estimate $\hat{\theta}$ is said to be unbiased if $\mathbb{E}[\hat{\theta}] = \theta_0$, where θ_0 is the true parameter value.

cf. Section 6.9, where in every other step (y_m, ξ_0) and (a, b, c) are fixed, respectively. When (y_m, ξ_0) is fixed, we have a standard system identification problem with no missing data and known initial value. When (a, b, c) is fixed, we have a linear LS problem for (y_m, ξ_0).

12.4.2 Maximum Likelihood Problem

We will state the ML formulation for missing data model estimation, which is unbiased. To this end, we will as before assume that e is a realization of a normal distribution with zero mean, and covariance equal to $\sigma^2 I$. Let $T_y = T_c^{-1} T_a$, $T_u = T_c^{-1} T_b$ and $T_i = T_c^{-1} \begin{bmatrix} I_n \\ 0 \end{bmatrix}$. Then we can write

$$e = T_y y - T_i \xi_0 - T_u u.$$

Also, we define $\mathcal{T}_m = T_y T_m^T$. Then it can be shown, [52, 113] that the ML problem is equivalent to

$$\text{minimize} \quad \frac{1}{2} \left\| \det \left(\mathcal{T}_m^T \mathcal{T}_m \right)^{\frac{1}{2n_o}} \left(T_y y - T_i \xi_0 - T_u u \right) \right\|_2^2 \tag{12.12}$$

with variables (θ, y_m), where n_o is the number of observed outputs. The difference as compared to (12.11) is the factor $\det \left(\mathcal{T}_m^T \mathcal{T}_m \right)^{\frac{1}{2n_o}}$, which makes the optimization problem much more challenging. It is still a separable nonlinear LS problem. For the case when also inputs are missing, see [52, 113].

Example 12.1 In this example, we consider identification of an ARMAX model for which $a = 0.7$, $b = (0.7, 0)$, and $c = 0.5$. The details of the experimental conditions are given in Exercise 12.1. In total, 40% of the data is missing. In Figure 12.1, the results of 100 experiments are presented for the estimates of a and c when using both the criterion in (12.11) and the one in (12.12). It is seen that the first criterion results in biased estimates, i.e. estimates that are not centered around the true values. This is not the case for the second criterion, which is the ML criterion.

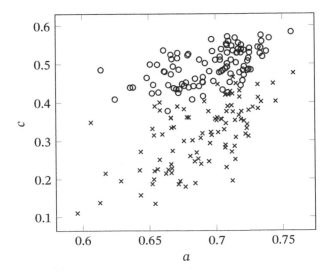

Figure 12.1 Plots showing the estimated values of a and c for 100 runs of system identification when data are missing. The crosses show the result when the criterion in (12.11) is used, and the circles shows the result when the criterion in (12.12) is used. The true values are $a = 0.7$ and $c = 0.5$.

12.5   Nuclear Norm system Identification

A common way to directly identify (12.5)–(12.6) is to use a so-called *subspace method*, see [68, 107]. The method is based on computing an SVD of a matrix built up from the collected data. More precisely, the method is based on the following Hankel matrix, see Section 2.5, of outputs

$$Y = \begin{bmatrix} y_0 & y_1 & \cdots & y_{M-1} \\ y_1 & y_2 & \cdots & y_M \\ \vdots & \vdots & \ddots & \vdots \\ y_{r-1} & y_r & \cdots & y_{M+r-2} \end{bmatrix}, \tag{12.13}$$

and a similarly defined input Hankel matrix U, where the entries are built up from values u_k. We choose M such that $N = M + r - 2$, where $N + 1$ is the number of data collected. The value of r has to be taken as greater than the state dimension n. For the case when $K = 0$ in (12.5), the following matrix is formed:

$$G = \frac{1}{N} Y \Pi^{\perp},$$

where

$$\Pi^{\perp} = I - U^T (UU^T)^{-1} U,$$

is a projection matrix onto the null-space of U. An SVD of G is computed to estimate the state dimension n, and then from a low-rank approximation of G, it is straightforward to compute the system matrices (A, B, C, D); see, e.g. [68]. Exercise12.3 provides some of the details. For the case when $K \neq 0$ so-called "instrument variables" can be used to modify G to avoid biased estimates, see [67, 68].

The number of singular values of G that are significantly larger than zero is the estimated dimension n of the state vector. It is often not easy to determine this value, since there is not a clear cut among the singular values when the data are corrupted with noise. One remedy to circumvent this problem is to consider *nuclear norm system identification* [67]. It turns out that this formulation seamlessly also can accommodate missing data in both inputs and outputs. It is based on the following optimization problem:

$$\text{minimize } \|G(\hat{y})\|_* + \lambda \|\hat{y} - y\|_2^2,$$

with variable $\hat{y}$, where $y \in \mathbb{R}^{pN}$ are the measured outputs stacked, $\lambda \in \mathbb{R}_{++}$ is a regularization parameter, and where $\hat{y} \in \mathbb{R}^{pN}$ is to be optimized such that it is close to y at the same time as the nuclear norm of $G : \mathbb{R}^{pN} \to \mathbb{R}^{pr \times N}$ is small, which is a heuristic for the rank of G, *cf.* Section 5.4. Here the most simple choice for G is $G(\hat{y}) = \hat{Y}\Pi^{\perp}$, where $\hat{Y}$ is defined similarly as Y in (12.13) above but using $\hat{y}$ instead of y.

The important observation is that G is linear in $\hat{y}$, and hence, the above optimization problem is a convex optimization problem, *cf.* (5.27). After the optimal $\hat{y}$ has been found, it is treated as the observed y in the subspace method, and hopefully, there is a more clear cut among the singular values such that it is easier to determine the state dimension n.

Another benefit with the nuclear norm formulation is that it is possible to address missing output data. The only modification that needs to be done is to consider

$$\text{minimize } \|G(\hat{y})\|_* + \lambda \|T_o (\hat{y} - y)\|_2^2,$$

with variable $\hat{y}$, where T_o as in Section 12.4 is a matrix that picks out the observed outputs. For the case of missing inputs and outputs, we refer to [67]. There also an efficient ADMM algorithm for solving the above optimization problem is presented.

12.6 Gaussian Processes for Identification

So far we have assumed that the system we would like to estimate a model for can be described within the model class we consider, i.e. that both the model and the true system are a linear system with state-dimension n. However, it can be shown that ARMAX systems can be arbitrarily well approximated with ARX models, see [69], if the model order n in the model is taken large enough. If both the number of data N and the model order n goes to infinity, and N faster than n, then the estimate is consistent. Since computing ARX models is equivalent to solving a linear LS problem, the solution can be obtained both efficiently and accurately. A drawback with the approach of using high model order is that even for a modest number of data, the variance of the estimate of the model parameters will be large. A remedy to this is to model the dynamical system as a Gaussian process, which by our previous discussions in Chapter 10 equivalently can be interpreted as regression in a Hilbert space or as a MAP estimate. For simplicity, we will in this section only consider the FIR case, but the extension to the ARX case is immediate.

12.6.1 MAP Estimate

We consider the regression model

$$y = \xi + T_b u + e,$$

which is a special case of the regression model derived in Section 12.3. It models a FIR system. Define

$$U = u_0 I + u_1 S + \cdots + u_N S^N,$$

where S is a lower triangular shift matrix. Now, we let U_{n+1} be the first $n + 1$ columns of U. Then it holds with $\theta = (b, \xi_0)$ that

$$y = \Phi\theta + e,$$

where

$$\Phi = \begin{bmatrix} U_{n+1} & J \end{bmatrix},$$

with $J = \begin{bmatrix} I & 0 \end{bmatrix}^T \in \mathbb{R}^{(N+1)\times n}$. This is clearly the same model as for the Gaussian process in Section 10.3 if we take $X = \Phi$ and $a = \theta$. We now assume that e is the outcome of a zero mean normally distributed random vector with covariance $\sigma^2 I$ and that θ is the outcome of a zero mean normally distributed random vector with covariance Σ and independent of e. It then follows that the estimate of θ is given by the conditional mean of θ given observations of y as

$$\hat{\theta} = \Sigma\Phi^T \left(\sigma^2 I + \Phi\Sigma\Phi^T\right)^{-1} y.$$

similarly as in Section 10.3. It can be shown that this is a consistent estimate if $\Sigma \in \mathbb{S}_{++}^N$, $\Phi\Phi^T/N$ converges to an invertible matrix and if $\Phi e/N$ converges to zero as N goes to infinity [26].

12.6.2 Empirical Bayes

There are many possible choices of $\Sigma \in \mathbb{S}_{++}^N$, and hence, it would be interesting to investigate if some are better than others. Assume that $\Sigma : \mathbb{R}^q \to \mathbb{S}_+^{2n+1}$, where $\Sigma(\eta)$ is describing a suitable subset of $\mathbb{S}_+^{2n+1}$ as the so-called *hyper parameter* η varies. Then a natural way to choose η is to maximize the likelihood function for the observations. For the above regression model, we have from Section 10.3

that y is the outcome of a zero mean normally distributed random vector with covariance $Z : \mathbb{R}^q \times \mathbb{R}_+ \to \mathbb{S}_{++}^{N+1}$ defined by

$$Z(\eta, \sigma^2) = \Phi\Sigma(\eta)\Phi^T + \sigma^2 I.$$

Therefore, the ML problem is equivalent to

$$\text{minimize } y^T Z(\eta, \sigma^2)^{-1} y + \ln \det Z(\eta, \sigma^2),$$

with variables (η, σ^2), where we also consider σ^2 as a hyper parameter. This approach is known as *Empirical Bayes* or *Type II Maximum Likelihood*. The advantage with this parameterization is that the ML criterion then is the difference between the two convex functions $g : \mathbb{R}^q \to \mathbb{R}$ and $h : \mathbb{R}^q \to \mathbb{R}$ defined by $g(\eta) = y^T Z(\eta, \sigma^2)^{-1} y$ and $h(\eta) = -\ln \det Z(\eta, \sigma^2)$, see exercises 4.5 and 4.7. Hence, the sequential convex optimization technique based on the majorization minimization principle applies, see Section 6.2.

How should then a suitable subset be chosen? Some insight can be obtained from the fact that for a FIR model, the parameter b is the impulse response of a linear dynamical system. One possible choice is to take $\Sigma_b : \mathbb{R}^3 \to \mathbb{S}_+^{n+1}$ defined by

$$\{\Sigma_b(\eta)\}_{j,k} = \lambda\alpha^{(k+j)/2}\rho^{|j-k|},$$

for $(j, k) \in \mathbb{N}_{n+1} \times \mathbb{N}_{n+1}$, where $\eta = (\lambda, \alpha, \rho)$ with $\lambda \in \mathbb{R}_+$, $0 \le \alpha < 1$, and $|\rho| \le 1$, see [26]. This is called a diagonal/correlated (DC) kernel.[3] We then let $\Sigma = \text{bdiag}(\Sigma_b, \Sigma_\xi)$, where $\Sigma_\xi \in \mathbb{S}_+^n$. Here α accounts for an exponential decay along the diagonal of Σ_b and ρ describes the correlation between neighboring impulse response coefficients. Exponential decay is expected for a stable linear system. There are several other choices for Σ_b discussed in the above reference. One interesting choice is to take $\Sigma(\eta) = \sum_{i=1}^q \eta_i\Sigma_i$, where $\Sigma_i \in \mathbb{S}_+^{2n+1}$ and $\eta \in \mathbb{R}_+^q$. This is motivated by the fact that for a linear dynamical system, the impulse response is the sum of the impulse responses of its partial fraction expansion. The different Σ_i could be obtained from fixed DC kernels modeling the different modes expected to be present in the model. For a survey on kernel methods in system identification, we refer the reader to [88].

Example 12.2 We consider a FIR model given by

$$y_k = \sum_{j=1}^{75} je^{-0.2j}u(k-j) + e_k.$$

We consider e_k to be realizations of i.i.d. Gaussian random variables with zero mean and unit variance. The input signal u_k is generated in a similar way. However, it is then low-pass filtered with a filter with cut-off frequency of 0.9. We generate data for $k \in \mathbb{N}_N$, where $N = 200$. We then estimate an FIR model of order $n = 50$ using Empirical Bayes with the DC kernel and another FIR model without using Empirical Bayes. After this, we generate new data denoted $(\bar{u}, \bar{y})$ in the same way as we generated the data (u, y) for estimation of the FIR model. We then generate $\hat{y}$ from

$$\hat{y}_k = \sum_{j=1}^n \hat{b}_j\bar{u}(k-j),$$

where $\hat{b}_j$ are the coefficients for the FIR models we estimated. This results in one $\hat{y}$ for the Empirical Bayes estimate and one for the estimate without using Empirical Bayes. We compare these $\hat{y}$ with one another and with $\bar{y}$ in Figure 12.2. We see that Empirical Bayes results in a much better model, since for that model $\bar{y}$ and $\hat{y}$ are much closer to one another.

---

3 Formally, the kernel is defined as $K_b : \mathbb{R}^{n+1} \times \mathbb{R}^{n+1} \to \mathbb{R}$, where $K_b = z^T\Sigma_b(\eta)\bar{z}$.

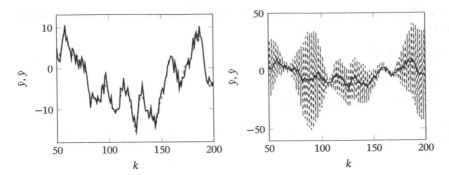

Figure 12.2 The left plot shows $\bar{y}_k$ (solid), and $\hat{y}$ (dashed), as function of k, for Empirical Bayes. The right plot shows $\bar{y}_k$ (solid), and $\hat{y}$ (dashed), as function of k, without Empirical Bayes.

12.7 Recurrent Neural Networks

We are now interested in obtaining a predictor for the nonlinear system described by (12.1–12.2). We consider the time-invariant case when there is no explicit dependence on k. There are many challenges. We know that for the linear and time-invariant case, there are infinitely many parameterizations, since the parameterization will depend on the choice of state. Also, we do not really know what the functions F and G should be unless we have some physical insight. Here we will resort to the fact that artificial neural networks ANNs can be used to approximate functions, as discussed in Section 10.7.

The idea in *recurrent neural networks* (RNNs) is to model the predictor as

$$x_{k+1} = F(x_k, u_k, \theta), \quad k \in \mathbb{Z}_{N-1}$$
$$\hat{y}_k = G(x_k, u_k, \theta), \quad k \in \mathbb{Z}_N,$$

where $F : \mathbb{R}^n \times \mathbb{R}^m \times \mathbb{R}^q \to \mathbb{R}^n$ and $G : \mathbb{R}^n \times \mathbb{R}^m \times \mathbb{R}^q \to \mathbb{R}^p$ are functions given by ANNs as described in Section 10.7. The vectors θ contains all the parameters that specify the affine propagation functions of the ANN. The word "recurrent" comes from the fact that the same ANN is used for each time step. It is not uncommon to only use a one-layer ANN to model F and G. Then it is possible to interpret the evolution of the state equation as a traditional ANN, for which the same weights are used in each layer. This interpretation is called *unrolling* of the RNN. There are also outputs for every layer. The fact that the same weight are used in every layer makes it impossible to use the multilayer back-propagation algorithm. However, it is of course still possible to compute the gradients and make use of the structure. A challenge with RNNs is that the resulting optimization problems might become ill-conditioned. In the learning community, this is called *vanishing gradient* or *exploding gradient*. A remedy to this is the so-called *long short-term memory* (LSTM) network, see [57]. For the application to system identification, see [70].

12.8 Temporal Convolutional Networks

Another way of obtaining a good predictor for a nonlinear time-invariant system is to generalize the linear ARX-model to a nonlinear ARX-model. This can be done by taking the predictor to be

$$\hat{y}_{k+1} = f(y_k, \dots, y_{k-n+1}, u_k, \dots, u_{k-n+1}; \theta), \quad n-1 \le k \le N-1,$$

where $f : \mathbb{R}^{2n} \times \mathbb{R}^q \to \mathbb{R}$ is a nonlinear function. For simplicity, we only consider the case of a scalar input signal and a scalar output signal. The results can be generalized to vector valued signals.

The vector θ contains all the parameters that define the predictor. The idea in *temporal convolutional networks* (TCNs) is to build up the function f using a tree structure, where each node in the tree is defined by a nonlinear ARX-model as well. The formal definition is as follows: let $x_k = (y_k, u_k)$ and let

$$\hat{y}_{k+1} = f^{(L)}\left(Z_k^{(L-1)}\right)$$

$$z_k^{(l)} = f^{(l)}\left(Z_k^{(l-1)}\right), \qquad l \in \mathbb{N}_{L-1}$$

$$z_k^{(0)} = x_k,$$

where

$$Z_k^{(l-1)} = \left(z_k^{(l-1)}, z_{k-d_l}^{(l-1)}, \ldots, z_{k-(\bar{n}-1)d_l}^{(l-1)}\right),$$

and where d_l is the so-called *dilation factor*. Typically, d_l increases exponentially with l, e.g. $d_l = 2^{l-1}$. We notice that $n - 1 = (\bar{n} - 1)d_L$ is the effective memory of the overall predictor. Each function $f^{(l)}$ can be defined in many different ways, and we refer to [4] for some examples. A very simple example would be to take it as a standard one-layer ANN. Because of the tree-structure of the TCN, it is possible to introduce further parallelism when computing gradients as compared to the back-propagation algorithm.

12.9 Experiment Design

We are interested in designing suitable input signals for system identification, and suitable in the sense that the covariance of the estimated model parameter is small in some sense. To fix the ideas we consider an FIR model given by

$$y_k = b_0 u_k + b_1 u_{k-1} + \ldots + b_n u_{k-n} + e_k, \quad k \in \mathbb{N}_N. \tag{12.14}$$

With the following definitions:

$$y = \begin{bmatrix} y_1 \\ \vdots \\ y_N \end{bmatrix}, \quad \varphi_k = \begin{bmatrix} u_k \\ \vdots \\ u_{k-n} \end{bmatrix}, \quad X = \begin{bmatrix} \varphi_1^T \\ \vdots \\ \varphi_N^T \end{bmatrix}, \quad e = \begin{bmatrix} e_1 \\ \vdots \\ e_N \end{bmatrix},$$

and $\theta = \begin{bmatrix} b_0 & \cdots & b_n \end{bmatrix}^T$ it holds that

$$y = X\theta + e.$$

We assume that e is the outcome of a normally distributed random variable E with zero mean and covariance $\sigma^2 I$, where $\sigma \in \mathbb{R}_{++}$ is the standard deviation. It is then straightforward[4] to show that the variance or covariance $\bar{P} \in \mathbb{S}_+^{n+1}$ for the LS estimate $\hat{\theta}$, which is the same as the ML estimate for this case, is given by

$$\bar{P} = \sigma^2 \left(X^T X\right)^{-1}.$$

From the above definition of X, we obtain the following SAA approximation

$$X^T X \approx N \begin{bmatrix} R_u(0) & \cdots & R_u(n) \\ \vdots & \ddots & \vdots \\ R_u(n) & \cdots & R_u(0) \end{bmatrix},$$

---

4 The estimate $\hat{\theta}$ satisfies the normal equations $X^T X \hat{\theta} = X^T Y$, where $Y = X\theta + E$ with θ being the true parameter value. Hence, $X^T X \mathbb{E}\hat{\theta} = X^T(X\theta + \mathbb{E}E) = X^T X\theta$. This shows that $\mathbb{E}\hat{\theta} = \theta$. Then the covariance is given by $\mathbb{E}\left[(\hat{\theta} - \mathbb{E}\hat{\theta})(\hat{\theta} - \mathbb{E}\hat{\theta})^T\right] = \mathbb{E}\left[(\hat{\theta} - \theta)(\hat{\theta} - \theta)^T\right] = \mathbb{E}\left[(X^T X)^{-1} X^T E E^T X(X^T X)^{-1}\right] = \sigma^2 (X^T X)^{-1}$.

where $R_u : \mathbb{Z} \to \mathbb{R}$ is the covariance function for the input signal of which u_k is a realization, see Sections 3.6 and 3.9, and (5.45). Our idea is now to find a good covariance function R_u for the input. We consider the vector $r \in \mathbb{R}^{n+1}$ defined as $r = \begin{bmatrix} R_u(0) & \cdots & R_u(n) \end{bmatrix}^T$ as a variable that we are going to find a good value for. We define the function $R : \mathbb{R}^{n+1} \to \mathbb{S}_+^{n+1}$ as

$$R(r) = N \begin{bmatrix} r_1 & \cdots & r_{n+1} \\ \vdots & \ddots & \vdots \\ r_{n+1} & \cdots & r_1 \end{bmatrix},$$

which is a linear function of r. We also define the function $P : \mathbb{R}^{n+1} \to \mathbb{S}_+^{n+1}$ as

$$P(r) = \sigma^2 R(r)^{-1}.$$

We have that $\bar{P} \approx P$ assuming that we are able to generate an input with the covariance function defined by r.

We now give different measures of what is a good covariance $P(r)$. The following quantities should be small:

A-optimality $\operatorname{tr} P(r)$
D-optimality $\ln \det P(r)$
E-optimality $\lambda_{\max}(P(r))$
L-optimality $\operatorname{tr}(WP(r))$, where $W \in \mathbb{S}_{++}^{n+1}$.

The reason for the name optimality above is that we would like to make the scalar functions of r above small. Then we will have a small covariance of the estimate $\hat{\theta}$. All optimality criteria can be related to the eigenvalues of $P(r)$, and the eigenvalues of a symmetric positive definite matrix are related to the length of the different principal axes of the confidence ellipsoids for the estimate that are given by $\{\theta \in \mathbb{R}^{n+1} | \ (\hat{\theta} - \theta)^T P(r)^{-1}(\hat{\theta} - \theta) \le \alpha\}$, where $\alpha \in \mathbb{R}_{++}$. Hence, A-optimality tries to make the sum of the lengths of principal axes as small as possible, and E-optimality tries to make the largest principal axis as small as possible. Regarding L-optimality this is a scaled version of A-optimality, and we will discuss why this is the case, and what it means later on. Since the volume of an ellipsoid is proportional to the product of the eigenvalues it follows that D-optimality tries to make the volume of the confidence ellipsoid as small as possible. Notice that we can make each of the criteria arbitrarily small by choosing r big enough, i.e. by using a high power in the signal u_k. In order to avoid this trivial and uninteresting solution, we will impose a bound on $r_1 \le L$, where $L \in \mathbb{R}_{++}$. Then the *signal-to-noise ratio* (SNR) will be L/σ^2.

12.9.1 The Positive Real Lemma

It is not the case that all r will result in covariance functions R_u that are possible to realize. We will here limit ourselves to covariance functions that can be obtained from the output of a *moving average* (MA) filter driven by white noise, i.e. let

$$U_k = V_k + c_1 V_{k-1} + \cdots + c_n V_{k-n}, \tag{12.15}$$

where V_k is independent of V_j for $j \ne k$, and where V_k has variance σ^2, $\sigma \in \mathbb{R}_+$, and $c_i \in \mathbb{R}$ for $1 \le i \le n$. The mean of V_k is zero. Then the covariance function $R_u : \mathbb{Z} \to \mathbb{R}$ is given by

$$R_u(k) = \mathbb{E}\big[U_l U_{l-k} \big] = \mathbb{E}\left[\sum_{i=0}^{n} c_i V_{l-i} \sum_{j=0}^{n} c_j V_{l-k-j} \right],$$

and we realize that $R_u(k) = 0$ for $k > n$. Here it holds that $c_0 = 1$. The Z-transform $\Phi_u : \mathbb{C} \to \mathbb{C}$ of the covariance function R_u is given by

$$\Phi_u(z) = \sum_{k=-n}^{n} R_u(k) z^{-k}.$$

A necessary and sufficient condition for R_u or equivalently r to be valid is that $\Phi_u(e^{i\omega}) \geq 0$ for all $\omega \in \mathbb{R}$. It is possible to characterize these r with a convex set. To this end, let

$$A = \begin{bmatrix} 0 & 0 \\ I_{n-1} & 0 \end{bmatrix}, \quad B = \begin{bmatrix} 1 \\ 0 \end{bmatrix}, \quad C = \begin{bmatrix} r_2 & \cdots & r_{n+1} \end{bmatrix}.$$

Define the function $\Psi : \mathbb{C} \to \mathbb{C}$ as

$$\Psi(z) = C(zI - A)^{-1}B + \frac{1}{2}r_1.$$

Since

$$(zI - A)\begin{bmatrix} z^{-1} \\ \vdots \\ z^{-n} \end{bmatrix} = B,$$

it follows that

$$\Psi(z) = \frac{1}{2}r_1 + r_2 z^{-1} + \cdots + r_{n+1}z^{-n} = \frac{1}{2}R_u(0) + R_u(1)z^{-1} + \cdots + R_u(n)z^{-n}.$$

We now have that

$$\Phi_u(z) = \Psi(z) + \Psi(1/z).$$

We will see that $\Phi_u(e^{i\omega}) = \Psi(e^{i\omega}) + \Psi(e^{-i\omega}) \geq 0$ for all $\omega \in \mathbb{R}$ is equivalent to the existence of a $Q \in \mathbb{S}^n$ such that the following constraint holds

$$K(Q,r) = \begin{bmatrix} Q - A^T Q A & C^T - A^T Q B \\ C - B^T Q A & r_1 - B^T Q B \end{bmatrix} \in \mathbb{S}_+^{n+1},$$

where $K : \mathbb{S}^n \times \mathbb{R}^{n+1} \to \mathbb{S}^{n+1}$. This equivalence is known as the *positive real lemma*. The set of Q and r that satisfies this constraint is convex since $\mathbb{S}_+^{n+1}$ is a convex cone, and the matrix K is affine in Q and r. We will show one direction of the proof of the positive real lemma. Assume that $K(Q,r) \in \mathbb{S}_+^{n+1}$. Notice that

$$K(Q,r) = \begin{bmatrix} Q & C^T \\ C & r_1 \end{bmatrix} - \begin{bmatrix} A & B \end{bmatrix}^T Q \begin{bmatrix} A & B \end{bmatrix},$$

and that

$$\begin{bmatrix} A & B \end{bmatrix}\begin{bmatrix} (zI - A)^{-1}B \\ 1 \end{bmatrix} = z(zI - A)^{-1}B.$$

From this, it is possible to show that

$$\begin{bmatrix} (z^{-1}I - A)^{-1}B \\ 1 \end{bmatrix}^T K(Q,r) \begin{bmatrix} (zI - A)^{-1}B \\ 1 \end{bmatrix} = \Phi_u(z),$$

from which the result follows that $\Phi_u(e^{i\omega}) \geq 0$ for all $\omega \in \mathbb{R}$. For a proof in the other direction see, e.g. [92].

12.9.2 D-Optimality

We will first look at D-optimality. It holds that $\ln \det P(r) = \ln \det R(r)^{-1} + 2(n + 1)\ln \sigma$ is a convex function of r. Hence, minimizing it is tractable. We get the following equivalent optimization problem, where we have removed the constant terms in the objective function:

$$\begin{aligned} \text{minimize} \quad & -\ln \det R(r) \\ \text{subject to} \quad & r_1 \leq L \\ & K(Q,r) \in \mathbb{S}_+^{n+1} \end{aligned}$$

with variables $r \in \mathbb{R}^{n+1}$ and $Q \in \mathbb{S}^n$. This is a conic optimization problem. We will see that the other optimization problems related to the other optimality criteria can also be cast as conic optimization problems.

12.9.3 E-Optimality

Minimizing a function $f : \mathbb{R}^n \to \mathbb{R}$ over some set $D \subset \mathbb{R}^n$ is equivalent to minimizing t over the epigraph of f. The minimal value of $f(x)$ will be equal to the minimal value of t. This is called the epigraph formulation of the optimization problem. Thus, minimizing $\lambda_{\max}(P(r))$ with respect to r is equivalent to minimizing t subject to $\lambda_{\max}(P(r)) \leq t$ or, equivalently, subject to $\lambda_{\min}(R(r)) \geq 1/t$. The smallest eigenvalue is greater than $1/t$ if and only if all eigenvalues are greater than $1/t$, and it can be shown that these inequalities hold if and only if $R(r) - 1/tI \in \mathbb{S}_+^{n+1}$. Also notice that minimizing t is the same as maximizing $1/t$. We now introduce a new variable $s = 1/t$. Then notice that maximizing s is the same as minimizing $-s$. Hence, the optimization problem for E-optimality can be stated as

$$\text{minimize} \quad -s$$
$$\text{subject to} \quad r_1 \leq L$$
$$R(r) - sI \in \mathbb{S}_+^{n+1}$$
$$K(Q,r) \in \mathbb{S}_+^{n+1}$$

with variables $r \in \mathbb{R}^{n+1}, s \in \mathbb{R}$ and $Q \in \mathbb{S}^n$, which is also a conic optimization problem.

12.9.4 A-Optimality

Similarly as the epigraph formulation, it holds that minimizing a sum of functions $\sum_{i=1}^m f_i(x)$, where $f_i : \mathbb{R}^n \to \mathbb{R}$ is equivalent to minimizing $\sum_{i=1}^m t_i$ subject to $f_i(x) \leq t_i$, where $t_i \in \mathbb{R}$ and where $i \in \mathbb{N}_m$, *cf.* Exercise 4.2. We now apply this to the A-optimality criterion and obtain the optimization problem:

$$\text{minimize} \quad \sum_{i=1}^{n+1} t_i$$
$$\text{subject to} \quad r_1 \leq L$$
$$\left(\sigma^2 R(r)^{-1}\right)_{ii} \leq t_i, \qquad i \in \mathbb{N}_{n+1}$$
$$K(Q,r) \in \mathbb{S}_+^{n+1}$$

with variables $r \in \mathbb{R}^{n+1}, t \in \mathbb{R}^{n+1}$ and $Q \in \mathbb{S}^n$. With $e_i \in \mathbb{R}^{n+1}, i \in \mathbb{N}_{n+1}$ being the standard basis vectors for $\mathbb{R}^{n+1}$, we can write the last constraints as

$$t_i - e_i^T \left(\sigma^2 R(r)^{-1}\right) e_i \geq 0, \qquad i \in \mathbb{N}_{n+1}.$$

By (2.58), this is equivalent to

$$\begin{bmatrix} t_i & e_i^T \\ e_i & \frac{1}{\sigma^2} R(r) \end{bmatrix} \in \mathbb{S}_+^{n+1}, \qquad i \in \mathbb{N}_{n+1},$$

and hence, it follows that the above optimization problem is also equivalent to a conic optimization problem.

12.9.5 L-Optimality

For L-optimality, we factorize W as $W = V^{-1}V^{-T}$ which can always be done with the help of, e.g. a Cholesky factorization, see (2.45). Now,

$$\text{tr}\,(WP(r)) = \text{tr}\left(V^{-T}\sigma^2 R(r)^{-1}V^{-1}\right) = \text{tr}\left(\sigma^2\left(VR(r)V^T\right)^{-1}\right),$$

and hence, we realize that we obtain the optimization problem for L-optimality by replacing $R(r)$ in the optimization problem for A-optimality with $VR(r)V^T$. Hence, this is just a weighted version of A-optimality with the weight matrix W.

We will discuss an application where L-optimality comes up naturally. It is not necessarily the case that we are interested in good values only of the parameters, but we might be more interested in obtaining a small Mean Square Error (MSE) when we use the model for prediction with new data. Hence, we define $\bar{X}$ similarly as X in the beginning of this section, but for the typical input signal for which we would like to have a good prediction. Then it can be shown[5] that the MSE for the predictor is

$$\sigma^2\text{tr}\left(\bar{X}(X^TX)^{-1}\bar{X}^T\right) + \sigma^2\text{tr}I,$$

and hence, the choice of $W = \bar{X}^T\bar{X}$ in L-optimal experiment design will result in minimizing the MSE for the predictor. We can interpret W/N as an estimate of the covariance $\mathbb{E}\left[\bar{U}\bar{U}^T\right]$, where $\bar{U}$ is the input for which we would like to have a small MSE.

12.9.6 Realization of Covariance Function

We now have several ways of finding optimal covariance functions R_u for the input signal u. Once it has been found we want to find an input signal u that is a realization of a stochastic process that has this covariance function. This is a so-called "realization problem." There are many ways to do this; see, e.g. the appendix of [118]. We will derive an MA-process that has the desired properties if its input is white noise $v(k) \in \mathbb{R}$. Remember that the Z-transform of the covariance function R_u is given by

$$\Phi_u(z) = \sum_{k=-n}^{n} R_u(k)z^{-k}.$$

Since R_u is an even function, we have that $\Phi_u(z) = \Phi_u(1/z)$. Hence, if z_i is a zero or pole of Φ_u so is $1/z_i$. Based on this the following *spectral factorization* follows, i.e. we may write

$$\Phi_u(z) = \kappa H(z)H(1/z),$$

where $\kappa \in \mathbb{R}$ is some constant, and where $H : \mathbb{C} \to \mathbb{C}$ is given by

$$H(z) = \frac{\Pi_{i=1}^n(z - z_i)}{\Pi_{i=1}^n(z - p_i)},$$

with $|z_i| \le 1$ and $|p_i| \le 1$. Notice that $z^n\Phi_u(z)$ is a polynomial, and hence,

$$z^n\Phi_u(z) = \kappa H(z)z^nH(1/z),$$

---

[5] Let the new data be generated from $\bar{Y} = \bar{X}\theta + \bar{E}$, where the random vector $\bar{E}$ has the same distribution as E. Also, let us estimate $\bar{Y}$ with $\hat{Y} = \bar{X}\hat{\theta}$, where $\hat{\theta}$ solves the normal equations $X^TX\hat{\theta} = X^TY$, where $Y = X\theta + E$. Then $\mathbb{E}\hat{Y} = \bar{X}\mathbb{E}\hat{\theta} = \bar{X}\theta$. Also $\hat{Y} - \mathbb{E}\hat{Y} = \bar{X}(\hat{\theta} - \mathbb{E}\hat{\theta}) = \bar{X}(X^TX)^{-1}X^TE$. Hence, $\text{tr}\,\mathbb{E}\left[(\bar{Y} - \hat{Y})(\bar{Y} - \hat{Y})^T\right] = \text{tr}\,\mathbb{E}\left[(\bar{Y} - \mathbb{E}\hat{Y} + \mathbb{E}\hat{Y} - \hat{Y})(\bar{Y} - \mathbb{E}\hat{Y} + \mathbb{E}\hat{Y} - \hat{Y})^T\right] = \sigma^2\text{tr}\left(\bar{X}(X^TX)^{-1}\bar{X}^T\right) + \sigma^2\text{tr}I.$

where the right-hand side will also be a polynomial. Let us now take $p_i = 0, i \in \mathbb{N}_n$. With $C : \mathbb{C} \to \mathbb{C}$ defined as the polynomial

$$C(z) = z^n + c_1 z^{n-1} + \cdots + c_n = \Pi_{i=1}^n (z - z_i),$$

we hence, have the following polynomial spectral factorization

$$z^n \Phi_u(z) = \kappa C(z) \tilde{C}(z),$$

where $\tilde{C} : \mathbb{C} \to \mathbb{C}$ is defined as $\tilde{C}(z) = z^n C(1/z)$. It follows that $\tilde{C}$ will have all its zeros outside the unit disk, since C has all its zeros inside the unit disk. Moreover, $H(z) = C(z)/z^n$. From this, it follows that the MA-process defined as

$$U_k = V_k + c_1 V_{k-1} + \cdots + c_n V_{k-n},$$

with V_k independent of V_j for $j \neq k$ with variance κ will have the covariance function R_u.

12.9.7 OE Model

It turns out that the solutions to the above optimization problems for optimal experiment design will all be trivial except for L-optimality. The optimal input will be white noise. However, for more general models than FIR models this is not the case. We will here consider a special case of an ARMAX model obtained by letting $T_c = T_a$ in (12.8), i.e.

$$T_a y = T_b u + T_a e.$$

Here we have assumed that $\xi = 0$. This model is an OE model. Since T_a is invertible we may write

$$y = T_a^{-1} T_b u + e.$$

We now consider T_a to be a function of $a = \begin{bmatrix} a_1 & \cdots & a_{n_a} \end{bmatrix}^T$ and T_b to be a function of $b = \begin{bmatrix} b_0 & \cdots & b_{n_b} \end{bmatrix}^T$ We also introduce the vector $\theta = \begin{bmatrix} a^T & b^T \end{bmatrix}$, and we define the function $T : \mathbb{R}^n \to \mathbb{R}^{N \times N}$ by $T(\theta) = T_a(a)^{-1} T_b(b)$, where $n = n_a + n_b + 1$. We realize that we have a nonlinear regression model

$$y = f(\theta) + e,$$

where $f : \mathbb{R}^n \to \mathbb{R}^N$ is given by $f(\theta) = T(\theta)u$. We now define

$$X = \frac{\partial f}{\partial \theta^T} \in \mathbb{R}^{N \times n}.$$

It can be shown that the covariance of the estimate of θ for ML estimation is given by $\sigma^2 (X^T X)^{-1}$ just as for the FIR model. We just have a different definition of X.[6] This is somewhat problematic, since the definition depends on the true parameter θ which we do not know. However, we will first make a preliminary estimate of θ without optimal inputs, and then we will use this estimate in the expression for X as if it was the true θ in order to figure out what is the optimal input to use. Then we will redo the estimation with this input to improve the estimate.

We partition X as $X = \begin{bmatrix} Y & Z \end{bmatrix}$, where

$$Y = \frac{\partial f}{\partial a^T} \in \mathbb{R}^{N \times n_a}, \qquad Z = \frac{\partial f}{\partial b^T} \in \mathbb{R}^{N \times (n_b + 1)}.$$

---

6 The inverse of this covariance matrix is known as the Fischer information matrix. Under mild conditions it holds that this covariance, as N goes to infinity, is the smallest possible covariance, after normalization with N, that can be obtained for any unbiased estimator, and this is called the Cramér-Rao lower bound.

We have that

$$\frac{\partial f}{\partial a_i} = -T_a^{-1} S^i T_a^{-1} T_b u, \quad 1 \le i \le n_a$$

$$\frac{\partial f}{\partial b_i} = T_a^{-1} S^i u, \quad 0 \le i \le n_b,$$

where $S^0 = I$. These expressions can be simplified since the shift matrix commutes with the Toeplitz matrices and their inverses. Actually, it can be shown that

$$Y = T_y U_y, \qquad Z = T_z U_z,$$

where $T_y = -T_a^{-1} T_a^{-1} T_b \in \mathbb{R}^{N \times N}$ and $T_z = T_a^{-1} \in \mathbb{R}^{N \times N}$ are lower triangular Toeplitz matrices, and where

$$U_y = \begin{bmatrix} u & Su & \cdots & S^{n_a - 1} u \end{bmatrix}, \qquad U_z = \begin{bmatrix} u & Su & \cdots & S^{n_b} u \end{bmatrix}.$$

We realize that U_y and Y_z are the first n_a and $n_b + 1$ columns of the lower triangular Toeplitz matrix

$$U = \begin{bmatrix} u & Su & \cdots & S^{N-1} u \end{bmatrix} \in \mathbb{R}^{N \times N}.$$

Since both T_y and T_z commute with this matrix, we have that

$$Y = U \tilde{T}_y, \qquad Z = U \tilde{T}_z,$$

where $\tilde{T}_y \in \mathbb{R}^{N \times n_a}$ and $\tilde{T}_z \in \mathbb{R}^{N \times (n_b+1)}$ are the first columns of T_y and T_z, respectively. Since we assume stability of the OE model, it follows that T_y and T_z are diagonally dominant, and hence, we may approximate Y and Z as

$$Y = \tilde{U} \tilde{T}_y, \qquad Z = \tilde{U} \tilde{T}_z,$$

where $\tilde{U}$ is a Toeplitz matrix with first column u and first row $\begin{bmatrix} u(0) & u(-1) & \cdots & u(-N+1) \end{bmatrix}$. Remember that $X = \begin{bmatrix} Y & Z \end{bmatrix}$, and hence,

$$X^T X = \begin{bmatrix} \tilde{T}_y & \tilde{T}_z \end{bmatrix}^T \tilde{U}^T \tilde{U} \begin{bmatrix} \tilde{T}_y & \tilde{T}_z \end{bmatrix},$$

where

$$\tilde{U}^T \tilde{U} \approx N \begin{bmatrix} R_u(0) & \cdots & R_u(N-1) \\ \vdots & \ddots & \vdots \\ R_u(N-1) & \cdots & R_u(0) \end{bmatrix}.$$

Thus, we can model this covariance matrix as a symmetric Toeplitz function $\bar{R} : \mathbb{R}^N \to \mathbb{S}^N$ defined as

$$\bar{R}(r) = N \begin{bmatrix} r_1 & \cdots & r_N \\ \vdots & \ddots & \vdots \\ r_N & \cdots & r_1 \end{bmatrix},$$

and hence, all entries of $X^T X$ will be linear in r. We now realize a difference as compared to the FIR-model case. There the vector r only had dimension $n + 1$, but here it has dimension N. We will limit ourselves to a lower dimension by considering inputs u which are such that $r_i = 0$ for $i > m$. Hence, the function $\bar{R}$ will be a banded matrix, and we will consider it to be a function of only $r \in \mathbb{R}^m$. We now define $R : \mathbb{R}^m \to \mathbb{S}^N$ as $R(r) = \begin{bmatrix} \tilde{T}_y & \tilde{Y}_u \end{bmatrix}^T \bar{R}(r) \begin{bmatrix} \tilde{T}_y & \tilde{Y}_u \end{bmatrix}$, which is a linear function of r. We finally define $P : \mathbb{R}^m \to \mathbb{S}_+^n$ as

$$P(r) = \sigma^2 R(r)^{-1},$$

similarly as for the FIR-model, and we can use the above optimization problem formulations to compute optimal experiments.

One might think that it is very costly to set up the optimization problems, since there are several matrix multiplications and inverses involved. However, it is the case that multiplication with both a banded Toeplitz matrix and the inverse of a banded Toeplitz matrix is equivalent to a filtering, and the multiplication with shift matrices is the same as delaying signals. Using sparse linear algebra is just as efficient as the filtering approach. We will now look at an example.

Example 12.3   We consider the OE model of this section with $n_a = n_b = n_k = 1$ where $a = a_1 = 0.95$, $b = b_0 = 0.1$ are the true parameters. The remaining parameters are $\sigma = 1$, $N = 300$, $L = 10$, $m = 10$. We will investigate D-optimality, and we are going to carry out 100 experiments for different realizations of e. We have generated e using `iddata` in MATLAB's Identification Toolbox, and then we have subtracted the mean and normalized such that the squared Euclidean norm of e is $L/\sigma = 10$. This data have been used to identify a preliminary model using `oe` in MATLAB's Identification Toolbox. Then this model has been used to carry out a D-optimal experiment. The output r of this optimization has been realized with an MA-model. We have then generated a new e using `iddata` and filtered it through the MA-filter to get an input for the second experiment. Also, for this input, the mean has been removed and a similar normalization as above has been carried out. The Bode-diagram of the transfer function of the MA-filter is shown in Figure 12.3. This is just one of all the 100 MA-filters that were computed, but they all look pretty much the same. We present the estimated values of a and b for the 100 experiments in the scatter-plots in Figure 12.4. We see that the optimal experiments result in a much smaller spread of the estimates, i.e. a smaller covariance, just as expected.

12.9.8   Experiment Design for Applications

As we have discussed before, it is not necessarily the case that we are interested in a small covariance of the parameter estimates. It all depends on what we are going to use the model for. In case

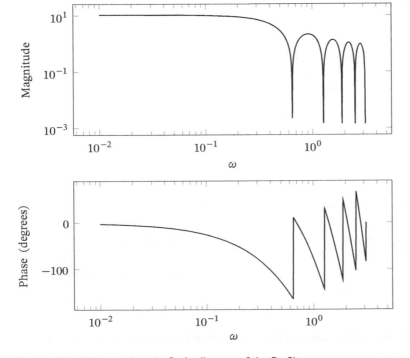

Figure 12.3   Plot showing the Bode-diagram of the Prefilter.

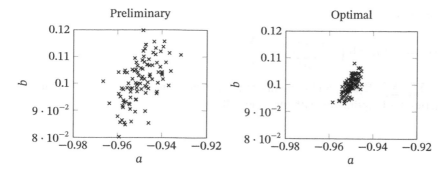

Figure 12.4 Scatter-plots showing the estimated values of a and b. The left figure shows the results from the preliminary nonoptimal experiment and the right figure shows the results from the optimal experiments.

one wants to use the model for prediction, then it is natural to try to make optimal experiments for minimizing the MSE of the predictor for the intended input signal. It turns out that also other application criteria related to the model can be interpreted as wanting to have a small MSE for a certain input signal. One example is if one wants to have a good model of the transfer function in a certain frequency band. Then this can be accomplished by using an input $\bar{u} \in \mathbb{R}^N$ that has most of its energy in this frequency band. We will model this with the covariance $\bar{\Sigma} = \mathbb{E}\left[\bar{U}\bar{U}^T\right]$, where we assume that $\mathbb{E}\bar{U} = 0$. Here $\bar{u}$ is the outcome of the random variable $\bar{U}$. We will as an example consider the OE-model in the previous section. The predictor is given by

$$\hat{Y}(\hat{\theta}) = T(\hat{\theta})\bar{U},$$

where $\hat{\theta}$ is the estimate of the model parameters θ. Let $\mu = \mathbb{E}\hat{\theta}$ and $\Sigma = \mathbb{E}\left[(\hat{\theta} - \mu)(\hat{\theta} - \mu)^T\right] = P(r)$, which are the mean and covariance of the ML estimate of the model parameters θ. We make a first-order Taylor series expansion of the ith column of $T(\hat{\theta})$:

$$T_i(\hat{\theta}) \approx T_i(\mu) + \frac{\partial T_i(\mu)}{\partial \theta^T}(\hat{\theta} - \mu), \qquad i \in \mathbb{N}_n.$$

We introduce the notation

$$L_i = \frac{\partial T_i(\mu)}{\partial \theta^T}, \qquad i \in \mathbb{N}_n,$$

for convenience. Since $\hat{\theta}$ and $\bar{U}$ are assumed to be independent it follows that

$$\mathbb{E}\hat{Y}(\hat{\theta}) \approx T(\mu)\mathbb{E}\bar{U} + \sum_{i=1}^{n} L_i \mathbb{E}\left[(\hat{\theta} - \mu)\bar{U}_i\right] = 0.$$

We also have

$$\operatorname{tr} \mathbb{E}\left[\hat{Y}(\hat{\theta})\hat{Y}(\hat{\theta})^T\right] \approx A + 2B + C,$$

where

$$A = \operatorname{tr}\left(T(\mu)\bar{\Sigma}T^T(\mu)\right)$$

$$B = \operatorname{tr} \mathbb{E}\left[\sum_{i=1}^{n} L_i(\hat{\theta} - \mu)\bar{U}_i\bar{U}^T T^T(\mu)\right] = 0$$

$$C = \operatorname{tr} \mathbb{E}\left[\sum_{i=1}^{n} L_i(\hat{\theta} - \mu)\bar{U}_i \sum_{j=1}^{n} \bar{U}_j(\hat{\theta} - \mu)^T L_j^T\right] = \operatorname{tr}\left(\Sigma \sum_{i=1}^{n}\sum_{j=1}^{n} \bar{\Sigma}_{ij} L_j^T L_i\right).$$

Hence, the MSE is given by

$$A + C = \text{tr}\left(T(\mu)\bar{\Sigma}T^T(\mu)\right) + \text{tr}\left(\Sigma\sum_{i=1}^{n}\sum_{j=1}^{n}\bar{\Sigma}_{ij}L_j^T L_i\right).$$

We now define $W = \sum_{i=1}^{n}\sum_{j=1}^{n}\bar{\Sigma}_{ij}L_j^T L_i$ and remember that $\Sigma = P(r)$, and we have shown that we have obtained a criterion for L-optimality with the weight W.

Exercises

12.1 Show that the optimization problems in (12.11) and (12.12) are the same when $T_a = T_c$.

12.2 We will investigate the variable projection method in Section 6.7 when applied to (12.12). We will take $x = y_m$ and $\alpha = (a, b, c)$. We will not consider ξ_0 as variable to estimate, i.e. we assume that it is zero, and we will generate data such that this holds true. We then write $F(x, \alpha) = \gamma\left(T_y y - T_u u\right)$, where $\gamma = \det\left(T_m^T T_m\right)^{\frac{1}{2n_0}}$. With $A(\alpha) = \gamma T_m$ and $b(\alpha) = \gamma(-T_o y_o + T_u u)$ it holds that $F(x, \alpha) = A(\alpha)x - b(\alpha)$.

(a) Show that P and $x(\alpha)$ as defined in Section 6.7 are given by

$$P = I - T_m T_m^\dagger$$
$$x(\alpha) = T_m^\dagger\left(-T_o y_o + T_u u\right),$$

where $T_m^\dagger = (T_m^T T_m)^{-1}T_m^T$.

(b) Show that

$$\frac{\partial A}{\partial \alpha_k}x - \frac{\partial b}{\partial \alpha_k} = \frac{\partial \gamma}{\partial \alpha_k}e + \gamma\frac{\partial e}{\partial \alpha_k},$$

where $e = T_y y - T_i \xi_0 - T_u u$. Notice that we do not consider $x = (y_m, \xi_0)$ to depend on α in this formula. However, when we use the formula we will despite this substitute with $x(\alpha)$ in order to evaluate e. This insight will simplify our code.

(c) Use the above results to implement the Kaufman Jacobian approximation for the variable projection method applied to the problem in (12.12). The function should return the nonlinear residual as its first output argument and the Jacobian approximation as its second output argument.
Hint: Expressions for the gradients $\partial e / \partial \alpha_k$ are given in Section 12.3. The partial derivative $\partial \gamma / \partial \alpha_k$ is derived in Exercise 2.21. Make sure to use a QR-factorization of T_m to make the code efficient and accurate, i.e. you should not implement the pseudo inverse of T_m in order to compute P and $x(\alpha)$, c.f. Exercise 2.22.

(d) Generate data for system identification experiments by taking $e \in \mathbb{R}^{N+1}$ to be a realization of zero mean independent normal distributed random variables with variance 0.5. Let $u \in \{-1, 1\}^{N+1}$ be generated by taking each component randomly from the set $\{-1, 1\}$ with equal probability for each value. Above let $N = 999$. Let $a = 0.7, b = (0.7, 0)$, and $c = 0.5$. This defines T_a, T_b, and T_c. From this, compute $y = T_a^{-1}(T_b u + T_c e)$. Define the matrix $\begin{bmatrix} T_o \\ T_m \end{bmatrix}$ by randomly shifting the rows of an identity matrix of dimension $N + 1$, and then let T_o contain the first 60% of the rows and T_m the remaining rows. This then defines $y_o = T_o y$ and $y_m = T_m y$. From the observed data (y_o, u), you then estimate $\alpha = (a, b, c)$ by running, e.g. a Levenberg–Marquardt algorithm with the Kaufman search direction and the residual ye. For this purpose, you can use lsqnonlin in MATLAB. The calling sequence is

```
OPTIONS = optimoptions('lsqnonlin','Algorithm',...
    'levenberg-marquardt','SpecifyObjectiveGradient',true);
x = lsqnonlin(@(x) missing_ML_kaufman(x,par),...
    zeros(3*n+1,1),[],[],OPTIONS);
```

You then need to write a function missing_ML_kaufman that as first output argument computes ye and as second output argument computes the Kaufman–Jacobian approximation. These computations should then be repeated with new data for in total 100 times. Make a two-dimensional scatter plot of the estimated values of a and c. You can then compare this with the results you obtain if you fix γ to one, i.e. solving (12.11). This means that the residual is e and that the Kaufman–Jacobian approximation is $P\frac{\partial e}{\partial a^T}$.

12.3 Consider a linear system

$$x_{k+1} = Ax_k + Bu_k$$
$$y_k = Cx_k + Du_k$$

for $k \in \mathbb{Z}_N$, where $x_k \in \mathbb{R}^n$, $u_k \in \mathbb{R}^m$, $y_k \in \mathbb{R}^p$, and where (A, B, C, D) are real-valued matrices of compatible dimensions. Consider Y defined as in (12.13) and also define U similarly from u_k.

(a) Show that

$$Y = \mathcal{O}X + \mathcal{T}U, \tag{12.16}$$

where

$$\mathcal{O} = \begin{bmatrix} C \\ AC \\ \vdots \\ A^{r-1}C \end{bmatrix}, \quad \mathcal{T} = \begin{bmatrix} D & 0 & 0 & \cdots & 0 \\ CB & D & 0 & \cdots & 0 \\ CAB & CB & D & & 0 \\ \vdots & & & \ddots & \ddots \\ CA^{r-2}B & CA^{r-3}B & \cdots & CB & D \end{bmatrix},$$

and where $X = \begin{bmatrix} x_0 & \cdots x_{M-1} \end{bmatrix}$.

(b) Assume that $\begin{bmatrix} X \\ U \end{bmatrix}$ has full row rank and that $\mathcal{O}$ has full column rank. Then use the results and definitions from Exercise 2.14 to conclude that $\mathcal{R}\mathcal{O} = \mathcal{R}L_{22}$. Let

$$L_{22} = \begin{bmatrix} U_1 & U_2 \end{bmatrix} \begin{bmatrix} \Sigma & 0 \\ 0 & 0 \end{bmatrix} \begin{bmatrix} V_1^T \\ V_2^T \end{bmatrix} = U_1 \Sigma V_1^T,$$

be an SVD. Show that there exists invertible state transformation $T \in \mathbb{R}^{n \times n}$, c.f. Exercise 2.12, such that $\bar{C} = CT = \bar{U}_1$, where $\bar{U}_1$ contains the first n rows of U_1, and that $\bar{A} = T^{-1}AT$ can be computed from U_1 by solving a linear system of equations.

(c) Show how $\bar{B} = T^{-1}B$ and D can be computed from (12.16) by solving a linear system of equations assuming that $\bar{C}$ and $\bar{A}$ are known.

12.4 Consider Example 12.2. You are asked to write a MATLAB code using the System Identification Toolbox that reproduces the result in the example. Play around with the value of the bandwidth. What happens if you use a bandwidth of one instead of 0.9 as is used in the example?

12.5 A sufficient and necessary condition for a vector

$$r = (r_1, \ldots, r_{n+1}),$$

to define a covariance function that can be described as the covariance of the output of the MA-filter (12.15) is the existence of a $Q \in \mathbb{S}_+^n$ such that the following constraint holds

$$K(Q, r) = \begin{bmatrix} Q - A^T Q A & C^T - A^T Q B \\ C - B^T Q A & r_1 - B^T Q B \end{bmatrix} \in \mathbb{S}_+^{n+1},$$

where

$$A = \begin{bmatrix} 0 & 0 \\ I_{n-1} & 0 \end{bmatrix}; \quad B = \begin{bmatrix} 1 \\ 0 \end{bmatrix}; \quad C = \begin{bmatrix} r_2 & \cdots & r_{n+1} \end{bmatrix}.$$

(a) Write a MATLAB function utilizing YALMIP that outputs the positive real lemma constraint with input variables being Q, and r. You should assume that the calling code has declared the input variables as Q = sdpvar(n) and r = sdpvar(n+1,1), respectively.

(b) Write a simple code to test your function by checking if $r = \begin{bmatrix} 3 & 21 \end{bmatrix}^T$ is valid. Also, try $r = \begin{bmatrix} 3 & 2 & -1 \end{bmatrix}^T$.

12.6 Write a MATLAB function `specfac` that given a vector $r \in \mathbb{R}^{n+1}$ containing values of a covariance function computes the vector $c \in \mathbb{R}^n$ containing the coefficients of the C-polynomial and the variance κ such that the MA-process defined in (12.15) has covariance as specific by r.

12.7 We consider estimating FIR models, i.e. models of the form

$$y(t) = b_0 u(t) + b_1 u(t-1) + \cdots + b_n u(t-n) + e(t).$$

(a) You should write a function that computes an L-optimal value of r given a positive definite weight matrix W, an upper limit L on r_1 and a standard deviation σ on the noise $e(t)$. Use the code from Exercise 12.5.a for the positive real lemma constraints.

(b) We now assume that the true system that we would like to estimate with an optimal input signal is a FIR model with $n = 3$. We assume that you would like to find an L-optimal input such that the MSE of the predictor is minimized for an input that is a sinusoidal given as $u(t) = 0.5 \sin 0.5t + 0.5 \sin 2t$. Write the code that computes the optimal covariance function r. Also, use the function `specfac` from Exercise 12.6 to compute the transfer function of the associated MA-filter and plot its Bode-diagram. What can you say about the input signal that is L-optimal?

Appendix A

A.1 Notation and Basic Definitions

A *set A* is a unordered collection of elements. It can contain finitely many elements or infinitely many elements. The number of elements in a set A is called its *cardinality*, and it is denoted by $|A|$. For two sets A and B, the *union* of A and B is $A \cup B = \{x \mid x \in A \text{ or } x \in B\}$, and the *intersection* of A and B is $A \cap B = \{x \mid x \in A \text{ and } x \in B\}$. The *set difference* between two sets A and B is $A \backslash B = \{x \in A \mid x \notin B\}$.

For two sets A and B, the *Cartesian product* of the two sets is $A \times B = \{(a, b) \mid a \in A, \ b \in B\}$. For any set A, we denote by $A^n = A \times \cdots \times A$ the set of n-dimensional *vectors* with entries in A.

We let $\bar{\mathbb{R}} = \mathbb{R} \cup \{-\infty, +\infty\}$ denote the set of *extended real* numbers. We introduce the notation

$$[a, b] = \{x \in \bar{\mathbb{R}} \mid a \leq x \leq b\},$$
$$(a, b) = \{x \in \bar{\mathbb{R}} \mid a < x < b\},$$
$$[a, b) = \{x \in \bar{\mathbb{R}} \mid a \leq x < b\},$$
$$(a, b] = \{x \in \bar{\mathbb{R}} \mid a < x \leq b\}.$$

For $A \subseteq \bar{\mathbb{R}}$, we define the *supremum* of A, which we denote by $\sup A$, as the smallest a such that $x \leq a$ for all $x \in A$. In case $a \in A$, we say that $a = \max A$, i.e. the *maximum* of A. If $a \notin A$, then A does not have a maximum element. For $A = [0, 1)$, we see that $\sup A = 1$, but A does not have a maximum element. Similarly, we define the *infimum* of A, denoted $\inf A$, as the largest a such that $x \geq a$ for all $x \in A$. We way that $a = \min A$ is the *minimum* element of A if $a \in A$, and if $a \notin A$, the set A does not have a minimum element.

Let f be a function $f : \mathcal{X} \to \mathcal{Y}$, where $\mathcal{X}$ is the *domain* of f and $\mathcal{Y}$ is its *codomain*. The function f assigns to each element x in the domain an element in the codomain denoted by $f(x)$. The set $\mathcal{Y}^{\mathcal{X}}$ is the set of all functions from $\mathcal{X}$ to $\mathcal{Y}$. When $\mathcal{X} = \mathbb{N}_n$, we frequently use $\mathcal{Y}^n$ to represent $\mathcal{Y}^{\mathbb{N}_n}$. Similarly, when $\mathcal{X} = \mathbb{Z}_+$, we frequently use $\mathcal{Y} \times \mathcal{Y} \times \cdots$ to represent $\mathcal{Y}^{\mathbb{Z}_+}$, i.e. we use infinite-dimensional vectors.

For a function $f : \mathcal{X} \to \mathcal{Y} \subseteq \bar{\mathbb{R}}$, we let the infimum of the function over a set $C \subseteq \mathcal{X}$ be defined as

$$\inf_{x \in C} f(x) = \inf \ \{f(x) \in \mathcal{Y} \mid x \in C\},$$

and similarly, we define the supremum of the function over a set $C \subseteq \mathcal{X}$ as

$$\sup_{x \in C} f(x) = \sup \ \{f(x) \in \mathcal{Y} \mid x \in C\}.$$

When the infimum is attained for some $x \in C$, we call it a minimum, and we have

$$\min_{x \in C} f(x) = \inf_{x \in C} f(x).$$

Optimization for Learning and Control, First Edition. Anders Hansson and Martin Andersen.
© 2023 John Wiley & Sons, Inc. Published 2023 by John Wiley & Sons, Inc.
Companion Website: www.wiley.com/go/opt4lc

We define the maximum of f analogously. Consider the function $\ln(x)$, which has domain $\mathbb{R}_{++}$. We have that the infimum over $\mathbb{R}_{++}$ is $-\infty$, which is not attained for an element of $\mathbb{R}_{++}$, and hence, the function does not have a minimum over its domain. However, it has a minimum over the set $[1, \infty)$. The infimum is attained for $x = 1$, and hence, the minimum value is equal to zero.

For a differentiable function $f : \mathcal{X} \to \mathbb{R}$, where $\mathcal{X}$ is an open subset of $\mathbb{R}$, we denote its *derivative* at $x \in \mathcal{X}$ with

$$\frac{df(x)}{dx}, \quad f'(x) \quad \text{or} \quad \dot{f}(x).$$

For a differentiable function $f : \mathcal{X} \to \mathbb{R}$, where $\mathcal{X}$ is an open subset of $\mathbb{R}^n$, we denote its *partial derivative* with respect to x_i at $x = (x_1, \ldots, x_n) \in \mathcal{X}$ with

$$\frac{\partial f(x)}{\partial x_i}.$$

For an integrable function $f : \mathcal{X} \to \mathbb{R}$, we denote the *definite integral* over $C \subseteq \mathcal{X} \subseteq \mathbb{R}^n$ with

$$\int_C f(x)dx.$$

When $C = C_1 \times \cdots \times C_n$, we may also write

$$\int_{C_1} \cdots \int_{C_n} f(x_1, \cdots, x_n)dx_1 \cdots dx_n.$$

When $C = [a, b] \subseteq \bar{\mathbb{R}}$, we often write

$$\int_a^b f(x)dx.$$

A.2 Software

Many of the problems in this book can be solved numerically with readily available software packages. The landscape of software for optimization, learning, and control is vast, so rather than attempting to provide a comprehensive list of software, we will provide only a brief overview of select software packages that are related to problems in this book. We will focus on two types of software: solvers and modeling tools. Generally speaking, solvers implement numerical methods for learning or optimization of some class of problems, and modeling tools allow the user to specify a wide range of problems in a solver-agnostic way using a high-level syntax.

A.2.1 Modeling Software

Modeling tools take a high-level description of an optimization problem as input and maps it to a format that is accepted by some solver through a number of transformations, e.g. epigraph reformulations, introduction of slack variables, etc. It then calls the solver and, if possible, recovers a solution to the problem specified by the user from the information returned by the solver. To illustrate the basic principle, suppose we would like to solve an instance of an support vector machine (SVM) training problem of the form

$$\begin{aligned}
\text{minimize} \quad & 1^T u + \gamma \|w\|_2, \\
\text{subject to} \quad & \text{diag}(y)(Cw + b) \geq 1 - u, \\
& u \geq 0,
\end{aligned}$$

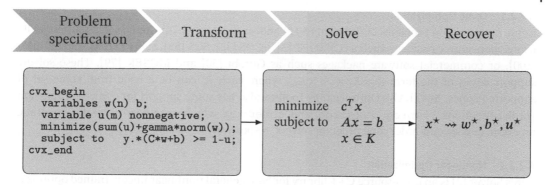

Figure A.1 Modeling tools allow the users to specify their problems using a high-level syntax. The problem is then transformed to a format that is accepted as some solver and, if possible, the information returned from the solver is used to recover a solution to the problem provided by the user.

with variables $w \in \mathbb{R}^n$, $b \in \mathbb{R}$, and $u \in \mathbb{R}^m$, problem data $C \in \mathbb{R}^{m \times n}$ and $b \in \{-1, 1\}^m$, and regularization parameter $\gamma > 0$. This problem is equivalent to an second-order cone program (SOCP), and it can be solved using one of many different software packages for conic optimization. However, the task of reformulating the problem so that it can be accepted by a given solver is often tedious and error-prone. Moreover, it is often necessary to start from scratch if we wish to use a another solver that requires the problem to be specified differently. Modeling tools automate this process, as illustrated in Figure A.1. The problem specification in the figure is based on the CVX modeling package for MATLAB, and it is clearly a self-explanatory and high-level representation of the mathematical description of the problem. Modeling tools also make it very easy to experiment with different problem formulations, but the user often has little or no control over transformations, which can make it difficult to exploit problem structure.

A.2.1.1 Disciplined Convex Programming

The CVX [47] modeling package for MATLAB has pioneered what is referred to as *disciplined convex programming*. It requires that user inputs a problem in a form that allows the software to verify convexity via a number of known composition rules. The problem is then reformulated as a conic optimization problem and passed to one of several possible solvers. The software packages CVXPY [31], Convex.jl [105], and CVXR [40] make similar modeling functionality available in the programming languages Python, Julia, and R, respectively.

A.2.1.2 YALMIP

The MATLAB toolbox YALMIP [71] is a modeling package that can be used with a wide range of solvers. YALMIP is an acronym for "Yet Another Linear Matrix Inequality Parser" and was initially developed for applications in control but has since grown into a full-fledged general-purpose modeling package with support for many classes of both convex and nonconvex problems.

A.2.1.3 JuMP

The Julia modeling package JuMP [36] is highly flexible modeling language for mathematical optimization, and it supports a large number of solvers through a generic solver-independent interface.

A.2.2 Optimization Software

We now provide an overview of optimization software for three broad classes of problems, namely conic optimization, nonlinear optimization, and stochastic optimization.

A.2.2.1 Conic Optimization

Conic optimization problems of the form (5.12) and/or the dual (5.16) can be solved numerically using open-source packages such as CVXOPT [3], ECOS [33], SCS [83], SDPT3 [104], and SeDuMi [100], or commercial software packages such as Gurobi [50] and MOSEK [79]. These solvers support cones of the form $K = K_1 \times \cdots \times K_m$, where each K_i can be a cone from some set of supported cones. MATLAB's Optimization Toolbox also has some support for conic optimization via the functions `linprog`, `quadprog`, and `coneprog`, which are linear program, quadratic program, and second-order cone program solvers, respectively.

A.2.2.2 Nonlinear Optimization

Ceres Solver [2] is an open-source C++ library for both general nonlinear unconstrained optimization problems as well as nonlinear least-squares problems with or without variable bounds. The solvers IPOPT [111] and SNOPT [42] are designed for smooth nonlinear optimization problems with constraints; IPOPT implements an interior-point method, whereas SNOPT is based on sequential quadratic programming. Examples of software packages for mixed-integer linear and nonlinear programming are BARON [101], CPLEX [59], Gurobi [50], and SCIP [108].

A.2.2.3 Stochastic Optimization

Gradient-based methods and stochastic methods are often a natural choice for learning problems that involve large quantities of training data. While these methods are typically conceptually simple and easy to implement, the computation of gradients or stochastic gradients is often much more complicated. Software packages such as TensorFlow [102] and PyTorch [87] address this issue by means of automatic differentiation, and both packages provide implementations of a wide range of stochastic optimization methods, which can be used for artificial neural network training and many other learning problems. MATLAB's Deep Learning Toolbox also includes a number of stochastic optimization methods.

A.2.3 Software for Control

We end with a brief overview of select software for optimal control, reinforcement learning, and system identification.

A.2.3.1 Optimal Control

Dynaprog is a MATLAB toolbox for solving finite horizon multistage deterministic decision problems using dynamic programming. The ACADO Toolkit is a software environment and algorithm collection for automatic control and dynamic optimization. It provides a general framework for using a variety of algorithms for direct optimal control, including model predictive control, state and parameter estimation, and robust optimization. The ACADO Toolkit is implemented as a self-contained C++ code and comes with a MATLAB interface. The object-oriented design allows for coupling of existing optimization packages and for extending it with user-written optimization routines. CasADi is an open-source tool for nonlinear optimization and algorithmic differentiation. It facilitates rapid and efficient implementation of different methods for numerical optimal control, both in an offline context and for online nonlinear model predictive control.

A.2.3.2 Reinforcement Learning

Gymnasium is an open source Python library for developing and comparing reinforcement learning algorithms by providing a standard application programming interface communication between learning algorithms and environments. The TensorFlow community has developed an extended version called TensorLayer, which provides some popular reinforcement learning modules that can easily be customized and assembled. DeepMind Lab is a Google 3D platform with customization for agent-based AI research. There is also the Reinforcement Learning Toolbox of MATLAB.

A.2.3.3   System Identification

One of the most popular software packages for system identification is arguably MATLAB's System Identification Toolbox. This toolbox has an interface to MATLAB's Deep Learning Toolbox, and hence, it is possible to seamlessly identify dynamical systems with models based on not only standard nonlinear regressors but also using ANNs.

References

1 P.-A. Absil, R. Mahoni, and R. Sepulchre. *Optimization Algorithms on Matrix Manifolds.* Springer, 2006.

2 S. Agarwal, K. Mierle, and The Ceres Solver Team. Ceres Solver, 3 2022. URL https://github .com/ceres-solver/ceres-solver.

3 M. S. Andersen, J. Dahl, and L. Vandenberghe. CVXOPT: A Python package for convex optimization, 2022. URL https://cvxopt.org.

4 C. Andersson, A. H. Ribeiro, K. Tiels, N. Wahlström, and T. B. Schön. Deep convolutional networks in system identification. In 2019 *IEEE 58th Conference on Decision and Control (CDC)*, 2019.

5 J. A. E. Andersson, J. Gillis, G. Horn, J. B. Rawlings, and M. Diehl. CasADi – A software framework for nonlinear optimization and optimal control. *Mathematical Programming Computation*, 11(1):1–36, 2019. doi: 10.1007/s12532-018-0139-4.

6 K. J. Åström and B. Wittenmark. *Computer Controlled Systems—Theory and Design.* Prentice-Hall, Inc., Englewood Cliffs, NJ, 1984.

7 K. J. Åström and B. Wittenmark. *Adaptive Control.* Addison-Wesley Longman Publishing Co., Inc., Reading, MA, 2nd edition, 1994. ISBN 0201558661.

8 P. Baldi and P. Sadowski. The dropout learning algorithm. *Artificial Intelligence*, 210:78–122, 2014. ISSN 0004-3702. doi: https://doi.org/10.1016/j.artint.2014.02.004. URL https://www .sciencedirect.com/science/article/pii/S0004370214000216.

9 A. G. Baydin, B. A. Pearlmutter, A. A. Radul, and J. M. Siskind. Automatic differentiation in machine learning: A survey. *Journal of Machine Learning Research*, 18(153):1–43, 2018. URL http://jmlr.org/papers/v18/17-468.html.

10 A. Beck. *First Order Methods in Optimization.* SIAM, Philadelphia, PA, 2017.

11 A. Beck and M. Teboulle. A fast iterative shrinkage-thresholding algorithm for linear inverse problems. *SIAM Journal on Imaging Sciences*, 2(1):183–202, 2009.

12 A. Beck and L. Tetruashvili. On the convergence of block coordinate descent type methods. *SIAM Journal on Optimization*, 23(4):2037–2060, January 2013. doi: 10.1137/120887679. URL https://doi.org/10.1137/120887679.

13 M. Benzi, G. H. Golub, and J. Liesen. Numerical solution of saddle point problems. *Acta Numerica*, 14:1–137, April 2005. doi: 10.1017/s0962492904000212. URL https://doi.org/10.1017/ s0962492904000212.

14 D. P. Bertsekas. *Dynamic Programming and Optimal Control—Volume I*. Athena Scientific, Belmont, MA, 1995.

Optimization for Learning and Control, First Edition. Anders Hansson and Martin Andersen.
© 2023 John Wiley & Sons, Inc. Published 2023 by John Wiley & Sons, Inc.
Companion Website: www.wiley.com/go/opt4lc

15 D. P. Bertsekas. *Constrained Optimization and Lagrange Multiplier Methods*. Athena Scientific, Belmont, MA, 1996.

16 D. P. Bertsekas. Incremental proximal methods for large scale convex optimization. *Mathematical Programming*, 129(2):163–195, June 2011. doi: 10.1007/s10107-011-0472-0.

17 D. P. Bertsekas. *Reinforcement Learning and Optimal Control*. Athena Scientific, Belmont, MA, 2019.

18 Ch. M. Bishop. *Pattern Recognition and Machine Learning*. Springer, 2006.

19 R. R. Bitmead, M. R. Gevers, I. R. Petersen, and R.J. Kaye. Monotonicity and stabilizability-properties of solutions of the Riccati difference equation: Propositions, lemmas, theorems, fallacious conjectures and counterexamples. *Systems & Control Letters*, 5(5):309–315, 1985. ISSN 0167-6911. doi: https://doi.org/10.1016/0167-6911(85)90027-1. URL https://www.sciencedirect.com/science/article/pii/0167691185900271.

20 S. Bittanti, A. J. Laub, and J. C. Willems. *The Riccati Equation*. Springer, 1991.

21 Å. Björk. *Numerical Methods for Least Squares Problems*. SIAM, 1996.

22 S. Boyd and L. Vandenberghe. *Convex Optimization*. Cambridge University Press, 2004.

23 D. A. Bristow, M. Tharayil, and A. G. Alleyne. A survey of iterative learning. *IEEE Control Systems*, 26:96–114, 07 2006. doi: 10.1109/MCS.2006.1636313.

24 A. E. Bryson and Y. C. Ho. *Applied Optimal Control*. Blaisdell Waltham, 1969.

25 J. R. Bunch and B. N. Parlett. Direct methods for solving symmetric indefinite systems of linear equations. *SIAM Journal on Numerical Analysis*, 8(4):639–655, December 1971. doi: 10.1137/0708060. URL https://doi.org/10.1137/0708060.

26 T. Chen, M. S. Andersen, L. Ljung, A. Chiuso, and G. Pillonetto. System identification via sparse multiple Kernel-based regularization using sequential convex optimization. *IEEE Transactions on Automatic Control*, 59(11):2933–2945, 2014.

27 A. R. Conn, N. I. M. Gould, and P. L. Toint. *Trust Region Methods*. 01 2000. ISBN 9780898719857.

28 A. C. C. Coolen. *Information Theory in Neural Networks—Lecture Notes of Course G31/CMNN14*. Department of Mathematics, King's College, 2002.

29 K. Crammer and Y. Singer. On the algorithmic implementation of multiclass kernel-based vector machines. *Journal of Machine Learning Research*, 2:265–292, 2001.

30 G. Cybenko. Approximation by superpositions of a sigmoidal function. *Mathematics of Control, Signals, and Systems*, 2:303–314, 1989.

31 S. Diamond and S. Boyd. CVXPY: A Python-embedded modeling language for convex optimization. *Journal of Machine Learning Research*, 17(83):1–5, 2016.

32 M. Diehl. *Real-Time Optimization for Large Scale Nonlinear Processes*. Ph.D. Thesis, Ruprächt-Karl-Universität, Heidelberg, June 2001.

33 A. Domahidi, E. Chu, and S. Boyd. ECOS: An SOCP solver for embedded systems. In *2013 European Control Conference*, pages 3071–3076, 2013. doi: 10.23919/ECC.2013.6669541.

34 P. Dommel and A. Pichler. Foundations of multistage stochastic programming, 2021.

35 J. Duchi, E. Hazan, and Y. Singer. Adaptive subgradient methods for online learning and stochastic optimization. *Journal of Machine Learning Research*, 12:2121–2159, 2011.

36 I. Dunning, J. Huchette, and M. Lubin. JuMP: A modeling language for mathematical optimization. *SIAM Review*, 59(2):295–320, 2017. doi: 10.1137/15M1020575.

37 M. Fazel. *Matrix Rank Minimization with Applications*. Ph.D. Thesis, Stanford University, February 2002.

38 A. Fischer and Ch. Igel. An introduction to restricted Boltzmann machines. In L. Alvarez, M. Mejail, L. Gomez, and J. Jacobo, editors, *Progress in Pattern Recognition, Image Analysis, Computer Vision, and Applications*, pages 14–36. Springer-Verlag, Berlin, Heidelberg, 2012.

39 R. A. Fisher. The use of multiple measurements in taxonomic problems. *Annals of Eugenics*, 7(2):179–188, 1936. doi: https://doi.org/10.1111/j.1469-1809.1936.tb02137.x. URL https://onlinelibrary.wiley.com/doi/abs/10.1111/j.1469-1809.1936.tb02137.x.

40 A. Fu, B. Narasimhan, and S. Boyd. CVXR: An R package for disciplined convex optimization. *Journal of Statistical Software*, 94(14), 2020. doi: 10.18637/jss.v094.i14. URL https://doi.org/10.18637/jss.v094.i14.

41 I. M. Gelfand and S.V. Fomin. *Calculus of Variations*. Prentice-Hall, Englewood Cliffs, NJ, 1963.

42 P. E. Gill, W. Murray, and M. A. Saunders. SNOPT: An SQP algorithm for large-scale constrained optimization. *SIAM Review*, 47(1):99–131, January 2005. doi: 10.1137/s0036144504446096. URL https://doi.org/10.1137/s0036144504446096.

43 G. Golub and V. Pereyra. Separable nonlinear least squares: The variable projection method and its applications. *Inverse Problems*, 19:R1–R26, 2003.

44 G. H. Golub and Ch. F. van Loan. *Matrix Computations*. The Johns Hopkins University Press, Baltimore, MD, 1996. 3rd edition.

45 I. Goodfellow, Y. Bengio, and A. Courville. *Deep Learning*. The MIT Press, Cambridge, MA, London, England, 2016.

46 R. M. Gower, M. Schmidt, F. Bach, and P. Richtarik. Variance-reduced methods for machine learning. *Proceedings of the IEEE*, 108(11):1968–1983, November 2020. doi: 10.1109/jproc.2020.3028013. URL https://doi.org/10.1109/jproc.2020.3028013.

47 M. Grant and S. Boyd. CVX: Matlab software for disciplined convex programming, version 2.1, March 2014. URL http://cvxr.com/cvx.

48 G. R. Grimmett and D. R. Stirzaker. *Probability and Random Processes*. Oxford University Press, New York, 1982.

49 S. Gunnarsson and M. Norrlöf. A short introduction to iterative learning control. Technical Report 1926, Linkoping University, Automatic Control, 1997.

50 Gurobi Optimization, LLC. Gurobi Optimizer Reference Manual, 2022. URL https://www.gurobi.com.

51 A. Hansson and P. Hagander. How to decompose semi-definite discrete-time algebraic Riccati equations. *European Journal of Control*, 5(2):245–258, 1999. ISSN 0947-3580. doi: https://doi.org/10.1016/S0947-3580(99)70159-7. URL https://www.sciencedirect.com/science/article/pii/S0947358099701597.

52 A. Hansson and R. Wallin. Maximum likelihood estimation of Gaussian models with missing data—Eight equivalent formulations. *Automatica*, 48:1955–1962, 2012.

53 C. R. Hargraves and S. W. Paris. Direct trajectory optimization using nonlinear programming and collocation. *Journal of Guidance Control and Dynamics*, 10(4):338–342, 1987.

54 T. Hastie, R. Tibshirani, and J. Friedman. *The Elements of Statistical Learning—Data Mining, Inference, and Prediction*. Springer, 2009.

55 J.-B. Hiriart-Urruty and C. Lemaréchal. *Fundamentals of Convex Analysis*. Springer, December 2001.

56 H. Hjalmarsson, M. Gevers, S. Gunnarsson, and O. Lequin. Iterative feedback tuning: Theory and applications. *IEEE Control Systems Magazine*, 18(4):26–41, 1998. doi: 10.1109/37.710876.

57 S. Hochtreiter and J. Schmidhuber. Long short-term memory. *Neural Computation*, 9(8):1735–1780, 1997.

58 B. Houska, H.J. Ferreau, and M. Diehl. ACADO toolkit – An open source framework for automatic control and dynamic optimization. *Optimal Control Applications and Methods*, 32(3):298–312, 2011.

59 IBM ILOG CPLEX. User's manual for CPLEX 20.1, 2021.

60 G. Karypis and V. Kumar. A fast and high quality multilevel scheme for partitioning irregular graphs. *SIAM Journal on Scientific Computing*, 20(1):359–392, 1999.

61 L. Kaufman. A variable projection method for solving separable nonlinear least squares problems. *BIT*, 15:49–75, 1975.

62 A. I. Khinchin. *Mathematical Foundations of Information Theory*. Dover Publications, 1957.

63 D. P. Kingma and J. Ba. Adam: A method for stochastic optimization, 2014.

64 A. N. Kolmogorov. On the representation of continuous functions of many variables by superposition of continuous functions of one variable and addition. *Doklady Akademii Nauk SSSR*, 114:953–956, 1957.

65 G. Lanckriet and B. K. Sriperumbudur. On the convergence of the concave–convex procedure. *Advances in Neural Information Processing Systems, 22*, 2009.

66 J. D. Lee, Y. Sun, and M. A. Saunders. Proximal Newton-type methods for minimizing composite functions. *SIAM Journal on Optimization*, 24(3):1420–1443, January 2014. doi: 10.1137/130921428.

67 Z. Liu, A. Hansson, and L. Vandenberghe. Nuclear norm system identification with missing inputs and outputs. *System & Control Letters*, 62(8):605–612, 2013.

68 L. Ljung. *System Identification—Theory for the User*. Prentice Hall, 1999.

69 L. Ljung and B. Wahlberg. Asymptotic properties of the least-squares method for estimating transfer functions and distrurbance spectra. *Advances in Applied Probability*, 24:412–440, 1992.

70 L. Ljung, C. Andersson, K. Tiels, and T. Schön. Deep learning and system identification. In *Proceedings of the 21st IFAC World Congress*, Berlin, pages 1175–1181, 2020.

71 J. Löfberg. YALMIP: A toolbox for modeling and optimization in MATLAB. In *Proceedings of the CACSD Conference*, Taipei, Taiwan, 2004.

72 G. G. Lorentz. The 13th problem of Hilbert. In F. Browder, editor, *Mathematical Developments Arising from Hilbert's Problems*, pages 419–430. American Mathematics Society, Providence, RI, 1976.

73 D. G. Luenberger. *Optimization by Vector Space Methods*. John Wiley & Sons, Inc., New York, 1969.

74 D. G. Luenberger. *Linear and Nonlinear Programming*. Addison-Wesley Publishing Company, 1984.

75 B. Ma. *An Improved Algorithm for Solving Constrained Optimal Control Problems*. Ph.D. thesis, Institute for Systems Research, University of Maryland, 1994.

76 L. Mirsky. A trace inequality of John von Neumann. *Monatshefte für Mathematik*, 79(4): 303–306, December 1975. doi: 10.1007/bf01647331. URL https://doi.org/10.1007/bf01647331.

77 K. L. Moore. *Iterative Learning Control for Deterministic Systems*. Industrial Control. Springer-Verlag, Berlin, 1993.

78 J. J. Moreé. The Levenberg-Marquardt algorithm: Implementation and theory. In G. A. Watson, editor, *Numerical Analysis. Lecture Notes in Mathematics*, vol. 630. Springer-Verlag, Berlin, Heidelberg, 1978.

79 MOSEK ApS. The MOSEK optimization suite. Version 10.0, 2022. URL http://mosek.com.

80 Y. E. Nesterov. A method for solving the convex programming problem with convergence rate $o(1/k^2)$. *Doklady Akademii Nauk SSSR*, 269:543–547, 1983.

81 Y. Nesterov. *Lectures on Convex Optimization*. Springer Optimization and its Applications. Springer International Publishing, 2nd edition, 2018.

82 Y. Nesterov and B. T. Polyak. Cubic regularization of Newton method and its global performance. *Mathematical Programming*, 108(1):177–205, April 2006. doi: 10.1007/s10107-006-0706-8. URL https://doi.org/10.1007/s10107-006-0706-8.

83 B. O'Donoghue, E. Chu, N. Parikh, and S. Boyd. Conic optimization via operator splitting and homogeneous self-dual embedding. *Journal of Optimization Theory and Applications*, 169(3): 1042–1068, June 2016. URL http://stanford.edu/boyd/papers/scs.html.

84 J. Omura. On the Viterbi decoding algorithm. *IEEE Transactions on Information Theory*, 15(1):177–179, 1969. doi: 10.1109/TIT.1969.1054239.

85 N. Parikh and S. Boyd. Proximal algorithms. *Foundations and Trends in Optimization*, 1(3):123–231, 2013.

86 T. A. Parks. *Reducible Nonlinear Programming Problems*. Ph.D. thesis, Rice University, 1985.

87 A. Paszke, S. Gross, F. Massa, A. Lerer, J. Bradbury, G. Chanan, T. Killeen, Z. Lin, N. Gimelshein, L. Antiga, A. Desmaison, A. Köpf, E. Yang, Z. DeVito, M. Raison, A. Tejani, S. Chilamkurthy, B. Steiner, L. Fang, J. Bai, and S. Chintala. *PyTorch: An Imperative Style, High-Performance Deep Learning Library*. Curran Associates Inc., Red Hook, NY, 2019.

88 G. Pillonetto, F. Dinuzzo, T. Chen, G. de Nicola, and L. Ljung. Kernel methods in system identification, machine learning and function estimation: A survey. *Automatica*, 50:657–682, 2014.

89 B. T. Polyak. Some methods of speeding up the convergence of iteration methods. *USSR Computational Mathematics and Mathematical Physics*, 4(5):1–17, January 1964. doi: 10.1016/0041-5553(64)90137-5.

90 M. J. D. Powell. On search directions for minimization algorithms. *Mathematical Programming*, 4(1):193–201, December 1973. doi: 10.1007/bf01584660. URL https://doi.org/10.1007/bf01584660.

91 J. C. Pratt. Sequential minimal optimization: A fast algorithm for training support vector machines. Technical report, Microsoft Research, April 1998.

92 A. Rantzer. On the Kalman-Yakobovich-Popov lemm. *System & Control Letters*, 28(1):7–10, 1996.

93 A. V. Rao. A survey of numerical methods for optimal control. *Advances in the Astronautical Sciences*, 135(1):497–528, 2010.

94 C. E. Rasmussen and C. K. I. Williams. *Gaussian Processes for Machine Learning*. MIT Press, 2006.

95 S. J. Reddi, S. Kale, and S. Kumar. On the convergence of Adam and beyond. In *International Conference on Learning Representations*, 2018. URL https://openreview.net/forum?id=ryQu7f-RZ.

96 H. Robbins and S. Monro. A stochastic approximation method. *The Annals of Mathematical Statistics*, 22(3):400–407, 9 1951. doi: 10.1214/aoms/1177729586.

97 W. J. Rugh. *Linear System Theory*. Prentice Hall, Englewood Cliffs, NJ, 1996.

98 A. N. Shiryayev. *Probability*. Springer-Verlag, New York, 1984.

99 J Sjöberg and M Viberg. Separable non-linear least squares minimization–possible improvements for neural net fitting. In *IEEE Workshop in Neural Networks for SignalProcessing*, 1997.

100 J. F. Sturm. Using SeDuMi 1.02, a Matlab toolbox for optimization over symmetric cones. *Optimization Methods and Software*, 11(1–4):625–653, January 1999. doi: 10.1080/10556789908805766. URL https://doi.org/10.1080/10556789908805766.

101 M. Tawarmalani and N. V. Sahinidis. A polyhedral branch-and-cut approach to global optimization. *Mathematical Programming*, 103:225–249, 2005.

102 TensorFlow. URL https://www.tensorflow.org/. Software available from tensorflow.org.

103 T. Tieleman and G. Hinton. Lecture 6.5—RmsProp: Divide the gradient by a running average of its recent magnitude. COURSERA: Neural Networks for Machine Learning, 2012.

104 K. C. Toh, M. J. Todd, and R. H. Tütüncü. SDPT3 — a Matlab software package for semidefinite programming, version 1.3. *Optimization Methods and Software*, 11(1–4):545–581, January 1999. doi: 10.1080/10556789908805762. URL https://doi.org/10.1080/10556789908805762.

105 M. Udell, K. Mohan, D. Zeng, J. Hong, S. Diamond, and S. Boyd. Convex optimization in Julia. In *SC14 Workshop on High Performance Technical Computing in Dynamic Languages*, 2014.

106 L. Vandenberghe and M. S. Andersen. Chordal graphs and semidefinite optimization. *Foundations and Trends® in Optimization*, 1(4):241–433, 2015. ISSN 2167-3888. doi: 10.1561/2400000006. URL http://dx.doi.org/10.1561/2400000006.

107 M. Verhaegen and V. Verdult. *Filtering and System Identification*. Cambridge University Press, Cambridge, UK, 2007.

108 S. Vigerske and A. Gleixner. SCIP: Global optimization of mixed-integer nonlinear programs in a branch-and-cut framework. *Optimization Methods and Software*, 33(3):563–593, 2018. doi: 10.1080/10556788.2017.1335312.

109 A. Viterbi. Error bounds for convolutional codes and an asymptotically optimum decoding algorithm. *IEEE Transactions on Information Theory*, 13(2):260–269, 1967. doi: 10.1109/TIT.1967.1054010.

110 A. J. Viterbi and J. K. Omura. *Principles of Digital Communication and Coding*. McGraw-Hill Book Company, New York, 1979.

111 A. Wächter and L. T. Biegler. On the implementation of an interior-point filter line-search algorithm for large-scale nonlinear programming. *Mathematical Programming*, 106(1):25–57, April 2005. doi: 10.1007/s10107-004-0559-y. URL https://doi.org/10.1007/s10107-004-0559-y.

112 M. J Wainwright and M. I. Jordan. Graphical models, exponential families, and variational inference. *Foundations and Trends in Machine Learning*, 1(1–2):1–305, 2008.

113 R. Wallin and A. Hansson. Maximum likelihood estimatin of linear SISO models subject to missing output data and missing input data. *International Journal of Control*, 87(11):2354–2364, 2014.

114 C.J.C.H. Watkins. *Learning from Delayed Rewards*. Ph.D. Thesis, Cambridge University, Cambridge, 1989.

115 S. J. Wright. *Primal-Dual Interior-Point Methods*. SIAM, 1997. ISBN 978-0-89871-382-4.

116 S. J. Wright. Coordinate descent algorithms. *Mathematical Programming*, 151(1):3–34, March 2015. doi: 10.1007/s10107-015-0892-3. URL https://doi.org/10.1007/s10107-015-0892-3.

117 C. F. J. Wu. On the convergence properties of the EM algorithm. *Annals of Statistics*, 11(1):95–103, 1983.

118 S. P. Wu, S. Boyd, and L. Vandenberghe. FIR filter design via spectral factorization and convex optimization. In B. Datta, editor, *Applied Computational Control, Signal and Communications*. Birkhauser, Boston, MA, 1997.

119 M. Yannakakis. Computing the minimum fill-in is NP-complete. *SIAM Journal on Algebraic Discrete Methods*, 2(1):77–79, March 1981. doi: 10.1137/0602010. URL https://doi.org/10.1137/0602010.

120 C. Zhang, S. Bengio, M. Hardt, B. Recht, and O. Vinyals. Understanding deep learning requires rethinking generalization, 2017. https://doi.org/10.48550/arXiv.1611.03530

121 H. Zhang and W. W. Hager. A nonmonotone line search technique and its application to unconstrained optimization. *SIAM Journal on Optimization*, 14(4):1043–1056, January 2004. doi: 10.1137/s1052623403428208. URL https://doi.org/10.1137/s1052623403428208.

Index

Printed and bound by CPI Group (UK) Ltd, Croydon, CR0 4YY

27/10/2024

14580679-0003